VOLUME NINETY FIVE

ADVANCES IN PARASITOLOGY

Echinococcus and Echinococcosis, Part A

VOLUME NINETY FIVE

ADVANCES IN PARASITOLOGY

Echinococcus and Echinococcosis, Part A

Edited by

R.C.A. THOMPSON
School of Veterinary and Life Sciences, Murdoch University, Murdoch Australia

P. DEPLAZES
Institute of Parasitology, University of Zürich, Zürich, Switzerland

A.J. LYMBERY
Freshwater Fish Group and Fish Health Unit, School of Veterinary and Life Sciences Murdoch University, Murdoch, Australia

ELSEVIER

ACADEMIC PRESS
An imprint of Elsevier
elsevier.com

Academic Press is an imprint of Elsevier
125 London Wall, London EC2Y 5AS, United Kingdom
The Boulevard, Langford Lane, Kidlington, Oxford OX5 1GB, United Kingdom
50 Hampshire Street, 5th Floor, Cambridge, MA 02139, United States
525 B Street, Suite 1800, San Diego, CA 92101-4495, United States

First edition 2017

ISBN: 978-0-12-811471-1
ISSN: 0065-308X

For information on all Academic Press publications visit our website at https://www.elsevier.com/books-and-journals

Publisher: Zoe Kruze
Acquisition Editor: Alex White
Editorial Project Manager: Helene Kabes
Production Project Manager: Vignesh Tamil
Designer: Alan Studholme

Typeset by TNQ Books and Journals

CONTENTS

VOLUME 95

PART A

VOLUME 96

PART B

CONTRIBUTORS

VOLUME 95
PART A

C.A. Alvarez Rojas
The University of Melbourne, Parkville, VIC, Australia

D. Antolova
Institute of Parasitology SAS, Kosice, Slovak Republic

K. Brehm
University of Würzburg, Würzburg, Germany

P.S. Craig
University of Salford, Greater Manchester, United Kingdom

G. Cringoli
University of Naples Federico II, Napoli, Italy

M. de la Rue
University of Santa Maria, Santa Maria RS, Brazil

P. Deplazes
University of Zürich, Zurich, Switzerland

J. Eckert
University of Zurich, Zurich, Switzerland

P. Giraudoux
University of Franche-Comté and Institut Universitaire de France, Besancon, France

M.F. Harandi
Research centre of Hydatid Disease in Iran, Kerman University of Medical Sciences, Kerman, Iran

D. Jenkins
Charles Sturt University, Wagga Wagga, NSW, Australia

E.J. Jenkins
University of Saskatchewan, Saskatoon, SK, Canada

U. Koziol
University of Würzburg, Würzburg, Germany; Universidad de la República, Montevideo, Uruguay

S. Lahmar
National School of Veterinary Medicine, Sidi Thabet, Tunisia

A.J. Lymbery
Murdoch University, Murdoch, WA, Australia

J. Magambo
Meru University of Science and Technology, Meru, Kenya

A. Massolo
University of Calgary, Calgary, Alberta, Canada

L. Rinaldi
University of Naples Federico II, Napoli, Italy

T. Romig
University of Hohenheim, Stuttgart, Germany

J.M. Schurer
University of Saskatchewan, Saskatoon, SK, Canada; University of Washington, Seattle, WA, United States

K. Takahashi
Rakuno Gakuen University, Ebetsu, Japan

R.C.A. Thompson
Murdoch University, Murdoch, WA, Australia

P.R. Torgerson
University of Zürich, Zurich, Switzerland

M. Wassermann
University of Hohenheim, Stuttgart, Germany

VOLUME 96
PART B

O. Akhan
Hacettepe University, Ankara, Turkey

C. Budke
Texas A&M University, College Station, TX, United States

A. Casulli
World Health Organization Collaborating Centre for the epidemiology, detection and control of cystic and alveolar echinococcosis, Istituto Superiore di Sanità, Rome, Italy

F.J. Conraths
Friedrich-Loeffler-Institut, Federal Research Institute for Animal Health, Institute of Epidemiology, Greifswald-Insel Riems, Germany

P.S. Craig
University of Salford, Greater Manchester, United Kingdom

B. Gottstein
University of Bern, Bern, Switzerland

D. Hegglin
University of Zurich, Zurich, Switzerland

P. Kern
University Hospital of Ulm, Ulm, Germany

M.W. Lightowlers
University of Melbourne, Werribee, VIC, Australia

A. Menezes da Silva
Estrada da Luz, Lisboa, Portugal

N. Müller
Institute of Parasitology, Vetsuisse Faculty, University of Bern, Bern, Switzerland

B. Müllhaupt
University Hospital of Zurich, Zürich, Switzerland

E. Ortona
Istituto Superiore di Sanità, Rome, Italy

M. Siles-Lucas
IRNASA-CSIC, Salamanca, Spain

A. Siracusano
Istituto Superiore di Sanità, Rome, Italy

P. Soboslay
University of Tübingen, Tübingen, Germany

P.R. Torgerson
University of Zurich, Zurich, Switzerland

K.A. Vizcaychipi
National Institute of Infectious Diseases, Buenos Aires, Argentina

D.A. Vuitton
WHO Collaborating Centre, Université de Franche-Comté, Besançon, France

J. Wang
University of Bern, Bern, Switzerland

Q. Wang
Sichuan Provincial Center for Disease Control and Prevention, Chengdu, China

PREFACE

Our aim in this thematic volume of *Advances in Parasitology* is to provide a complete synthesis of what is known about the cestode parasite *Echinococcus* and the diseases it causes, echinococcosis (hydatid disease). It builds on the success of two previous volumes 'The Biology of *Echinococcus* and Hydatid disease' and '*Echinococcus* and Hydatid Disease' published by Allen and Unwin and CAB International, respectively, and details the major advances that have taken place since. The 10 chapters demonstrate that in addition to its medical, veterinary and economic significance, *Echinococcus* is an intriguing biological phenomenon. They detail the major advances that have taken place during the last 20 years, particularly in our understanding of taxonomy, genetic variation, developmental biology, host–parasite relationships, geographic distribution and host range, diagnosis, control and clinical management. In addition, we have included a chapter covering historical aspects of echinococcosis that highlights major contributions to knowledge about the parasite and the diseases it causes. In this respect, four doyens of the field and contributors to the previous volume have passed away in recent years: Desmond Smyth, Michael Gemmell, Robert Rausch and Rudolf Ammann. We are therefore pleased to have the opportunity that this volume provides to detail and highlight the major contributions they have made to the field.

Although major advances have been made in research on *Echinococcus* and echinococcosis, many questions remain, particularly in the areas of developmental biology and host–parasite relationships. Control efforts have had limited impact globally, and *Echinococcus* is an emerging problem in some parts of the world. As with many zoonoses, control is hampered by anthropogenic factors that influence both domestic and wild cycles of transmission.

The overriding theme of the book is that a comprehensive understanding of the biology of *Echinococcus* is essential for the effective treatment and control of echinococcosis. The links between laboratory knowledge and field applications are emphasized throughout the book. Consequently, we hope that research workers, teachers and students of parasitology, clinicians and field workers will find this work an indispensable source of information, but that it will also provide a model for the integration of basic and applied research in parasitology.

Andrew Thompson, Peter Deplazes and Alan Lymbery

CHAPTER ONE

Historical Aspects of Echinococcosis

J. Eckert*[,1], R.C.A. Thompson[§]
*University of Zurich, Zurich, Switzerland
[§]Murdoch University, Murdoch, WA, Australia
[1]Corresponding author: E-mail: johannes.eckert@bluewin.ch

Contents

Advances in Parasitology, Volume 95
ISSN 0065-308X
http://dx.doi.org/10.1016/bs.apar.2016.07.003

Abstract

Echinococcosis is a zoonosis whose history dates back to antiquity. This article provides an overview on the general history of echinococcosis, including the elucidation of *Echinococcus* life cycles and the long controversy on the aetiology of the cystic and alveolar forms of echinococcosis (CE and AE), lasting about 100 years since the middle of the 19th century. Furthermore, selected historical aspects of some fields of echinococcosis research are discussed and compared with our current knowledge, such as geographic distribution and epidemiology of CE (*Echinococcus granulosus*) and AE (*Echinococcus multilocularis*), clinical aspects and pathology, diagnosis in humans and animals, treatment (with focus on chemotherapy), control and basic research. A short paragraph is devoted to the neotropical forms of echinococcosis, caused by *Echinococcus vogeli* and *Echinococcus oligarthrus*. In this context the achievements of some ancestral pioneers of echinococcosis research are particularly highlighted and appreciated. Finally, the role of associations, international organizations (World Health Organization and others) and international working groups in echinococcosis research and control is briefly outlined. The retrospective reveals both the admirable achievements of our ancestors and the scientific progress of more recent times. But, it also shows the gaps in our knowledge, skills and resources that we need to control or even eradicate echinococcosis.

1. INTRODUCTION

Although echinococcosis has a long history dating back to ancient times, it is still a relevant zoonosis today with considerable socioeconomic impact, affecting humans in many parts of the world (WHO, 2001a; Craig and Pawlowski, 2002; Eckert et al., 2011; Torgerson and Macpherson, 2011; Vuitton et al., 2015). Here, we highlight selected historical aspects of some fields of echinococcosis research, with examples of what contributions our ancestors have made to our current knowledge which has continuously grown, especially since the 19th century and what changes have occurred over the years. The reader is also referred to a chapter on

Table 1 Forms of echinococcosis

Forms of echinococcosis	Causative agent	Typical definitive hosts	Typical intermediate hosts	Aberrant hosts
Cystic (CE)	*Echinococcus granulosus* and some closely related species (see chapter: "Biology and Systematics of *Echinococcus*", this volume)	Dog, wolf, dingo, jackal other canids	Sheep, goat, cattle, pig, horse, wild ungulates, marsupials etc.	Humans, monkeys, many other mammals
Alveolar (AE)	*Echinococcus multilocularis*	Fox species, wolf, raccoon dog, domestic dog, cat	Small herbivorous mammals, predominantly rodents	Humans, monkeys, dog, pig, horse etc.
Polycystic (PE)	*Echinococcus vogeli*	Bush dog, domestic dog	Paca, agouti	Humans, monkeys
Unicystic (UE)	*Echinococcus oligarthrus*	Wild cats (Felidae)	Agoutis, paca, opossums	Humans

Further information on the echinococcosis forms listed in the table can be obtained from reviews (Thompson and Lymbery, 1995; WHO, 2001a; Eckert and Deplazes, 2004; D'Alessandro and Rausch, 2008, and others).

echinococcosis published by Grove in 'A History of Human Helminthology' (Grove, 1990).

Echinococcosis is a zoonosis involving carnivores as definitive and a broad spectrum of mammalian species as intermediate hosts. Humans, monkeys and some other mammals can be affected as aberrant hosts. Currently, four forms of echinococcosis are distinguished (WHO, 2001a; D'Alessandro and Rausch, 2008) (Table 1). Cystic and alveolar echinococcosis (CE and AE), the most important forms with the widest geographic ranges, are the focus of this article. CE is also called hydatid disease (Greek: *hydatis*: water bladder).

2. *ECHINOCOCCUS GRANULOSUS* AND *ECHINOCOCCUS MULTILOCULARIS*

2.1 General historical aspects

2.1.1 Early knowledge of hydatids

The metacestodes of *E. granulosus* are bladders (cysts) of variable and often large size filled with clear liquid and are called 'hydatids.' The first indications of hydatids date back to antiquity and stem from **Hippocrates** (~460–377 BC) who wrote in his aphorisms (VII, 55): 'In those whose water stuffed liver opens into the omentum, the belly is filled with water, and they die.' (Neisser,

1877; Fuchs, 1895). **Galen** (129–~200 BC) regarded the liver as the main site of hydatids and mentioned their occurrence in slaughter animals (Hosemann, 1928). Around 50 AD, **Aretaeus** (or Aretaios) of Cappadocia described in his work 'De causis et signis morborum' different clinical entities of people. He noted that in patients with ascites, numerous small, fluid-filled blisters may be present in the abdomen and that some liquid emerges when an abdominal puncture is attempted with a trocar (Neisser, 1877). In the following periods, the presence of hydatids in animals and humans was repeatedly reported in the literature, for example, in the 16th and 17th century. **Wolckerus** described an alleged abscess, from which 300 hydrous bladders deflated (Neisser, 1877; Langenbuch, 1890). In 1679 **Théophile Bonet** (1620–89) in Geneva published a summary of the pathological knowledge in a work entitled 'Sepulchretum sive anatomia practica' ('burial ground or practical anatomy') (Ackerknecht, 1989), containing hints on some patients harbouring hydatids (Langenbuch, 1890).

2.1.2 Knowledge on the nature of hydatids

Until the early modern age, the true nature of hydatids remained unknown, and they were regarded before that time as degenerated glands, as accumulations of serum or mucus between laminar cell layers or as descendants of so-called 'milk vessels' or blood vessels (Langenbuch, 1890). First indications of the animal nature of metacestodes arose from observations of **Francesco Redi** (1626–97) who recognized in 1684 in Florence that cysticerci (metacestodes of *Taenia* spp.) are able to move like animals (Grove, 1990). In the following year 1685, **Philip Jacob Hartmann** (1648–1707)—Professor of Medicine and History of Medicine at the University of Königsberg/Germany (now Kaliningrad/Russia)—confirmed the animal nature of cysticerci, describing a small, spherical structure that was connected with the metacestode bladder. Apparently, he had described a Cysticercus tenuicollis with scolex, the metacestode of *Taenia hydatigena* (Enigk, 1986). These observations were unknown to **Edward Tyson** (1650–1708), a doctor in Oxford. He also had observed the motility of C. tenuicollis and reported in 1687 to the Royal Society that 'hydatids in animals are a sort of living creatures,' which he called 'lumbrici hydropici,' thus providing a clue for a possible link with 'worm' parasites (Grove, 1990). **Peter Simon Pallas** (1741–1811) born in Berlin, allocated the hydatids as a discrete group to the 'bladder worms' and described in his medical dissertation in 1760 (submitted to the University of Leiden/The Netherlands) 'small bodies' (brood capsules) on the inner wall of the bladders (Neisser, 1877; Enigk, 1986). In

Figure 1 J.A.E. Goeze (1731–93). *Enigk, K., 1986. Geschichte der Helminthologie im deutschsprachigen Raum. G. Fischer, Stuttgart. ISBN 3-437-20350-9. With permission of Karl Enigk Foundation, DSZ-Deutsches Stiftungszentrum, Essen, Germany, 19.01.2016.*

Figure 2 C.A. Rudolphi (1771–1832). *Leuckart, R., 1863. Die menschlichen Parasiten und die von ihnen herrührenden Krankheiten. Band 1 und 2, Leipzig, C.F. Winter'sche Verlagshandlung. Photo: J. Eckert.*

these bodies, the priest **John August Ephraim Goeze** (1731–93) in Quedlinburg/Germany recognized in 1782 tapeworm scoleces (Enigk, 1986). In 1801 **Carl Asmund Rudolphi** (1771–1832) (Dr phil. 1773 and Dr med. 1795 at the University of Greifswald/Germany) introduced the name '*Echinococcus*' to zoology (Rudolphi, 1801) (Figs 1 and 2).

2.1.3 Elucidation of the life cycle of Echinococcus granulosus

2.1.3.1 Development in final hosts

In 1688 **Johann Jacob Wepfer** (1620–95), physician in Schaffhausen/Switzerland, had observed that a tapeworm stage occurring in the liver of

Figure 3 J.J. Wepfer (1620–95). *With permission of Stadtarchiv Schaffhausen, Switzerland, 07.01.2016.*

mice (subsequently called *Strobilocercus fasciolaris*) exhibited features of an intestinal cat tapeworm, which was then called 'latis lumbricus intestinorm' (today *Taenia taeniaeformis*). Thus Wepfer was the first to make a link between the intestinal stage of tapeworms and its cystic stage developing in internal organs (Enigk, 1986; Grove, 1990) (Fig. 3).

Approximately 150 years later, in 1842, the Dane **Johannes J.S. Steenstrup** (1813–97)—Lecturer of Zoology and Mineralogy at the Academy in Sorø on Zealand and later Professor at the University of Copenhagen—formulated his theory on alternations of generations. In this sense, he regarded the cystic stages as early forms in the development of helminths that were unknown to him (Grove, 1990). In 1845 **Felix Dujardin** concurred with this view in France. He infected dogs and cats experimentally with cysticerci, but he was reluctant to publish his results, which resulted in Küchenmeister in Germany preempting him with publication of his data (Enigk, 1986).

After graduation in 1846 as a medical doctor at the University of Leipzig/Germany, **Gottlob Friedrich Heinrich Küchenmeister** (1821–90) settled down as a medical practitioner and obstetrician in Zittau and from 1859 in Dresden (Enigk, 1986). In 1851 during his time in Zittau, he fed Cysticercus pisiformis (metacestodes of *Taenia pisiformis*) to four red foxes and isolated from their intestines young tapeworms with scoleces identical with those of the cysticerci (Küchenmeister, 1851; Enigk, 1986). With this, Küchenmeister had detected the life cycle of taeniid tapeworms, but this basic observation by a medical practitioner was heavily criticized by **Carl Theodor Ernst von Siebold** (1804–85) who was Professor of

Figure 4 G.F.H. Küchenmeister (1821–90). *Enigk, K., 1986. Geschichte der Helminthologie im deutschsprachigen Raum. G. Fischer, Stuttgart. ISBN 3-437-20350-9. With permission of Karl Enigk Foundation, DSZ-Deutsches Stiftungszentrum, Essen, Germany, 19.01.2016.*

Figure 5 C.T.E. von Siebold (1804–85). *Enigk, K., 1986. Geschichte der Helminthologie im deutschsprachigen Raum. G. Fischer, Stuttgart. ISBN 3-437-20350-9. With permission of Karl Enigk Foundation, DSZ-Deutsches Stiftungszentrum, Essen, Germany, 19.01.2016.*

Physiology at the University of Breslau/Germany (now Wrocław/Poland) since 1850 (Enigk, 1986). Later, however, von Siebold followed the example of Küchenmeister. In 1852 he infected several dogs with protoscoleces from hydatid cysts and found in their small intestines tapeworms that were a few millimetres long at 27 day post infection (p.i.) composed of a scolex and two to three proglottids (von Siebold, 1853). For the first time, he had obtained experimentally *E. granulosus* which had been described previously by Batch, 1786, Zeder, 1803, and Rudolphi, 1801 (cit. in Enigk, 1986). Soon thereafter, these findings were confirmed by several authors (Figs 4 and 5).

Figure 6 R. Leuckart (1822–98). *Enigk, K., 1986. Geschichte der Helminthologie im deutschsprachigen Raum. G. Fischer, Stuttgart. ISBN 3-437-20350-9. With permission of Karl Enigk Foundation, DSZ-Deutsches Stiftungszentrum, Essen, Germany, 19.01.2016.*

However, it remained still unknown, how metacestodes in intermediate hosts develop from intestinal tapeworms of carnivores. Küchenmeister (1853) infected dogs with metacestodes of *Taenia multiceps*, isolated the tapeworm eggs and transferred them to a sheep, in which he later found young metacestodes. Finally **Rudolf Leuckart** (1822–98) infected piglets with eggs of '*Taenia echinococcus*,' obtained from a dog and detected 4 weeks later in the livers of these animals small vesicles (0.25 to 0.35 mm) showing a laminated layer ('glashelle Kapsel'). He also depicted and described in detail the further development of these stages including the formation of brood capsules and protoscoleces (Leuckart, 1863).

Rudolf Leuckart, born in Helmstedt/Germany, had studied medicine in Göttingen. He was first associate Professor of Zoology in Giessen and then appointed in 1869 as a Professor of Comparative Anatomy at the University of Leipzig. By his profound and broad-based zoological and parasitological work, R. Leuckart had a long-lasting impact on the development of biological sciences and parasitology in the 19th century and beyond (Enigk, 1986) (Figs 6 and 7).

2.1.3.2 Development in intermediate hosts, cyst formation and secondary echinococcosis

After Leuckart's observations (see above), it was discussed for some time how the oncospheres migrate in the intermediate host's body. A few authors had speculated that oncospheres, hatched in the gastrointestinal tract from the eggs, may be incorporated passively like 'foreign bodies' (Hosemann,

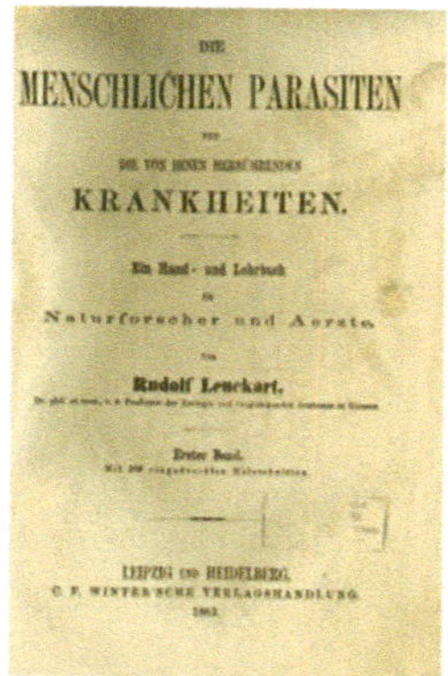

Figure 7 Book of Leuckart (1863): title page. *Photo: J. Eckert.*

1928). Dévé in France and Dew in Australia were the first to show that the oncopheres penetrated actively the intestinal wall, entered small portal veins and were subsequently carried in the bloodstream to the liver. Both Dew (1925) and Dévé (in Hosemann, 1928) had observed larvae in portal veins of the liver a few hours after experimental infection of pigs. It was concluded that most larvae remain in the liver, but a few of them may pass the liver 'filter' and are transported to the lung or other organs. This view was consistent with the sites of cysts in the human body, predominantly localized in the liver and less frequently in other organs [according to Dévé (in Hosemann, 1928): 74.5% liver, 8.6% lung and pleura, 2.3% spleen, 2.1% kidneys, 6.2% muscles].

Another important aspect of practical and basic interest was the aetiology of secondary echinococcosis which may occur in body cavities of human patients following cyst rupture. In 1898 von Alexinsky (in Hosemann, 1928) had experimentally demonstrated that intraperitoneal injection of 'hydatid sand' (containing protoscoleces) resulted in cyst formation. Several years later, Dévé confirmed these findings by injecting hydatid sand intraperitoneally to rabbits with the result that 23 of 32 of his experiments were positive. Furthermore, Dévé could clearly demonstrate by histological studies that cysts can arise from protoscoleces (Dévé, in Hosemann; Dévé, 1902, 1946).

Félix Dévé (1872–1951) studied medicine in Paris. In 1990 he began his long series of investigations on echinococcosis and presented in 1901 his thesis on secondary echinococcosis (Dévé, 1901). In 1942 he was appointed as Professor of Clinical Medicine in Rouen. He published three books and more than 300 articles on hydatid disease (Grove, 1990).

Harold Robert Dew (1891–1962) studied medicine at the University of Melbourne and graduated in 1914. In 1920 he became a fellow of the

Figure 8 R.L. Virchow (1821–1902). *Public domain: http://ihm.nlm.nih.gov/images/B25667.*

Royal College of Surgeons, and in 1923 he joined the surgical staff of the Royal Melbourne Hospital and was appointed as an assistant director of the Walter Eliza Hall Institute of Medical Research. Later he was appointed as the first full-time Professor of Surgery in Australia by the University of Sydney (Grove, 1990). An impressive document of his scientific activities and achievements is his monograph: 'Hydatid disease. Its pathology, diagnosis and treatment' (Dew, 1928).

2.1.4 Initial findings on human alveolar echinococcosis

In 1852 **Ludwig Buhl** (1816–80), pathologist at the University of Munich/Germany, had diagnosed in a human liver an uncommon liver tumour which he initially called 'alveolarcolloid' ('colloid cancer') but recognized it in 1854 as a deformed *Echinococcus* lesion (Buhl, 1855). In 1855 **Rudolf Ludwig Virchow** (1821–1902) presented to the 'Physicalisch—Medicinische Gesellschaft' in Würzburg a detailed report on 'Die multiloculäre, ulcerierende Echinokokkengeschwulst der Leber' (multilocular, ulcerating *Echinococcus* tumour), referring to cases described by Buhl and some other authors (Virchow, 1856). At that time, Virchow was Professor of Pathology at the University of Würzburg where he developed his famous concept of cellular pathology (Enigk, 1986) (Fig. 8).

2.1.5 Controversy on the aetiology of cystic and alveolar echinococcosis

With the identification of the 'alveolarkolloid' as a form of echinococcosis two disease patterns were recognized in the mid-19th century, which are called today cystic echinococcosis (CE) and alveolar echinococcosis (AE), respectively. However, it was unclear whether these two forms were caused by a single or by two different *Echinococcus* species. In the following period, a lively debate developed in which 'unicists' faced a group of 'dualists.' The unicists, among them L. Buhl, R. Virchow, G. F. H. Küchenmeister, R. Leuckart, F. Devé and H. R. Dew, regarded *E. granulosus* (then called *Taenia echinococcus* or *T. echinococcus cysticus*) as the cause of human CE and AE but assumed that structure and growth characteristics of the metacestodes are widely variable, depending on mechanical and physiological factors in the human host tissue (Hosemann et al., 1928). These and other arguments did not convince the dualists who embraced different *Echinococcus* species as causative agents. **A. Morin** (1875) was apparently one of the first who considered a dualistic conception in his thesis at the University of Berne/Switzerland (Morin, 1875). His view was supported by other authors, including **Adolf Posselt** (1867—1936), Professor of Internal Medicine at the University of Innsbruck/Austria (Fig. 9).

In 1901/1902 Posselt was the first to provide clear experimental evidence that human AE is caused by *E. multilocularis* (then called *Taenia echinococcus alveolaris*). He isolated from a human patient alveolar parasite tissue containing numerous protoscoleces and infected a parasite-free dog. After 49 days, he found in the intestine numerous small tapeworms with typical morphological features of the adult stages of *E. multilocularis* (Posselt, 1928). In addition to the results of the feeding experiment, Posselt presented many other plausible arguments (parasite morphology, pathology, histology, clinical signs, geographic distribution etc.) in support of the dualistic conception,

Figure 9 A. Posselt (1867—1916). *Aspöck H (ed.) (2002): Amöben, Bandwürmer, Zecken... (p. 354). Denisia 6. ISBN 3-85474-088-3. With permission of Prof Dr H. Aspöck, Institut für Spezifische Prophylaxe und Tropenmedizin, Medizinische Universität Wien, Austria, 09.01.2016.*

summarized in a classical treatize 'Der Alveolarechinokokkus und seine Chirurgie' (the alveolar *echinococcus* and its surgery) (Posselt, 1928). Photos in this publication indicated that the structure of the gravid uterus of *T. echinococcus alveolaris* clearly differs from that of *T. echinococcus cysticus* (= *E. granulosus*) from Australia, and drawings showed morphological differences of the scolex hooks (Posselt, 1928). Fifty six years later, Vogel (1957) could reexamine the specimens from Posselt's feeding experiment and recognized 'all characteristics' of *T. echinococcus alveolaris* (= *E. multilocularis*). The controversy between 'unicists' and 'dualsits' had persisted for approximately 100 years. Even in 1953, Dew (cited in Vogel, 1957) has stated with regard to the various forms of echinococcosis in humans and domestic animals: 'If the essential unity of the hydatid species is admitted, and I think all workers now admit this, all the above forms, simple and bizarre, are different forms assumed by the larval stage of the same parasite.' However, a new chapter of the story had already been opened in 1951 by the work of Rausch and Schiller in Alaska (see Section 2.1.6).

2.1.6 Elucidation of the life cycles of Echinococcus sibiricensis and Echinococcus multilocularis

With the successful infection of a dog with fertile alveolar metacestode material isolated from a human patient in 1901/1902 Posselt had already elucidated part of the life cycle of *E. multilocularis* (Posselt, 1928). However, many questions remained open until **Robert Rausch** and **E. L. Schiller** published the results of their pioneering studies on alveolar echinococcosis of Inuits in Alaska (Rausch and Schiller, 1951, 1954, 1956). They identified arctic foxes (*Alopex lagopus*) and sledge dogs as definitive hosts of a new *Echinococcus* species which they described as *E. sibiricensis*, found alveolar metacestodes in rodents (field vole and red-backed vole), succeeded in infecting experimentally microtine rodents with eggs of the newly identified *Echinococcus* species, and associated these findings with AE cases in the Inuit population (Rausch and Schiller, 1951, 1954, 1956). **Robert L. Rausch** (1921–2012) was then a member of the United States Public Health Service at the Arctic Health Research Center in Alaska since 1948 and since 1978 Professor at the Washington State University in Seattle.

In the mid-1950s **Johannes Vogel** (1900–80), helminthologist at the former Tropical Institute in Hamburg/Germany (inspired by the findings of Rausch and Schiller, as he mentioned in one his publications) conducted epidemiological studies on *E. multilocularis* in Southern Germany (Vogel, 1955, 1957). In the region of the 'Swabian Alb,' he found 4 of 10 red foxes infected with *E. multilocularis* and identified voles (*Microtus arvalis*) as natural intermediate hosts. Furthermore, he infected successfully foxes, dogs and

Figure 10 R.L. Rausch (1921–2012). *100 Jahre Veterinärmedizinische Fakultät der Universität Zürich (p. 178). With permission of Dean, Vetsuisse Faculty, Zürich, Switzerland, 12.01.2016.*

cats with metacestodes from rodents, and rodents (e.g., *M. arvalis*, *Sigmodon hispidus*) with *Echinococcus* eggs isolated from foxes. In an extensive morphological study, he could not find significant differences between specimens of the European *E. multilocularis* and the Alaskan *E. sibiricensis* and proposed for priority reasons the name *E. multilocularis* Leuckart 1863 for both of these parasites (Vogel, 1955, 1957). With this, a 100-year dispute on the aetiology of CE and AE was clarified (Rausch, 1986).

Robert Rausch (1921–2012)[1] acquired a broad education in zoology, entomology, veterinary medicine, parasitology, bacteriology and wildlife biology. He began his research career in 1948 at the US Public Health Service at the Arctic Research Center in Alaska, where he spent 27 years and became Chief of the Infectious Disease Section. After 3 years as Professor of Parasitology at the University of Saskatchewan, he was appointed in 1978 as Professor at the Washington State University in Seattle, School of Medicine, Department of Comparative Medicine. After his retirement in 1992, he continued his work on cestodes and other parasites for many years. His wife, **Virginia Rausch**, deserves great appreciation for the close scientific cooperation with her husband and her constant support (Figs 10 and 11).

Johannes (Hans) Vogel (1900–80), born in Dresden/Germany, had studied natural sciences in Jena and medicine in Jena and Hamburg from 1919 until 1927. In 1927 he became assistant at the Division of

[1] Source: Hoberg, E.P. (2014). In memoriam. Robert Lloyd Rausch—a life in nature and field biology 1921–2012. Journal of Parasitology 100 (4), 547–552.

Figure 11 J. Vogel (1900–80). *Enigk, K., 1986. Geschichte der Helminthologie im deutschsprachigen Raum. G. Fischer, Stuttgart. ISBN 3-437-20350-9. With permission of Karl Enigk Foundation, DSZ-Deutsches Stiftungszentrum, Essen, Germany, 19.01.2016.*

Helminthology at the Bernhard-Nocht-Institut für Schiffs-und Tropenkranheiten (now Bernhard-Nocht-Institut für Tropenmedizin), and in 1935 he was promoted to Head of the Division as a successor of F. G. H. Fülleborn. From 1963 to 1968, he served as the Director of the Institute (Enigk, 1986). H. Vogel has made major contributions to parasitology, predominantly in cestode and trematode research.

J. Vogel and R. Rausch had cooperated in a collegial manner and acquired a high international scientific and personal reputation through their outstanding work on echinococcosis and other parasitic diseases. More details on the historical development can be found in Enigk's book 'Geschichte der Helminthologie im deutschsprachigen Raum' (history of helminthology in the German speaking area) (Enigk, 1986) and in a review by Tappe et al. (2010a).

3. SPECIFIC HISTORICAL ASPECTS

3.1 Geographic distribution and epidemiology

By the end of the 19th century and the beginning of the 20th century, a remarkable body of knowledge had accumulated on the worldwide geographic distribution and prevalence of echinococcosis in humans and animals, but the overall picture had considerable gaps.

3.1.1 Echinococcus granulosus and cystic echinococcosis

3.1.1.1 Distribution

Early knowledge of the occurrence of CE in humans was published by **Egbert Schwarz**, Professor at the University of Rostock/Germany, who had

DIE
ECHINOKOKKENKRANKHEIT

VON

Prof. Dr. G. HOSEMANN
Prof. Dr. E. SCHWARZ, Prof. Dr. J. C. LEHMANN
und Prof. Dr. A. POSSELT

MIT EINEM GELEITWORT
VON
GEH.-RAT PROF. DR. W. MÜLLER

MIT [illegible] ABBILDUNGEN UND 2 FARBIGEN TAFELN

VERLAG VON FERDINAND ENKE IN STUTTGART
1·9·2·8

Figure 12 Book of Hosemann et al. (1928): title page. *Photo: J. Eckert.*

reviewed the international literature of the period from the middle of the 19th century until the 1920s, published in a remarkable monograph 'Die Echinokokkenkrankheit' edited by Hosemann et al. (1928) (Fig. 12). Schwarz (1928) presented information on human CE in South America, with high prevalences in Argentina, Paraguay, Uruguay and Brazil. In Argentina, 970 human cases were registered in 1901. Only sporadic cases were reported in some other South American countries (Chile, Bolivia, Peru, Ecuador, Columbia, Venezuela), in Mexico and North America (United States of America, Canada). Little information was available on the general situation in Africa, but rather high prevalences were known to occur in southern Africa and Algeria, whereas only sporadic cases were recorded in other countries of northern Africa (Morocco, Tunisia, Egypt) as well as in regions of eastern Africa.

In the northern hemisphere, Iceland with its low number of inhabitants (~64,000 in 1849) had very high prevalences of human CE with a range of 104 to 235 cases each year in the period 1896 to 1903, and declining numbers of 105 to 33 in the period 1904 to 1920 (Schwarz, 1928). In other parts of Europe, the CE prevalences were categorized by Schwarz (1928) as medium to high in regions of France, England, Germany, parts of the Netherlands, the Mediterranean region (Portugal, Spain, Italy), Hungary, Dalmatia[2], Serbia, Greece and some other Balkan states. For example, northern Germany had a high prevalence with 1.98% of human CE cases among 4250 autopsies in Rostock in the period 1861 to 1905. Sporadic cases were reported in

[2] Historical region of western Balkan.

Belgium, Scotland, Ireland, the Scandinavian area (Denmark, Sweden, Norway), Finland and the Baltic region, Austria and some of the Balkan states (Bulgaria, Romania) (Schwarz, 1928). In Russia, areas in the southern and eastern parts were regarded as especially affected by CE. Apparently, little information existed on the situation in Asia, and only medium prevalences in India and sporadic occurrence in China and Japan are mentioned by Schwarz (1928). On the other hand, the high prevalences in mainland Australia, Tasmania and New Zealand are described in detail. With reference to J. D. Thomas, Schwarz (1928) underlined the great importance of CE in Australia, especially in Victoria, and reported that 200 people died of CE within 14 years (14 per year) in the 1860s, and 52 people in 1884 (Schwarz, 1928).

In summary, Schwarz (1928) had classified several regions as highly endemic for *E. granulosus* and human CE, especially Argentina, Australia and New Zealand, but also parts of northwestern and southern Africa, Iceland, large parts of Europe, except Scandinavian countries, and eastern Russia. Furthermore, he provided some data on CE in livestock and of *E. granulosus* in definitive hosts (e.g., 80% infected dingos in Australia). He pointed out that echinococcosis occurs more or less everywhere and that its detection depends on the awareness of the disease. Studies in the following decades have confirmed Schwarz's statement showing that *E. granulosus* and CE have a wide range in the northern and southern hemisphere (Matossian et al., 1977; Andersen et al., 1993, 1997; Schantz et al., 1995; WHO, 2001a). According to Torgerson and Macpherson (2011), there is a persistently high burden of CE in many parts of the world with an estimated one million or more people currently suffering from CE globally and financial losses of $2 billion caused by CE in global livestock populations. In some regions, the prevalence is low (e.g., western Europe), whereas other areas are highly affected (e.g., regions in central Asia and China) (Torgerson and Shaikenov 2004, Torgerson and Macpherson, 2011; Torgerson et al., 2011).

3.1.1.2 Epidemiology

In the 1980s/1990s, experimental studies by **M. Gemmell** and coworkers generated important new basic knowledge for understanding the epidemiology of taeniid cestodes and breaking the 'epidemiological code' with the aid of mathematical modelling. This knowledge and data from other authors have been reviewed in several excellent articles (Gemmel and Lawson, 1986; Gemmell et al., 1986; Roberts et al., 1986; Gemmell and Roberts, 1995; Gemmell, 1997; Gemmell et al., 2001). Gemmell and coworkers were the first to apply the concept of the basic reproductive rate (R_0) to studies on the transmission of *Echinococcus*. They analyzed and quantified the contributions made by the parasites, the definitive and intermediate hosts and the environment to transmission dynamics. The results revealed inter alia that *E. granulosus* and other taeniid cestodes (*T. hydatigena* and *Taenia ovis*) have an overdispersed distribution with only a small number of definitive and intermediate hosts (dogs and sheep, respectively) harbouring a large number of parasites. Compared with *Taenia* species, *E. granulosus* has a much lower biotic potential (= capacity of organisms to reproduce to an extent that enduring survival of the species is secured). As acquired immunity to *E. granulosus* in dogs is weak or lacking, it does not play a role in regulating the parasite population. In contrast, immunity to superinfection by *E. granulosus*, *T. hydatigena* and *T. ovis* can be acquired or induced in sheep. Therefore the immune status of intermediate hosts can be a constraint on the parasite population but only under high infection pressure. The parasite population is also influenced by environmental factors, such as climate and egg-dispersal mechanisms. According to Gemmell et al. (2001) for understanding the transmission dynamics and for planning control programmes, the following factors are of great significance: (1) biotic potential of the parasite in the definitive host, (2) acquired immunity as a density-dependent constraint by the intermediate host, and (3) climate as a density-dependent constraint in the free-living egg phase (Fig. 13).

Considerable merit is due to **Michael Gemmell** (1926–2003), who with his team in Dunedin/New Zealand who collated and analyzed 30 years

Figure 13 M.A. Gemmell (1926–2003). *Congress photo, collection J. Eckert.*

of their experimental infection and transmission data on *Taenia* spp. and *E. granulosus* infecting livestock and dogs. For the first time, they were able to build a quantified approach to measure the dynamics of taeniid transmission between mammalian definitive and intermediate hosts. These transmission studies formed the foundation for understanding the epidemiology of *E. granulosus* and represented an important contribution to the development of detailed concepts for planning, implementation and evaluation of control interventions. M. Gemmell, born in London, studied veterinary medicine at the University of Sydney where he graduated in 1950. In 1958 he became the Director of the Hydatid Research Unit of the Otago Medical School in Dunedin where he worked until the late 1980s. During his career, he was responsible for several major ground-breaking developments, he was an enthusiastic scientist and a competent advisor for national institutions and international organizations [World Health Organization (WHO), Pan American Health Organization (PAHO), Food and Agriculture Organization of the United Nations (FAO) etc.] and has certainly influenced generations of scientists of various fields through his ideas and excellent work[3]. Studies in Gemmell's area of research are now continued by younger generations (e.g., Torgerson, 2003, 2006; Budke et al., 2005).

3.1.2 *Echinococcus multilocularis* and alveolar echinococcosis

Compared to *E. granulosus*, the epidemiology of *E. multilocularis* is more complex due to its predominantly sylvatic life cycle involving foxes and other wild carnivores as definitive hosts and a large number of small mammals (mainly rodents) as intermediate hosts. Therefore historically, the determination of epidemiological key factors has proven to be more difficult.

3.1.2.1 Human cases of alveolar echinococcosis

After the unequivocal identification of the first human cases of AE in southern Germany in the mid-1850s (Buhl, 1855; Virchow, 1856), further cases were reported in Germany (Vierodt, 1886) and adjacent areas, for example in Switzerland (1858) (Dardel, 1927). Posselt (1928) listed a total of 651[4] AE cases, which he had collected until spring 1928, with the following geographic distribution: Germany: 168, Switzerland: 164, Austria: 96, other

[3] Torgerson, P., Eckert, J. (2003): Obituary: Micheal A. Gemmell (1926–2003) (unpublished).

[4] The total number of cases indicated in Posselt's publication is only 600, apparently due to a calculation error in a table.

Das Blasenwurmleiden
in der Schweiz

Von
Dr. med. Gustave Dardel

Mit Vorwort von
Prof. Dr. A. Posselt

Bern · Verlag A. Francke A.-G. · 1927

Figure 14 Dissertation of G. Dardel (1927): title page. *Photo: J. Eckert.*

alpine regions and Mähren[5]: 6, France and Italy: each 3, North America: 2, and Russia: 209. Of 440 European cases, 428 (97%) had been diagnosed at that time in Germany, Switzerland and Austria. As early as 1900, Posselt had drawn attention to differences in the geographic ranges of CE and AE in Europe and defined southern Germany, Switzerland, Austrian alpine regions and certain regions in Russia as 'classical distribution areas' of AE (Posselt, 1928). Approximately 4% of the cases were found outside this area, namely in Germany up to Hannover and Berlin, in the Baltic region (St. Petersburg/Russia, Tartu/Estonia) and in Warszawa/Poland (Posselt, 1928; Schwarz, 1928). In Russia, AE cases were known to occur in regions around Moscow, Kazan and Tomsk (Posselt, 1928). In Alaska, autochthonous human AE cases were diagnosed in 1947 (Rausch and Schiller, 1951) and in Japan in 1926 (Katsurashima in Suzuki et al., 1993). Posselt (1928) assumed an almost complete separation of the distribution area of AE and CE, but his colleague Schwarz (1928) did not support this view in its strict form. By the end of the 1920s, autochthonous human AE cases were documented only in four central European countries (Germany, Switzerland, Austria, and France), as well as in Russia and Japan (Posselt, 1928; Katsurashima in Suzuki et al., 1993) (Fig. 14).

In the following decades, human AE cases were recorded in many more countries (Eckert, 1996; WHO, 2001a; Vuitton et al., 2003, 2015). Vuitton et al. (2003) listed at least 28 countries with reported human AE cases, including two in North America, 13 in Eurasia and the Middle East, two

[5] Mähren = Moravia, eastern part of the Czech Republic.

in Asia, and 11 in (western and central) Europe. Of 559 human AE cases registered by the European Echinococcosis Registry during the period 1996 to 2000 in nine European countries, 539 (96%) were diagnosed in the well-known 'classical' endemic area (Austria, Germany, Switzerland) and in France (Kern et al., 2003). Within this area, systematic active case finding studies have only been performed in Switzerland. In this country, the nationwide annual average numbers of new human AE cases were initially low with 0.6, 3.1 and 3.0 cases in the periods 1855 to 1900, 1901 to 1924 and 1926 to 1955, respectively. In the following period 1956 to 2000, these values were higher and varied between 6.6 and 10.0 cases, corresponding to annual incidences per 100,000 population between 0.10 and 0.16 (reviewed in Eckert et al., 1995; Schweiger et al., 2007). These data could be interpreted as a situation of 'endemic stability.' However, a distinct increase to 19.2 cases per year was observed in the period 2001 to 2005, resulting in an incidence rate of 0.26, possibly associated with increasing fox populations in rural and urban areas (Schweiger et al., 2007). A similar stable situation existed on Hokkaido/Japan where in the period 1937 to 1997, the average number of new cases per year varied between 4.2 and 11.2 (reviewed in WHO, 2001a).

The People's Republic of China is a more recently recognized focus with the first human AE cases detected at the end of the 1950s and endemic areas in eight provinces or autonomous regions and high prevalence rates (Vuitton et al., 2003). The global burden of AE has been estimated at over 600,000 Disability-Adjusted Life Years) (Torgerson and Macpherson, 2011).

3.1.2.2 *Echinococcus multilocularis* in definitive and intermediate hosts

Knowledge of the geographic range of AE was initially based on human cases. Dardel (1927) in Switzerland associated human AE with dogs but mentioned that foxes, hares and cattle could possibly play a role in the transmission cycle. Rare cases of multilocular liver echinococcosis in livestock, predominantly in cattle, were initially misdiagnosed as AE by several authors but later identified as CE (Huber, 1861 in Tappe et al., 2010a). The potential role of dogs as definitive hosts of *E. multilocularis* was first substantiated 1901/1902 by the experimental infection of a dog with metacestodes obtained from human AE case (Posselt, 1928).

In the 1950s, arctic foxes (*A. lagopus*) and sledge dogs were identified in Alaska as definitive hosts of *E. multilocularis* (then called *E. sibiricensis*), with voles as intermediate hosts (Rausch, 1951, Rausch and Schiller, 1954, 1956).

The major role played by foxes in the life cycle and epidemiology of *E. multilocularis* was subsequently documented by studies in Europe (Vogel, 1955, 1957, 1961), United States of America (North Dakota) (Rausch and Richards, 1971), the former Soviet Union (Lukashenko, 1971) and Hokkaido/Japan (Yamashita, 1963 cit.in Zehyle, 1982). For example, in North Dakota, 67 (70%) of 96 red foxes were carriers of *E. multilocularis* (Rausch and Richards, 1971) and 23% in Hokkaido (Yamashita, 1963 in Zeyhle, 1982). In some of the early European publications, the following prevalence rates of *E. multilocularis* in red foxes were recorded: eastern Switzerland 36% (8/22) (Bouvier et al., 1957), each 40% (4/10) in northeastern Switzerland and southern Germany (Swabian Alb) (Vogel, 1955, 1961), 5% (8/167) in France (Coudert et al., 1970), and 13.5% (598/4441) in southwestern Germany (Zehyle, 1982). Concurrently, rodent species (e.g., field voles in Germany) were identified as natural intermediate hosts (Vogel, 1961).

In studies performed in various geographic regions since the mid-1950s further definitive host species (wolf, coyote, raccoon dog, corsac fox, Tibet fox etc.) of *E. multilocularis* and numerous natural intermediate host species (small mammals, mainly arvicolid and cricetid rodents) were identified (reviewed in Rausch, 1986, Schantz et al., 1995; WHO, 2001a; Vuitton et al., 2003). Furthermore, life cycle patterns were described (Rausch, 1986, 1995) (see below), as well as the sylvatic cycle (wild carnivores—wild intermediate hosts) and the synanthropic cycle (domestic dog—wild intermediate hosts) recognized as epidemiologically relevant (WHO, 2001a). Significant contributions to the epidemiological knowledge were made by studies on the dynamics of intermediate host populations and their relationships to landscape characters (Giraudoux et al., 2002).

3.1.2.3 Expansion or new detection of endemic areas?

By the end of the 1980s, *E. multilocularis* was known to occur in foxes in Central Europe only in four countries, including Austria, France, Germany and Switzerland (Stössel, 1989; Eckert, 1996). Further examinations of foxes (mostly large numbers) performed since 1989 revealed that the European endemic area of *E. multilocularis*, as determined by the presence of *E. multilocularis* in definitive hosts (predominantly foxes) is much larger than previously anticipated. By the mid-1990s, the known geographic range of *E. multilocularis* included the 'endemic region in Central Europe, most of northern Eurasia, form Bulgaria and Turkey through most of Russia, and the newly independent nations of the former Soviet Union, extending eastward to several of the Japanese islands. In North America the cestode is found throughout the northern tundra zone and in a discontinuous zone in the south' (Schantz et al., 1995). How the image of the known endemic area in Europe has changed is evidenced by the fact that *E. multilocularis* has now been recorded in definitive hosts in at least 22 of 47 European countries (status 2016, see chapter "Epidemiology", this volume).

Evidence of *E. multilocularis* spreading to previously nonendemic regions has been reported from Japan and North America. In Japan, the parasite was apparently introduced by means of foxes from the Kuriles to Rebun Island off the northern coast of Hokkaido where it was known to occur since 1936. In the period 1966 to 1971, infected humans and animals were found in the eastern part of Hokkaido, and in 1992 *E. multilocularis* was considered present throughout Hokkaido (Kamiya et al., 2004). In North America, the known endemic zone has extended from the northern tundra zone through parts of Canada further south to central states of the United States of America. It was assumed that arctic foxes migrating from the tundra southward were the means by which *E. multilocularis* became established in red foxes and rodents in Canada and subsequently in central North America (Rausch, 1967b, Rausch and Fay, 2002). However, recent studies suggest a more complex situation. For example, a European strain of *E. multilocularis* has been identified in Canada, which might have been introduced with foxes (for commercial use) or dogs imported from Europe (Massolo et al., 2014).

As mentioned above, in Europe, the known geographic range of *E. multilocularis* in foxes has extended considerably since the 1980s, and cases of human AE have been found in regions previously not recognized as endemic, for example in Poland and Lithuania. To explain this situation, various factors have been proposed, including *E. multilocularis* dispersal by means of fox migrations in recent years, increase of fox populations, changes of landscape characters, and increased disease awareness, misdiagnosis and underreporting of human AE cases, and the use inadequate techniques for diagnosing *E. multilocularis* in foxes and other definitive hosts. An excellent overview of this discussion is presented in review articles of Vuitton et al. (2003, 2015).

There is no doubt that by routine necropsy of foxes only with macroscopic inspection of the intestinal mucosa, infections with *E. multilocularis* can be easily overlooked, especially if worm numbers are low. This is supported by the fact that in Europe the parasite has been detected in many regions soon after the employment of necropsy techniques targeted to *E. multilocularis*. Most likely, an increased disease awareness has played also a role in the identification of previously unknown endemic areas.

Well-documented indicators for fox migrations since the end of the 1980s were the increasing fox numbers in cities in Europe, Canada and Japan (Deplazes et al., 2002) and the establishment of the life cycle including foxes and rodents in the urban environment (Deplazes et al., 2002, 2004).

Interesting and epidemiologically relevant questions are how long the recently detected endemic regions have existed unnoticed and when they might have been established. Knapp et al. (2009a) studied the genetic diversity of *E. multilocularis* in Europe using the microsatelite marker EmsB in association with matching the fox hosts geographical positions. A central core of the European focus was identified in Switzerland and the Swabian Jura (Germany) flanked by neighbouring regions where *E. multilocularis* exhibits a lower genetic diversity than that in the centre. The authors concluded that *E. multilocularis* has needed more than a few decades to migrate distances of more than 1000 km; 'thus in those countries, the apparent emergence of human AE is more likely due to an active search as a consequence of disease awareness and only secondarily due to an increase of parasite prevalence' (Knapp et al., 2009a).

3.1.2.4 Life cycle patterns of *Echinococcus granulosus* and *Echinococcus multilocularis*

In the past, at least 16 *Echinococcus* species were described (Verster, 1965; Ohbayashi, 1993), but subsequently only four of them were recognized as valid (*E. granulosus*, *E. multilocularis*, *Echinococcus oligarthrus*, *Echinococcus vogeli*) (Verster, 1965; Rausch, 1967a, Thompson, 1986a,b). The fact that 10 subspecies were morphologically recognized in *E. granulosus* and only three in *E. multilocularis* was an indication of a considerable intraspecific variation in *E. granulosus*. Studies of populations of *E. granulosus* recovered in different regions have demonstrated intraspecific variations in final and intermediate host assemblages as well as in morphological and other characteristics (Thompson, 1986a,b 1995). Meanwhile intensive studies by various research groups have identified several strains of *E. granulosus* differing in their host preferences, infectivity to humans, and other characteristics, as described in detail in chapter "Biology and Systematics of *Echinococcus*' (this volume). Less intraspecific variability has been described in *E. multilocularis*. Respective research in the last five to six decades has deeply changed and enriched the understanding of epidemiological interconnections and has an impact for disease prevention and control.

3.2 Clinical aspects and pathology

3.2.1 Echinococcosis in humans

The literature published at the turn of the 19th to 20th century is a rich source of information on clinical aspects and pathology of human echinococcosis. This may partially be due to the fact that autopsies of humans were formerly frequently performed, and some authors had collected and evaluated large numbers of cases. For example, in Munich/Germany, 14,830 autopsies were made in the period 1854 to 1887 (436 per year) (Schwarz, 1928), and Neisser (1877) listed in his thesis 968 echinococcosis cases, compiled from the literature and hospital reports, many of them with case histories. **Albert Neisser** (1855—1916) was born in Schweidnitz, Germany (now Swidnica, Poland). He studied medicine at the University of Breslau (now Wrocław), and published in 1877 his dissertation on echinococcosis ('Die Echinococcen-Krankheit'). In 1879 he discovered the causative agent of gonorrhoea (*Neisseria gonorrhoeae*) and was appointed to Professor of Dermatology at the University of Wrocław in 1907 (Figs 15 and 16).

At that time, clinical signs and pathologies of echinococcosis were quite well known, and CE and AE were distinguished as clinical entities although it was still uncertain whether these two forms of human echinococcosis are caused by a single or two different *Echinococcus* species. Neisser (1877), Langenbuch (1890) and Lehmann (1928) presented detailed descriptions of clinical signs and pathologies of CE, and Posselt (1928) contributed a comprehensive review of AE. The differences between both forms were clearly documented, and many of the previously known characteristic

Figure 15 A. Neisser (1855—1916). *Wikipedia—die freie Enzyklopädie.*

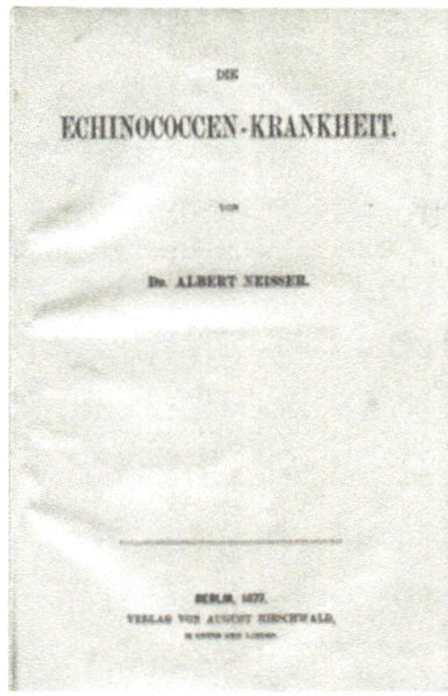
DIE
ECHINOCOCCEN-KRANKHEIT.
Dr. ALBERT NEISSER.

Figure 16 Book of A. Neisser (1877): title page. *Photo: J. Eckert.*

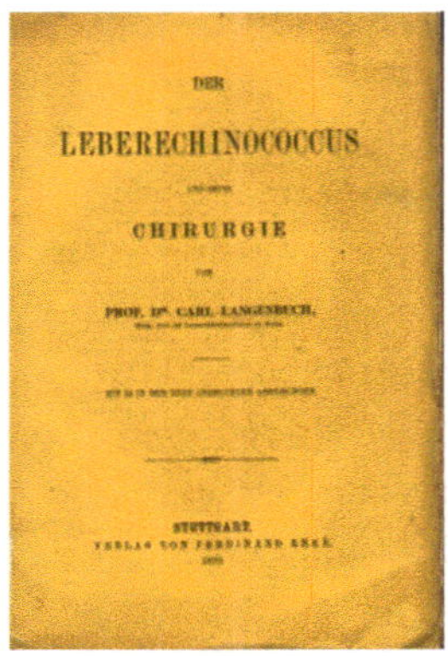
DER
LEBERECHINOCOCCUS
CHIRURGIE
STUTTGART.

Figure 17 Book of Langenbuch (1890): title page. *Photo: J. Eckert.*

features are still valid (Ammann and Eckert, 1996; Pawlowski et al., 2001; Kern et al., 2004; Junghanss et al., 2008; Brunetti et al., 2010) (Fig. 17).

Over the years, clinical research and exchange of information has led to a better understanding of human echinococcosis, a more precise definition of clinical cases, and an international classification of disease entities. Such classification systems are relevant for improving the diagnosis and the prerequisites for therapeutic interventions. The WHO Informal Working Groups on Echinococcosis (IWGE) (see Section 5) developed an international classification of ultrasound images of abdominal CE (WHO-IWGE, 2003). This grading system allows clinicians to identify five different types of (liver) cysts (e.g., uni- or multilocular, active or inactive), to follow the development of cysts, to recommend specific procedures of intervention and to apply standardized criteria in assessing the evolution of cysts after puncture–aspiration–injection–reaspiration (PAIR) or chemotherapy (WHO-IWGE, 2003; Junghanss et al., 2008). Sensitivity and specificity of ultrasound (US) examinations for abdominal cysts are high and reported to be at least 93% to 98% or 88% to 90%, respectively (Macpherson et al., 2003; WHO-IWGE, 2003). Based on the International Classification of Diseases and Related Health Problems (ICD 10), CE cases can be subclassified into several disease entities (Brunetti et al., 2010). The European Network for Concerted Surveillance of Alveolar Echinococcosis (see Section 5) and the WHO-IWGE have proposed a classification system for AE which includes imaging findings on the localization of the parasite in the liver, extrahepatic involvement of neighbouring organs, and absence or presence of distant metastases (PNMs) (Kern et al., 2006). As in CE, definition of AE cases follows the ICD system mentioned above (Brunetti et al., 2010). Regarding details and other clinical aspects, the reader is referred to the reviews of Junghanss et al. (2008) Brunetti et al. (2010) and chapter "Clinical management...", this volume.

In the old literature, the tumour-like proliferation and metastasis formation of *E. multilocularis* metacestodes in the human body was well recognized, but the mechanisms were unknown (Posselt, 1928). The proliferation of the parasite was explained by Jahn (1927) with diverticle formation of the vesicle wall (germinal and laminated layer) resulting in a network of many interconnected vesicles of various diameters and forms. Rausch (1954) infected voles (*Microtus pennsylvanicus*) orally with *Echinococcus* eggs obtained from arctic foxes and studied the development of the alveolar metacestodes and the hosts' histological reactions from 20 h to 170 days p.i. Metacestodes developed primarily in the liver; they grew rapidly and formed aggregates of cysts usually less than 10 mm in diameter. In long-lasting infections, metastatic foci were observed in various organs. At 14 days p.i., when a subgerminal 'laminated layer' was not yet discernable, Rausch observed 'numerous isolated masses of germinal tissue which are the forerunners of new vesicles.' Regarding metastasis formation, Rausch (1954) thought that the dilatation of blood vessels in the liver resulting from larval growths 'apparently is adequate to allow bits of larval tissue to pass into the hepatic vain and thence into the systemic circulation.' According to Lukashenko (1975), parasite proliferation is due to internal fission of vesicles in two or more parts which form daughter vesicles after separation. Vogel (1978) reviewed the literature (back to Leuckart in 1886 and others) and discussed various hypotheses regarding the growth of *E. multilocularis* metacestodes. In his own studies with experimentally infected voles (*M. arvalis*), he had identified by light microscopy small protuberances of the germinal layer, devoid of a central cavity and a laminated layer, that he regarded as 'juvenile form of the vesicular stage excellently suited to enter narrow inter spaces.'

Following Vogel's studies, Eckert et al. (1983) observed that the subcutaneous transplantation of *E. multilocularis* metacestode tissue into the neck region of rodents (*Meriones unguiculatus*) resulted not only in parasite growth but also metastasis formation first in regional lymph nodes and subsequently in the lungs, indicating parasite spreading via the lymph and blood system. In further electron microscopical studies, slender solid cell columns (buds) protruding from the germinal layer were detected, that were devoid of a laminated layer. These root-like protrusions infiltrated the surrounding host tissue and transformed later into tube-like structures with a central cavity, covered by a laminated layer (Mehlhorn et al., 1983). Thus the previous observations of Rausch (1954) and Vogel (1978) could be confirmed and extended. To our knowledge, there are no reports on such structures in metacestode tissue obtained from humans. It can be assumed that proliferation involving the root-like protuberances only occurs in very young, actively proliferating parts of the metacestode. Tappe et al. (2010b) reconstructed three-dimensional images of histological sections of metacestode material from AE patients and observed a root-like network of interconnected tubules.

In the old literature, the immunological interplay between metacestodes of *E. granulosus* and *E. multilocularis* and their hosts was discussed in terms of the role of parasite antigens in stimulating humoral antibody and cellular skin reactions as well as anaphylactic events after sudden release of hydatid cyst fluid (Lehmann, 1928). Furthermore, the chemical composition of cyst fluid was discussed in some detail (Neisser, 1877). Recent developments of the immunobiology of the *Echinococcus* infection are described in previous reviews (e.g., Heath, 1995; Lightowlers and Gottstein, 1995; Gottstein and Hemphill, 2008; Mejri et al., 2010; Siracusano et al., 2012; Zhang et al., 2012) and in chapter "Immunology...", this volume.

3.2.2 Echinococcosis in animals

A vast older literature exists on CE in many intermediate and aberrant host species, including natural history of the cysts, their organ localization, clinical manifestations, prevalence in various countries and regions etc. (Dardel, 1927; Hosemann, 1928; Thompson and Allsopp, 1988; WHO, 2001a; see also databanks). Since the 1990s numerous reports on naturally acquired AE in a wide spectrum of mammalian accidental hosts have been published, including nutria, wild boar, domestic pig, horse, dog and various genera of monkeys. Severe clinical implications and lethal cases of AE were observed in dogs and monkeys (reviewed in WHO, 2001a; Deplazes and Eckert, 2001). For information on AE infections in natural intermediate hosts and the wide spectrum of intermediate host species, see reviews (WHO, 2001a; Rausch, 1986, 1995, Vuitton et al., 2003).

3.3 Diagnosis of echinococcosis

3.3.1 Diagnosis in humans

Formerly, the clinical diagnosis of echinococcosis in humans was almost exclusively dependent on symptoms which could be discovered by inspection and palpation (e.g., distended abdomen, palpable fluctuating cysts) (Langenbuch, 1890; Lehmann, 1928). Although the clinical manifestations, caused by cysts of *E. granulosus* in various organ systems, were quite well known to authors in the second half of the 19th century, differential diagnosis (tumours, liver abscess etc.) was difficult in many cases. Therefore over the years, attempts were made to introduce and apply improved diagnostic methods.

3.3.1.1 Diagnostic puncture

Neisser (1877) underlined the importance of puncture as a diagnostic tool and presented a table with detailed information on the differential diagnosis of CE, ascites, ovarian cysts, hydronephrotic lesions etc., including data on specific weight and composition of the puncture fluid. Diagnostic puncture, which had already been occasionally applied in the ancient world and the middle ages, was widely used in the 19th and the beginning of the 20th century. This method was regarded as useful and mostly harmless, but the risks were also recognized (dissemination of protoscoleces, allergic reactions, bacterial infections etc.) (Langenbuch, 1890; Hosemann, 1928). According to Lehmann (1928), puncture was a malpractice and 'prohibited in all circumstances.' Still today, fine-needle puncture is used occasionally (e.g., cysts or unclear lesions in seronegative persons).

3.3.1.2 Immunodiagnosis

Immunodiagnosis of echinococcosis in humans dates from the beginning of the 20th century. For the detection of circulating anti-*Echinococcus* antibodies Ghedini, Weinberg and Parvu (Weinberg, 1909; Lehmann, 1928) developed a complement fixation test (also known as Ghedini–Weinberg test), and Fleig and Lisbonne (1907 in Lehmann, 1928) a precipitation test while Casoni and Botteri (in Lehmann, 1928) invented an intradermal test (later known as Casoni test). From the 1950s, these tests were complemented or gradually replaced by better methods, such as indirect haemagglutination test, bentonite and latex agglutination tests, immunoprecipitation and immunoelectrophresis. Some of these tests were quite sensitive in detecting CE, especially of the liver, but a general problem was the low degree of specificity (Kagan, 1968; Varela-Diaz and Coltorti, 1976; Rickard and Lightowlers, 1986). Later on, the repertoire of antibody tests was extended by further procedures, including the indirect fluorescent antibody test, the enzyme-linked immunosorbent assay (ELISA) and some secondary tests. The use of purified or recombinant antigens in modern testing procedures significantly improved the reliability of diagnostic results (Lightwolers and Gottstein, 1995; WHO, 2001a) further information see chapters "Immunology..." and "Laboratory diagnosis", this volume.

3.3.1.3 Imaging procedures

Around 1900 radiography was introduced as a new diagnostic technique, followed since the late 1960s by ultrasonography, computed tomography, angiography, cholangiography, magnetic resonance imaging, and in the early 1990s, by positron emission tomography. These imaging procedures presented new options for diagnosing of CE and AE in humans as documented in an overview by von Sinner and Lewall (2001). The introduction of US for the diagnosis of human abdominal echinococcosis in the early 1970s was of special relevance, not only for the clinical diagnosis as described above, but also for mass screening of populations (see below).

3.3.1.4 Mass screening of human populations

The use of portable ultrasound scanners for mass screening of populations since the mid-1980s and early 1990s was a great step forward. Community-based mass screenings for CE have been performed in several countries in Africa, South America and Asia (China), including remote areas and large groups of people (up to 20,220 in one of the studies) (Macpherson et al., 2003). In these studies, ultrasound mass screening has proven a reliable and relatively cheap method for demonstrating the true extent of human CE (Macpherson and Milner, 2003; Macpherson et al., 2003). Screening surveys for human AE have also been conducted in endemic areas, for example, in China and France (Craig et al., 1996). However, serological tests are required in many cases to confirm the aetiology of the lesions (Macpherson and Milner, 2003; Macpherson et al., 2003). In some of these programmes, ELISAs, in combination with western blot analyses, were used for detection of specific serum antibodies. In view of the low prevalence of human AE in most of the endemic areas, it is essential to use only test systems which are highly sensitive and specific (reviewed in WHO, 2001a). The special value of ultrasound mass screening is that cases of CE and AE can be detected in an early stage thus improving the chances for effective treatment and a better prognosis.

3.3.2 Diagnosis of echinococcosis in animals

3.3.2.1 Definitive hosts

In the past, the diagnosis of the intestinal *E. granulosus* infection in living dogs was highly unreliable because the eggs of *Echinococcus* and *Taenia* species are morphologically indistinguishable, and spontaneously and irregularly eliminated small *Echinococcus* proglottids can easily be overlooked. Better results were obtained by examination of faecal samples collected after purgation of dogs with arecoline hydrobromide. This method was used as a diagnostic aid in many of the control programmes and in epidemiological studies but can now be replaced by the coproantigen ELISA (see below). The examination of the small intestine for mature or immature *E. granulosus* stages at necropsy is restricted for obvious reasons to small numbers of dogs. The purgation and necropsy techniques have been described and reviewed in various publications, including WHO documents (WHO, 1984b, 2001a).

Necropsy was the only option for diagnosing intestinal *E. multilocularis* in foxes and other definitive hosts. However, at routine necropsies with macroscopic inspection of the mucosa, the parasites can be easily missed, especially when worm numbers are low. Therefore since the late 1970s in many studies, either the intestinal scraping technique (IST) or the sedimentation and counting technique (SCT) was employed after the intestines or carcasses had been kept deep frozen at −80°C for at least 4 days (routinely 7 days for carcasses) in order to kill parasite eggs and to reduce or exclude a potential infection risk for laboratory personnel (Eckert et al., 2001; Mathis and Deplazes, 2004). The IST has been widely used in European epidemiological studies on *E. multilocularis* in foxes, but it may underestimate the true prevalence by about 20% (Mathis and Deplazes, 2004). The SCT (or modifications of it) has usually a higher sensitivity (over 90%) (Hofer et al., 2000; Mathis and Deplazes, 2004). Both the IST and SCT are labour intensive and cost intensive (Mathis and Deplazes, 2004).

In the 1990s, significant progress was achieved by the development of methods for (1) detecting *Echinococcus*—specific antigens by ELISAs in faecal samples (Allan et al., 1992; Deplazes et al., 1992; and others reviewed in Allan and Craig, 2006) or (2) DNA in faecal material or in isolated eggs (Bretagne et al., 1993; Mathis et al., 1996; Deplazes et al., 2003; Mathis and Deplazes, 2004, 2006; Trachsel et al., 2007) by molecular techniques. These methods detect intestinal infections with *Echinococcus* species with high sensitivity and specificity (reviewed in Allan and Craig, 2006; Mathis and Deplazes, 2004) and can be applied for in vivo (faecal samples) and postmortem (intestinal content) examinations of animals. Furthermore, DNA detection allows the identification of *Echinococcus* species and strains (Stefanic et al., 2004) and the species-specific detection of *Echinococcus* eggs in the environment (faecal samples, soil) or in other materials (e.g., vegetables) (Mathis and Deplazes, 2004).

3.3.2.2 Intermediate and aberrant host animals

Early lesions caused by *Echinococcus* species in organs of animals are sometimes difficult to identify macroscopically or histologically. In such cases, DNA detection is a valuable diagnostic aid. Furthermore, serological examinations for specific antibodies have gained significance in the diagnosis of CE or AE in certain animals, such as monkeys or dogs.

3.4 Treatment of echinococcosis

3.4.1 Treatment of human echinococcosis

3.4.1.1 Cyst puncture

As early as in ancient times and the middle ages, doctors had tried to inactivate hydatids by **puncture** or **minor surgery** (see above diagnosis). In the mid-19th century, the French physicians **Récamier** and **Moissenet** began to employ this method specifically for treatment of CE (Langenbuch, 1890). *Echinococcus* cysts were punctured with a cannula, and the fluid was aspirated with a syringe in order to harm the parasite. Cysts were also drained by means of a trochar (Langenbuch, 1890). After **Boinet** in France and **Weber** in New York had recommended in 1851 to inject tincture of iodine into punctured cysts, solutions of many other substances were used in this indication (usually after aspiration of some cyst fluid) with the intention of increasing the detrimental effects on hydatids. The list of substances (and natural fluids) is long and included 'filix mas' ('worm fern,' an anthelmintic, containing filicic acid), ox bile, chlorinated water, copper sulphate, β-naphthol, boric acid, salicylic acid, mercury (II) chloride (sublimate), alcohol, formalin (1%) and formalin–glycerin (Neisser, 1877; Langenbuch, 1890; Lehmann, 1928). According to Neisser (1877), of 160 patients with abdominal hydatids, 97 (61%) were cured after puncture (single or repeated), aspiration and injection (substances not defined). The risks of puncture (spillage of hydatid fluid with subsequent dissemination of protoscoleces, anaphylactic shock, toxicity of the injected chemicals) were already well known in the 19th century. However, until recent years, formalin was injected into cysts for intraoperative killing of protoscoleces. In 1996 a WHO expert group deemed formalin unsafe and recommended that its use be stopped (WHO, 1996).

3.4.1.2 Puncture–aspiration–injection–reaspiration

The old principle of puncture experienced a revival within an improved and thoroughly controlled procedure, which was developed in the mid-1980s as the so-called **PAIR method**. PAIR includes the following steps: (1) percutaneous cyst puncture under ultrasonographic guidance, (2) aspiration of a substantial portion of cyst fluid, (3) injection of a parasitocidal solution (20% sodium chloride solution or preferably 95% ethanol; approximately an equivalent of one-third of the amount aspirated), and (4) reaspiration of the fluid content after 5 min in case of ethanol injection and at least 15 min if sodium chloride solution is used (Ben Amor et al., 1986; Filice and Brunetti, 1997; Brunetti et al., 2010). In 2001 the WHO Informal Working Group on Echinococcosis (WHO-IWGE, see below) published guidelines for the indication and use of PAIR as well as on its benefits and risks (WHO, 2001b; Brunetti et al., 2010). PAIR is considered a minimal invasive and alternative technique for surgery or chemotherapy of CE patients that harbour certain types of hepatic cysts defined according to an international classification system (WHO-IWGE, 2003). For further information on benefits and risk of PAIR, see Junghanss et al. (2008).

3.4.1.3 Surgery

Surgery has always played a prominent role in the therapy of echinococcosis. With regard to CE, Lehmann stated in 1928 'that treatment of the *Echinococcus* today is fundamentally purely surgical.' Accordingly, detailed descriptions of different surgical techniques can be found in the older literature (e.g., Langenbuch, 1890; Lehmann, 1928) and in the following periods (e.g., Morris and Richards, 1992; Uchino et al., 1993) up to liver transplantation (Koch et al., 2003). The use of surgery for treatment of CE and AE as single (potentially curative) method or in combination with chemotherapy and adjuvant procedures, as well as indications and contraindications is discussed in several reviews prepared by groups of international experts (Pawlowski et al., 2001; Junghanss et al., 2008; Brunetti et al., 2010).

3.4.1.4 Chemotherapy

In the literature of the 19th and the early 20th century, many attempts of systemic chemotherapy of CE and AE are described, reviewed by Langenbuch (1890) and Posselt (1928). They included, for example, 'heroic' treatments with emetics which were intended to cause cyst rupture and draining of the fluid into a body cavity. Laxatives, concentrated sodium chloride solutions or potassium iodide, turpentine, kalium iodide, or mercury (I) chloride (calomel) and other substances were used for systemic treatment. Langenbuch (1890) mentioned that by calomel treatment, 'man became more poisoned than the worm.' Later approaches of chemotherapy with atoxyl, neosalvarsan, arsenobenzole and thymol compounds failed to show success (Thiodét, 1954/55, Burkhardt, 1981). Thymol injections were still used in the 1960s, but there was no sound evidence for their efficacy against CE or AE neither in patients nor in experimental animals (Burkhardt, 1981).

3.4.1.4.1 Chemotherapy with benzimidazoles. Until the mid-1970s, many attempts to find antiparasitic drugs against experimental larval echinococcosis in rodents were unsuccessful. The tide turned in 1974 after Thienpont et al. (1974), researchers at Janssen Pharmaceutica Beerse, Belgium, had detected the efficacy of mebendazole against metacestodes of *T. taeniaeformis* in mice. Scientists in Australia described a high efficacy of this drug against metacestodes of *T. pisiformis* in rabbits and of *Mesocestoides corti* and *E. granulosus* in mice (Heath and Chevis, 1974; Heath et al., 1975). Authors in Russia (Krotov et al., 1974) and the United States of America (Campbell et al., 1975)[6] reported that the tumorous growth of *E. multilocularis* metacestodes in rodents can be inhibited by treatment with increased doses of mebendazole.

Further detailed experimental studies revealed that prolonged oral treatment of rodents with albendazole, fenbendazole, flubendazole and mebendazole significantly (mostly >90%) inhibited the proliferation of *E. multilocularis* metacestodes, damaged the parasite structure, prevented metastasis formation and prolonged the survival time of the treated animals, but usually did not kill the parasites. On the other hand, in the same model, cysts of *E. granulosus* could be killed (Eckert and Pohlenz, 1976; Eckert et al., 1978; Burkhardt, 1981; Schantz et al., 1982; Eckert, 1986). Concurrently,

[6] W.C. Campbell (formerly Merck Institute for Therapeutic Research, Rahway, New Jersey, United States of America), one of the laureates of the Nobel Prize Medicine and Physiology 2015.

pharmacological studies on the bioavailability of high oral doses of benzimidazoles and the presumably effective serum drug levels were performed in experimental animals and humans (Witassek et al., 1981; Luder et al., 1986). The results formed the basis for monitoring of serum drug levels in patients and adaptation of the oral doses.

Among the early reports on the use of high oral mebendazole doses against AE in humans were those of Akovbiantz et al. (1977) in Switzerland and Wilson et al. (1978) in Alaska, and against CE communications of Bekhti et al. (1977) in Belgium, Danis et al. (1977) in France, Beard et al. (1978) in Australia and Al-Moslih et al. (1978) in Iraq (for further references, see Schantz et al., 1982). In 1980 an international workshop on 'Chemotherapy of Larval Echinococcosis in Animals and Humans' was organized by the Janssen Foundation at Janssen Pharmaceutica, Beerse, Belgium, which was particularly supported by Dr Paul A. Janssen and his coworkers (Schantz et al., 1982). Since most of the early studies included only single or small numbers of patients and had given rather inconsistent results, the Swiss Study Group on Echinococcosis (steering committee members: A. Akovbiantz, R. Ammann, J. Bircher, J. Eckert) suggested that the WHO Parasitic Disease Programme (WHO/PDP) at WHO in Geneva coordinate international studies on the treatment of human echinococcosis with benzimidazoles (WHO/PDP, 1981). The project was supported by Dr A. Davis (Director, PDP), Dr Z. Matyas (Chief, Veterinary Public Health [VPH]) and Dr Z. Pawlowski (Senior Medical Officer, PDP) and launched in 1981, based on a uniform protocol (WHO/PDP, 1981, 1984). In 1986 the first results of a multicentric trial were published, which included 85 CE and 54 AE patients treated with mebendazole, albendazole or flubendazole at clinical centres in Anchorage, Beirut, Besançon, Paris, Rome, Sofia and Zurich. Pharmaceutical companies (Janssen Pharmaceutica, Smith, Kline & French) and several University institutes and hospitals cooperated in the studies (Davis et al., 1986). In the following years, various aspects of chemotherapy of human CE and AE were subjects of discussions at workshops or conferences and numerous publications (reviewed in Schantz et al., 1982; Eckert, 1986; Sato et al., 1993; Ammann and Eckert, 1995; Horton, 1997; WHO, 2001a; Horton, 2003; Eckert and Deplazes, 2004; Kern, 2004, 2006; Junghanss et al., 2008, and others). The continuous interest of WHO (Dr F.-X. Meslin, Dr T. Fujikura, Dr Z. Pawlowski, Dr L. Savioli) into echinococcosis and the international cooperation of research groups within the framework of 'WHO Informal Working Groups on Echinococcosis' (see below) proved to be stimulating and important for the further development of chemotherapy. Methods and results were discussed and evaluated at several international meetings, for example, 1983 in Geneva, Switzerland (WHO, 1984a); 1990 in Anchorage, Alaska; 1992 in Besançon, France (WHO/CDS/VPH, 1992); 1993 in Beijing, China (WHO/VPH, 1993); 1994 in Al-Ain, United Arab Emirates (WHO, 1996) and 1995 in Hokkaido, Japan. In 1996 these efforts resulted in the publication of 'Guidelines for treatment of cystic and alveolar echinococcosis in humans' prepared by numerous experts of the WHO-IWGE (WHO, 1996). An updated guideline was published in 2010 (Brunetti et al., 2010). Results on the benefits and problems of chemotherapy of human CE and AE have been reviewed in various articles (e.g., Ammann et al., 1999; Pawlowski et al., 2001; Eckert and Deplazes, 2004; Kern, 2006, 2010; Eckert et al., 2011; chapter "Clinical management...", this volume).

3.4.1.4.2 Alveolar echinococcosis. Historically, AE had a high lethality rate in untreated patients of 90% within 10 years from the onset of clinical symptoms and virtually 100% within 15 years (Ammann and Eckert, 1996). In a Swiss study, the analysis of data of 329 patients over a period of 35 years revealed that current case management and treatments (improved diagnosis and surgery, chemotherapy with mebendazole or albendazole, and other measures) substantially improved the life expectancy of AE patients compared to the 1970s. Whereas the average life expectancy of a male AE patient in 1970 was 6.2 years, it increased to 25.1 years in 2005 (Torgerson et al., 2008). Vuitton (2009) concluded that 'benzimidazoles have deeply modified the management of cystic echinococcosis patients and life expectancy of alveolar echinococcosis patients.'

Due to the activities and endurance of **Rudolf Ammann** (1926–2015), Professor of Gastroenterology at the University Hospital Zürich, and a group of colleagues and coworkers (see list of references), a large group of AE patients under treatment and medical care was observed in Switzerland over more than three decades (Ammann et al., 1999, 2004; Kadry et al., 2005; Torgerson et al., 2008). A sign of R. Ammann's tireless passion for echinococcosis research is that the results of his last long-term study were published 3 months before his death (Ammann et al., 2015).

3.4.1.4.3 Cystic echinococcosis. Chemotherapy of CE patients with albendazole or mebendazole is often only partially effective, and rarely curative with complete regression of the cysts. Results for over 2000 well-controlled cases treated with benzimidazoles and evaluated for up to 12 months have shown that 10% to 30% of the cysts die (cure),

50% to 70% respond (degeneration or size reduction of cysts), and 20% to 30% do not exhibit morphological changes (Horton, 1997; Todorov et al., 1992; Pawlowski et al., 2001; Horton, 2003; Junghanss et al., 2008). Based on a meta-analysis, including 711 patients, it was estimated that even 2 years after initiation of treatment, 40% of the cysts are still active or become active again (Stojkovic et al., 2009).

Although chemotherapy of human AE and CE with benzimidazoles represents significant progress, these drugs are not fully satisfactory. In recent years, several substances have been tested in vitro or in laboratory rodents, but so far no drugs superior to benzimidazoles have been developed for use in humans (Hemphill et al., 2010).

3.4.2 Chemotherapy of intestinal Echinococcus infections in carnivores

One of the key measures to prevent CE in humans is mass treatment of dogs against *E. granulosus* (see below). Up to 1977, several anthelmintics were used in this way, including arecoline hydrobromide, bunamidine hydrochloride, niclosamide, and nitroscanate (Gemmell, 1978). Since arecoline hydrobromide acts as a purgative and often eliminates only a proportion of the *Echinococcus* burden, its value in control programmes and as a diagnostic tool (see above) is limited. The other anticestodal drugs mentioned above had certain limitations and were not fully satisfactory for various reasons. A significant advance was the introduction of praziquantel. Its efficacy against intestinal cestodes was announced in 1975 by Thomas et al., scientists of the Bayer Company, Germany. At a single dose, this well-tolerated drug has a very high and reliable efficacy against mature and immature intestinal stages of *Taenia* species as well as of *E. granulosus* and *E. multilocularis*. Since 1977 it has been successfully and widely used for treating individual dogs or cats infected with *Echinococcus* species and in control campaigns against *E. granulosus* in dogs and *E. multilocularis* in dogs and foxes (see below).

3.5 Control and Prevention

3.5.1 Echinococcus granulosus

Shortly after the elucidation of the life cycle of *E. granulosus* in 1853, first proposals were made for controlling this parasite. For example, Küchenmeister (1855) in Germany suggested official regulations that organs of slaughter animals containing 'bladder worms' should not be left for dogs to eat as food but should be destroyed. In view of the environmental contamination with *Echinococcus* eggs, he recommended caution with unboiled drinking water, raw root vegetables or fruit windfall and demanded from governments educational programmes on echinococcosis. In Iceland, which had previously a high incidence of human CE, **Dr Arthur Leared**, an English physician who visited the country in 1862, recommended the destruction of hydatid cysts and the treatment of dogs with anthelmintics (Grove, 1990; Thakur, 2002).

The first long-term control campaign was started in **Iceland** in 1864 with an educational programme introduced by **Harald Krabbe** (1831–1917), then assistant at the Royal and Agricultural College in Copenhagen, Denmark (Enigk, 1986). The first legislation to control hydatid disease in Iceland was established in 1869, including registration of all dogs, payment of tax on nonworking dogs, destruction of hydatid cysts and infected viscera and several other measures, such as restrictions to dog populations, and annual treatment of dogs, initially with areca nut (containing arecoline as an active ingredient), from 1930 with arecoline hydrobromide, and from 1977 with praziquantel (Beard, 1973; Thakur, 2002; PAHO, 2002). The programme lasted 110 years from 1869 to 1979. Iceland is now free of *E. granulosus*; the last human case of CE was diagnosed in 1960.

Since 1959, further long-term and large-scale (nation-wide or regional) control campaigns have been performed, including those in New Zealand (1959–97) (Gemmell, 1978, 1987, 1990; Heath et al., 2002), Tasmania (Meldrum and McConnell, 1968; Thompson, 2002), Falkland Islands (1965–77) (PAHO, 2002), and Cyprus (1971–94) (Polydorou, 1984, 1994; Economides and Christofi, 2002; Christofi, 2011). These programmes were based on general measures outlined by Atwater (1969) in the textbook 'Veterinary Medicine and Human Health,' edited by **Calvin W. Schwabe** (1927–2006), Professor of Epidemiology at the School of Veterinary Medicine and School of Medicine, University of California, Davis, United States of America (Schwabe, 1969). He was widely known as the father of veterinary epidemiology and directed as a consultant the WHO global hydatid research and control programme in 1960 (Kass et al., 2006). Specific plans and strategies for the control of *E. granulosus* were substantially elaborated by **M.A. Gemmell** and his team in New Zealand, summarized by

Atwater (1969) and in WHO guidelines (WHO, 1981, 1984b). Essential components of these programs were legislation, the establishment of a control authority, long-term funding, collection of base-line data (human cases of CE, infection rates of dogs and intermediate hosts), collection and evaluation of continuing data, educational measures and technical control measures. These programs employed several key control measures, such as sanitary education, surveillance of food animals, construction and improvement of slaughterhouses, meat inspection, adequate disposal of slaughter offal, quarantine of premises with infected dogs and livestock, registration of dogs, demonstration of the infection in dogs, and dog dosing with anticestodal drugs at regular intervals (e.g., four to eight times a year with praziquantel). The programme in Cyprus included a strong policy on killing of infected ownerless and owned dogs (PAHO, 2002).

The structures and results of these programs were reviewed on several occasions, for example, up to 1974 by the Pan American Zoonoses Center and WHO (Gemmell and Varela-Diaz, 1980), by PAHO in 1994 and 1999 (PAHO, 1994, 2002), Larrieu and Pérez Palacios, 2002; Craig and Larrieu (2006) and WHO-IGWE (2011). Furthermore, related data are documented in many publications (e.g., Gemmell and Lawson, 1986; Gemmell, 1987; Schantz et al., 1995; WHO, 2001a). Over many years, these programs resulted in a significant reduction of the infection rates of dogs, sheep or other livestock and humans. Until 1999 the programs in New Zealand, Tasmania and the Falkland Islands had achieved the consolidation or maintenance of eradication phase (PAHO, 2002). In Cyprus, eradication was claimed in 1985, but control had to be reintroduced in 1994 (PAHO, 2002) and appeared to be successful (the last infected dog and intermediate hosts were found in 1996 and 2010, respectively) (Christofi, 2011). Since the 1970s, several South American countries have initiated control programs, including regions in Argentina, Uruguay, Brazil, Chile and Peru (Gemmell and Lawson, 1986; PAHO, 1994). The programs in Argentina and Chile were restricted to small parts of the endemic areas and are classified as partially successful; others have been discontinued or modified from the original design (Gavidia, 2011). Control programs have also been initiated in other regions, for example, more recently in Kyrgyzstan (Abdykerimov, 2011) and China (Wang, 2011). In China, control programmes were implemented in 2006 in 10 counties in the Sichuan Province and extended in

2010 to 170 counties in seven provinces or regions (Sichuan, Xinjiang, Inner Mongolia, Gansu, Qinghai, Ningxia, and Tibet) (Wang, 2011). In Sichuan Province, the proportion of diagnosed CE patients dropped from 2.4% (4247/178,358) in 2008 to 0.3% (717/,237,399) in 2010, and the rates of coproantigen-positive dogs in the same period from 18% to 15.9% (Wang, 2011).

The successful historical examples show that control campaigns against *E. granulosus* are long-term actions which require a high expenditure of time and financial resources. Therefore in recent programmes, efforts are made to introduce innovations and improvements, such as mass population screening (ultrasound examinations, serology) and treatment of human CE cases in order to increase the awareness and health status of local people, and vaccination of livestock with the new EG95 vaccine (Heath et al., 2003; Craig, 2011; Lightowlers, 2006, 2011; Torgerson, 2011; Wang, 2011). Mathematical models are useful for estimating the disease burden for the community and for evaluating various control options (Torgerson, 2003, 2006).

3.5.2 *Echinococcus multilocularis*

Control of *E. multilocularis* is very difficult because the primary cycle is sylvatic, typically with foxes as definitive and rodents as intermediate hosts. Rausch et al. (1990) performed a 10-year field trial in an Alaskan village where dogs in a synanthropic cycle had access to rodents as intermediate hosts. All dogs of the village were treated monthly with praziquantel (5 mg/kg body weight). By this intervention, the infection prevalence in rodents was reduced from 29% to 5%, but the infection rate rebounded quickly toward pretreatment levels when treatment was discontinued (Wilson, cit. in Schantz et al., 1995). In two large field trials in Germany, comprising areas of 3400 to 5000 km^2, the prevalence of *E. multilocularis* in sylvatic cycles could be significantly reduced (15%/67% and 3%/26%, respectively) by regular delivery of praziquantel-containing baits to foxes (Romig et al., 1999; Hansen et al., 2003; Romig, 2011). However, the prevalence rebounded to precontrol levels after termination of baiting (Romig, 2011). A significant reduction of the *E. multilocularis* prevalence in wild foxes by baiting with praziquantel was also documented in Hokkaido, Japan (Ito et al., 2003; Takahashi et al., 2013). On the other hand, it has been shown that fox baiting in small (~1 km^2), highly endemic urban areas can contribute to the reduction of environmental contamination with *E. multilocularis* eggs (Hegglin et al., 2003). However, the currently available data suggest that sustainable effects can only be achieved by permanent intervention schemes. In order to reduce the potential infection risk for pet owners, regular praziquantel treatment of dogs and cats which catch wild rodents is recommended in endemic areas.

3.6 Aspects of basic research

In his famous textbook on parasites of humans, Leuckart (1863) has described with impressive precision the anatomy and development of cestodes, including juvenile and adult stages of '*Taenia echinococcus*' and the development of hydatid cysts with formation of brood capsules and protoscoleces. Subsequent generations of authors have referred to this basic information (e.g., Hosemann, 1928). Further studies on the structure, biology and physiology of *E. granulosus* and other species were performed later, increasingly since the 1950s, and have been reviewed by Smyth (1964), Thompson (1986b), McManus and Bryant (1986), Thompson (1995), McManus and Bryant (1995) and others.

3.6.1 Echinococcus granulosus

Professor **James Desmond Smyth** (1917–99) was one of the eminent pioneers in the field of basic echinococcosis research. He was born in Dublin and obtained the degrees of BA and BSc in 1940 at the University of Dublin. Important positions in his career were Professor of Experimental Biology at Trinity College in Dublin (1955), Foundation Professor of Zoology in the School of General Studies at the National University, Canberra/Australia (1959–70) and Professor of Parasitology at Imperial College of Science and Technology, London (1970–82) (Bryant and Barwick, 2016). His major contributions to research on echinococcosis include the establishment of systems for the in vitro cultivation of *E. granulosus*, basic studies on the development of the adult parasite (germinal and somatic differentiation, proglottisation, segmentation etc.) and elucidation of external factors influencing parasite development (Smyth, 1964). Considerable progress has been achieved by the axenic cultivation of *E. granulosus* (sheep strain) with the reproducible development of protoscoleces to segmented, sexual mature stages without egg production. Furthermore, development of eggs or protoscoleces to sterile cysts was achieved (Smyth and Davies, 1974; Howell,

Figure 18 J.D. Smyth (1917–99). *Collection R.C.A. Thompson.*

1986; Howell and Smyth, 1995). Of special interest were Smyth's studies and considerations on the anatomical, biochemical and physiological host factors which might be involved in the parasite's host and intermediate host specificity (Smyth, 1968). Smyth (1964) stated that speciation of *Echinococcus* is a complex matter, and he pointed to the possibility that different 'races,' 'strains' or 'subspecies' may exist (see chapter: "Biology and Systematics of *Echinococcus*", this volume). Instructive and innovative presentations of the biochemical–physiological and immunological parasite–host interactions were characteristic of his excellent text books (Smyth, 1962). Smyth's contributions provided an important basis for subsequent studies on host specificity, establishment of the parasites in the definitive host, activities of *Echinococcus* parasites at the intestinal interface etc. (Thompson, 1986b, 1995; Thompson and Lymbery, 2013) (Fig. 18).

3.6.2 Echinococcus multilocularis

In contrast to *E. granulosus*, in vitro cultivation of intestinal stages of *E. multilocularis* resulted in highly variable results (reviewed by Howell, 1986; Howell and Smyth, 1995). However, uniform development of protoscoleces to immature stages in vitro could be obtained (Thompson et al., 1990) as well as gravid worms following partial development in the definitive host (Thompson and Eckert, 1982). Rausch and Jentoft (1957) were apparently the first who reported the development of small vesicles of *E. multilocularis* from minced metacestode tissue (isolated from rodents) in complex media in vitro. They believed that the maintenance of larval tissue in vitro may permit metabolic and other studies 'hitherto not practicable.' A significant step in this direction was the maintenance of *E. multilocularis* metacestode tissue blocks (0.1 to 0.3 g wet weight) in vitro with Eagle's minimal essential medium with 10% foetal bovine serum and antibiotics (Ramp and Eckert, 1986, 1987). Various isolates retained the proliferative capacity in the intermediate host for up to 40 weeks. This system was further developed for long-term in vitro production of large numbers of vesicles (Hemphill and Gottstein, 1995), which can be used for in vitro drug screening and other purposes (Siles-Lucas and Hemphill, 2002). From such vesicles, primary *Echinococcus* cells that are devoid of host cells could be isolated and perpetuated in vitro (Spiliotis and Brehm, 2008; Brehm and Spiliotis, 2008). These achievements contributed to the reduction of experimental animals otherwise needed for strain maintenance in the laboratory. This applies also for cryopreservation of *E. multilocularis* tissue blocks which can be preserved in a viable stage in liquid nitrogen for many years (Eckert and Ramp, 1985).

Little exact information existed on the development of *E. multilocularis* in definitive hosts. Vogel (1957) infected experimentally 10 dogs, 4 red foxes and 6 cats with *E. multilocularis* with metacestode material from rodents or of human origin and obtained highly variable numbers of adult intestinal stages in all dogs and foxes and in 5 of the cats. However, a quantitative evaluation could not be done. Such studies were performed by Kapel et al. (2006) who infected red foxes, raccoon dogs, domestic dogs and domestic cats and determined the susceptibility and reproductive potential of these species. Furthermore, a detailed study of the development of the parasites in the various hosts was done (Thompson et al., 2006).

4. NEOTROPICAL *ECHINOCOCCUS* SPECIES

4.1 Taxonomy

Several *Echinococcus* species, including *E. granulosus* s.l., *E. oligarthrus* and *E. vogeli*, occur in the neotropical region of Central and South America. It is assumed that *E. granulosus* was introduced with animals from Europe at the beginning of the 16th century (D'Alessandro and Rausch, 2008). For a long time, this species was considered to be the sole cause of echinococcosis in South America, although further *Echinococcus* species were known to occur in animals originating from this region (D'Alessandro et al., 1979), namely *E. oligarthrus* (Lühe, 1910) and *E. cruzi* (Brumpt and Joyeux,

1924). In the 1960s the work of Lothar Szidat (1892–1973) stimulated new interest in this field as he had described three new species, namely *Echinococcus patagonicus*, *Echinococcus pampeanus* and *Echinococcus cepanzoi* (Szidat, 1971). Of these, *E. patagonicus* and *E. cepanzoi* were considered conspecific with *E. granulosus* (Schantz et al., 1975, 1976) and *E. cruzi* with *E. oligarthrus* (Verster, 1965; Rausch et al., 1984); *E. pampeanus* is most similar to *E. oligarthrus* (Rausch and Bernstein, 1972). After the description of *E. vogeli* by Rausch and Bernstein (1972), this species, together with *E. granulosus*, *E. multilocularis* and *E. oligarthrus*, was accepted as valid species (Rausch and Bernstein, 1972; Thompson, 1986b). In recent years, six genotypes of the *E. granulosus* complex have been identified in South American countries and were allocated to the newly established species *E. granulosus* s.s., *Echinococcus ortleppi*, and *Echinococcus canadensis* (Cucher et al., 2015; see also chapter "Biology and Systematics of *Echinococcus*", this volume). *E. granulosus* s.l. is not further considered in this section.

E. oligarthrus and *E. vogeli* are considered to be indigenous to South America and of ancient origin because their metacestodes typically occur in terrestrial rodents belonging to the group of Hystricognathi (e.g., pacas, agoutis). Such rodents were dominant in South America from ~22 to 5 million years before present (D'Alessandro and Rausch, 2008). Both species have a peculiar history. **Johann Natterer** (1781–1843), an Austrian scientist had collected helminths from a puma [*Felis* (= *Puma*) *concolor*] in Brazil and brought the material back to Vienna in 1836 (Tappe et al., 2008). **Karl Moritz Diesing** (1800–67), from 1827 custos at the 'Hofmuseum' in Vienna, classified small cestodes from the puma initially as *Taenia crassicollis* and later as *Taenia oligarthra*. **Maximilian F.L. Lühe** (1870–1916), from 1909 Professor at the University of Königsberg (now Kaliningrad) (Enigk, 1986) studied Natter's material and concluded that *T. echinococcus* and *T. oligarthra* are closely related (Lühe, 1910). Finally, **Thomas W.M.** Cameron (1894–1980), London School of Hygiene and Tropical Medicine, discovered the adult tapeworm in a jaguarundi (*Felis yagouaroundi*) (which had died in the London Zoo) and described it in detail as *E. oligarthrus* (Cameron, 1926). The story of the other indigenous species is also linked to a zoo animal as it was isolated in 1970 by Rausch and Bernstein (1972) from a bush dog (*Speothos venaticus*) captured in Ecuador and kept in the Los Angeles Zoo. This parasite differed from other *Echinococcus* species and was described as new species, *E. vogeli*, in 'recognition of the contributions to the understanding of the taxonomy of *Echinococcus* species made by Professor Hans Vogel...' (Rausch and Bernstein, 1972, see also page 14).

4.2 Life cycles and epidemiology

The natural definitive host of *E. vogeli* is the bush dog from which this species was isolated for the first time by Rausch and Bernstein (1972). Adult cestodes have also been found in a naturally infected domestic dog (D'Alessandro et al., 1981) and in experimentally infected dogs (see below). Pacas (*Cuniculus paca*) and agoutis (*Dasyprocta* spp.) are the only known natural intermediate hosts. Natural metacestode infections occur also in aberrant hosts, such as humans and monkeys (in zoos) and are known as polycystic echinococcosis (Rausch and D'Alessandro, 2002; D'Alessandro and Rausch, 2008).

The life cycle of *E. vogeli* was elucidated in 1955 by G.E. Vogelsang and J. Barnola in Venezuela and in 1975 by AD'Alessandro in Columbia, who infected dogs experimentally with cysts from an agouti (*Dasyprocta agouti*) and a paca (*C. paca*), respectively, and obtained strobilar stages of this species (Rausch et al., 1978). In further studies, Rausch et al. (1978) had reared strobilar stages of *E. vogeli* by infecting dogs with metacestodes isolated from human patients, naturally infected pacas and an agouti.

In the life cycle of *E. oligarthrus*, wild felids (puma, jaguar, jaguarundi etc.) act as definitive hosts and rodents (agoutis, pacas, opossums etc.) as intermediate hosts. The strobilar stage develops to maturity in experimentally infected domestic cats, and several rodent species are susceptible to experimental metacestode infections (Thatcher and Sousa, 1968; Sousa and Thatcher, 1969; Rausch and D'Alessandro, 2002). In humans, the metacestode stages cause the unicystic form of echinococcosis.

After the description of the strobilar stage of *E. oligarthrus*, the metacestode stage remained unknown for a long time, although in 1914, the French parasitologists, Brumpt (1877–1951) and Joyeux (1881–1966), had found multilocular metacesteodes in agoutis in Brazil which they tentatively named *E. cruzi*, but they also considered an association with *T. oligarthra* (Brumpt and Joyeux, 1924; Tappe et al., 2008). Many years later, Thatcher and Sousa (1968) studied a multilocular metacestode stage obtained from a nutria (*Myocastor coypus*) born in a United States Zoo. This stage was previously identified as *E. granulo*sus, but Thatcher expressed the view that the nutria could be a natural host of *E. oligarthrus*.

A reliable diagnosis of the larval stages of *E. vogeli* and *E. oligarthrus* was difficult until Rausch et al. (1978) demonstrated that the dimensions of rostellar hooks of protoscoleces provide a means for the discrimination. Today, molecular methods allow a species-specific differential diagnosis (Tappe et al., 2008).

4.3 Human cases and pathology

The first cases of 'alveolar hydatid disease' were recorded in 1903 and in following years in Argentina by Viñas (in Rausch et al., 1978; D'Alessandro et al., 1995; Tappe et al., 2008), but it remained unclear for a long time, whether such cases are caused by atypical metacestodes of *E. granulosus* or other *Echinococcus* species (Rausch et al., 1978). Finally, at the end of the 1970s, studies by Rausch, D'Alessandro and coworkers (reviewed in Rausch and D'Alessandro, 2002; D'Alessandro and Rausch, 2008) demonstrated that the metacestodes developing in humans were predominantly that of *E. vogeli*. By March 2007, in 12 Latin American countries, 172 human cases of neotropical echinococcosis had been recorded (D'Alessandro and Rausch 2008), and at least eight further cases were reported thereafter (Siqueira et al., 2007, 2010; Knapp et al., 2009b). Most of the cases were caused by *E. vogeli* (for details see reviews by Rausch and D'Alessandro, 2002; D'Alessandro and Rausch, 2008).

Of special interest regarding the pathogenicity of *E. vogeli* and *E. oligarthrus* in natural intermediate hosts and in humans are differences in structure and proliferation of the metacestode stages, as described by Rausch et al., 1981, Rausch and Alessandro 1999). Further information on this aspect, on clinical characteristics, diagnosis, epidemiology etc. can be found in excellent reviews of Rausch and D'Alessandro (2002) and D'Alessandro and Rausch (2008). These scientists deserve special recognition for their great contributions to the research into neotropical echinococcosis. Robert Rausch made most of his contributions when he was Professor at the University of Washington, Seattle, Washington (see also p. 12). A.D'Alessandro was a member of Tulane University, New Orleans, Louisiana, and its Center for Medical Research in Cali, Columbia.

5. THE ROLE OF ASSOCIATIONS, INTERNATIONAL ORGANIZATIONS, AND INTERNATIONAL WORKING GROUPS IN ECHINOCOCCOSIS RESEARCH AND CONTROL

International associations and organizations have played and still play a significant role in the field of echinococcosis by facilitating international exchange of knowledge and promoting cooperation in research and control programmes. In 1941 the 'Asociación International de Hidatilogia' was founded in South America, later known as 'International Association of Hydatidology (IAH)' and since 2015 as 'World Association of

Echinococcosis' (WAE). Since its foundation this association has organized regular scientific meetings ('Jornadas Internationales'), predominantly in South American countries, and international congresses in various countries around the world. In 1991 the 50th anniversary of the IAH was celebrated with a special congress in Rome. Congress abstracts or other communications were published in 'Archivos Internationales de Hidatidosis.' More information on the WAE can be obtained from http://www.echinoworld.org.

As early as 1948, the Pan American Health Organization (PAHO) and the World Health Organization (WHO) had published a resolution—based on a proposal submitted by representatives of Argentina, Brazil, Paraguay and Uruguay—recommending to the public health authorities of the American countries that they intensify the epidemiological investigation of hydatidosis and issue laws and regulations directed toward prevention and control of hydatidosis (PAHO, 1948). In 1978 the 31st **WHO World Health Assembly** recognized the need for elaboration of strategies and methods for the control of zoonoses and food-borne diseases and adopted a resolution on 'Prevention and control of zoonoses and foodborne disease due to animal products' (WHO, 1978). With reference to this resolution, 'Guidelines for surveillance, prevention and control of echinococcosis/hydatidosis' were prepared in cooperation between FAO/UNEP/WHO[7] and published in two editions (WHO, 1981, 1984b), followed in 2001 by a 'WHO/OIE[8]' 'Manual on echinococcosis in humans and animals: a public health problem of global concern' (WHO, 2001a). These documents were prepared by large groups of international experts. WHO has developed a worldwide **network of zoonoses centres** in order to 'provide essential technical cooperation to country health programmes with respect to zoonoses and related foodborne diseases' (WHO, 1984b). Some of these centres are specifically concerned with echinococcosis, for example, in France (Université de Franche-Comté, Besançon), Italy (University of Pavia) and China (Xinjiang Medical University first Affiliated Hospital, and Xinjiang Center for Control and Prevention, Urumqi). Some activities of WHO and related organizations are mentioned in sections above, others are documented in various publications of which only some are included here as examples (Varela-Diaz and Coltorti, 1976; Gemmell and Varela-Diaz, 1980; PAHO, 1994; PAHO, 2002; WHO-IGWE, 2011). In 2013 the 66th **WHO World Health**

7 FAO: Food and Agriculture Organization of the United Nations. UNEP: United Nations Environment Programme.

8 OIE: Office International des Epizooties (World Organisation of Animal Health).

Assembly adopted a resolution (WHA66.12) which calls member states to improve the health and social well-being of affected populations. Based on this resolution, a list of 'neglected tropical diseases' was published which also contains echinococcosis, although this disease is not restricted to the tropics (WHO, 2013).

Based on a proposal by Dr K. Bögel (then chief of the Veterinary Public Health Unit (VPH), WHO), 'WHO Informal Working Groups on Echinococcosis' (WHO-IWGE) were conceived in 1985 and officially established in 1987 with the intention to promote international exchange of knowledge and cooperation between groups working in various fields of echinococcosis (Meslin and Vuitton, 2011). The IWGE consisted of a coordinating group (chairman J. Eckert, University of Zürich, Switzerland, 1987 to 1995, vice-chairmen Lord E.J.L. Soulsby, University of Cambridge, United Kingdom and P.M. Schantz, Centres for Disease Control, Atlanta, United States of America) and the following subgroups (chairmen in prarentheses): (1) Biology and strain variation (R.C.A. Thompson, Murdoch University, Murdoch, Western Australia); (2) Immunology (M.D. Rickard, University of Melbourne, Werribee, Australia; (3) Immunodiagnosis (B. Gottstein, University of Berne, Switzerland); (4) Medical aspects (Z.S. Pawlowski, Clinic of Parasitic and Tropical Diseases, Poznań, Poland; (5) Epidemiology and Control (M.A. Gemmell, Hydatid Research Unit, Dunedin, New Zealand); and (6) Chemotherapy (J. Eckert, Zürich). Each of the subgroups had invited several international experts for cooperation. In 1995, Dr F.-X. Meslin (WHO/VPH) decided to transform all subgroups to a single group and to designate a coordinator for a 4-year term (Vuitton, 1997; Meslin and Vuitton, 2011). These were D.A. Vuitton (University Hospital Besançon, France, 1995 to 1999), P. Schantz (Centers for Disease Control, Atlanta, United States of America, 2000 to 2004), Ph. Craig (University of Salford, Manchester, United Kingdom, 2005 to 2010) and P. Kern (Department of Internal Medicine, University of Ulm, Germany, 2011 to 2015) (Meslin and Vuitton, 2011). Under the umbrella of this working group, a number of sections for special topics and tasks were established.

The WHO-IWGE held many expert meetings (for example[9], in Montreal, Canada, 1987; Zürich and Geneva, Switzerland, 1988; East-Berlin, former German Democratic Republic, 1989; Stuttgart-Hohenheim,

[9] Selected sources WHO reports: Montreal: WHO/CDS/VPH 87.72; Zürich: WHO/CDS/VPH88. 78; Geneva: WHO/CDS/VPH 88.79; Stuttgart-Hohenheim: WHO/CDS/VPH 89.85; Besançon: WHO/CDS/VPH 93.118; Beijing: WHO/VPH 93.131.

Germany, 1989; Besançon, France, 1992; Beijing, China, 1993; Al-Ain, United Arab Emirates, 1994; Geneva, Switzerland, 2011); they organized sessions on specific research topics at each World Congress of Hydatidology from 1990 to 2015 and published guidelines on treatment of CE and AE (WHO, 1996), PAIR (WHO, 2001b), PNM classification of alveolar echinococcosis (Kern et al., 2006), a consensus paper on the diagnosis and treatment of CE and AE in humans (Brunetti et al., 2010) and a WHO/OIE manual on echinococcosis in humans and animals (WHO, 2001a) (Figs 19 and 20).

Another example of international cooperation in the field of echinococcosis is the 'European Network for Concerted Surveillance of Alveolar Echinococcosis,' established in 1998 with the aims: (1) to determine the prevalence of human cases and identify risk factors as well as prognostic

Figure 19 ***Meeting of the WHO Informal Meeting of Working Groups on Echinococcosis Research, 13–16.09.1988, World Health Organization, Geneva.*** From left: first row: Mrs Fujikura (Japan), D. Heath (New Zealand), P. Schantz (United States of America), M. Rickard (Australia), C. Macpherson (United Kingdom), T. Fujikura (Japan, WHO Geneva); second row: Jiang Cipeng (China), E.J.L. Soulsby (United Kingdom), M. Gemmell (New Zealand), C. Arme (United Kingdom), B. Gottstein (Switzerland), Z. Pawlowski partially hidden (Poland, WHO Geneva), M. Lightowlers (Australia), A. Thompson (Australia). Further group member J. Eckert (Switzerland, photographer). *Original: J. Eckert.*

Figure 20 ***Meeting of the WHO Informal Working Group on Echinocococcosis in Al-Ain, United Arab Emirates (UAE) October 1994.*** From left: first row: M. Kamiya, partially hidden (Japan); Vice Dean, Medical Faculty, Al-Ain (UAE); J. Eckert (Switzerland, chairman); local secretary of the meeting; D.A. Vuitton (France); J. Pawlowska (Poland); Y. Kamiya (Japan); A. McLedingham (Dean Medical Faculty, Al-Ain, UAE); F.K. Dar (Medical Faculty, Al Ain, vice-chairman). Second/third row: N.N partially hidden; N.N.; N.N.; G. N. Alwar (Lebanon); T. Todorov (Bulgaria); N.N.; P. Kern (Germany); Z. Pawlowski (Poland, vice-chairman); N.N.; H. Wen (China); N.N.; C.E. Tanner (Canada); L. Savioli (WHO, Geneva); P. S. Craig (United Kingdom); J. Horton (United Kingdom); N.N.; De Rycke (Belgium); W.N. von Sinner (Saudi Arabia). The authors regret that not all persons (N.N.) could be identified. *Congress photo, collection J. Eckert.*

factors and (2) to determine the role of final hosts (foxes, dogs and cats) as possible sources of infection for humans. This network—funded by the European Commission and other sources—was coordinated by D.A. Vuitton (University of Franche-Comté, Besançon, France) and P. Kern (Ulm University, Ulm, Germany) for human epidemiology ('EurEchinoReg'), and by P. Giraudoux (University of Franche-Comté, Besançon, France) and T. Romig (University of Hohenheim, Germany) for animal epidemiology ('EchinoRisk'). It stimulated the establishment of national reference centres and concerted data collection on human cases of AE (Kern et al., 2003; Tamarozzi et al., 2015). The network on human epidemiology has been reactivated in 2014 to 2015 to set up a fully integrated European Alveolar Echinococcosis Database 'EurEchino' in order to collect AE cases online all over Europe (Charbonnier et al., 2014). In parallel, the EC-funded project 'Heracles,' an international endeavour to coordinate studies on CE in Europe and Turkey, was established in 2013, coordinated by A. Casulli (Istituto Superiore di Sanità, Rome, Italy); it also includes an epidemiological collection of CE cases all over Europe (http://www.heracles-fp7.eu/erce.html).

After the major endemic area for CE and AE was disclosed in Western China at the end of the 1980s, internationally supported projects were launched, thanks to the European Commission then to the National Institutes of Health/National Sciences Foundation of the United States of America (TransTech projects, coordinated by P.S. Craig, Salford University and P. Giraudoux, Université de Franche-Comté, Besançon, France, and locally supported by Xinjiang University, Lanzhou Medical University, Sichuan Center for Control and Prevention, and Ningxia University, People's Republic of China). These projects were the basis for the Chinese National Survey of Echinococcosis in the 2000s and the Chinese National Echinococcosis Control Programme launched by the People's Republic of China in 2010 (Chinese Ministry of Health, 2007).

ACKNOWLEDGEMENTS

The authors are grateful to Professor Dominique Vuitton for her valuable information and advice regarding the role of associations, international organizations and international working groups on echinococcosis.

REFERENCES

Abdykerimov, K., 2011. Control of *Echinococcus granulosus* and Cystic Echinococcosis. Control in Kyrgyzstan. Report of the WHO Informal Working Group on cystic and alveolar echinococcosis surveillance, prevention and control, with the participation of the Food and Agriculture Organization of the United Nations and the World Organisation for Animal Health. Geneva, World Health Organization, (WHO/HTM/NTD/NZD/2011.2), ISBN 978 92 4 150292 4.

Ackerknecht, E.H., 1989. Geschichte der Medizin, sixth ed. Enke-Verlag, Stuttgart.

Akovbiantz, A., Eckert, J., Hess, U., Schmid, M., 1977. Clinical experience with mebendazole treatment of human echinococcosis. XI Int. Congr. Hydatidosis Athens, 30.05.-2.6. Abstract.

Allan, J.C., Craig, P.S., 2006. Coproantigens in taeniasis and echinococcosis. Parasitol. Int. 55 (Suppl.), S75–S80.

Allan, J.C., Craig, P.S., Garcia Noval, J., Mencos, F., Liu, D., Wang, Y., Wen, H., Zhou, P., Stringer, R., Rogan, M., Zeyhle, E., 1992. Coproantigen detection for immunodiagnosis of echinococcosis and taeniasis in dogs and humans. Parasitology 104 (Pt 2), 347–356.

Al-Moslih, M.I., Al-Rawas, A.Y., Rassam, S., Al-Dabagh, M.A., Al-Janabi, T.A., Shafik, M.A., Al-Ani, M.A., 1978. Treatment of hydatidosis. Br. Med. J. 2, 1435.

Ammann, R., Eckert, J., 1995. Clinical diagnosis and treatment of echinococcosis in humans. In: Thompson, R.C.A., Lymbery, A.J. (Eds.), *Echinococcus* and Hydatid Disease. CAB International Wallingford Oxon, pp. 411–463. ISBN:0 851 198 910 1.

Ammann, R.W., Eckert, J., 1996. Cestodes: *Echinococcus*. Gastroenterol. Clin. N. Am. 25, 655–689.

Ammann, R.W., Fleiner-Hoffmann, A., Eckert, J., Schweizerische Echinokokkose-Studiengruppe (SESG), 1999. Schweizerische Studie für Chemotherapie der alveolären Echinokokkose - Rückblick auf ein 20 jähriges klinisches Forschungsprojekt. Schweiz. Med. Wochenschr. 129, 323–332.

Ammann, R.W., Renner, E.C., Gottstein, B., Grimm, F., Eckert, J., Renner, E.L., Swiss Echinococcosis Study Group, 2004. Immunosurveillance of alveolar echinococcosis by specific humoral and cellular immune tests: long-term analysis of the Swiss chemotherapy trial (1976–2001). J. Hepatol. 41, 551–559.

Ammann, R.W., Stumpe, K.D., Grimm, F., Deplazes, P., Huber, S., Bertogg, K., Fischer, D.R., Müllhaupt, B., 2015. Outcome after discontinuing long-term benzimidazole treatment in 11 patients with non-resectable alveolar echinococcosis with negative FDG-PET/CT and anti-EmII/3-10 serology. PLoS Negl. Trop. Dis. 9 (9), e0003964.

Andersen, F.L., Chai, J., Liu, F. (Eds.), 1993. Compendium on Cystic Echinococcosis with Special Reference to the Xinjiang Uygur Autonomous Region, The Peoples Republic of China. Brigham Young University Provo, UT 84602, USA.

Andersen, F.L., Ouhlli, H., Kachani, M. (Eds.), 1997. Compendium on Cystic Echinococcosis in Africa and Middle Eastern Countries With Special Reference to Morocco. Brigham Young University, Provo, UT 84602, USA.

Atwater, R., 1969. Prevention, control, and eradication of disease. In: Schwabe, C.W. (Ed.), Veterinary Medicine and Human Health, second ed. Williams and Wilkins Company, Baltimore, pp. 374–440.

Beard, T.C., 1973. The elimination of echinococcosis from Iceland. Bull. World Health Organ. 48, 653–660.

Beard, T.C., Rickard, M.D., Goodman, H.T., 1978. Medical treatment for hydatids. Med. J. Aust. (i), 633–635.

Bekhti, A., Schaaps, J.-P., Capron, M., Dessaint, J.P., Santoro, F., Capron, A., 1977. Treatment of hepatic hydatid disease with mebendazole: preliminary results in four cases. Br. Med. J. 2, 1047–1051.

Ben Amor, N., Gargouri, M., Gharbi, H.A., Golvan, Y.J., Ayachi, K., Kchouck, H., 1986. Essai de traitement par ponction des kystes hydatiques abdominaux inoperables. Ann. de Parasitol. Humaine Comparée 61, 689–692.

Bouvier, G., Burgisser, H., Schneider, P.A., 1957. Observavtions sue le maladies du gibier, des oiseaux et de poissons. Schweiz. Arch. für Tierheilkd. 99, 461–477.

Brehm, K., Spiliotis, M., 2008. Recent advances in the in vitro cultivation and genetic manipulation of *Echinococcus multilocularis* metacestodes and germinal cells. Exp. Parasitol. 119, 506–515.

Bretagne, S., Guillou, J.P., Morand, M., Houin, R., 1993. Detection of *Echinococcus multilocularis* DNA in fox faeces using DNA amplification. Parasitology 106 (Pt 2), 193–199.

Brumpt, E., Joyeux, C., 1924. Description d'un nouvel echinococcque: *Echinococcus cruzi* n.sp. Ann. Parasitol. 2 (3), 226–231.

Brunetti, E., Kern, P., Vuitton, D.A., Writing Panel for the WHO-IWGE, 2010. Expert consensus for the diagnosis and treatment of cystic and alveolar echinococcosis in humans. Acta Trop. 114, 1–16.

Bryant, C., Barwick, R.E., 2016. Smyth, James Desmond (1917–1999). Obituaries Australia, National Centre of Bibliography, Australian National University. http://oa.anu.edu.au.

Budke, C.M., Jiamin, Q., Craig, P.S., Torgerson, P.R., 2005. Modeling the transmission of *Echinococcus granulosus* and *Echinococcus multilocularis* in dogs for a high endemic region of the Tibetan plateau. Int. J. Parasitol. 35, 163–170.

Buhl, L., 1855. Ueber die zusammengesetzte Echinokokkengeschwulst der Leber. Briefliche Mitteilungen, vorgelegt von R. Virchow. Verhandlungen der physikalisch-medizinischen Gesellschaft zu Würzburg, vol. 6, pp. 428–429.

Burkhardt, B., 1981. Beiträge zur experimentellen Chemotherapie der larvalen Echinokokkose mit Untersuchungen zur biologischen Verfügbarkeit von Mebendazol bei Nagetieren (Dissertation). Veterinär-Medizinische Fakultät, Universität Zürich.

Campbell, W.C., McCracken, R.O., Blair, L., 1975. Effect of parenterally injected benzimidazole compounds on *Echinococcus multilocularis* and *Taenia crassiceps* metacestodes in laboratory animals. J. Parasitol. 61, 844–852.

Cameron, T.W.M., 1926. Observations on the genus *Echinococcus* Rudolphi 1801. J. Helminthol. 4, 13–22.

Charbonnier, A., Knapp, J., Demonmerot, F., Bresson-Hadni, S., Raoul, F., Grenouillet, F., Millon, L., Vuitton, D.A., Damy, S., 2014. A new data management system for the French National Registry of human alveolar echinococcosis cases. Parasite 21, 69.

Chinese Ministry of Health, 2007. The Office of the Ministry of Health. Technical Scheme of Hydatid Disease Control and Prevention Notice (In Chinese).

Coudert, J., Euzéby, J., Garin, J.P., 1970. Fréquence de *E. multilocularis* chez le renard commun (*Vulpes vulpes*) dans le secteur nord-est de la France. Lyon Med. 224, 293–298.

Craig, P., 2011. Control of *Echinococcus granulosus* and Cystic Echinococcosis. General Aspects. WHO Report (WHO/HTM/NTD/NZD/2011.2), ISBN 978 92 4 150292 4 (full title of the report see Abdykerimov, 2011).

Craig, P.S., Larrieu, E., 2006. Control of cystic echinococcosis/hydatidosis: 1863–2002. Adv. Parasitol. 61, 443–508.

Craig, P., Pawlowski, Z., 2002. Cestode Zoonoses: Echinococcosis and Cysticercosis. IOS Press, Amsterdam, ISSN: 1566–7693.

Craig, P.S., Rogan, M.T., Allan, J.C., 1996. Detection, screening and community epidemiology of taeniid cestode zoonoses: cystic echinococcosis, alveolar echinococcosis and neurocysticercosis. Adv. Parasitol. 38, 169–250.

Christofi, G., 2011. Control of *Echinococcus granulosus* and Cystic Echinococcosis Control in Cyprus and the Mediterranean Area. WHO Report (WHO/HTM/NTD/NZD/2011.2), ISBN 978 92 4 150292 4 (full title of the report see Abdykerimov, 2011).

Cucher, M.A., Macchiaroli, N., Baldi, G., Camicia, F., Prada, L., Maldonado, L., Ávila, H.G., Fox, A., Gutiérrez, A., Negro, P., López, R., Jensen, O., Rosenzvit, M., Kamenetzky, L., 2015. Cystic echinococcosis in South America: systematic review of species and genotypes of *Echinococcus granulosus* sensu lato in humans and natural domestic hosts. Trop. Med. Int. Health 21, 166–175.

D'Alessandro, A., Rausch, R.L., 2008. New aspects of neotropical polycystic (*Echinococcus vogeli*) and unicystic (*Echinococcus oligarthrus*) echinococcosis. Clin. Microbiol. Rev. 21, 380–401.

D'Alessandro, A., Ramirez, L.E., Chapadeiro, E., Lopes, E.R., de Mesquita, P.M., 1995. Second recorded case of human infection by *Echinococcus oligarthrus*. Am. J. Trop. Med. Hyg. 52, 29–33.

D'Alessandro, A., Rausch, R.L., Cuello, C., Aristizabal, N., 1979. *Echinococcus vogeli* in man, with a review of polycystic hydatid disease in Colombia and neighboring countries. Am. J. Trop. Med. Hyg. 28, 303–317.

D'Alessandro, A., Rausch, R.L., Morales, G.A., Collet, S., Angel, D., 1981. *Echinococcus* infections in Colombian animals. Am. J. Trop. Med. Hyg. 30, 1263–1276.

Danis, M., Brücker, G., Gentilini, M., Richard-Lenoble, D., Smith, M., 1977. Treatment of hepatic hydatid disease. Br. Med. J. 2, 1356.

Dardel, G., 1927. Das Blasenwurmleiden in der Schweiz. Verlag A. Francke, Bern.
Davis, A., Pawlowski, Z.S., Dixon, H., 1986. Multicentric clinical trials of benzimidazolecarbamates in human echinococcosis. Bull. World Health Organ. 64, 383–388.
Deplazes, P., Eckert, J., 2001. Veterinary aspects of alveolar echinococcosis-a zoonosis of public health significance. Veterinary Parasitol. 98, 65–87.
Deplazes, P., Dinkel, A., Mathis, A., 2003. Molecular tools for studies on the transmission biology of *Echinococcus multilocularis*. Parasitology 127 (Suppl.), S53–S61.
Deplazes, P., Gloor, S., Stieger, C., Hegglin, D., 2002. Urban transmission of *Echinococcus multilocularis*. In: Craig, P., Pawlowski, Z. (Eds.), Cestode Zoonoses: Echinococcosis and Cysticercosis. IOS Press, Amsterdam, pp. 287–297. ISSN: 1566–7693.
Deplazes, P., Gottstein, B., Eckert, J., Jenkins, D.J., Ewald, D., Jimenez-Palacios, S., 1992. Detection of *Echinococcus* coproantigens by enzyme-linked immunosorbent assay in dogs, dingoes and foxes. Parasitol. Res. 78, 303–308.
Deplazes, P., Hegglin, D., Gloor, S., Romig, T., 2004. Wilderness in the city: the urbanization of *Echinococcus multilocularis*. Trends Parasitol. 20, 77–84.
Dévé, F., 1901. De L'echinococcose Sécondaire. These Inaugurale, Paris.
Dévé, F., 1902. De al greffe hydatique obtune par l'inoculation de scolex. Bull. Mémoires de la Société de Chir. de Paris 28, 905–910.
Dévé, F., 1946. L'echinococcose Secondaire. Massen et Co, Paris.
Dew, H.R., 1925. The histogenesis of the hydatid parasite (*Taenia echinococcus*) in the pig. The development of the *Taenia echinococcus* from embryo to hydatid cyst. Med. J. Aust. 1925 (i), 101–110.
Dew, H.R., 1928. Hydatid Disease. Its Pathology, Diagnosis and Treatment. Australian Medical Publishing Company, Sydney.
Eckert, J., 1986. Prospects for treatment of the metacestode stage of *Echinococcus*. In: Thompson, R.C.A. (Ed.), The Biology of *Echinococcus* and Hydatid Disease. Allen and Unwin, London, pp. 250–284.
Eckert, J., 1996. *Echinococcus multilocularis* and alveolar echinococcosis in Europe (except parts of eastern Europe). In: Uchino, J., Sato, N. (Eds.), Alveolar Echinococcosis. Strategy for Eradication of Alveolar Echinococcosis of the Liver. Fuji Shoin, Sapporo, Japan, pp. 27–43.
Eckert, J., Deplazes, P., 2004. Biological, epidemiological, and clinical aspects of echinococcosis, a zoonosis of increasing concern. Clin. Microbiol. Rev. 17, 107–135.
Eckert, J., Pohlenz, J., 1976. Zur Wirkung von Mebendazol auf Metazestoden von *Mesocestoides corti* und *Echinococcus multiocularis*. Tropenmedizin Parasitol. 27, 247–262.
Eckert, J., Ramp, T., 1985. Cryopreservation of *Echinococcus multilocularis* metacestodes and subsequent proliferation in rodents (*Meriones*). Z. für Parasitenkd. 71, 777–787.
Eckert, J., Barandun, G., Pohlenz, J., 1978. Chemotherapie der larvalen Echinokokkose bei Labortieren. Schweiz. Med. Wochenschr. 108, 1104–1112.
Eckert, J., Deplazes, P., Craig, P.S., Gemmell, M.A., Gottstein, B., Heath, D., Jenkins, D.J., Kamyia, M., Lightowler, M., 2001. Echinococcosis in animals: clinical aspects, diagnosis and treatment. In: Eckert, J., Gemmell, M.A., Meslin, F.-X., Pawlowski, Z. (Eds.), Manual on Echinococcosis in Humans and Animals: A Public Health Problem of Global Concern. World Organisation for Animal Health, Paris, pp. 72–99.
Eckert, J., Deplazes, P., Kern, P., 2011. Alveolar echinococcosis (*Echinococcus multilocularis*) and neotropical forms of echinococcosis (*Echinococcus vogeli* and *Echinococcus oligathrus*). In: Palmer, S.R., Lord Soulsby, Toregson, P.R., Brown, D.W.G. (Eds.), Oxford Textbook of Zoonoses, second ed. Oxford University Press, Oxford, ISBN 978-0-19-857002-8, pp. 669–699.
Eckert, J., Jacquier, P., Baumann, D., Raeber, P.-A., 1995. Echinokokkose des Menschen in der Schweiz, 1984–1992. Schweiz. Med. Wochenschr. 125, 1989–1998.

Eckert, J., Thompson, R.C.A., Mehlhorn, H., 1983. Proliferation and metastases formation of larval *Echinococcus multilocularis*. I. Animal model, macroscopical and histological findings. Z. für Parasitenkd. 69, 737–748.

Enigk, K., 1986. Geschichte der Helminthologie im deutschsprachigen Raum. G. Fischer, Stuttgart, ISBN 3-437-20350-9.

Economides, P., Christofi, G., 2002. Experience gained and evaluation of the echinococcosis/hydatidosis eradication programme in Cyprus. In: PAHO, pp. 51–59.

Filice, C., Brunetti, E., 1997. Use of PAIR in human cystic echinococcosis. Acta Trop. 64, 95–107.

Fuchs, R., 1895. Hippokrates. Sämtliche Werke ins Deutsche übersetzt und ausführlich commentiert. Verlag, Dr. H. Lüneburg, München.

Gavidia, C., 2011. Control of *Echinococcus Granulosus* and Cystic Echinococcosis. Control in South America. WHO Report (WHO/HTM/NTD/NZD/2011.2), ISBN 978 92 4 150292 4 (full title of the report see Abdykerimov, 2011).

Gemmell, M.A., 1978. The Styx Field Trial. Effect of treatment of the definitive host for tapeworms on larval forms in the intermediate host. Bull. World Health Organ. 56, 433–443.

Gemmell, M.A., 1987. A critical approach to the concepts of control and eradication of echinococcosis/hydatidosis and taeniasis/cysticercosis. Int. J. Parasitol. 17, 465–472.

Gemmell, M.A., 1990. Australasian contributions to an understanding of the epidemiology and control of hydatid disease caused by *Echinococcus granulosus*-past, present and future. Int. J. Parasitol. 20, 431–456.

Gemmell, M.A., 1997. Quantifying the transmission dynamics of the family Taeniidae with particular reference to *Echinococcus* spp.: an update. In: Andersen, F.L., Ouhlli, H., Kachani, M. (Eds.), Compendium on Cystic Echinococcosis in Africa and Middle Eastern Countries with Special Reference to Morocco. Brigham Young University, Provo, UT 84602, USA, pp. 54–84.

Gemmell, M.A., Lawson, J.R., 1986. Epidemiology and control of hydatid disease. In: Thompson, R.C.A. (Ed.), The Biology of *Echinococcus* and Hydatid Disease. Allen & Unwin, London, ISBN 0-04-591020-0, pp. 189–216.

Gemmell, M.A., Roberts, M.G., 1995. Modelling *Echinococcus* life cycles. In: Thompson, R.C.A., Lymbery, A.J. (Eds.), *Echinococcus* and Hydatid Disease. CAB International, Wallingford, Oxon, ISBN 0 85198 910 1, pp. 333–354.

Gemmell, M.A., Varela-Diaz, V.M., 1980. Hydatidosis/Echinococcosis. Review of Programs for Control up to 1974. Series of Scientific and Technical Monographs C.P.Z. - 8. Pan American Zoonoses Center, Buenos Aires, Argentina.

Gemmell, M.A., Lawson, J.R., Roberts, M.G., Kerin, B.R., Mason, C.J., 1986. Population dynamics in chinococcosis and cysticercosis: comparison of the response of *Echinococcus granulosus*, *Taenia hydatigena* and *T. ovis* to control. Parasitology 93 (Pt 2), 357–369.

Gemmell, M.A., Roberts, M.G., Beard, T.C., Lawson, J.R., 2001. Quantitative epidemiology and transmission biology with specials reference to *Echinococcus granulosus*. In: Eckert, J., Gemmell, M.A., Meslin, F.-X., Pawlowski, Z.S. (Eds.), WHO/OIE Manual on Echinococcosis in Humans and Animals: A Public Health Problem of Global Concern. World Health Organisation for Animal Health, Paris, ISBN 92-9044-522-X, pp. 143–163.

Giraudoux, P., Delattre, P., Takahashi, K., Raoul, F., Quéré, J.-P., Craig, P., Vuitton, D., 2002. Transmission ecology of *Echinococcus multilocularis* in wildlife: what can be learned from comparative studies and multiscale approaches? In: Craig, P., Pawlowski, Z. (Eds.), Cestode Zoonoses: Echinococcosis and Cysticercosis. IOS Press, Amsterdam, pp. 251–266. ISSN: 1566–7693.

Gottstein, B., Hemphill, A., 2008. *Echinococcus multilocularis*: the parasite-host interplay. Exp. Parasitol. 119, 447–452.

Grove, D.I., 1990. A History of Human Helminthology. CAB International, Wallingford, Oxon.
Hansen, F., Tackmann, K., Jeltsch, F., Wissel, C., Thulke, H.H., 2003. Controlling *Echinococcus multilocularis*-ecological implications of field trials. Prev. Veterinary Med. 60, 91–105.
Heath, D.D., 1995. Immunology of *Echinococcus* infections. In: Thompson, R.C.A., Lymbery, A.J. (Eds.), *Echinococcus* and Hydatid Disease. CAB International, Wallingford, Oxon, ISBN 0 85198 910 1, pp. 183–199.
Heath, D.D., Chevis, R.A.F., 1974. Mebendazole and hydatid cysts. Lancet 1974, 218–219.
Heath, D.D., Christie, M., Chevis, R.A.F., 1975. The lethal effect of mebendazole on secondary *Echinococcus granulosus*, cysticerci of *Taenia pisiformis* and tetrathyridia of *Mesocestoides corti*. Parasitology 70, 273–285.
Heath, D.D., Jensen, O., Lightowlers, M.W., 2003. Progress in control of hydatidosis using vaccination – a review of formulation and delivery of the vaccine and recommendations for practical use in control programs. Acta Trop. 85, 133–143.
Heath, D.D., Pharo, H., van der Log, P., 2002. Hydatids close to eradication in New Zealand. In: PAHO (2002), pp. 47–50.
Hegglin, D., Ward, P.I., Deplazes, P., 2003. Anthelmintic baiting of foxes against urban contamination with *Echinococcus multilocularis*. Emerg. Infect. Dis. 9, 1266–1272.
Hemphill, A., Gottstein, B., 1995. Immunology and morphology studies on the proliferation of in vitro cultivated *Echinococcus multilocularis* metacestodes. Parasitol. Res. 81, 605–614.
Hemphill, A., Stadelmann, B., Scholl, S., Müller, J., Spiliotis, M., Müller, N., Gottstein, B., Siles-Lucas, M., 2010. *Echinococcus* metacestodes as laboratory models for the screening of drugs against cestodes and trematodes. Parasitology 137, 569–587.
Hofer, S., Gloor, S., Müller, U., Mathis, A., Hegglin, D., Deplazes, P., 2000. High prevalence of *Echinococcus multilocularis* in urban red foxes (*Vulpes vulpes*) and voles (*Arvicola terrestris*) in the city of Zürich, Switzerland. Parasitology 120 (Pt 2), 135–142.
Horton, R.J., 1997. Albendazole in treatment of human cystic echinococcosis: 12 years of experience. Acta Trop. 64, 79–93.
Horton, J., 2003. Albendazole for the treatment of echinococcosis. Fundam. Clin. Pharmacol. 17, 205–212.
Hosemann, G., Schwarz, E., Lehmann, J.C., Posselt, A. (Eds.), 1928. Die Echinokokken-Krankheit. F. Enke, Stuttgart.
Hosemann, G., 1928. Morphologie und Biologie des *Echiococcus cysticus*. In: Hosemann, G., Schwarz, E., Lehmann, J.C., Posselt, A. (Eds.), Die Echinokokken-Krankheit. F. Enke, Stuttgart, pp. 1–66.
Howell, M.J., 1986. Cultivation of *Echinococcus* species in vitro. In: Thompson, R.C.A. (Ed.), The Biology of *Echinococcus* and Hydatid Disease. Allen & Unwin, London, ISBN 0-04-591020-0, pp. 143–163.
Howell, M.J., Smyth, J.D., 1995. Maintenance and cultivation of *Echinococcus* species in vitro and in vivo. In: Thompson, R.C.A., Lymbery, A.J. (Eds.), Echinococcus and Hydatid Disease. CAB International, Wallingford, Oxon, ISBN 0-85198-910-1, pp. 201–232.
Ito, A., Romig, T., Takahashi, K., 2003. Perspective on control options for *Echinococcus multilocularis* with particular reference to Japan. Parasitology 127 (Suppl.), S159–S172.
Jahn, M.L., 1927. Über den Wachstumstypus des *Echinococcus alveolaris* und die durch ihn bedingten Reaktionsformen des Wirtsgewebes. Beiträge zur Pathol. Anatomie uzur Allg. Pathol. 76, 1–35.
Junghanss, T., da Silva, A.M., Horton, J., Chiodini, P.L., Brunetti, E., 2008. Clinical management of cystic echinococcosis: state of the art, problems, and perspectives. Am. J. Trop. Med. Hyg. 79, 301–311.

Kadry, Z., Renner, E.C., Bachmann, L.M., Attigah, N., Renner, E.L., Ammann, R.W., Clavien, P.A., 2005. Evaluation of treatment and long-term follow-up in patients with hepatic alveolar echinococcosis. Br. J. Surg. 92, 1110–1116.

Kagan, I.G., 1968. A review of serological tests for the diagnosis of hydatid disease. Bull. World Health Organ. 39, 25–37.

Kamiya, M., Nonaka, N., Sumiya, G., Oku, Y., 2004. Effective countermeasures against alveolar echinococcosis in red fox population of Hokkaido, Japan. In: Torgerson, P.R., Shaikenov, B. (Eds.), *Echinococcus* in Central Asia: Problems and Solutions. Publishing House "Dauir", Zurich and Alamty, pp. 273–282.

Kapel, C.M., Torgerson, P.R., Thompson, R.C., Deplazes, P., 2006. Reproductive potential of *Echinococcus multilocularis* in experimentally infected foxes, dogs, raccoon dogs and cats. Int. J. Parasitol. 36, 79–86.

Kass, P.H., McCapes, R.H., Pritchard, W.R., 2006. In: memoriam Calvin W. Schwabe. http.//senate.universityofcalifornia.edu/inmemoriam/calvinschwabe.htm.

Kern, P., 2004. *Echinococcus granulosus* infection: definitions, medical treatment and outcome. In: Torgerson, P., Shaikenov, B. (Eds.), Echinococcosis in Central Asia: Problems and Solutions. Publishing House "Dauir", Zurich & Almaty, ISBN 9965-517-92-4, pp. 166–175.

Kern, P., 2006. Medical treatment of echinococcosis under the guidance of good clinical practice (GCP/ICH). Parasitol. Int. 55 (Suppl.), S273–S282.

Kern, P., 2010. Clinical features and treatment of alveolar echinococcosis. Curr. Opin. Infect. Dis. 23, 505–512.

Kern, P., Bardonnet, K., Renner, E., Auer, H., Pawlowski, Z., Ammann, R.W., Vuitton, D.A., Kern, P., 2003. European echinococcosis registry: human alveolar echinococcosis, Europe, 1982–2000. Emerg. Infect. Dis. 9, 343–349.

Kern, P., Wen, H., Sato, N., Vuitton, D.A., Gruener, B., Shao, Y., Delabrousse, E., Kratzer, W., Bresson-Hadni, S., 2006. WHO classification of alveolar echinococcosis: principles and application. Parasitol. Int. 55 (Suppl.), S283–S287.

Knapp, J., Bart, J.M., Giraudoux, P., Glowatzki, M.L., Breyer, I., Raoul, F., Deplazes, P., Duscher, G., Martinek, K., Dubinsky, P., Guislain, M.H., Cliquet, F., Romig, T., Malczewski, A., Gottstein, B., Piarroux, R., 2009a. Genetic diversity of the cestode *Echinococcus multilocularis* in red foxes at a continental scale in Europe. PLoS Negl. Trop. Dis. 3 (6), e452.

Knapp, J., Chirica, M., Simonnet, C., Grenouillet, F., Bart, J.-M., Sako, Y., Itoh, S., Nakao, M., Ito, A., Millon, L., 2009b. *Echinococcus vogeli* infection in a hunter, French Guiana. Emerg. Infect. Dis. 15, 2029–2031.

Koch, S., Bresson-Hadni, S., Miguet, J.P., Crumbach, J.P., Gillet, M., Mantion, G.A., Heyd, B., Vuitton, D.A., Minello, A., Kurtz, S., European Collaborating Clinicians, 2003. Experience of liver transplantation for incurable alveolar echinococcosis: a 45-case European collaborative report. Transplantation 75, 856–863.

Krotov, A.I., Chernyaeva, A.I., Kovalenko, F.P., Bayandina, D.G., Budanova, I.S., Kuznetsova, O.E., Voskoboinik, L.V., 1974. Experimental therapy of alveococcosis. II. Effectiveness of some antinematode agents in the alveococcosis of laboratory animals (russisch). Meditsinskaya Parazitol. i Parazit. Bolezni 43, 314–319.

Küchenmeister, F., 1851. Vorläufige Mittheilung über *Cysticercus pisiformis* der Kaninchen. Z. für Klin. Med. 2, 240.

Küchenmeister, F., 1853. Experimente über die Entstehung der Cestoden 2 ter Stufe, zunächst des Coenurus cerebralis. Z. für Klin. Med. 4, 448–451.

Küchenmeister, F., 1855. Die in und an dem Körper des Menschen vorkommenden Parasiten. Verlag Teubner, Leipzig.

Langenbuch, C., 1890. Der Leberechinococcus und seine Chirurgie. Verlag Ferdinand Enke, Stuttgart.

Larrieu, E., Pérez Palacios, A., 2002. Prospects of hydatidosis control in continental areas. In: PAHO, pp. 106–115.

Lehmann, J.C., 1928. Allgemeine Pathologie und Klinik der Echinokokkenkrankheit. In: Hosemann, G., Schwarz, E., Lehmann, J.C., Posselt, A. (Eds.), Die Echinokokken-Kankheit. F. Enke, Stuttgart, pp. 115–304.

Leuckart, R., 1863. Die menschlichen Parasiten und die von ihnen herrührenden Krankheiten. Leipzig, C.F. Winter'sche Verlagshandlung. Band 1 und 2.

Lightowlers, M.W., 2006. Vaccines against cysticercosis and hydatidosis: foundations in taeniid cestode immunology. Parasitol. Int. 55, S39–S43.

Lightowlers, M.W., 2011. Advances in Programme Execution and a Proposal for an Integrated Pilot Project for Control of Cystic Echinococcosis. EG95 Vaccine and Immunization and Evaluation of New Strategies Regimes. WHO Report (WHO/HTM/NTD/NZD/2011.2), ISBN 978 92 4 150292 4 (full title of the report see Abdykerimov, 2011).

Lightowlers, M.W., Gottstein, B., 1995. Echinococcosis/hydatidosis: antigens, immunological and molecular diagnosis. In: Thompson, R.C.A., Lymbery, A.J. (Eds.), *Echinococcus* and Hydatid Disease. CAB International, Wallingford, Oxon, ISBN 0 85198 910 1, pp. 183–199, 355–410.

Luder, P.J., Siffert, B., Witassek, F., Meister, F., Bircher, J., 1986. Treatment of hydatid disease with high oral doses of mebendazole. Long-term follow-up of plasma mebendazole levels and drug interactions. Eur. J. Clin. Pharmacol. 31, 443–448.

Lühe, M., 1910. Cystotänien südamerikanischer Feliden. Zool. Jahrb. Suppl. 12, 687–710.

Lukashenko, N.P., 1971. Problems of epidemiology and prophylaxis of alveococcosis (multilocular echinococcosis): a general review – with particular reference to the U.S.S.R. Int. J. Parasitol. 1, 125–134.

Lukashenko, N.P., 1975. Alveokokkoz. Izdatelsatvo "Medicina", Moscow (in Russian).

Macpherson, C.N., Milner, R., 2003. Performance characteristics and quality control of community based ultrasound surveys for cystic and alveolar echinococcosis. Acta Trop. 85, 203–209.

Macpherson, C.N., Bartholomot, B., Frider, B., 2003. Application of ultrasound in diagnosis, treatment, epidemiology, public health and control of *Echinococcus granulosus* and *E. multilocularis*. Parasitology 127, S21–S35.

Massolo, A., Liccioli, S., Budke, C., Klein, C., 2014. *Echinococcus multilocularis* in North America: the great unknown. Parasite 21, 73.

Mathis, A., Deplazes, P., 2004. Diagnosis of *Echinococcus granulosus* and *E. multilocularis* in animals and identification in environment samples. In: Torgerson, P., Shaikenov, B. (Eds.), Echinococcosis in Central Asia: Problems and Solutions. Publishing house "Dauir", Zurich & Almaty, ISBN 9965-517-92-4, pp. 149–165.

Mathis, A., Deplazes, P., 2006. Copro-DNA tests for diagnosis of animal taeniid cestodes. Parasitol. Int. 55 (Suppl.), S87–S90.

Mathis, A., Deplazes, P., Eckert, J., 1996. An improved test system for PCR-based specific detection of *Echinococcus multilocularis* eggs. J. Helminthol. 70, 219–222.

Matossian, R.M., Rickard, M.D., Smyth, J.D., 1977. Hydatidosis: a global problem of increasing importance. Bull. World Health Organ. 55, 499–507.

McManus, D.P., Bryant, C., 1986. Biochemistry and physiology of *Echinococcus*. In: Thompson, R.C.A. (Ed.), The Biology of *Echinococcus* and Hydatid Disease. Allen & Unwin, London, ISBN 0-04-591020-0, pp. 114–142.

McManus, D.P., Bryant, C., 1995. Biochemistry, physiology and molecular biology of *Echinococcus*. In: Thompson, R.C.A., Lymbery, A.J. (Eds.), *Echinococcus* and Hydatid Disease. CAB International, Wallingford, Oxon, ISBN 0-85198-910-1, pp. 135–181.

Mejri, N., Hemphill, A., Gottstein, B., 2010. Triggering and modulation of the host-parasite interplay by *Echinococcus multilocularis*: a review. Parasitology 137, 557–568.

Meldrum, G.K., McConnell, J.D., 1968. The control of hydatid disease in Tasmania. Aust. Veterinary J. 44, 212–217.

Mehlhorn, H., Eckert, J., Thompson, R.C.A., 1983. Proliferation and metastases formation of larval *Echinococcus multilocularis*: II.Ultrastructural investigations. Z. für Parasitenkd. 69, 749–763.

Meslin, F.X., Vuitton, D., 2011. History of the Informal Working Group on Echinococcosis. WHO Report (WHO/HTM/NTD/NZD/2011.2), ISBN 978 92 4 150292 4 (full title of the report see Abdykerimov, 2011).

Morin, A., 1875. Deux cas de tumeur à échinocoques multiloculaires. Thèse. Berne.

Morris, D.L., Richards, K.S., 1992. Hydatid Disease: Current Medical and Surgical Management. Butterworth-Heinemann, Oxford, ISBN 0 7506 13 79 3.

Neisser, A., 1877. Die Echinococcen-Krankheit. Verlag August Hirschwald. Berlin.

Ohbayashi, M., 1993. Parasitology. In: Uchino, J., Sato, N. (Eds.), Alveolar Echinococcosis of the Liver. Hokkaido University School of Medicine, Sapporo, Japan, pp. 21–32. INSS: 0385-6089.

PAHO, 1948. Resolution CD2.R8 Hydatidosis. 2nd Directing Council. Pan American Health Organization/World Health Organization, Washington, DC.

PAHO, 1994. In: Ruiz, A., Schantz, P., Arámbulo III, P. (Eds.), Proceedings of the Scientific Working Group on the Advances in the Prevention, Control and Treatment of Hydatidosis. Pan American Health Organization, Washington, DC, 26-28 October 1994, Montevideo, Uruguay, pp. 1–306. PAHO/HCP/HCV/95/01.

PAHO, 2002. Report of the PAHO/WHO working group on perspectives and possibilities of control and eradication of hydatidosis. In: XIX International Congress of Hydatidology, San Carlos de Bariloche, Rio Negro, Argentina, 20 to 24 September, 1999. Pan American Health Organization, Washington, DC, USA, ISBN 92 75 12420 5.

Pawlowski, Z., Eckert, J., Vuitton, D.A., Ammann, R.W., Kern, P., Craig, P.S., Dar, K.F., De Rosa, F., Filice, C., Gottstein, B., Grimm, F., Macpherson, C.N.L., Sato, N., Todorov, T., Uchino, J., von Sinner, W., Wen, H., 2001. Echinococcosis in humans: clinical aspects, diagnosis and treatment. In: Eckert, J., Gemmell, M.A., Meslin, F.-X., Pawlowski, Z. (Eds.), WHO/OIE Manual on Echinococcosis in Humans and Animals: A Public Health Problem of Global Concern. World Health Organisation for Animal Health, Paris, ISBN 92-9044-522-X, pp. 20–71.

Polydorou, K., 1984. How echinococcosis was conquered in Cyprus. World Health Forum (WHO) 5, 160–165.

Polydorou, K., 1994. Echinococcosis/hydatidosis eradication in Cyprus. In: Ruiz, A., Schantz, P., Arámbulo III, P. (Eds.), Proceedings of the Scientific Working Group on the Advances in the Prevention, Control and Treatment of Hydatidosis, 26–28 October 1994, Montevideo, Uruguay. Pan American Health Organization, pp. 113–122.

Posselt, A., 1928. Der Alveolarechinokokkus und seine Chirurgie. In: Hosemann, G., Schwarz, E., Lehmann, J.C., Posselt, A. (Eds.), Die Echinokokken-Krankheit. F. Enke, Stuttgart, pp. 305–418.

Ramp, T., Eckert, J., 1986. Long-term in vitro maintenance of *Echinococcus multilocularis* metacestode tissue blocks and subsequent proliferation in the intermediate host (*Meriones unguiculatus*). Institute of Parasitology, University of Zürich. Unpublished report.

Ramp, T., Eckert, J., 1987. Long-term maintenance in vitro of *Echinococcus multilocularis* metacestode tissue blocks and subsequent proliferation in the intermediate host (*Meriones unguiculatus*). Zentralblatt für Bakteriologie und Hygiene A, 267, p. 304.

Rausch, R., 1954. Studies on the helminth fauna of Alaska. XX. The histogenesis of the alveolar larva of *Echinococcus* species. J. Infect. Dis. 94, 178–186.

Rausch, R.L., 1967a. A consideration of infraspecific categories in the genus *Echinococcus* Rudolphi, 1801 (Cestoda: Taeniidae). J. Parasitol. 53, 484–491.

Rausch, R.L., 1967b. On the ecology and distribution of *Echinococcus* spp. (Cestoda: Taeniidae) and characteristics of their development in the intermediate host. Ann. de Parasitol. Humaine Comparée 42, 16–93.

Rausch, R.L., 1986. Life-cycle patterns and geographic distribution of *Echinococcus* species. In: Thompson, R.C.A. (Ed.), The Biology of *Echinococcus* and Hydatid Disease. Allen and Unwin, London, ISBN 0-04-591020-0, pp. 44–80.

Rausch, R.L., 1995. Life-cycle patterns and geographic distribution of *Echinococcus* species. In: Thompson, R.C.A., Lymbery, A.J. (Eds.), *Echinococcus* and Hydatid Disease. CAB International, Wallingford, Oxon, ISBN 0 85198 910 1, pp. 89–134.

Rausch, R.L., Bernstein, J.J., 1972. *Echinococcus vogeli* sp. n. (Cestoda: Taeniidae) from the bush dog, *Speothos venaticus* (Lund). Z. für Tropenmedizin Parasitol. 23, 25–34.

Rausch, R.L., D'Alessandro, A., 1999. Histogenesis in the metacestode of *Echinococcus vogeli* and mechanism of pathogenesis in polycystic hydatid disease. J. Parasitol. 85, 410–418.

Rausch, R.L., D' Alessando, A., 2002. The epidemiology of echinococcosis caused by Echinococcus oligarthrus and E. vogeli in the Neotropics. In: Craig, P., Pawlowski, Z. (Eds.), Cestode zoonoses: echinooccosis and cysticercosis. IOS Press, Amsterdam, pp. 107–113. ISSN: 1566–7693.

Rausch, R.L., Fay, F.H., 2002. Epidemiology of alveolar echinococcosis, with reference to St. Lawrence Island, Bering Sea. In: Craig, P., Pawlowski, Z. (Eds.), Cestode Zoonoses: Echinococcosis and Cysticercosis. IOS Press, Amsterdam, pp. 309–325, 1566–7693.

Rausch, R., Jentoft, V.L., 1957. Studies on the helminth fauna of Alaska. XXXI. Observations on the propagation of the larval *Echinococcus multilocularis* Leuckart, 1863, in vitro. J. Parasitol. 43, 1–8.

Rausch, R.L., Richards, S.H., 1971. Observations on parasite-host relationships of *Echinococcus multilocularis* Leuckart, 1863, in North Dakota. Can. J. Zool. 49, 1317–1330.

Rausch, R.L., Schiller, E.L., 1951. Hydatid disease (echinococcosis) in Alaska and the importance of rodent intermediate hosts. Science 113, 57–58.

Rausch, R.L., Schiller, E.L., 1954. Studies on the helminth fauna of Alaska. XXIV. *Echinococcus sibiricensis* n. sp., from St. Lawrence Island. J. Parasitol. 40, 659–662.

Rausch, R.L., Schiller, E.L., 1956. Studies on the helminth fauna of Alaska. XXV. The ecology and public health significance of *Echinococcus sibiricensis* Rausch & Schiller, 1954, on St. Lawrence Island. Parasitology 46, 395–419.

Rausch, R.L., D'Alessandro, A., Ohbayashi, M., 1984. The taxonomic status of *Echinococcus cruzi* Brumpt and Joyeux, 1924 (Cestoda: Taeniidae) from an agouti (Rodentia: Dasyproctidae) in Brazil. J. Parasitol. 70, 295–302.

Rausch, R.L., D'Alessandro, A., Rausch, V.R., 1981. Characteristics of the larval *Echinococcus vogeli* (Rausch & Bernstein 1972) in the natural intermediate host, the paca, *Cuniculus paca* L. (Rodentia: Dasyproctidae). Am. J. Trop. Med. Hyg. 30, 1043–1052.

Rausch, R.L., Rausch, V.R., D'Alessandro, A., 1978. Discrimination of the larval stages of *Echinococcus oligarthrus* (Diesing, 1863) and *E. vogeli* Rausch and Bernstein, 1972 (Cestoda: Taeniidae). Am. J. Trop. Med. Hyg. 27, 1195–1202.

Rausch, R.L., Wilson, J.F., Schantz, P.M., 1990. A programme to reduce the risk of infection by *Echinococcus multilocularis*: the use of praziquantel to control the cestode in a village in the hyperendemic region of Alaska. Ann. Trop. Med. Parasitol. 84, 239–250.

Roberts, M.G., Lawson, J.R., Gemmell, M.A., 1986. Population dynamics in echinococcosis and cysticercosis: mathematical model of the life-cycle of *Echinococcus granulosus*. Parasitology 92, 621–641.

Rickard, M.D., Lightowlers, M.W., 1986. Immunodiagnosis of hydatid disease. In: Thompson, R.C.A. (Ed.), The Biology of *Echinococcus* and Hydatid Disease. George Allen & Unwin, London, pp. 217–249. ISBN: 0-04-591020.0.

Romig, T., 2011. Control of *Echinococcus granulosus* and Cystic Echinococcosis. Control of *Echinococcus multilocularis*. WHO Report (WHO/HTM/NTD/NZD/2011.2), ISBN 978 92 4 150292 4 (full title of the report see Abdykerimov, 2011).

Romig, T., Bilger, B., Merli, M., Dinkel, A., Lucius, R., Mackenstedt, U., 1999. Bekämpfung von *Echinococcus multilocularis* in einem Hochendemiegebiet Süddeutschlands. In: Neuere Methoden und Ergebnisse zur Epidemiologie von Parasitosen. Deutsche Veterinärmedizinische Gesellschaft, Giessen, pp. 172–183.

Rudolphi, C.A., 1801. Beobachtungen über die Eingeweidewürmer. Arch. für Zool. Zootomie 2, 1–65.

Sato, N., Uchino, J., Suzuki, K., 1993. Chemotherapy. In: Uchino, J., Sato, N. (Eds.), Alveolar Echinococcosis of the Liver. Hokkaido University School of Medicine, Sapporo, Japan, pp. 151–166. INSS 0385-6089.

Schantz, P.M., van den Bossche, H., Eckert, J., 1982. Chemotherapy for larval echinococcosis in animals and humans: report of a workshop. Z. für Parasitenkd. 67, 5–26.

Schantz, P.M., Chai, J., Craig, P.S., Eckert, J., Jenkins, D.J., Macpherson, C.N.L., Thakur, A., 1995. Epidemiology and control of hydatid disease. In: Thompson, R.C.A., Lymbery, A.J. (Eds.), *Echinococcus* and Hydatid Disease. CAB International, Wallingford, Oxon, ISBN 0 85198 910 1, pp. 233–331.

Schantz, P.M., Cruz-Reyes, A., Coll,i, C., Lord, R.D., 1975. Sylvatic echinococcosis in Argentina. I. On the morphology and biology of strobilar *Echinococcus granulosus* (Batsch, 1786) from domestic and sylvatic animal hosts. Tropenmedizin Parasitol. 26, 334–344.

Schantz, P.M., Colli, C., Cruz-Reyes, A., Prezioso, U., 1976. Sylvatic echinococcosis in Argentina. II. Susceptibility of wild carnivores to *Echinococcus granulosus* (Batsch, 1786) and host-induced morphological variation. Tropenmedizin Parasitol. 27, 70–88.

Siracusano, A., Delunardo, F., Teggi, A., Ortona, E., 2012. Host-parasite relationship in cystic echinococcosis: an evolving story. Clin. Dev. Immunol. 2012, 639362.

Smyth, J.D., 1962. Introduction to Animal Parasitology. The English University Press, London.

Smyth, J.D., 1964. The biology of the hydatid organisms. Adv. Parasitol. 2, 169–219.

Smyth, J.D., 1968. In vitro studies and host-specificity in *Echinococcus*. Bull. World Health Organ. 39, 5–12.

Smyth, J.D., Davies, Z., 1974. In vitro culture of the strobilar state of *Echinococcus granulosus* (sheep strain): a review of basic problems and results. Int. J. Parasitol. 4, 631–644.

Stössel, T., 1989. Literaturübersicht zur Häufigkeit und geographischen Verbreitung der Echinokokkose bei Menschen und Tieren in Ländern der EG und EFTA (Dissertation Medizinische Fakultät). Universität Zürich.

Schwabe, C.W., 1969. Veterinary Medicine and Human Health, second ed. Williams and Wilkins Company, Baltimore.

Schwarz, E., 1928. Die geographische Verbreitung der Echinokokkenkrankheit. In: Hosemann, G., Schwarz, E., Lehmann, J.C., Posselt, A. (Eds.), Die Echinokokken-Krankheit. F. Enke, Stuttgart, pp. 67–114.

Schweiger, A., Ammann, R.W., Candinas, D., Clavien, P.A., Eckert, J., Gottstein, B., Halkic, N., Müllhaupt, B., Prinz, B.M., Reichen, J., Tarr, P.E., Torgerson, P.R., Deplazes, P., 2007. Human alveolar echinococcosis after fox population increase, Switzerland. Emerg. Infect. Dis. 13, 878–882.

Siles-Lucas, M., Hemphill, A., 2002. Cestode parasites: application of in vivo and in vitro models for studies on the host-parasite relationship. Adv. Parasitol. 51, 133–230.

Siqueira, N.G., de Almeida, F.B., Chalub, S.R., Machado-Silva, J.R., Rodrigues-Silva, R., 2007. Successful outcome of hepatic polycystic echinococcosis managed with surgery and chemotherapy. Trans. R. Soc. Trop. Med. Hyg. 101, 624–626.

Siqueira, N.G., Almeida, F.B., Suzuki, Y.A., Lima, R.N., Machado-Silva, J.R., Rodrigues-Silva, R., 2010. Atypical polycystic echinococcosis without liver involvement in Brazilian patients. Trans. R. Soc. Trop. Med. Hyg. 104, 230–233.
Sousa, O.E., Thatcher, V.E., 1969. Observations on the life-cycle of *Echinococcus oligarthrus* (Diesing, 1863) in the Republic of Panama. Ann. Trop. Med. Parasitol. 63, 165–175.
Spiliotis, M., Brehm, K., 2008. Axenic in vitro cultivation of *Echinococcus multilocularis* metacestodes vesicles and the generation of primary cell cultures. In: Rupp, S., Sohn, K. (Eds.), Host-pathogen Interactions. Huma Press, Toronto, ISBN 978-1-58829-886-7, pp. 245–262.
Stefanić, S., Shaikenov, B.S., Deplazes, P., Dinkel, A., Torgerson, P.R., Mathis, A., 2004. Polymerase chain reaction for detection of patent infections of *Echinococcus granulosus* ("sheep strain") in naturally infected dogs. Parasitol. Res. 92, 347–351.
Stojkovic, M., Zwahlen, M., Teggi, A., Vutova, K., Cretu, C.M., Virdone, R., Nicolaidou, P., Cobanoglu, N., Junghanss, T., 2009. Treatment response of cystic echinococcosis to benzimidazoles: a systematic review. PLoS Negl. Trop. Dis. 3, e524.
Suzuki, K., Sato, N., Uchino, J., 1993. Epidemiology. In: Uchino, J., Sato, N. (Eds.), Alveolar Echinococcosis of the Liver, vol. 30. Hokkaido University Medical Library Series, pp. 1–20. INSS 0385-6089.
Szidat, L., 1971. Neue Aspekte des Echinococcose-Problems. Angew. Parasitol. 12, 133–143.
Takahashi, K., Uraguchi, K., Hatakeyama, H., Giraudoux, P., Romig, T., 2013. Efficacy of anthelmintic baiting of foxes against *Echinococcus multilocularis* in northern Japan. Veterinary Parasitol. 198, 122–126.
Thakur, A.S., 2002. Eradication of hydatid disease in Iceland. In: PAHO, pp. 32–38.
Tamarozzi, F., Rossi, P., Galati, F., Mariconti, M., Nicoletti, G.J., Rinaldi, F., Casulli, A., Pozio, E., Brunetti, E., 2015. The Italian registry of cystic echinococcosis (RIEC): the first prospective registry with a European future. Eurosurveillance 20 (18) pii:21115.
Tappe, D., Kern, P., Frosch, M., Kern, P., 2010a. A hundred years of controversy about the taxonomic status of *Echinococcus* species. Acta Trop. 115, 167–174.
Tappe, D., Stich, A., Frosch, M., 2008. Emergence of polycystic neotropical echinococcosis. Emerg. Infect. Dis. 14, 292–297.
Tappe, D., Zidowitz, S., Demmer, P., Kern, P., Barth, T.F., Frosch, M., 2010b. Three-dimensional reconstruction of *Echinococcus multilocularis* larval growth in human hepatic tissue reveals complex growth patterns. Am. J. Trop. Med. Hyg. 82, 126–127.
Thatcher, V.E., Sousa, O.E., 1968. *Echinococcus oligarthrus* (Diesing, 1863) developing in a United States zoo. J. Parasitol. 54, 847–848.
Thienpont, D., Vanparijs, O., Hermans, L., 1974. Anthelmintic activity of mebendazole against Cysticercus fasciolaris. J. Parasitol. 60, 1052–1053.
Thiodét, J.J., 1954/55. Die Chemotherapie der Echinococcenerkrankung. Therapiewoche 5, 366–368.
Thomas, H., Gönnert, R., Pohlke, R., Seubert, J., 1975. A new compound against adult tapeworms. In: Proceedings, 7th International Conference of the World Association for the Advancement of Veterinary Parasitology, Thessaloniki (Abstract No. 51).
Thompson, R.C.A. (Ed.), 1986a. The Biology and Systematics of *Echinococcus* and Hydatid Disease. Allen & Unwin, London, ISBN 0-04-591020-0.
Thompson, R.C.A., 1986b. Biology and systematics of *Echinococcus*. In: Thompson, R.C.A. (Ed.), The Biology of *Echinococcus* and Hydatid Disease. Allen & Unwin, London, ISBN 0-04-591020-0, pp. 5–43.
Thompson, R.C.A., 1995. Biology and systematics of *Echinococcus*. In: Thompson, R.C.A., Lymbery, A.J. (Eds.), *Echinococcus* and Hydatid Disease. CAB International, Wallingford, Oxon, ISBN 0-85198-910-1, pp. 1–50.
Thompson, R.C.A., 2002. Hydatidosis eradication within insular systems. In: PAHO, pp. 39–46.

Thompson, R.C.A., Allsopp, C.E., 1988. Hydatidosis: Veterinary Perspectives and Annotated Bibliography. CAB International, Wallingford, ISBN 0 85 198 610 2.
Thompson, R.C., Eckert, J., 1982. The production of eggs by *Echinococcus multilocularis* in the laboratory following in vivo and in vitro development. Z. für Parasitenkd. 68, 227–234.
Thompson, R.C.A., Lymbery, A.J., 1995. *Echinococcus* and Hydatid Disease. In: CAB International, Wallingford, Oxon, ISBN 0 85198 910 1.
Thompson, R.C.A., Lymbery, A.J., 2013. Let's not forget the thinkers. Trends Parasitol. 29, 581–584.
Thompson, R.C., Deplazes, P., Eckert, J., 1990. Uniform strobilar development of *Echinococcus multilocularis* in vitro from protoscolex to immature stages. J. Parasitol. 76, 240–247.
Thompson, R.C., Kapel, C.M., Hobbs, R.P., Deplazes, P., 2006. Comparative development of *Echinococcus multilocularis* in its definitive hosts. Parasitology 132 (Pt 5), 709–716.
Todorov, T., Vutova, K., Mechkov, G., Georgiev, P., Petkov, D., Tonchev, Z., Nedelkov, G., 1992. Chemotherapy of human cystic echinococcosis: comparative efficacy of mebendazole and albendazole. Ann. Trop. Med. Parasitol. 86, 59–66.
Torgerson, P.R., 2003. The use of mathematical models to simulate control options for echinococcosis. Acta Trop. 85, 211–221.
Torgerson, P.R., 2006. Mathematical models for the control of cystic echinococcosis. Parasitol. Int. 55 (Suppl.), S253–S258.
Torgerson, P., 2011. Advances in Programme Execution and a Proposal for an Integrated Pilot Project for Control of Cystic Echinococcosis. Integrated Approaches for Cystic Echinococcosis Control. WHO Report (WHO/HTM/NTD/NZD/2011.2), ISBN 978 92 4 150292 4 (full title of the report see Abdykerimov, 2011).
Torgerson, P.R., Macpherson, C.N., 2011. The socioeconomic burden of parasitic zoonoses: global trends. Veterinary Parasitol. 182, 79–95.
Torgerson, P.R., Shaikenov, B., 2004. Echinococcosis in Central Asia: Problems and Solutions. Publishing house "Dauir", Zurich and Alamty.
Torgerson, P.R., Macpherson, C.N., Vuitton, D.A., 2011. Cystic echinococcosis. In: Palmer, S.R., Lord Soulsby, Torgerson, P.R., Brown, D.W. (Eds.), Oxford Textbook of Zoonoses, second ed. Oxford University Press, Oxford, ISBN 978-0-19-857002-8, pp. 650–688.
Torgerson, P.R., Schweiger, A., Deplazes, P., Pohar, M., Reichen, J., Ammann, R.W., Tarr, P.E., Halkik, N., Müllhaupt, B., 2008. Alveolar echinococcosis: from a deadly disease to a well-controlled infection. Relative survival and economic analysis in Switzerland over the last 35 years. J. Hepatol. 49, 72–77.
Trachsel, D., Deplazes, P., Mathis, A., 2007. Identification of taeniid eggs in the faeces from carnivores based on multiplex PCR using targets in mitochondrial DNA. Parasitology 134 (Pt 6), 911–920.
Uchino, J., Sato, N., Nakajima, Y., Matsushita, M., Takahashi, M., Une, Y., 1993. Treatment. In: Uchino, J., Sato, N. (Eds.), Alveolar Echinococcosis of the Liver. Hokkaido University School of Medicine, Sapporo, Japan, pp. 137–149. INSS 0385-6089.
Varela-Díaz, V.M., Coltorti, E.A., 1976. Human Hydatidosis. Immunodiagnostic Techniques. Series of Scientific and Technical Monographs C.P.Z. - 7. Pan American Zoonoses Center, Buenos Aires, Argentina.
Verster, A.J.M., 1965. Review of *Echinococcus* species in South Africa. Onderstepoort J. Veterinary Res. 32, 7–118.
Vierodt, H., 1886. Abhandlung über den multilokulären Echinococcus. Akademische Verlagsbuchhandlung. von J.C.B. Mohr, Freiburg.
Virchow, R., 1856. Die multiloculäre, ulcerirende Echinokokkengeschwulst der Leber. Verhandlungen der physicalisch-medicinischen Ges. Würzburg. 6, 84–95.
Vogel, H., 1955. Über den Entwicklungszyklus und die Artzugehörigkeit des europäischen Alveolar echinococcus. Dtsch. Med. Wochenschr. 80, 931–932.

Vogel, H., 1961. Biologie und Parasitologie des Alveolarechinokokkus. Chir. Prax. 61, 407–412.
Vogel, H., 1957. Über den *Echinococcus multilocularis* Süddeutschlands. l. Das Bandwurmstadium von Stämmen menschlicher und tierischer Herkunft. Z. für Tropenmedizin Parasitol. 8, 404–454.
Vogel, H., 1978. Wie wächst der Alveolarechinokokkus? Tropenmedizin Parasitol. 29, 1–11.
Von Sieboldt, C.T.E., 1853. Ueber die Verwandlung der *Echinococcus*-Brut in Taenien. Z. für Wiss. Zool. 4, 409–425.
Von Sinner, W., Lewall, D., 2001. Hydatid disease (echinococcosis). In: Palmer, P.E.S., Reeder, M.M. (Eds.), The Imaging of Tropical Diseases, vol. 1. Springer, Berlin, ISBN 3-540-56028-9, pp. 205–353.
Vuitton, D.A., 1997. The WHO informal working group on echinococcosis. Coordinating board of the WHO-IWGE. Parassitologia 39, 349–353.
Vuitton, D.A., 2009. Benzimidazoles for the treatment of cystic and alveolar echinococcosis: what is the consensus? Expert Rev. Anti Infect. Ther. 7, 145–149.
Vuitton, D.A., Demonmerot, F., Knapp, J., Richou, C., Grenouillet, F., Chauchet, A., Vuitton, L., Bresson-Hadni, S., Millon, L., 2015. Clinical epidemiology of human AE in Europe. Veterinary Parasitol. 213, 110–120.
Vuitton, D.A., Zhou, H., Bresson-Hadni, S., Wang, Q., Piarroux, M., Raoul, F., Giraudoux, P., 2003. Epidemiology of alveolar echinococcosis with particular reference to China and Europe. Parasitology 127 (Suppl.), S87–S107.
Wang, Q., 2011. Control of *Echinococcus granulosus* and Cystic Echinococcosis. Control in China. WHO Report (WHO/HTM/NTD/NZD/2011.2), ISBN 978 92 4 150292 4 (full title of the report see Abdykerimov, 2011).
Weinberg, M., 1909. Séro-diagnostic de l'echinococcose. Ann. l'Institut Pasteur 23, 472–502.
WHO, 1978. Resolution WHA 31.48: Prevention and Control of Zoonoses and Foodborne Disease Due to Animal Products. 31st World Health Assembly, May 1978. World HealthOrganization, Geneva.
WHO, 1981. Eckert, J., Gemmell, M.A., Soulsby, E.J.L. (Eds.), Guidelines for surveillance Prevention and Control of Echinococcosis/Hydatidosis. World Health Organization, Geneva, pp. 1e148
WHO, 1984a. Treatment of Human Echinococcosis. Report of an informal meeting, Geneva, 15–16 December 1983. World Health Organization, Geneva.
WHO, 1984b. Eckert, J., Gemmell, M.A., Matyas, Z., Soulsby, E.J.L. (Eds.), Guidelines for Surveillance, Prevention and Control of Echinococcosis/Hydatidosis, second ed. World Health Organization, Geneva, pp. 1–148. VPH/81.28.
WHO, 1996. WHO Informal Working Group on Echinococcosis. Guidelines for treatment of cystic and alveolar echinococcosis. Bull. World Health Organ. 74, 231–242.
WHO, 2001a. Eckert, J., Gemmell, M.A., Meslin, F.-X., Pawlowski, Z. (Eds.), WHO/OIE Manual on Echinococcosis in Humans and Animals: A Public Health Problem of Global Concern. World Health Organisation for Animal Health, Paris, ISBN 92-9044-522-X.
WHO, 2001b. PAIR: Puncture, Aspiration, Injection, Re-aspiration. An Option for the Treatment of Cystic Echinococcosis. WHO/CDS/CSR/APH/2001.6. World Health Organization, Geneva.
WHO, 2013. Neglected Tropical Diseases. www.who.int.neglected_diseases/diseases/en.
WHO-IWGE, 2003. WHO Informal Working Group: international classification of ultrasound images in cystic echinococcosis for application in clinical and field epidemiological settings. Acta Trop. 85, 253–261.
WHO-IWGE, 2011. WHO InformalWorking Group. Report of the WHO InformalWorking Group on cystic and alveolar echinococcosis, surveillance, prevention and control, with the participation of the Food and Agriculture Organization of the United

Nations and the World Organisation for Animal Health. 22–23 June, 2011. Department of Control of Neglected Tropical Diseases. World Health Organization, Geneva.
WHO/PDP, 1981. Treatment of Human Echinococcosis. Report of an infomal meeting, 29 June -1 July 1981. PDP/82.1. World Health Organization, Geneva.
WHO/PDP, 1984. Treatment of Human Echinococcosis. Report of an informal meeting, 15–16 December 1983. PDP. World Health Organization, Geneva.
WHO/VPH, 1993. Report of WHO Working Group Meeting on Echinococcosis Research and Control, Beijing, China, 12 October 1993. VPH/93.131. World Health Organization, Geneva.
WHO/CDS/VPH, 1992. Report of the WHO Working Group Meeting on Clinical Medicine and Chemotherapy of Alveolar and Cystic Echinococcosis. Besancon, France, 10 October 1992. WHO/CDS/VPH/93.118. World Health Organization, Geneva.
Wilson, J.F., Davidson, M., Rausch, R.L., 1978. A clinical trial of mebendazole in the treatment of alveolar hydatid disease. Am. Rev. Respir. Dis. 118, 747–757.
Witassek, F., Burkhardt, B., Eckert, J., Bircher, J., 1981. Chemotherapy of alveolar echinococcosis. Comparison of plasma mebendazole concentrations in animals and man. Eur. J. Clin. Pharmacol. 20, 427–433.
Zeyhle, E., 1982. Die Verbreitung von *Echinococcus multilocularis* in Südwestdeutschland. In: Bähr, R. (Ed.), Probleme der Echinokokkose unter Berücksichtigung parasitologischer und klinischer Aspekte. Verlag Hans Huber, Bern, ISBN 3-456-81183-7, pp. 26–33.
Zhang, W., Wen, H., Li, J., Lin, R., McManus, D.P., 2012. Immunology and immunodiagnosis of cystic echinococcosis: an update. Clin. Dev. Immunol. 2012, 101895.

CHAPTER TWO

Biology and Systematics of *Echinococcus*

R.C.A. Thompson
Murdoch University, Murdoch, WA, Australia
E-mail: a.thompson@murdoch.edu.au

Contents

Abstract

The biology of *Echinococcus*, the causative agent of echinococcosis (hydatid disease) is reviewed with emphasis on the developmental biology of the adult and metacestode stages of the parasite. Major advances include determining the origin, structure and functional activities of the laminated layer and its relationship with the germinal layer;

Advances in Parasitology, Volume 95
ISSN 0065-308X
http://dx.doi.org/10.1016/bs.apar.2016.07.001

and the isolation, in vitro establishment and characterization of the multipotential germinal cells. Future challenges are to identify the mechanisms that provide *Echinococcus* with its unique developmental plasticity and the nature of activities at the parasite-host interface, particularly in the definitive host. The revised taxonomy of *Echinococcus* is presented and the solid nomenclature it provides will be essential in understanding the epidemiology of echinococcosis.

1. INTRODUCTION

Echinococcus Rudolphi, 1801 (see Chapter 1), is a small endoparasitic flatworm belonging to the Class Cestoda (Table 1). It is a 'true tapeworm' (Subclass Eucestoda) and as such exhibits the features that characterize this group (Table 1; Fig. 1). It has no gut and all metabolic interchange takes

Table 1 Classification of *Echinococcus*

Phylum Platyhelminthes
Soft-bodied, triploblastic and acoelomate; dorsoventrally flattened with cellular outer body covering; excretory system protonephridial
Class Cestoda
Endoparasites; gut absent; outer body covering a living syncytial tegument with microtriches
Subclass Eucestoda
True tapeworms; adults characteristically with elongated body (strobila) consisting of linear sets of reproductive organs (proglottids); specialized anterior attachment organ, a scolex; hermaphrodite with indirect life cycles
Order Cyclophyllidea
Scolex with four muscular suckers and a rostellum usually armed with hooks strobila consisting of proglottids in various stages of development and each proglottid clearly demarcated by external segmentation; eggs round, not operculate, containing nonciliated six-hooked oncosphere
Family Taeniidae
Adults in small intestine of carnivores and humans; intermediate hosts all mammalian; scolex with rostellum usually armed with double row of hooks; genitalia unpaired in each proglottid with marginal genital pore irregularly alternating; eggs with radially striated hardened 'shell' (embryophore) metacestode a cysticercus, coenurus, **hydatid** or strobilocercus

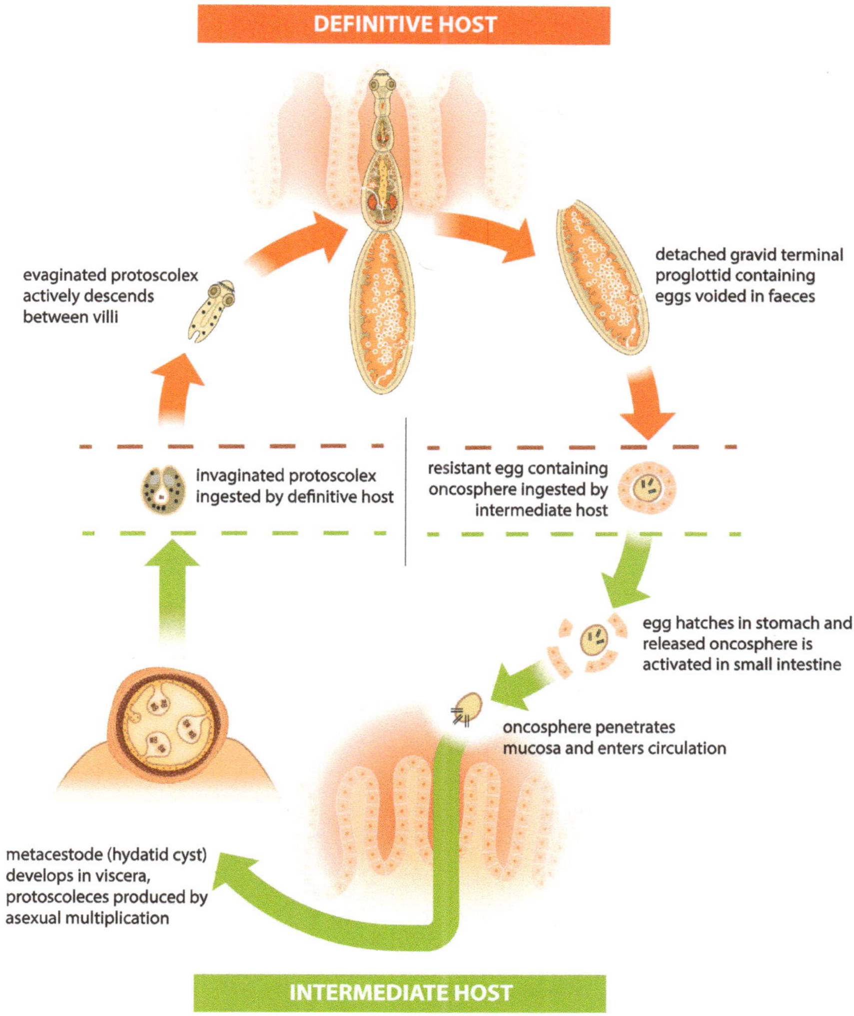

Figure 1 Basic life cycle of *Echinococcus*.

place across the syncytial outer covering, the tegument. Anteriorly, the adult possesses a specialized attachment organ, the scolex, which has two rows of hooks on the rostellum, and four muscular suckers (Fig. 2). A narrow 'neck' region separates the scolex from the rest of the body, or strobila, which is 'segmented' and consists of a number of reproductive units (proglottids) (Fig. 2). *Echinococcus* has an indirect, two host life cycle in which the sexually reproducing adult is hermaphrodite and the larval metacestode stage, the hydatid cyst, proliferates asexually (Fig. 1).

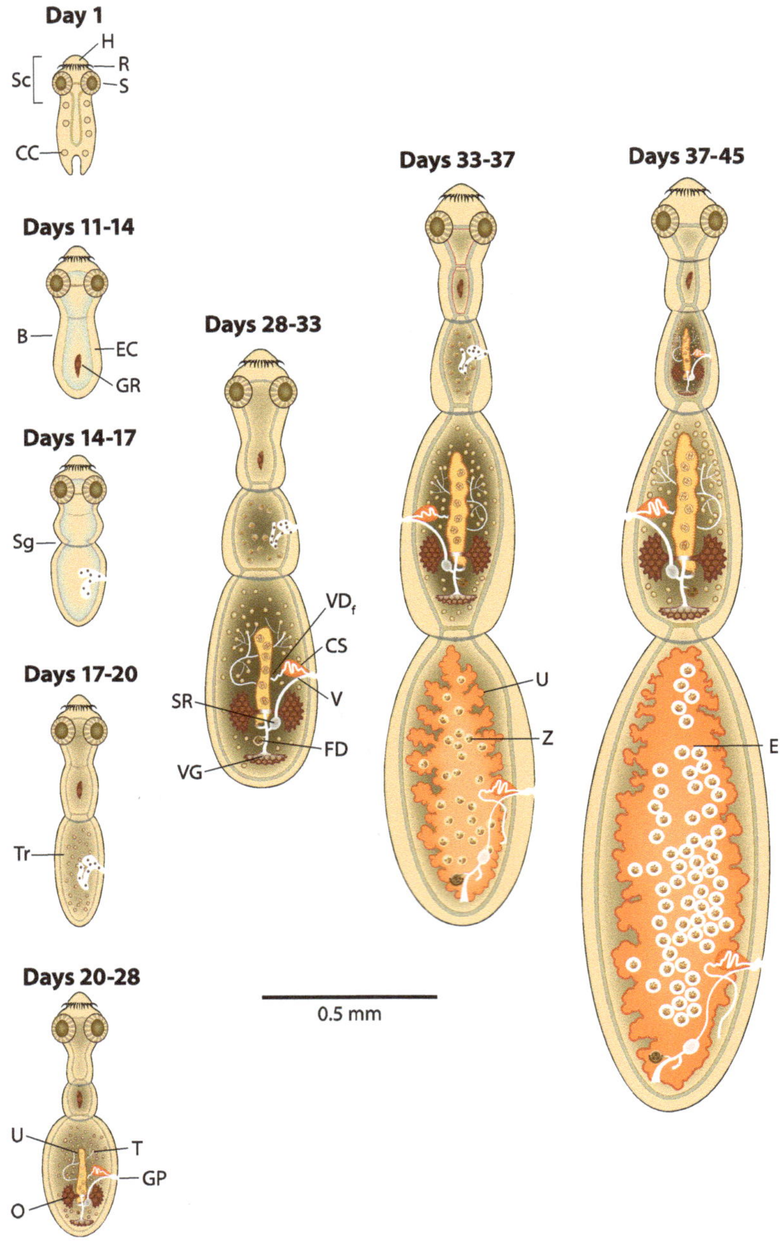

Figure 2 Stages of development of adult *Echinococcus granulosus* in the definitive host. (The periods at which various stages appear may vary and are dependent on 'strain'/isolate of parasite and various host factors). Day 1: Protoscolex has evaginated and elongated; contains numerous calcareous corpuscles. Days 11–14: calcareous corpuscles

Apart from its size, only a few millimetres long, and the possession of rarely more than five proglottids, *Echinococcus* is a typical taeniid cestode requiring two mammalian hosts to complete its life cycle (Fig. 1); a carnivorous definitive host in which the adult cestode develops in the small intestine, and a herbivorous or omnivorous intermediate host in which the metacestode develops, usually in the viscera. Unlike *Taenia*, the metacestode exhibits a low degree of host specificity and has a much greater reproductive potential. The definitive host is always a carnivore. The metacestode is a fluid-filled cystic structure that undergoes asexual multiplication to produce large numbers of scolices, termed protoscoleces. There may be several thousand protoscoleces within a single cyst, and each one is capable of developing into a sexually mature adult worm. Following sexual reproduction adult worms produce fertilized eggs, each containing a single embryo (oncosphere). Proglottids containing fully developed eggs are voided with the faeces of the definitive host. They may attach to the perianal region of the definitive host or contaminate the environment. The eggs that are released from proglottids are surrounded by a thick, resistant outer covering and are capable of surviving in the environment for extended periods.

have disappeared; lateral excretory canals are conspicuous; genital rudiment present denoting formation of first proglottid; constriction and clear area below the neck ('Banding') marks the site of the first segment. Days 14–17: genital rudiment has divided into two and extends unilaterally; first segment fully formed. Days 17–20: rudimentary testes appear in first proglottid; initial stages in formation of second proglottid. Days 20–28: two-segmented worm; male genitalia – testes, cirrus and vas deferens – have developed; female genitalia – ovary, Mehlis' gland and vitelline gland – still developing; uterus appears as a streak; both cirrus and vagina open to exterior via lateral genital pore. Days 28–33: male and female genitalia in terminal proglottid fully mature; uterus still dilating; penultimate proglottid has developing genitalia; either a band or third segment appears. Days 33–37: ovulation and fertilization in terminal proglottid; fully dilated uterus contains dividing zygotes; male and female genitalia degenerating in terminal proglottid; mature genitalia in penultimate proglottid and developing genitalia in ante-penultimate proglottid; strobila divided by three or four segments. Days 37–45: gravid with embryonated eggs in uterus of terminal proglottid – eggs have fully formed embryophore ('thick-shelled') and contain embryo (oncosphere); zygotes in uterus of penultimate proglottid and maturing genitalia in ante-penultimate proglottid; strobila divided by three, four or five segments. *B*, band; *CC*, calcareous corpuscles; *CS*, cirrus sac; *E*, embryonated eggs; *EC*, excretory canal; *FD*, female reproductive ducts; *GP*, genital pore; *GR*, genital rudiment; *H*, Hooks, *O*, ovary; *R*, rostellum; *S*, sucker; *Sc*, scolex; *Sg*, segment; *SR*, seminal receptacle; *Tr*, rudimentary testes; *T*, testes; *U*, uterus; *V*, vagina; VD_f, vas deferens; *VG*, vitelline gland; *Z*, zygotes. Scale bar, 0.5 mm. *Based on an original drawing by L.M. Kumaratilake.*

Numerous species of herbivorous or omnivorous intermediate hosts are susceptible to infection with the metacestode following accidental ingestion of the eggs.

2. TERMINOLOGY

Infection with *Echinococcus* may be naturally transmitted between humans and other animals. It thus claims membership of the most significant group of communicable diseases, the zoonoses. The clinical and economic significance of the parasite are almost completely confined to infection with the metacestode. Hydatid disease, hydatidosis and echinococcosis are all terms used to refer to infection with the metacestode. Strictly speaking, the terms hydatid disease and hydatidosis should be restricted to infection with the metacestode, and echinococcosis to infection with the adult stage. This is the convention with *Taenia* infections in which the terms cysticercosis and taeniasis apply to infection with the metacestode (cysticercus) and adult, respectively. However, more recently a consensus has been reached to use the term echinococcosis for infections with the metacestode of *Echinococcus* to clarify the distinctness between the diseases in humans caused by *Echinococcus granulosus* and *Echinococcus multilocularis*, cystic echinococcosis (CE) and alveolar echinococcosis (AE), respectively (Table 2). *Echinococcus oligarthra* and *Echinococcus vogeli* both cause polycystic echinococcosis (PE) in humans.

The term strain was introduced during the period of taxonomic uncertainty to refer to intraspecific variants of *Echinococcus* with defined phenotypic and subsequently genotypic characteristics (Lymbery and Thompson, 2012; Thompson and Lymbery, 1990, 1996). The term strain has largely been replaced by genotype as this enabled a numerical system to be developed to refer to the different strains (Table 2). However, the majority of strains/genotypes are now recognized as distinct species and the revised classification is shown in Table 2.

3. TAXONOMY

3.1 Species, strains and species

There has been a long history of taxonomic confusion at the species level in the genus *Echinococcus*. This has been reviewed extensively over the years and it is not intended to reiterate the history here. In summary,

Table 2 Current taxonomy of *Echinococcus*

Species	Strain/genotype	Known intermediate hosts	Known definitive hosts	Infectivity to humans	Disease
Echinococcus granulosus	Sheep/G1	Sheep (cattle, pigs, camels, goats, macropods)	Dog, fox, dingo, jackal and hyena	Yes	CE
	Tasmanian sheep/G2	Sheep (cattle?)	Dog, fox	Yes	CE
	Buffalo/G3	Buffalo (cattle?)	Dog, fox?	Yes	CE
Echinococcus equinus	Horse/G4	Horses and other equines	Dog	Probably not	CE?
Echinococcus ortleppi	Cattle/G5	Cattle	Dog	Yes	CE
Echinococcus canadensis	Cervids/G8,G10	Cervids	Wolves, dog	Yes	CE
Echinococcus intermedius	Camel/Pig/G6/G7	Camels, pigs, sheep	Dog	Yes	CE
Echinococcus felidis	Lion/?	Warthog, (zebra, wildebeest, bushpig, buffalo, various antelope, giraffe Hippopotamus?)	Lion	?	-
Echinococcus multilocularis	Some isolate variation	Rodents, domestic and wild pig, dog, monkey, (horse?)	Fox, dog, cat, wolf, racoon-dog, coyote	Yes	AE
Echinococcus shiquicus	?	Pika and ?	Tibetan fox and?	?	AE?
Echinococcus vogeli	None reported	Rodents	Bush dog	Yes	PE
Echinococcus oligarthra	None reported	Rodents	Wild felids	Yes	PE

AE, alveolar echinococcosis; *CE*, cystic echinococcosis; *PE*, polycystic echinococcosis.
Data from Thompson, R.C.A., Lymbery, A.J., Constantine, C.C., 1995. Variation in *Echinococcus*: towards a taxonomic revision of the genus. Adv. Parasitol. 35, 145–176; Thompson, R.C.A., McManus, D.P., 2002. Towards a taxonomic revision of the genus *Echinococcus*. Trends Parasitol. 18, 452–457; Jenkins, D.J., Romig, T., Thompson, R.C.A., 2005. Emergence/re-emergence of *Echinococcus* spp.—a global update. Int. J. Parasitol. 35, 1205–1219; Thompson, R.C.A., 2008. The taxonomy, phylogeny and transmission of *Echinococcus*. Exp. Parasitol. 119, 439–446; Thompson, R.C.A., Jenkins, D.J., 2014. *Echinococcus* as a model system. Int. J. Parasitol. 44, 865–877.

many species have been described and just as many invalidated on sound taxonomic grounds. What has been clear, however, from the earliest descriptions of the parasite is that the genus exhibits considerable variability at the species level in terms of host specificity, morphology, antigenicity, development rate, and cycles of transmission (Lymbery and Thompson, 2012; Thompson and Lymbery, 1988; Thompson and McManus, 2001, 2002; Thompson, 2008). However, concerns about the genetic basis of phenotypic differences, particularly with respect to morphology, intermediate host specificity and evidence of reproductive isolation, have been the principle reasons for questioning the taxonomic status of some species (also see Chapter 3).

One of the most important observations in the recent history of *Echinococcus* taxonomy was made as a result of studies on the in vitro cultivation of the parasite, in which protoscoleces collected from hydatid cysts in horses failed to develop in the same way as those of sheep origin. Protoscoleces from horses evaginated and increased in length but failed to undergo proglottisation or segmentation, even though they were grown in exactly the same diphasic medium (Smyth and Davies, 1974a). This fairly simple observation resulted in radical shifts in our understanding of the epidemiology of echinococcosis and transmission of the aetiological agents as well as their taxonomy and phylogenetic relationships (Howell and Smyth, 1995; Thompson and Lymbery, 2013; Thompson and Jenkins, 2014). The results demonstrated that there were fundamental physiological differences between *E. granulosus* of sheep and horse origin and the coining of the term 'physiological strain differences' (Smyth and Davies, 1974b; Smyth, 1982). This had a broad influence beyond *Echinococcus*, and in particular the importance of combining phenotypic and genetic differences in the characterization and description of parasites at the intraspecific level (Lymbery and Thompson, 2012; Thompson and Lymbery, 1990, 1996; Thompson and Jenkins, 2014).

The observation of physiological differences between the two parasites of sheep and horse origin complemented earlier epidemiological and taxonomic studies on *Echinococcus* of horse origin (Williams and Sweatman, 1963). These demonstrated morphological differences between the two forms that were considered to be taxonomically significant, and to reflect differences in host specificity and their life cycles. The sympatric occurrence of distinct sheep and horse dog cycles in several European countries (Gonzalez et al., 2002; Thompson and Smyth, 1975; Thompson, 2001) supported the existence of two separate host-adapted species. In addition, epidemiological evidence not only demonstrated distinct differences in intermediate host specificity but also that, unlike the sheep strain (= *E. granulosus*), the horse strain (= *Echinococcus equinus*; Table 2) does not appear to be infective to humans (Thompson and Lymbery, 1988; Thompson, 1995, 2008).

The outcomes of this research caused an attitudinal shift in studies on *Echinococcus* that were not constrained by taxonomic issues with the growing realization that a 'strain' was an acceptable term when describing variability at the phenotypic, and subsequently genotypic levels within species of parasites (Lymbery and Thompson, 2012; Thompson and Lymbery, 1990, 1996). As such, research on the horse and sheep strains of *E. granulosus* led to similar studies on *Echinococcus* of cattle, pig, camel and cervid origin with the description and characterization of several new strains/genotypes (Thompson and McManus, 2002; Thompson, 2008, Table 2). These studies not only confirmed the existence of a number of host-adapted life cycles in different parts of the world but also provided additional data on developmental differences between strains which may impact on control (Lymbery and Thompson, 2012; Thompson, 2001, 2008; Thompson and Lymbery, 1988; Thompson and McManus, 2002). These informal groupings were retained for many years but with the advent of molecular characterization they were shown to be genetically distinct (Thompson and McManus, 2001). PCR-based techniques using a variety of genetic loci, and sequencing of nuclear and mitochondrial DNA, coupled with molecular epidemiological studies in endemic areas, confirmed the genetic and morphological distinctness of the host-adapted strains and revealed phylogenetic relationships which support a robust, meaningful taxonomy based on a previously documented nomenclature (Table 2; Bowles et al., 1994; Cruz-Reyes et al., 2007; Harandi et al., 2002; Huttner et al., 2009; Jenkins et al., 2005 Lavikainen et al., 2003, 2006; Moks et al., 2008; Nakao et al., 2013; Pednekar et al., 2009; Romig et al., 2006, 2015; Saarma et al., 2009; Thompson et al., 1995, 2006a, 2014; Thompson, 2001, 2008; Thompson and McManus, 2002; Thompson and Jenkins, 2014; Tigre et al., 2016). Interestingly, the nomenclature used for these 'species' conforms to that proposed by observational parasitologists in the 1920s–60s, before molecular tools were available to confirm and support their morphological descriptions and epidemiological observations (Thompson et al., 1995, 2014; Thompson and McManus, 2002; Thompson, 2008). Importantly, these molecular epidemiological studies have given confidence to the morphological characters used for species discrimination, which now offer a simple, cost-effective means of parasite identification in endemic foci where the application of molecular tools may not be practical or cost effective (Harandi et al., 2002, 2012; Lymbery and Thompson, 2012; Sharbatkhori et al., 2011).

Interestingly, in terms of considering the life cycles of these host-adapted species, Rausch (1997) considered that a uniform, typical larval structure, with long survival without degenerative change and high protoscolex production, are characteristic of metacestodes of recognized species in their natural intermediate hosts, and this is the case with *E. granulosus, E. multilocularis, Echinococcus equinus, Echinococcus ortleppi, Echinococcus canadensis, Echinococcus intermedius* and *Echinococcus felidis* (Table 2). In the future, it is possible that new species of *Echinococcus* will be discovered, particularly as more studies investigate the parasites of wildlife in areas such as Africa and China. Molecular tools will undoubtedly play a role in these studies, as for example the finding a novel genotype of *E. granulosus* in a human patient in Ethiopia (Wassermann et al., 2016). It will be important to ensure that descriptions of new strains or species do not rely solely on molecular data and that as much biological information is obtained to support the epidemiological significance of such discoveries.

4. EPIDEMIOLOGICAL SIGNIFICANCE OF INTRA- AND INTERSPECIFIC VARIATION

Differences in host specificity between strains and species of *Echinococcus* is clearly important in areas where there are different host assemblages and cycles of transmission (Lymbery and Thompson, 2012; see Chapter 5). Developmental differences may also impact on the control of *Echinococcus* in different regions, particularly if the regular drug treatment of dogs is used to prevent transmission with the frequency of treatment less than the time required for worms to reach patency. Research has demonstrated that a standard treatment frequency may not 'break the cycle of transmission' since the maturation rate and onset of egg production varies between strains and species of *Echinococcus* (Lymbery and Thompson, 2012; Thompson, 2001; Thompson and McManus, 2001, 2002). Similarly, differences in development and maturation between species of definitive hosts must be taken into account in areas where more than one species is involved in transmission, e.g., *E. multilocularis* in foxes, dogs and cats (Kapel et al., 2006; Thompson et al., 2006b). Evidence of differences in infectivity for humans, clinical manifestations and pathogenicity are also clearly important (Lymbery and Thompson, 2012; Thompson, 2001). As described earlier, all available epidemiological evidence supports the conclusion that humans are not susceptible to infection with *Echinococcus equinus*. With *E. canadensis*, it has long been thought that the clinical consequences of infection in humans are negligible compared to infection with *E. granulosus* (Thompson, 2015). In part, this may be due to the long progression of the disease in humans, often without symptoms, and the nonspecificity of symptoms when they do occur. However, the limitations of serological tests used to diagnose cystic infections caused by *E. canadensis* have contributed to human cases being underdiagnosed (Jenkins et al., 2013; Thompson et al., 2014; Thomposn, 2015). There has been a reliance on tests developed for *E. granulosus* and there are known to be antigenic differences between *E. canadensis* and *E. granulosus* (Jenkins et al., 2011, 2013; Schurer et al., 2013, 2014). The two genotypes of *E. canadensis* also appear to vary in virulence in humans with G8 more pathogenic than previously considered, with two severe cases recently reported (Jenkins et al., 2011; Thompson, 2015). Some species of *Echinococcus* develop very differently in different species of intermediate host. *Echinococcus granulosus* produces viable, fertile, cysts in sheep whereas in cattle and pigs, cysts are usually sterile and these hosts play little role in transmission of *E. granulosus*. *Echinococcus granulosus* preferentially affects the lungs of wild macropod marsupials and establishes clinically significant infections in contrast to the seemingly benign infections that develop in the liver and lungs of sheep infected with this species (Barnes et al., 2011; Thompson, 2013).

As mentioned earlier, differences in the antigenic characteristics between species of *Echinococcus* will have a bearing on the development of immunodiagnostic procedures in different countries (Cameron, 1960; Gottstein et al., 1983; Huldt et al., 1973; Jenkins et al., 2013; Lightowlers et al., 1984; Thompson, 2015; see also Chapter 9). It was proposed that an obvious corollary to this must be the assumption that a vaccine developed against one particular species or strain of *Echinococcus* may not protect against infection with another strain (Thompson, 1995, 2001). This has now been confirmed with recent investigations showing that the EG95 vaccine antigen developed to protect against infection with *E. granulosus* is immunologically different in *Echinococcus intermedius* (Alvarez Rojas et al., 2014). It may therefore be necessary to develop 'genotype-specific' vaccines in the future (Alvarez Rojas et al., 2014). It will also be interesting to see whether species and strains of *Echinococcus* vary in their response to particular chemotherapeutic regimes. Albendazole is the most widely used drug to treat CE and AE, and as with other helminths β-tubulin is believed to be the target. Recent research has demonstrated some variation in the β-tubulin gene of *E. granulosus* (Pan et al., 2011). Further, Brehm and Koziol (2014) have shown that in *E. multilocularis*, germinal cells express a β-tubulin isoform with limited affinity to benzimidazoles (See Chapter 4).

5. DEVELOPMENT OF *ECHINOCOCCUS*

Most of what we know about the developmental biology of *Echinococcus* and the host-parasite relationship has been obtained from studies on the metacestode. This is because it is relatively straightforward to maintain larval stages in axenic culture, and the availability of rodent models for maintaining cystic infections following either primary or secondary infection (Howell and Smyth, 1995). The practical and ethical difficulties, as well as safety concerns, of undertaking in vivo studies in the definitive hosts of *Echinococcus* have been a major limiting factor in studying the development of the adult parasite. If it was not for the pioneering studies of Smyth (Howell and Smyth, 1995; Smyth et al., 1966; Smyth and Davies, 1974a; Thompson and Lymbery, 2013) in developing in vitro systems for studying the developmental biology of both larval and adult stages, the underlying principles of differentiation, host-parasite relationships and evolutionary biology (Smyth, 1969; Smyth et al., 1966) are unlikely to have been discovered at a formative era in parasitology in which hypotheses were developed that could influence future, emerging research in the genomics area (Thompson and Lymbery, 2013). As such, Smyth (1969) saw the potential of exploiting *Echinococcus* as a novel model system for studying parasitism as distinct from a model for studies on evaluating anthelmintics or other antiparasitics. He thus expanded the definition of a 'model' to embrace studies on all the biological activities that supported the parasite life style of *Echinococcus*, as well as the potential of this model system for broader studies of a more fundamental nature in biology, for example, developmental plasticity and stem cells (heterogeneous morphogenesis), gene regulation and evolutionary developmental biology (Cucher et al., 2011; Smyth, 1969, 1987; Thompson and Lymbery, 2013; Thompson and Jenkins, 2014; see later and Chapter 4).

5.1 Adult

5.1.1 Establishment in the definitive host

The definitive host acquires infection by ingesting viable protoscoleces. These may be ingested still within the hydatid cyst and the masticatory actions of the host will assist in tearing open the cyst and freeing the brood capsules. The process of excystment — removal of brood capsule and other cystic tissue — is further assisted by the action of pepsin in the stomach. Prior to ingestion, the apical region of the protoscolex (suckers, rostellum and hooks) is invaginated within the mucopolysaccharide-coated basal region of the protoscolex tegument (Fig. 1). This protects the scolex until it is stimulated to evaginate. In dogs experimentally infected with evaginated protoscoleces of *E. granulosus*, far fewer worms establish than in dogs infected with invaginated protoscoleces (Thompson, 1995).

The precise nature of the stimulus for evagination is not known. Protoscoleces are sensitive to environmental changes and evaginate in response to variations in temperature and osmotic pressure, and to agitation (Thompson, 1995). However, in an intact hydatid cyst, protoscoleces will remain viable, with the majority invaginated for several days depending on the temperature (Thompson, 1995), thus enhancing transmission in sylvatic cycles reliant on predation or scavenging by definitive hosts. Aerobic conditions appear to be essential for evagination but specific enzymes or bile are not essential although the rate of evagination is increased in the presence of bile (Smyth, 1967, 1969).

In the definitive host, the time required for evagination is variable, with the majority of protoscoleces evaginated after 6 h but complete evagination takes up to 3 days (Thompson, 1977). Following evagination, protoscoleces are initially very active as they have to rapidly locate and attach to the mucosal surface in the crypts of Lieberkuhn to avoid being swept out of the small intestine, with some actually within the crypts by 6 h after infection (Thompson, 1977). As such, motility is enhanced by a well-developed nervous system and glycogen energy reserves (Brownlee et al., 1994; Camicia et al., 2013; Hemer et al., 2014; Smyth, 1967). The evaginated protoscoleces are rich in glycogen which acts as an energy reserve although this is rapidly used up, usually within 3 h (Smyth, 1967). Activity then declines as energy reserves are replenished, accounting for a lag phase in growth during the first 3 days of infection in dogs (Thompson, 1977). Developing worms attach mainly by grasping substantial plugs of tissue with their suckers (Smyth et al., 1969; Thompson et al., 1979; Thompson and Eckert, 1983, Fig. 3). The hooks only superficially penetrate the mucosal epithelium but their shape ensures that they act as anchors to assist in preventing the worm being dislodged.

The intestinal distribution of the adult worm appears to be similar for *E. granulosus* and *E. multilocularis* with worms dispersed unevenly along

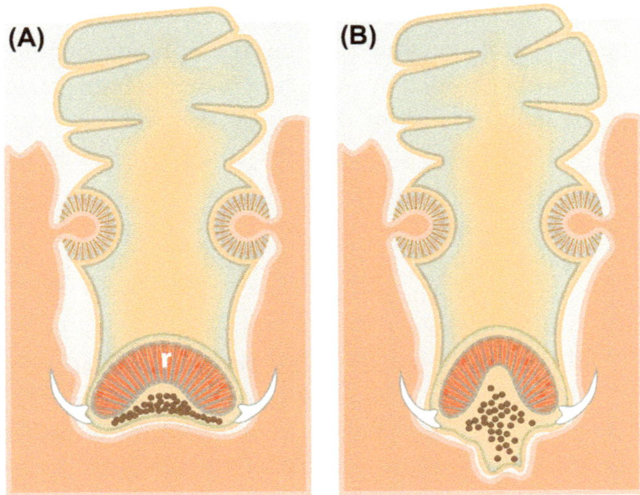

Figure 3 Diagram illustrating adult *Echinococcus* in situ in the intestine, with suckers on the scolex grasping the epithelium at the base of the villi. Rostellum is deeply inserted into a crypt of Lieberkühn (A and B) and extensibility of the apical rostellar region is shown in (B). Rostellar gland with secretory material is anterior to the rostellar pad (r). *From Thompson, R.C.A., Jenkins, D.J., 2014. Echinococcus as a model system. Int. J. Parasitol. 44, 865–877.*

the intestine with most worms located in the proximal regions (Constantine et al., 1998; Thompson et al., 2006b). It is not known to what extent the developing adult *Echinococcus* migrates within the small intestine, although it does move between adjacent villi. Constantine et al. (1998) found that adult *E. granulosus* in dogs did change location in the gut to form aggregates which led to differences in worm density along the intestine. Whether this is a consequence of attraction between worms and/or to particular microenvironmental sites is unclear. These local high densities of worms may promote rapid development but as the worms grow, competitive interactions or local immune responses by the host inhibit development with a change from cytosolic to more energetically efficient mitochondrial metabolism (Constantine et al., 1998).

5.1.2 Activities at the interface

Observations that contact of the scolex of *E. granulosus* with a proteinaceous base stimulated strobilate development led to studies on the nature of the interface both in vitro and in vivo and the discovery of the rostellar gland which comprises a modified group of tegumental cells situated in the apical

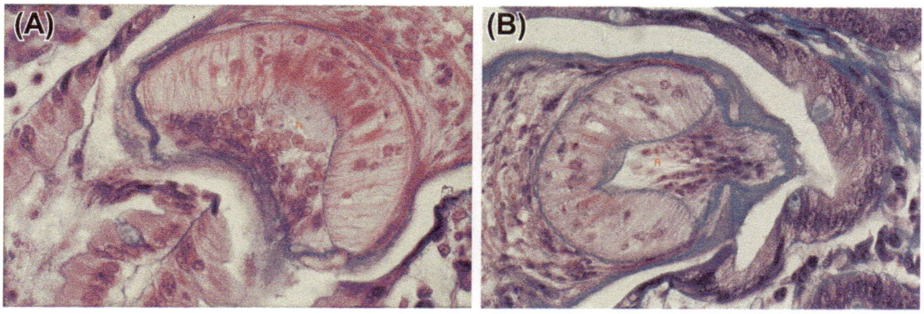

Figure 4 Section of mucosal wall of a dog's small intestine showing rostellum of *Echinococcus granulosus*, deeply inserted in a crypt of Lieberkuhn showing apical rostellar gland (R) retracted (A) and extended (B). Section stained with Martius Scarlet Blue ×60.

rostellum (Smyth, 1964a; Smyth et al., 1969; Thompson et al., 1979; Thompson and Eckert, 1983, Figs 3 and 4). However, subsequent studies demonstrated that *E. multilocularis* can differentiate sexually in monophasic media (without a solid serum base) (Howell and Smyth, 1995; Smyth and Davies, 1975; Smyth, 1979) suggesting that the stimulus for strobilate development must be more complex than previously thought (Constantine et al., 1998; Thompson et al., 1990, 2006b).

Although an adult *Echinococcus* may alter its position and move up and down and between adjacent villi during development, this may not occur once the worm reaches maturity (Constantine et al., 1998). Between 20 and 35 days after infection, respectively, *E. multilocularis* and *E. granulosus* are found in a position characteristic of the mature worm. The rostellum is deeply inserted into a crypt of Lieberkuhn with the mobile apical rostellar region usually fully extended, the hooks superficially penetrating the mucosal epithelium and the suckers grasping the epithelium at the base of the villi (Smyth et al., 1969; Thompson et al., 1979; Thompson and Eckert, 1983, Figs 3 and 4). Invasion of the crypts of Lieberkuhn by the mature worm is of particular physiological significance to *Echinococcus*. It is a characteristic not shared by other adult taeniids, which achieve only a relatively superficial attachment to the mucosa of the definitive host (Beveridge and Rickard, 1975; Featherstone, 1971), presumably because of their greater size. *Echinococcus* has a very mobile and extensible apical rostellar region. Extension of this region into the crypts coincides with the commencement of secretory activity of the rostellar gland (Figs 3 and 4) and the release of secretory material by a holocrine process into the interface between parasite and host (Smyth, 1964a,b; Smyth et al., 1969; Thompson et al., 1979; Thompson and Eckert, 1983; Thompson and Jenkins, 2014). The secretion is proteinaceous containing cystine and lipid. It is not known if there are one or more proteins secreted but Siles-Lucas et al. (2000) demonstrated that the secretion contains a regulatory protein (14-3-3) that is released into the host-parasite interface. The origin and site of synthesis of the secretion has not been determined, although large amounts occur in both the perinuclear and distal cytoplasm of the tegument as well as in the tegumental nuclei (Thompson et al., 1979; Herbaut et al., 1988). Recently, Kouguchi et al. (2013) used a surface glycoprotein from *E. multilocularis* as a vaccine in dogs which induced significant protection when administered via a mucosal route and demonstrated antibodies raised by their vaccine on the surface of the suckers, rostella and hooks. More research is required to determine whether this glycoprotein is related to the secretions from the rostellar gland.

The crypts of Lieberkuhn may represent a site of particular nutritional significance for mature *Echinococcus*. Nutrients could be derived from the lysis of host cells but there is no evidence that the rostellar gland secretion is histolytic or has any enzymatic activity. An important factor to be considered is the timing of secretory activity, which coincides with a levelling off in growth of the worm at around 30 days after infection and the commencement of egg production (Thompson et al., 1979, 2006b). The secretion may therefore be associated with the maturation of ova and/or subsequent release of the gravid proglottid (apolysis). Gland activity could be recurrent with a cycle of activity associated with the maturation and release of each proglottid. The rostellar gland of *Echinococcus* is seemingly unique to *Echinococcus* (Thompson and Jenkins, 2014). Although rostellar secretions have been described in larval *Taenia crassiceps* (Krasnoshchekov and Pluzhnnikov, 1981), no gland has been described. Rostellar glands have been described in other adult cestodes, particularly proteocephalids but their function is also unclear and structurally they are different to the rostellar gland in *Echinococcus* (McCullough and Fairweather, 1989; Zd'arska and Nebesarova, 2003). Modified glandular parts of the scolex tegument have been described in some other cestodes and the most-favoured role for their secretions is one of attachment (Hayunga, 1979; Ohman-James, 1973; Richards and Arme, 1981; Sawada, 1973; Specian and Lumsden, 1981). In *Echinococcus* perhaps firm attachment is a prerequisite for apolysis, since unattached gravid worms in vitro do not shed their terminal proglottids (Thompson and Eckert, 1982). An adhesive function to assist in retention of the worms most adequately accounts for the special location of the rostellar secretory cells, site of release, and timing if apolysis necessitates particularly firm positioning.

Mature *E. granulosus* possesses two morphologically distinct types of microthrix (Thompson et al., 1982; Irshadullah et al., 1990). On the strobila, they are bladelike and rigid for most of their length and probably serve to keep the absorptive surface of the parasite and host apart, thus maintaining a free flow of nutrients at the interface between the two absorptive surfaces. On the apical rostellum and scolex the microtriches are long, slender filamentous types apparently flexible for most of their length, thus allowing the scolex and rostellum to achieve close contact with the host, perhaps to enhance adhesion (Mettrick and Podesta, 1974; Thompson et al., 1979).

One other possible function for the rostellar gland secretion is that of protection. It is feasible that the secretion may protect the worm either by inhibiting or inactivating host digestive enzymes or by interfering with the host's immune effector mechanisms (see also Chapter 7). However, it would be reasonable to expect such a protective mechanism to operate throughout the life of the adult worm, unless, as seems to be the case, it is only after maturity that a permanent and very intimate association is achieved between parasite and host. Recent studies have demonstrated a Kunitz-type protease inhibitor (EgKI-2) in the *E. granulosus* genome, which is highly expressed in the adult worm and may play a protective role in preventing proteolytic enzyme attack thereby ensuring survival in the definitive host (Ranasinghe et al., 2015). Further research is required to determine whether EgKI-2 is a component of the rostellar gland secretions.

Echinococcus seldom engenders a morphologically apparent host response, although occasionally in heavy infections, there may be an excessive production of mucus (Thompson, 1995). As emphasized by Heath (1995), the scolex is in intimate contact with the systemic circulation even in the Payer's patches and would appear to maintain its privileged integrity by suppression of cytotoxic and effector cell activity in the region of the scolex. The host tissue that is grasped by the suckers is usually necrotic, but the hooks cause little damage (Thompson et al., 1979). Observations at the ultrastructural level have shown that hook damage is restricted to columnar cells with an associated loss of some host microvilli (Thompson et al., 1979). The epithelium of parasitised crypts is commonly flattened and there may be occasional rupture of a crypt wall with release of host cells into the crypt (Smyth et al., 1969). Adult worms have been observed to invade the lamina propria, but this appears to be a rare event. No substantial pathology or evidence of a host cellular reaction has been observed in infections with adult *E. granulosus* or *E. multilocularis* (Thompson et al., 1979; Thompson and Eckert, 1983).

However, the presence of the adult worm does not go unnoticed by the host and a specific humoral response with the production of circulating IgG antibodies does occur (Jenkins and Rickard, 1986). Deplazes et al. (1993) also demonstrated local humoral, IgG and IgA, and cellular reactions in the intestine of dogs experimentally infected with *E. granulosus*, emphasizing the importance of Peyer's patches in localized, specific immune responses.

The intimate association of the rostellar gland and its secretions suggests a role(s) that enhances the host-parasite relationship in favour of the parasite, which may be regulatory, nutritional and/or protective. The relationship between rostellar gland activity and localized humoral and cellular reactions (Deplazes et al., 1993) is not known but such localized reactions demonstrate stimulation of host immune effector mechanisms. The rostellar gland secretory molecules would seem to be obvious candidates for exploitation in vaccine studies since a focus on prophylaxis of the definitive host may be more attractive than the intermediate host, particularly for the control of *E. multilocularis* (Thompson, 1995; Thompson and Jenkins, 2014).

There is clearly a need for more studies on the interface and attachment of adult *Echinococcus*, as well as other cyclophyllidean cestodes (Pospekhova and Bondarenko, 2014; Thompson and Jenkins, 2014).

5.1.3 Differentiation

The development of the adult parasite involves germinal and somatic differentiation and can be divided into the following processes: proglottisation; maturation; growth; segmentation (Thompson, 1995; Thompson et al., 2006b). Germinal differentiation comprises proglottisation, which refers to the sequential formation of new reproductive units (proglottids), and the maturation of the proglottids. Somatic differentiation consists of growth, i.e., increase in size, and the somatic delineation of each proglottid by segmentation (strobilisation). Segmentation in cestodes is not to be confused with true mesodermal segmentation (metamerism) which occurs by distal growth not proximally as in cestodes (Freeman, 1973). In some cestodes, including *Echinococcus* (see later), proglottisation may occur without segmentation. Thus both terms are necessary for a full comprehension of the process of development (Freeman, 1973), but should not be referred to interchangeably. Segmentation in cestodes, including *Echinococcus*, does not involve the formation of any separatory structure or 'interproglottid' membrane between adjoining proglottids (Mehlhorn et al., 1981). The demarcation of each proglottid is purely an external phenomenon caused by an infolding of the tegument which gives rise to the characteristic constricted appearance.

It also appears that the microtriches in the infolded regions of the tegument may be linked together thus stabilizing the infoldings (Mehlhorn et al., 1981).

The four developmental processes described earlier take place independently. This has been demonstrated by studies on *E. granulosus* and *E. multilocularis* in vitro (Smyth, 1971; Smyth and Davies, 1975; Smyth and Barrett, 1979), and *E. granulosus* in vivo (Thompson, 1977; Constantine et al., 1998; Thompson et al., 2006b). Further, in the adult parasite, somatic and germinal differentiation are independently associated with a transition from cytosolic to mitochondrial energy metabolism (Constantine et al., 1998). This very complicated process of cytodifferentiation was considered to indicate the possible existence of several primitive cell lines as in *Hymenolepis diminuta* (Sulgostowska, 1972, 1974). However, preliminary studies on cytodifferentiation in adult *E granulosus* suggested that only one primitive cell type exists located in the neck region of the adult worm (Gustafsson, 1976) (see later and Chapter 4). In vitro studies on *Echinococcus* demonstrated that this so-called 'germinative' (germinal) cell was also extremely sensitive to environmental and/or nutritive conditions (Smyth and Barrett, 1979; Howell and Smyth, 1995). In some cultures mainly germinal cells were produced, leading to proglottisation and maturation but no segmentation, whereas in other cultures more somatic cells were produced leading to growth without sexual maturation. These observations from in vitro studies have been complemented by studies in vivo, comparing the development of adult *E. multilocularis* in foxes, raccoon dogs, cats and dogs. Thompson et al. (2006b) compared developmental processes in the different definitive hosts, and by examining germinal and somatic differentiation, confirmed that these processes can be influenced by their environment; in this case the small intestine of different carnivore host species. In cats, the investment by worms in the somatic processes of growth and segmentation was not complemented in terms of maturation, in contrast to foxes, dogs and raccoon dogs, demonstrating the fine balance that exists which can easily be upset if environmental factors are not correct (Thompson et al., 2006b).

5.1.4 Sequential development

The newly evaginated protoscolex contains an abundance of calcareous corpuscles (Fig. 2), which consist of an organic base and inorganic material (Smyth, 1969). They are of cellular origin with the characteristic concentric layers of mineral deposition increasing with age (Ohnishi and Kutsumi, 1991; Pawlowski et al., 1988). They develop from living cells and two different mechanisms of formation coexist with corpuscles originating from the nucleus or cytoplasm. Their function may be that of a buffering system or a source of inorganic ions, CO_2 and phosphates (Smyth, 1969), but their transitory nature suggests an association with cell death (Thompson, 1995) and recently they have been shown to be associated with autophagy and catabolic processes (Loos et al., 2014). Within 3–4 days after infection, the lateral excretory canals of the young worm are clearly evident and by the end of the first week a posterior excretory bladder is seen (Smyth and Davies, 1974a). The excretory system of *Echinococcus*, like all other cestodes, is based on the platyhelminth protonephridial system with the lateral excretory canals (Fig. 2) acting as collecting ducts for numerous flame cells distributed throughout the parenchyma. However, the physiology of excretion has not been investigated. Evidence that the excretory ducts of some pseudophyllidean cestodes are capable of absorption (Lindroos and Gardberg, 1982) raises the possibility that the excretory system of *Echinococcus* could also function as a distributive system.

The sequence of development described later and illustrated in Fig. 2 refers to *E. granulosus*. Although it is essentially the same in other species, the rate of development varies, particularly in relation to growth, onset of egg production and number of proglottids produced (Kapel et al., 2006; Thompson et al., 2006b). The first sign of proglottisation is the appearance of a genital rudiment or anlagen which may appear as early as 11 days after infection, separated from the scolex by a clear band. By 14 days the first proglottid is clearly evident as a darkly staining body demarcated from the scolex by the transverse infolding of the tegument which delineates the first segment. Within 1–2 days a lateral branch forms from the genital rudiment which will eventually open to the exterior via the genital pore. Subsequent stages of maturation follow the general cestode pattern and are summarized in Fig. 2. Growth, as determined by total worm length, exhibits a steady log linear increase throughout the first 35 days of infection apart from a lag period during the first 3 days (Thompson, 1995). Growth also levels off prior to egg production (Kapel et al., 2006; Thompson, 1995; Thompson et al., 2006b).

5.1.5 Sexual reproduction

Mature *Echinococcus* is hermaphrodite (Fig. 2) and capable of both self- and cross-insemination (Smyth and Smyth, 1969) although it is predominantly self-fertilizing (Lymbery et al., 1997; Lymbery and Thompson, 2012; see Chapter 3). It is not known whether cross-insemination between two individuals takes place. Hermaphroditism combined with self-insemination is obviously an advantage to a small worm such as *Echinococcus*, which might find it difficult to find another worm, particularly in light infections. Furthermore, such a reproductive mechanism has a significant evolutionary potential (see Chapter 3). The requirements for self-insemination in *Echinococcus* appear to be extremely complex as suggested by the repeated failure to achieve fertilization in vitro (see Howell and Smyth, 1995; Smyth and Davies, 1974a; Smyth, 1979; Thompson and Jenkins, 2014). It has been suggested that there may be specific or nonspecific factors in the intestinal secretions of the definitive host which activate the cirrus to commence its copulatory movements and that without such stimulation self-insemination may not occur (Smyth, 1982).

5.1.6 Egg production and subsequent development

The initial onset of egg production varies between species and even between strains. In *E. granulosus* it ranges from 34 to 58 days, whereas *E. multilocularis* has a far more rapid rate of maturation with egg production commencing between 28 and 35 days after infection (Kapel et al., 2006; Thompson and Eckert, 1982; Thompson et al., 1984; Thompson et al., 2006b).

Although development up to the initial onset of egg production has been extensively studied, there have been few studies of subsequent development. Thompson et al. (2006b) studied the maturation of adult *E. multilocualris* in experimentally infected foxes, dogs, raccoon dogs and cats, at 35, 63 and 90 days postinfection. They found that egg production was a continuous process throughout the 90 day period (Fig. 5). The number of eggs produced is uncertain, with reports varying between 100 and 1500 per proglottid (Heath and Lawrence, 1991; Rausch, 1975; Thompson and Eckert, 1982), with *E. multilocularis* producing fewer eggs per proglottid than *E. granulosus*. A study by Kapel et al. (2006) found that approximately 114, 42 and 27 eggs per worm were excreted in the faeces of dogs, raccoon dogs and foxes, respectively, experimentally infected with *E. multilocularis*, over a 90-day period. However, it is not known how often species of *Echinococcus* produce gravid proglottids. Based on the rate of development during the first 40 days of infection, it has been estimated that gravid proglottids of *E. granulosus* are produced and detached every 7—14 days (Schantz, 1982; Smyth, 1964a). However, without further investigation it is impossible to conclude whether the rate of proglottisation after apolysis in *Echinococcus* is constant or declines. It is also not known how long the adult parasite may survive in the definitive host. It has been reported that adult worms become

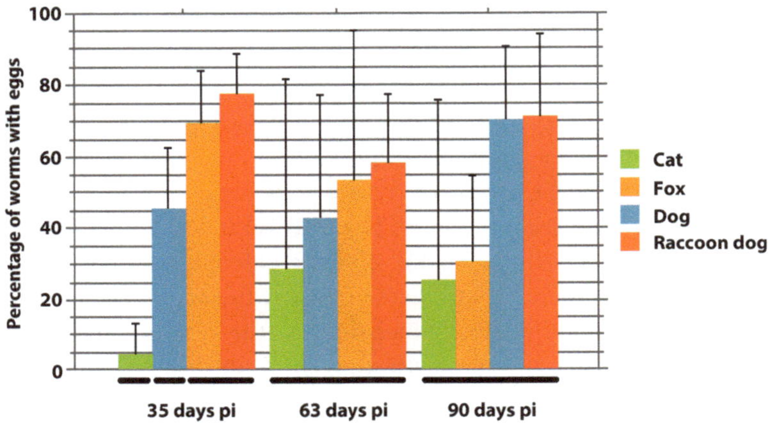

Figure 5 Mean percentage of *Echinococcus multilocularis* worms with eggs calculated for each infected host individual at 35, 63 and 90 days postinfection for each of the four host species. *From Thompson, R.C.A., Kapel, C.M., Hobbs, R.P., Deplazes, P., 2006b. Comparative development of Echinococcus multilocularis in its definitive hosts. Parasitology 19, 1–8.*

senescent after 6–20 months, although worms may live for 2 years or longer (Schantz, 1982). In the absence of accurate information on the rate of production and release of gravid proglottids and the life span of the adult parasite, it is impossible to accurately determine the reproductive potential of *Echinococcus* in the definitive host.

5.2 Egg

Taeniid eggs are spherical to ellipsoid in shape and usually range in size from 30 to 50 μm and from 22 to 44 μm in their two diameters. They are morphologically indistinguishable at the light microscope level and ultrastructural studies of the eggs of *E. granulosus, E. multilocularis* and various *Taenia* species have shown that they possess similar structures consisting of several layers and membranes (Fig. 6) (Morseth, 1965; Sakamoto, 1981; Swiderski, 1982). The embryophore is the principal layer affording physical protection to the embryo, or oncosphere, since the vitelline layer ('egg shell' or outer envelope) is passively removed from the egg before it is liberated. The embryophore is relatively thick and impermeable, consisting of polygonal blocks composed of an inert keratin-like protein, which are held together by a cementing substance (Morseth, 1966; Nieland, 1968; Sakamoto, 1981).

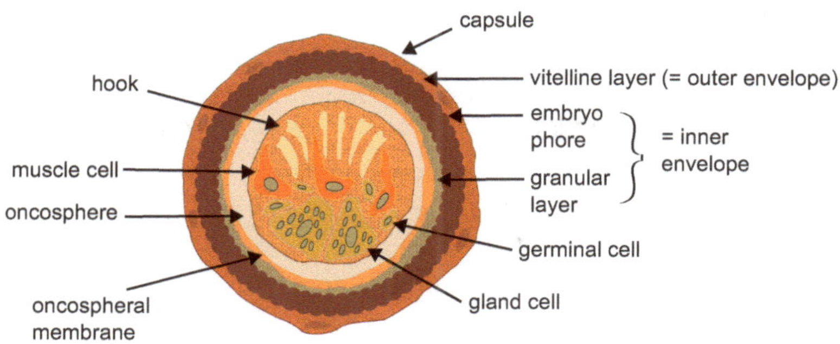

Figure 6 Diagram of the egg of *Echinococcus. Redrawn from Thompson, R.C.A., 1995. Biology and systematics of Echinococcus. In: Thompson, R.C.A., Lymbery, A.J., (Eds.), Echinococcus and Hydatid Disease. CAB International, Wallingford, Oxon, UK, pp. 1–50.*

When released from the definitive host, the egg of *Echinococcus* is presumed to be fully embryonated and infective to a suitable intermediate host. However, taeniid eggs at the time of expulsion are probably at different stages of maturation and immature eggs may mature in the environment under appropriate conditions (Gemmell and Roberts, 1995).

Echinococcus eggs are extremely resistant enabling them to withstand a wide range of environmental temperatures for many months (Gemmell et al., 1986; Schantz et al., 1995; Thevenet et al., 2005; Veit et al., 1995). Dessication is lethal and the end points for temperature are approximately +40°C to −70°C (Gemmell and Roberts, 1995). However, the availability of moisture is a limiting factor in survival and recent research has shown the eggs of *E. multiloculalris* suspended in water could survive for 2 h after exposure to a temperature of +65°C (Federer et al., 2015).

5.2.1 Hatching and activation

When ingested by a suitable intermediate host, viable eggs of *Echinococcus* hatch in the stomach and small intestine. Hatching is a two-stage process involving (1) the passive disaggregation of the embryophoric blocks in the stomach and intestine and (2) the activation of the oncosphere and its liberation from the oncospheral membrane (reviewed by Holcman and Heath, 1997; Jabbar et al., 2010; Lethbridge, 1980). Disaggregation of the embryophoric blocks appears to require the action of proteolytic enzymes, including pepsin and pancreatin, in the stomach and/or intestine but does not depend on any one specific enzyme. The oncosphere plays no part in disaggregation of the embryophore and remains essentially dormant until activated. Evidence suggests the oncosphere may be stimulated to free itself from the oncospheral membrane following changes in membrane permeability brought about by the surface active properties of bile salts. This led to the proposal that bile may play a part in determining intermediate host specificity, since its composition varies between different species of vertebrate (Smyth, 1969). However, the situation is certainly not as straightforward since eggs of *E. granulosus* were shown to hatch in extraintestinal sites including the lung, liver and peritoneal cavity of sheep and rodents inoculated experimentally by tracheostomy or intraperitoneal injection (Blood and Lelijveld, 1969; Borrie et al., 1965; Colli and Williams, 1972; Kumaratilake and Thompson, 1981; Williams and Colli, 1970). Eggs inoculated into the peritoneal cavity were rapidly surrounded by adhering neutrophils and macrophages which probably released hydrolytic enzymes causing dissolution of the embryophore. Eggs of *Echinococcus* can also be hatched and activated in vitro using chemicals and enzymes not derived from a particular species of intermediate host. Thus hatching requirements do not depend on the physiological characteristics of the definitive host gut and are not specific. Consequently, factors which regulate whether eggs of a particular taeniid species will or will not develop in a particular intermediate host must operate on the oncosphere either during the invasive or establishment phases (Thompson, 1995).

5.2.2 Penetration and tissue localization

The liberated, activated oncosphere exhibits intricate rhythmic movements involving the body and hooks, the coordinated movement of the latter effected by a complex muscular system (Swiderski, 1983). The so-called 'penetration glands' are also prominent at this stage. Studies in sheep and rabbits have shown that the oncospheres of *E. granulosus* penetrate the tips of the villi in the jejunal and upper ileal region of the small intestine (Heath, 1971). The oncospheres initially attach to the microvillous border of the villi, presumably using their hooks as anchors. Studies on several taeniid species, including *E. granulosus*, have shown that oncospheres rapidly migrate through the epithelial border of the villi, reaching the lamina propria within 3–120 min after hatching (reviewed by Lethbridge, 1980; Jabbar et al., 2010). Penetration appears to involve hook and body movements presumably assisted by the penetration gland secretions. Stainable material in the penetration glands is totally extruded from between the hooks at the time, and in the place where the oncosphere is actively engaged in penetration (reviewed by Fairweather and Threadgold, 1981; Jabbar et al., 2010). Degeneration of host tissue also occurs in the vicinity of the invading oncosphere (Heath, 1971). It is therefore assumed that penetration gland secretions must aid the penetration process by causing lysis of host tissue. However, the putative enzymatic nature of the secretion has yet to be established, although oncospheral penetration glands are the source of the EG95 antigen used to vaccinate against CE (Jabbar et al., 2011). Alternatively, penetration may be purely mechanical, involving hook and body movements. The secretion may have other functions such as to assist adhesion, act as a lubricant or afford protection against host digestive enzymes or immunological factors (Fairweather and Threadgold, 1981; Lethbridge, 1980). Ultrastructural studies (Swiderski, 1983) demonstrated that the oncosphere of *E. granulosus* has three types of gland cells (Fig. 6), thus several secretions with different functions may be produced during penetration. Harris et al. (1989) found that not all penetration gland secretions are necessarily shed during penetration and much secretory material is retained in the oncospheral epithelium where it appears to be involved in the formation of transitory microvilli which disappear within 6 days and may have a digestive function. As in the adult worm, the oncosphere has also been shown to release a Kunitz-type protease inhibitor, EgKI-1, which is highly expressed in the oncosphere and is a potent chymotrypsin and neutrophil elastase inhibitor that binds calcium and reduces neutrophil infiltration (Ranasinghe et al., 2015). EgKI-1 may be involved in host immune evasion by inhibiting neutrophil elastase and cathepsin G once this stage is exposed to the mammalian blood system (Ranasinghe et al., 2015).

The factors that determine the final localization of the metacestode of *Echinococcus* in a given host are not clear but probably include anatomical and physiological characteristics of the host as well as the species and strain of parasite. Heath (1971) provided strong circumstantial evidence that oncospheres of *E. granulosus* are capable of completing a lymphatic or venous migration. He further postulated that since the lymphatic lacteals of the villus differed in size between different hosts, the size of the oncosphere in relation to the venules and lacteals in various animals may determine the distribution of cysts between the liver and lungs. It has also been suggested that the microvilli on the surface of the developing metacestode may assist in initial retention in the liver and lungs (Harris et al., 1989).

5.2.3 Postoncospheral development

Once the oncosphere attains a site of predilection, postoncospheral development proceeds leading to the formation of the metacestode. The oncosphere of *Echinococcus* very rapidly undergoes a series of reorganizational events during the first 14 days, involving cellular proliferation, degeneration of oncospheral hooks, muscular atrophy, vesicularization and central cavity formation, and development of both germinal and laminated layers (Heath and Lawrence, 1976; Rausch, 1954; Sakamoto and Sugimura, 1970). Slais (1973) demonstrated that postoncospheral development was initiated by the growth and division of primary germinal cells. Slais (1973) and Swiderski (1983) described five pairs of these cells in the posterior pole of the oncosphere. Although the complexity and plasticity of developmental processes in the adult and metacestode stages initially led to the belief that several primitive cell lines must exist (see earlier), all available evidence supports the existence of only one primitive morphological cell type, as a pool of uncommitted, undifferentiated multipotent germinal, or stem, cells in both the adult and metacestode (Gustafsson, 1976; Koziol et al., 2014; Smyth, 1969; Thompson et al., 1990; Thompson, 1995; Thompson and Lymbery, 2013; Thompson and Jenkins, 2014), although there are subpopulations with different gene expression patterns (Koziol et al., 2014; and see Chapter 4). The germinal cells are a component of the syncytial germinal layer of the metacestode and neck region of the adult worm. Ultrastructural studies reveal unremarkable rounded cells of variable size of around 4 μm (Albani et al., 2010; Gustafsson, 1976; Mehlhorn et al., 1983). Cell proliferation derives from the continuous replicative activity of these dividing stem cells located in the germinal layer of the metacestode or neck region of the adult worm (Galindo et al., 2003; Gustafsson, 1976). They have considerable proliferative potential (Eckert et al., 1983; Galindo et al., 2003; Martínez et al., 2005; Mehlhorn et al., 1983) and are the only proliferating cells in *Echinococcus* (Koziol et al., 2014). This is particularly well illustrated by the capacity of the parasite for indefinite perpetuation in the larval stage by the passage of protoscoleces or germinal layer material in rodents (secondary hydatidosis; Howell and Smyth, 1995). In AE caused by the metacestode of *E. multilocularis*, the proliferating larval parasite has an infiltrative capacity to establish distant foci of infection due to the distribution via blood or lymph of detached germinal cells (Ali-Khan et al., 1983; Ammann and Eckert, 1996; Eckert et al., 1983; Mehlhorn et al., 1983, Fig. 7; and see later).

Problems with host cell contamination dogged early attempts to establish germinal cell lines of *E. granulosus* and *E. multilocularis* (reviewed in Howell and Smyth, 1995). In addition, their isolation from the germinal layer, and their in vitro propagation, could have been hampered by the fact that the germinal layer is a syncytium. However, the establishment and

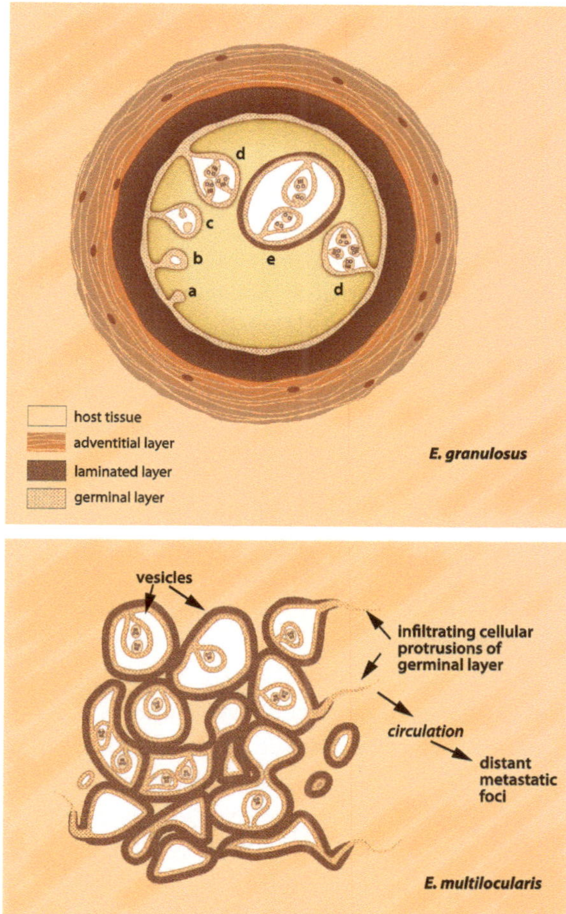

Figure 7 Diagram illustrating the structural differences between the metacestodes of *Echinococcus granulosus* (a–d stages in development of protoscoleces and brood capsule, and e daughter cyst) and *Echinococcus multilocularis*. *From Thompson, R.C.A., Jenkins, D.J., 2014. Echinococcus as a model system. Int. J. Parasitol. 44, 865–877.*

long-term perpetuation of *Echinococcus* germinal cells has now been achieved for both species (Albani et al., 2010; Spiliotis and Brehm, 2009; Spiliotis et al., 2008; Yamashita et al., 1997). The germinal cells behave very much like classical stem cells with the formation of cell aggregates and clusters with cavity formation, and there is cytological evidence of transformation (Albani et al., 2013; Spiliotis et al., 2008; see Chapter 4).

5.3 Metacestode

Metacestodes of the four species of *Echinococcus* have certain basic features in common which can be illustrated by a detailed examination of *E. granulosus*. Differences exhibited by the other three species will then be discussed.

5.3.1 Structure

5.3.1.1 *Echinococcus granulosus*

The fully developed metacestode of *E. granulosus* is typically unilocular, subspherical in shape, fluid-filled and exhibits the least complex structure of the four species (Cameron and Webster, 1969; Moro and Schantz, 2009; Rausch et al., 1981; Schantz, 1982; Thompson, 2001). The cyst consists of an inner germinal or nucleated layer supported externally by a tough, elastic, acellular laminated layer of variable thickness, surrounded by a host-produced fibrous adventitial layer (Fig. 7). Typically *E. granulosus* produces a single-chambered unilocular cyst in which growth is expansive by concentric enlargement. Asexual proliferation of the germinal layer and brood capsule formation takes place entirely endogenously. Pouching of the cyst walls may occur giving rise to secondary chambers communicating with the central cavity (Vanek, 1980). Sometimes the central cavity may be partly separated from the secondary chambers by incomplete septa. Occasionally cysts may abut and coalesce, forming groups or clusters of small cysts of different size. In some hosts, particularly human, where unusually large cysts may develop, daughter cysts may form within the primary cyst (Moro and Schantz, 2009; Thompson, 2001, Fig. 7).

The germinal layer is similar in structure to the tegument of the adult worm, consisting of a distal cytoplasmic syncytium from which microtriches project into the overlaying laminated layer (Bortoletti and Ferretti, 1973, 1978; Lascano et al., 1975; Morseth, 1967). The cell bodies and nuclei comprise the perinuclear, or proliferative cell layer, which contains several cell types including tegumental, muscle, glycogen-storing and undifferentiated cells. The tegumental cells are multinucleated, which may be indicative of their rapid growth (Rodriguez-Caabeiro and Casado, 1988). Cytoplasmic connections between the two layers maintain continuity. The undifferentiated cells of the perinuclear layer are proliferative and are responsible for the formation of brood capsules which originate endogenously as small nuclear masses, or buds, which proliferate towards the cystic cavity (Fig. 7; Slais, 1973; Thompson, 1976), whereas in *E. multilocularis* recent evidence suggests they arise from an invagination of the germinal layer (Koziol et al.,

2016). Brood capsules enlarge, vacuolate and become stalked. Within their lumen, a repetition of the asexual budding process takes place, leading to the production of numerous protoscoleces. The formation of protoscoleces is asynchronous with a number of different developmental stages being present in a brood capsule at the same time (see Thompson, 1995). Fully developed protoscoleces are characterized by the possession of hooks on the invaginated rostellum. The spines of microtriches are the precursors of the hook blades which become enveloped by the rostellar tegument with subsequent formation of the guard and handle (Rogan and Richards, 1987) with subsequent addition of hook material in the adult worm (Hobbs et al., 1990). Hook formation must be subject to environmental factors of host origin given the variability in hook number and size in the same species of *Echinococcus* from different hosts.

In addition to its proliferative activity, the germinal layer is involved in secretory activity and in this respect, Monteiro et al. (2010) identified several molecules in hydatid cyst fluid that could play a role in host evasion that presumably were secreted by the germinal layer. Irigoin et al. (2001) showed that myo-Inositol hexakisphosphate (IP(6)) which is present in the germinal and laminated layers of *E. granulosus* inhibits complement activation and Breijo et al. (2008) indicated that the establishment and survival of the hydatid cyst is associated with the control of complement and, consequently, of local inflammation.

The delicate germinal layer is supported externally by the acellular and elastic laminated layer. All species of *Echinococcus* are characterized by the possession of a laminated layer which, because it is periodic acid-Schiff positive (Kilejian et al., 1961), provides a useful diagnostic marker. The laminated layer also undoubtedly assists in supporting the cyst and allows an often considerable intracystic tension to develop (Cameron and Webster, 1969; Slais, 1973). It is a remarkable and specialized interface in the intermediate host providing a physiochemical barrier with apparent multi-functionality and a structure whose biosynthesis has become a model system for carbohydrate chemistry (Diaz et al., 2011a,b; Parkinson et al., 2012; Thompson and Jenkins, 2014). The laminated layer comprises a specialised extracellular matrix unique to *Echinococcus* (Fig. 7), whose synthesis is a major metabolic activity of the much thinner germinal layer (Diaz et al., 2011a; Lin et al., 2012; Parkinson et al., 2012). It is a carbohydrate-protein complex of highly glycosylated mucin glycoproteins (Kilejian and Schwabe, 1971). Ultrastructurally it can be seen to be a microfibrillate, three-dimensional meshwork matrix in which aggregates of electron-dense material occur (Richards et al., 1983). The origin of the laminated layer was controversial for some time, but electron microscopy and in vitro studies have unequivocally demonstrated that it is entirely of parasite (germinal layer) origin (see Thompson, 1995).

Considerable metabolic activity in the germinal layer is required to synthesize and maintain the interfacial barrier of the laminated layer (Parkinson et al., 2012). The role of the laminated layer would appear to be one of protection by modulating the host-parasite interface, since cyst survival is dependent upon its integrity (Gottstein et al., 2002; Stadelmann et al., 2010). Whether this is purely physical or if there is selective permeability is not known. It may protect the cyst from immunological attack by offering an immunologically inert barrier which can deny access to host defence cells (Coltorti and Varela-Diaz, 1974). Immunoglobulin, however, can pass through the laminated layer but the capacity to regulate penetration of macromolecules into the cyst appears to be a function of the germinal rather than the laminated layer (Coltorti and Varela-Diaz, 1974).

Smyth (1969) commented on the significance of the presence of a human blood group P-like substance in the laminated layer of *Echinococcus* and its significance as a model system in better understanding the immunological basis of the host-parasite relationship. This P1 blood-antigen motif has since attracted much attention and has been further characterized as a protein-carbohydrate, trisaccharide/mucin complex containing galactosamine, yet no biological function has been described to date (Lin et al., 2012). Recently, however, Nicolao et al. (2014) described the expression of an ATP-dependent transporter, P-glycoprotein in *E. granulosus* (Eg-Pgp). How this may relate to the laminated layer is not clear but given its clinical significance in humans as a transporter involved in the efflux of a wide variety of lipophilic substrates, such as toxins and xenobiotics, and its role in the ineffective therapeutic treatment of cancer cells and microbial pathogens (Nicolao et al., 2014), it seems likely to have a role in enhancing the survival of the *Echinococcus* metacestode. Recent research has shown that the laminated layer protects *E. granulosus* against the nitric oxide protective response of the host by increasing arginase activity in macrophages, which counteracts the nitric oxide production (Amri and Touil-Boukoffa, 2015). Stadelmann et al. (2010) investigated the molecular and functional characterization of *E. multilocularis* phosphoglucose isomerase (EmPGI), which is a component of the laminated layer, and proposed that besides its role in glycolysis, EmPGI could also act as a factor that stimulates parasite growth and potentially induces the formation of novel blood vessels around the developing metacestode in vivo. Noya et al. (2014) also demonstrated that mucin-like peptides from *E. granulosus* induce antitumour activity. In studies on the identification and characterization of Emp53, the homologue of human tumour suppressor p53, from *E. multilocularis*, Cheng et al. (2015) concluded that since the parasite develops in host organs, it must have evolved a stress defence system, which could involve Emp53, to cope with various genotoxic and cellular stresses that may cause DNA damage and genomic instability.

The host fibrous capsule (adventitial layer) which typically surrounds fully developed, viable cysts of *E. granulosus*, is the product of a three-layered host cellular inflammatory reaction initiated in the early stages of postoncospheral development (Cameron and Webster, 1969; Slais and Vanek, 1980; Smyth and Heath, 1970). The initial intensity of this reaction varies between hosts and governs the fate of the developing metacestode. If too intense it will cause the degeneration and eventual death of the parasite, whereas in suitable intermediate hosts the initial reaction resolves, leaving a fibrous capsule. The latter situation is common where a stable host-parasite relationship has evolved (Rausch, 1997), as appears to be the case, for example, between *Echinococcus equinus* and the horse (Roneus et al., 1982; Thompson, 1977) and *E. granulosus* in sheep. In contrast, *E. granulosus* rarely produces protoscoleces in cattle and the inflammatory response does not resolve causing the destruction of the developing cyst (Thompson, 2008).

5.3.1.2 *Echinococcus multilocularis*

The metacestode of *E. multilocularis* is the most complex and develops quite differently to that of *E. granulosus* (Braithwaite et al., 1985; D'Alessandro et al., 1979; FAO, 1982; Ohbayashi et al., 1971; Wilson and Rausch, 1980). It is a multivesicular, infiltrating structure with no limiting host–tissue barrier (adventitial layer), consisting of numerous small vesicles embedded in a dense stroma of connective tissue (Thompson, 1995, Fig. 7). The larval mass usually contains a semisolid matrix rather than fluid. Proliferation occurs both endogenously and exogenously and is attributable to the undifferentiated cells of the germinal layer (Mehlhorn et al., 1983; Moro and Schantz, 2009; Sakamoto and Sugimura, 1970). The metacestode consists of a network of filamentous solid cellular protrusions of the germinal layer which are responsible for infiltrating growth (Fig. 7) transforming into tubelike and cystic structures (Eckert et al., 1983; Mehlhorn et al., 1983; Vogel, 1978). Furthermore, the detachment of germinal cells from infiltrating cellular protrusions and their subsequent distribution via the lymph or blood can give rise to the distant metastatic foci characteristic of *E. multilocularis* (Ali-Khan et al., 1983; Eckert et al., 1983; Mehlhorn et al., 1983; Thompson, 2001).

Since the first report of the metacestode of *E. multilocularis* in extraintestinal sites in dogs and cats in 1990 from Germany, there have been a growing number of cases reported from Europe and Canada (Corsini et al., 2015; Deplazes et al., 1997; Geisel et al., 1990; Losson and Coignoul, 1997; Pergrine et al., 2012; and reviewed in Weiss et al., 2010). It is not known whether such infections resulted directly from the ingestion of eggs or indirectly by autoinfection as a result of a previously acquired worm burden, but they illustrate the unusual developmental potential of *E. multilocularis* (Thompson, 2001).

5.3.1.3 *Echinococcus vogeli* and *Echinococcus oligarthra*

The metacestodes of *E. vogeli* and *E. oligarthra* (= *E. oligarthrus*) have been less studied but exhibit developmental and structural characteristics considered intermediate to those of *E. granulosus* and *E. multilocularis* (Moro and Schantz, 2009; Rausch et al., 1981). The metacestodes of both species are termed polycystic since they are characterized by the internal division of fluid-filled cysts to form multichambered cyst masses (D'Alessandro et al., 1979; Morales et al., 1979; Gottstein and Hemphill, 1997; Moro and Schantz, 2009; Rausch et al., 1981). *Echinococcus vogeli* produces cysts varying greatly in size from 2 to 80 mm, which may occur singly, in small groups, or occasionally in dense aggregations in which each cyst is enclosed by its separate adventitia. In *E. vogeli,* endogenous proliferation and convolution of both germinal and laminated layers leads to the formation of secondary subdivisions of the primary vesicle with production of brood capsules and protoscoleces in the resultant chambers, which are often interconnected. In *E. oligarthra*, there is less subdivision into secondary chambers and the laminated layer is much thinner than that of *E. vogeli* (Sousa and Thatcher, 1969; Rausch et al., 1981). Exogenous proliferation has been reported in both species but, at least in *E. vogeli*, it appears to be abnormal and restricted compared to *E. multilocularis*, and does not occur in the natural intermediate host (Gottstein and Hemphill, 1997).

5.3.2 Asexual reproduction and differentiation

The asexual reproduction exhibited by species of *Echinococcus* has a potential unsurpassed by other tapeworms and is of particular evolutionary significance (see Chapter 3). The metacestode has a potentially unlimited sequential generative capacity (reviewed by Whitfield and Evans, 1983). Although some germinal cells initiate the production of new brood capsules and protoscoleces, a pool of uncommitted, undifferentiated, germinal cells remain. This fact makes possible the indefinite perpetuation of larval *Echinococcus* in rodents by repeated intraperitoneal passage of protoscoleces or germinal layer material (secondary hydatidosis), and the development in humans and other animals of secondary cysts following the rupture of a primary cyst (see Chapter 10). Thus undifferentiated cells retained in the germinal layer and protoscolex are capable of initiating new cycles of asexual multiplication. Apart from being able to initiate the production of new protoscoleces, a protoscolex has a dual capability (heterogeneous morphogenesis) since if ingested by a suitable definitive host it will develop into an adult worm (Cucher et al., 2011; Thompson, 1995, 2001). However, even the adult worm must retain some undifferentiated multipotential germinal cells, since adult worms can de-differentiate in a cystic direction under unfavourable conditions (Smyth, 1969).

5.3.3 Rate of development

Although in both *E. granulosus* and *E. multilocularis*, initial reorganization of the oncosphere and formation of the germinal and laminated layers occurs rapidly, usually within the first 14 days (see earlier), the rate of subsequent development differs markedly. In *E. granulosus*, it is slow and variable and dependent on a number of factors including the strain of parasite, the species and strain of host and the degree of infection. Heath (1973) concluded that *E. granulosus* cysts increase in diameter by between 1 and 5 cm per year depending on factors yet unresolved. The time taken for brood capsule formation is also extremely variable. The earliest recorded is 195 days in mice following oral infection with eggs (Colli and Schantz, 1974). In pigs, 10–12 months has been reported (Slais, 1980), whereas in sheep reports range from 10 months to 4 years (Heath, 1973; Gemmell et al., 1986). The production of brood capsules and protoscoleces does not seem to depend on cyst size and in mice it has been found that it is not always the largest cysts which develop protoscoleces (Colli and Schantz, 1974). In horses, fertile cysts as small as 2 mm in diameter have been reported (Edwards, 1981). The life span of hydatid cysts of *E. granulosus* can be as long as 16 years in horses (Roneus et al., 1982) and 53 years in humans (Spruance, 1974).

In contrast to *E. granulosus*, *E. multilocularis* develops rapidly in its natural intermediate host, producing protoscoleces in only 2–4 months, an adaptation to the short-lived arvicoline rodents it utilizes (Rausch, 1975;

Woolsey et al., 2015). Thereafter, proliferation of vesicles is curtailed and there is little if any further increase in size (Rausch and Wilson, 1973). In humans, growth is very different. Proliferation continues indefinitely although there are few if any protoscoleces produced (Moro and Schantz, 2009; Rausch and Wilson, 1973). The larval mass proliferates peripherally and at the same time regressive changes occur centrally (Ammann and Eckert, 1996; Moro and Schantz, 2009). Thus a progressively enlarging mass of necrotic tissue with a relatively thin zone of viable proliferating parasite is produced. The term 'alveolar hydatid' was initially used to describe this form of growth, which is not a feature of the development in natural intermediate host species.

6. PERSPECTIVES FOR THE FUTURE

In terms of the biology of *Echinococcus*, there have been major advances made since I last expressed some hopes and predictions (Thompson, 1995). Most notable have been (1) determining the origin, structure and functional activities of the laminated layer and its relationship with the germinal layer; and (2) the isolation, in vitro establishment and characterization of the multipotential germinal cells (see Chapter 4). The latter is a particularly remarkable achievement and along with the sequence data that is now available, we are well placed to build on Desmond Smyth's legacy (Thompson and Lymbery, 2013) and identify the mechanisms that provide *Echinococcus* with its unique developmental plasticity, as well as providing answers to questions such as what governs host specificity and the nature of activities at the parasite-host interface, particularly in the definitive host. For example, discovering the nature and function(s) of rostellar gland secretions in the adult worm would be a major breakthrough in understanding the host-parasite relationship. However, this presents a challenge given the lack of an ethically acceptable model for maintaining the adult parasite in vivo, given the difficulties in undertaking experimental infections in the definitive hosts. However, in vitro techniques are established for growing the adult parasite from protoscolex to maturity and with the advanced imaging and analytical tools now available, they should be exploited to investigate the functional aspects of the rostellar gland. 'New tools' per se, will be valuable in discovering novel aspects of the biology of *Echinococcus* but unless we 'rediscover' the fundamental issues that were laid down by earlier workers we may miss clues to guide data mining.

The last two decades have proved particularly productive in terms of elucidating the phylogenetic relationships of species and intraspecific variants within the genus *Echinococcus* (see Chapter 3). The development and application of appropriate molecular tools have been instrumental in this regard and have underscored a taxonomic revision of the genus. We now have a sound nomenclature, which is essential in understanding the epidemiology of echinococcosis. The revised taxonomy has been built on the basis of phenotypic (morphology, host specificity) and genotypic data. It is important that we maintain such a holistic approach in characterising species and strains in the future and do not rely solely on molecular data alone.

ACKNOWLEDGEMENT

With thanks to Mark Preston, brilliant Graphic Designer at Murdoch Design for producing all the figures.

REFERENCES

Albani, C.M., Elissondo, M.C., Cumino, A.C., Chisari, A., Denegri, G.M., 2010. Primary cell culture of *Echinococcus granulosus* developed from the cystic germinal layer: biological and functional characterization. Int. J. Parasitol. 40, 1269–1275.

Albani, C.M., Cumino, A.C., Elissondo, M.C., Denegri, G.M., 2013. Development of a cell line from *Echinococcus granulosus* germinal layer. Acta Trop. 128, 124–129.

Ali-Khan, Z., Siboo, R., Gomersall, M., Faucher, M., 1983. Cystolytic events and the possible role of germinal cells in metastasis in chronic alveolar hydatidosis. Ann. Trop. Med. Parasitol. 77, 497–512.

Alvarez Rojas, C.A., Romig, T., Lightowlers, M.W., 2014. *Echinococcus granulosus* sensu lato genotypes infecting humans – review of current knowledge. Int. J. Parasitol. 44, 9–18.

Ammann, R.W., Eckert, J., 1996. Cestodes: Echinococcus. Gastroenterol. Clin. North Am. 25, 655–689.

Amri, M., Touil-Boukoffa, C., 2015. A protective effect of the laminated layer on *Echinococcus granulosus* survival dependent on upregulation of host arginase. Acta Trop. 149, 186–194.

Barnes, T.S., Hinds, L.D., Jenkins, D.J., Bielefeldt-Ohmann, H., Lightowlers, M.W., Coleman, G.T., 2011. Comparative pathology of pulmonary hydatid cysts in macropods and sheep. J. Comp. Pathol. 144, 113–122.

Beveridge, I., Rickard, M.D., 1975. The development of *Taenia pisiformis* in various definitive host species. Int. J. Parasitol. 5, 633–639.

Blood, B.D., Lelijveld, J.L., 1969. Studies on sylvatic echinococcosis in southern South America. Z. Tropenmed. Parasitol. 20, 475–482.

Borrie, J., Gemmell, M.A., Manktelow, B.W., 1965. An experimental approach to evaluate the potential risk of hydatid disease from inhalation of *Echinococcus* ova. Br. J. Surg. 52, 876–878.

Bortoletti, G., Ferretti, G., 1973. Investigation on larval forms of *Echinococcus granulosus* with electron microscope. Riv. Parassitol. 34, 89–110.

Bortoletti, G., Ferretti, G., 1978. Ultrastructural aspects of fertile and sterile cysts of *Echinococcus granulosus* developed in hosts of different species. Int. J. Parasitol. 8, 421–431.

Bowles, J., Blair, D., McManus, D.P., 1994. Molecular genetic characterisation of the cervid strain ("northern form") of *Echinococcus granulosus*. Parasitology 109, 215–221.

Braithwaite, P.A., Lomas, R., Thompson, R.C.A., 1985. Hydatid disease: the alveolar variety in Australia. A case report with comment on the toxicity of mebendazole. Aust. N. Z. J. Surg. 55, 519–523.

Brehm, K., Koziol, U., 2014. On the importance of targeting parasite stem cells in anti-echinococcosis drug development. Parasites 21, 72.

Breijo, M., Anesetti, G., Martínez, L., Sim, R.B., Ferreira, A.M., 2008. *Echinococcus granulosus*: the establishment of the metacestode is associated with control of complement-mediated early inflammation. Exp. Parasitol. 118, 188–196.

Brownlee, D.J., Fairweather, I., Johnston, C.F., Rogan, M.T., 1994. Immunocytochemical localization of serotonin (5-HT) in the nervous system of the hydatid organism, *Echinococcus granulosus* (Cestoda, Cyclophyllidea). Parasitology 109, 233–241.

Cameron, T.W.M., Webster, G.A., 1969. The histogenesis of the hydatid cyst (*Echinococcus* spp.). Part 1. Liver cysts in large mammals. Can. J. Zool. 47, 1405–1410.

Cameron, T.W.M., 1960. The incidence and diagnosis of hydatid cyst in Canada. *Echinococcus granulosus* var. canadensis. Parassitologia 2, 381–390.

Camicia, F., Herz, M., Prada, L.C., Kamenetzky, L., Simonetta, S.H., Cucher, M.A., Bianchi, J.I., Fernández, C., Brehm, K., Rosenzvit, M.C., 2013. The nervous and prenervous roles of serotonin in *Echinococcus* spp. Int. J. Parasitol. 43, 647–659.

Cheng, Z., Zhu, S., Wang, L., Liu, F., Tian, H., Pengsakul, T., Wang, Y., 2015. Identification and characterisation of Emp53, the homologue of human tumor suppressor p53, from *Echinococcus multilocularis*: its role in apoptosis and the oxidative stress response. Int. J. Parasitol. 45, 517–526.

Colli, C.W., Schantz, P.M., 1974. Growth and development of *Echinococcus granulosus* from embryophores in an abnormal host (*Mus musculus*). J. Parasitol. 60, 53–58.

Colli, C.W., Williams, J.F., 1972. Influence of temperature on the infectivity of eggs of *Echinococcus granulosus* in laboratory rodents. J. Parasitol. 58, 422–426.

Coltorti, E.A., Varela-Diaz, V.M., 1974. *Echinococcus granulosus*: penetration of macromolecules and their localization on the parasite membranes of cysts. Exp. Parasitol. 35, 225–231.

Constantine, C.C., Lymbery, A.J., Jenkins, D.J., Bennet-Jenkins, E.M., Behm, C.A., Thompson, R.C.A., 1998. Factors influencing the development and carbohydrate metabolism of *Echinococcus granulosis* in dogs. J. Parasitol. 84, 873–881.

Corsini, M., Geissbühler, U., Howard, J., Gottstein, B., Spreng, D., Frey, C.F., 2015. Clinical presentation, diagnosis, therapy and outcome of alveolar echinococcosis in dogs. Vet. Rec. 177, 569. http://dx.doi.org/10.1136/vr.103470.

Cruz-Reyes, A., Constantine, C.C., Boxell, A.C., Hobbs, R.P., Thompson, R.C.A., 2007. *Echinococcus granulosus* from Mexican pigs is the same strain as that in Polish pigs. J. Helminthol. 81, 287–292.

Cucher, M., Prada, L., Mourglia-Ettlin, G., Dematteis, S., Camicia, F., Asurmendi, S., Rosenzvit, M., 2011. Identification of *Echinococcus granulosus* microRNAs and their expression in different life cycle stages and parasite genotypes. Int. J. Parasitol. 41, 439–448.

D'Alessandro, A., Rausch, R.L., Cuello, C., Aristizabal, N., 1979. *Echinococcus vogeli* in man, with a review of polycystic hydatid disease in Colombia and neighboring countries. Am. J. Trop. Med. Hyg. 28, 303–317.

Deplazes, P., Thompson, R.C.A., Constantine, C.C., Penhale, W.J., 1993. Primary infection of dogs with *Echinococcus granulosus*: systemic and local (Peyer's patches) immune responses. Vet. Immunol. Immunopathol. 40, 171–184.

Deplazes, P., Arnold, P., Kaser-Hotz, B., 1997. Concurrent infections of the liver and intestine with *Echinococcus multilocularis* in dogs. Arch. Int. Hidatid. 32, 201–202.

Diaz, A., Casaravilla, C., Irigoin, F., Lin, G., Previato, J.O., Ferreira, F., 2011a. Understanding the laminated layer of larval *Echinococcus* I: structure. Trends Parasitol. 27, 204–213.

Diaz, A., Casaravilla, Allen, J.E., Sim, R.B., Ferreira, F., 2011b. Understanding the laminated layer of larval *Echinococcus* II: immunology. Trends Parasitol. 27, 263–272.

Eckert, J., Thompson, R.C.A., Mehlhorn, H., 1983. Proliferation and metastases formation of larval *Echinococcus multilocularis*. I. Animal model, macroscopical and histological findings. Z. ParasitKde 69, 737–748.

Edwards, G.T., 1981. Small fertile hydatid cysts in British horses. Vet. Rec. 108, 460–461.

Fairweather, I., Threadgold, L.T., 1981. *Hymenolepis nana*: the fine structure of the 'penetration gland' and nerve cells within the oncosphere. Parasitology 82, 445–458.

FAO, 1982. Echinococcosis/hydatidosis Surveillance, Prevention and Control: FAO/UNEP/WHO Guidelines. FAO Animal Production and Health Paper No. 29. Food and Agriculture Organisation of the United Nations, Rome.

Featherstone, D.W., 1971. *Taenia hydatigena*. II. Evagination of cysticerci and establishment in dogs. Exp. Parasitol. 29, 242–249.

Federer, K., Armua-Fernandez, M.T., Hoby, S., Wenker, C., Deplazes, P., 2015. In vivo viability of *Echinococcus multilocularis* eggs in a rodent model after different thermo-treatments. Exp. Parasitol. 154, 14–19.

Freeman, R.S., 1973. Ontogeny of cestodes and its bearing on their phylogeny and systematics. Adv. Parasitol. 11, 481–557.

Galindo, M., Paredes, R., Marchant, C., Mino, V., Galanti, N., 2003. Regionalization of DNA and protein synthesis in developing stages of the parasitic platyhelminth *Echinococcus granulosus*. J. Cell. Biochem. 90, 294–303.

Geisel, O., Barutzki, D., Minkus, G., 1990. Hunde als Finnentrager (Intermediarwirt) von *Echinococcus multilocularis* (Leuckart, 1863), Vogel 1957. Kleintierpraxis 35, 275–280.

Gemmell, M.A., Roberts, M.G., 1995. Modelling *Echinococcus* life cycles. In: Thompson, R.C.A., Lymbery, A.J. (Eds.), *Echinococcus* and Hydatid Disease. CAB International, Wallingford, pp. 333–354.

Gemmell, M.A., Lawson, J.R., Roberts, M.G., 1986. Population dynamics in echinococcosis and cysticercosis: biological parameters of *Echinococcus granulosus* in dogs and sheep. Parasitology 92, 599–620.

Gonzalez, L.M., Daniel-Mwambete, K., Montero, E., Rosenzvit, M.C., McManus, D.P., Garate, T., Cuesta-Bandera, C., 2002. Further molecular discrimination of Spanish strains of *Echinococcus granulosus*. Exp. Parasitol. 102, 46–56.

Gottstein, B., Hemphill, A., 1997. Immunopathology of echinococcosis. Chem. Immunol. 66, 177–208.

Gottstein, B., Eckert, J., Fey, H., 1983. Serological differentiation between *Echinococcus granulosus* and *E. multilocularis* infections in man. Z. ParasitKde 69, 347–356.

Gottstein, B., Dai, W.J., Walker, M., Stettler, M., Müller, N., Hemphill, A., 2002. An intact laminated layer is important for the establishment of secondary *Echinococcus multilocularis* infection. Parasitol. Res. 88, 822–828.

Gustafsson, M.K.S., 1976. Basic cell types in *Echinococcus granulosus* (Cestoda, Cyclophyllidea). Acta Zool. Fenn. 146, 1–16.

Harandi, F.M., Hobbs, R.P., Adams, P.J., Mobedi, I., Morgan-Ryan, U.M., Thompson, R.C.A., 2002. Molecular and morphological characterisation of *Echinococcus granulosus* of human and animal origin in Iran. Parasitology 125, 367–373.

Harandi, M.F., Haiialilo, E., Shokouhi, M., 2012. Larval hook length measurement for differentiating G1 and G6 genotypes of *Echinococcus granulosus* sensu lato. Turk. Parazitol. Derg. 36, 215–218.

Harris, A., Heath, D.D., Lawrence, S.B., Shaw, R.J., 1989. *Echinococcus granulosus*: ultrastructure of epithelial changes during the first 8 days of metacestode development in vitro. Int. J. Parasitol. 19, 621–629.

Hayunga, E.G., 1979. The structure and function of the scolex glands of three caryophyllid tapeworms. Proc. Helminthol. Soc. Wash. 46, 171–179.

Heath, D.D., Lawrence, S.B., 1976. *Echinococcus granulosus*: development in vitro from oncosphere to immature hydatid cyst. Parasitology 73, 417–423.
Heath, D.D., Lawrence, S.B., 1991. Daily egg-production of dogs infected with *Echinococcus granulosus*. Arch. Int. Hidatid. 30, 321–328.
Heath, D.D., 1971. The migration of oncospheres of *Taenia pisiformis, T. serialis* and *Echinococcus granulosus* within the intermediate host. Int. J. Parasitol. 1, 145–152.
Heath, D.D., 1973. The life cycle of *Echinococcus granulosus*. In: Brown, R.W., Salisbury, J.R., White, W.E. (Eds.), Recent Advances in Hydatid Disease. Hamilton Medical and Veterinary Association, Victoria Australia, pp. 7–18.
Heath, D.D., 1995. Immunology of *Echinococcus* infections. In: Thompson, R.C.A., Lymbery, A.J. (Eds.), *Echinococcus* and Hydatid Disease. CAB International, Wallingford, pp. 183–199.
Hemer, S., Konrad, C., Spiliotis, M., Koziol, U., Schaack, D., Förster, S., Gelmedin, V., Stadelmann, B., Dandekar, T., Hemphill, A., Brehm, K., 2014. Host insulin stimulates *Echinococcus multilocularis* insulin signalling pathways and larval development. BMC Biol. 12, 5.
Herbaut, C., Petavy, A.-F., Deblock, S., Gabrion, C., 1988. Nuclear inclusions of rostellar cells of *Echinococcus multilocularis* (Cestoda). Parasitol. Res. 74, 399–402.
Hobbs, R.P., Lymbery, A.J., Thompson, R.C.A., 1990. Rostellar hook morphology of *Echinococcus granulosus* (Batsch, 1786) from natural and experimental Australian hosts and its implications for strain recognition. Parasitology 101, 273–281.
Holcman, B., Heath, D.D., 1997. The early stages of *Echinococcus granulosus* development. Acta Trop. 64, 5–17.
Howell, M.J., Smyth, J.D., 1995. Cultivation. In: Thompson, R.C.A., Lymbery, A.J. (Eds.), *Echinococcus* and Hydatid Disease. CAB International, Wallingford, Oxon, UK, pp. 200–232.
Huldt, G., Johansson, G.S.O., Lantto, S., 1973. Echinococcosis in northern Scandinavia. Arch. Environ. Health 26, 35–40.
Huttner, M., Siefert, L., Mackenstedt, U., Romig, T., 2009. A survey of *Echinococcus* species in wild carnivores and livestock in East Africa. Int. J. Parasitol. 39, 1269–1276.
Irigoín, F., Ferreira, F., Fernández, C., Sim, R.B., Díaz, A., 2002. Myo-Inositol hexakisphosphate is a major component of an extracellular structure in the parasitic cestode *Echinococcus granulosus*. Biochem. J. 362, 297–304.
Irshadullah, M., Nizami, W.A., Ahmad, M., 1990. Polymorphism in the microtriches of adult *Echinococcus granulosus*: scanning electron microscopy. Zool. Anz. 224, 321–327.
Jabbar, A., Swiderski, Z., Mlocicki, D., Beveridge, I., Lightowlers, M.W., 2010. The ultrastructure of taeniid cestode oncospheres and localization of host-protective antigens. Parasitology 137, 521–535.
Jabbar, A., Jenkins, D.J., Crawford, S., Walduck, A.K., Gauci, C.G., Lightowlers, M.W., 2011. Oncospheral penetration glands are the source of the EG95 vaccine antigen against cystic hydatid disease. Parasitology 138, 89–99.
Jenkins, D.J., Rickard, M.D., 1986. Specific antibody responses in dogs experimentally infected with *Echinococcus granulosus*. Am. J. Trop. Med. Hyg. 35, 345–349.
Jenkins, D.J., Romig, T., Thompson, R.C.A., 2005. Emergence/re-emergence of *Echinococcus* spp.—a global update. Int. J. Parasitol. 35, 1205–1219.
Jenkins, E.J., Schurer, J.M., Gesy, K.M., 2011. Old problems on a new playing field: helminth zoonoses transmitted among dogs, wildlife, and people in a changing northern climate. Vet. Parasitol. 182, 54–69.
Jenkins, E.J., Castrodale, L.J., de Rosemond, S.J.C., Dixon, B.R., Elmore, S.A., Gesy, K.M., Hoberg, L., Polley, E.P., Polley, L., Schurer, J.M., Simard, M., Thompson, R.C.A., 2013. Tradition and transition: parasitic zoonoses of people and animals in Alaska, northern Canada, and Greenland. Adv. Parasitol. 82, 33–204.

Kapel, C.M., Torgerson, P.R., Thompson, R.C.A., Deplazes, P., 2006. Reproductive potential of *Echinococcus multilocularis* in experimentally infected foxes, dogs, raccoon dogs and cats. Int. J. Parasitol. 36, 79–86.

Kilejian, A., Schwabe, C.W., 1971. Studies on the polysaccharides of the *Echinococcus granulosus* cyst, with observations on a possible mechanism for laminated membrane formation. Comp. Biochem. Physiol. 40B, 25–36.

Kilejian, A., Schinazi, L.A., Schwabe, C.W., 1961. Host-parasite relationships in echinococcosis. V. Histochemical observations on *Echinococcus granulosus*. J. Parasitol. 47, 181–188.

Kouguchi, H., Matsumoto, J., Nakao, R., Yamano, K., Oku, Y., Yagi, K., 2013. Characterization of a surface glycoprotein from *Echinococcus multilocularis* and its mucosal vaccine potential in dogs. PLoS One 8, e69821.

Koziol, U., Rauschendorfer, T., Rodríguez, L.Z., Krohne, G., Brehm, K., 2014. The unique stem cell system of the immortal larva of the human parasite *Echinococcus multilocularis*. EvoDevo 5, 10.

Koziol, U., Jarero, F., Olson, P.D., Brehm, K., 2016. Comparative analysis of Wnt expression identifies a highly conserved developmental transition in flatworms. BMC Biol. 14, 10.

Krasnoshchekov, G.P., Pluzhnnikov, L.T., 1981. Rostellar organ of *Taenia crassiceps* (Cestoda: Taeniidae) larvae. Parazitologia 15, 519–524. In Russian.

Kumaratilake, L.M., Thompson, R.C.A., 1981. Maintenance of the life cycle of *Echinococcus granulosus* in the laboratory following in vivo and in vitro development. Z. ParasitKde 65, 103–106.

Lascano, E.F., Coltorti, E.A., Varela-Diaz, V.M., 1975. Fine structure of the germinal membrane of *Echinococcus granulosus* cysts. J. Parasitol. 61, 853–860.

Lavikainen, A., Lehtinen, M.J., Meri, T., Hirvelä-Koski, V., Meri, S., 2003. Molecular genetic characterization of the Fennoscandian cervid strain, a new genotypic group (G10) of *Echinococcus granulosus*. Parasitology 127, 207–215.

Lavikainen, A., Lehtinen, M.J., laaksonen, S., Agren, E., Oksanen, A., Meri, S., 2006. Molecular genetic characterization of *Echinococcus* isolates of cervid origin from Finland and Sweden. Parasitology 133, 565–570.

Lethbridge, R.C., 1980. The biology of the oncosphere of cyclophyllidean cestodes. Helminthol. Abstr. A 49, 59–72.

Lightowlers, M.W., Rickard, M.D., Honey, R.D., Obendorf, D.L., Mitchell, G.F., 1984. Serological diagnosis of *Echinococcus granulosus* infection in sheep using cyst fluid antigen processed by antibody affinity chromatography. Aust. Vet. J. 61, 101–108.

Lin, G., Todeschini, A.R., Koizumi, A., Neves, J.L., González, H., Dematteis, S., Hada, Previato, J.O., Ferreira, F., Mendonça-Previato, L., Díaz, A., 2012. Further structural characterization of the *Echinococcus granulosus* laminated layer carbohydrates: the blood-antigen P1-motif gives rise to branches at different points of the O-glycan chains. Glycobiology 23, 438–452.

Lindroos, P., Gardberg, T., 1982. The excretory system of *Diphyllobothrium dendriticum* (Nitzsch 1824) plerocercoids as revealed by an injection technique. Z. ParasitKde 67, 289–297.

Loos, J.A., Caparros, P.A., Nicolao, M.C., Denegri, G.M., Cumino, A.C., 2014. Identification and pharmacological induction of autophagy in the larval stages of *Echinococcus granulosus*: an active catabolic process in calcareous corpuscles. Int. J. Parasitol. 44, 415–427.

Losson, B.J., Coignoul, F., 1997. Larval *Echinococcus multilocularis* infection in a dog. Vet. Rec. 141, 49–50.

Lymbery, A.J., Thompson, R.C.A., 2012. The molecular epidemiology of parasite infections: tools and applications. Mol. Biochem. Parasitol. 181, 102–116.

Lymbery, A.J., Constantine, C.C., Thompson, R.C.A., 1997. Self-fertilization without genomic or structuring in a parasitic tapeworm. Evolution 51, 289–294.

Martínez, C., Paredes, R., Stock, R.P., Saralegui, A., Andreu, M., Cabezón, C., Ehrlich, R., Galanti, N., 2005. Cellular organization and appearance of differentiated structures in developing stages of the parasitic platyhelminth *Echinococcus granulosus*. J. Cell. Biochem. 94, 327–335.
McCullough, J.S., Fairweather, I., 1989. The fine structure and possible functions of scolex gland cells in *Trilocularia acanthiaevulgaris* (Cestoda, Tetraphyllidea). Parasitol. Res. 75, 575–582.
Mehlhorn, H., Becker, B., Andrews, P., Thomas, H., 1981. On the nature of the proglottids of cestodes: a light and electron microscopic study on *Taenia, Hymenolepis* and *Echinococcus*. Z. ParasitKde 65, 243–259.
Mehlhorn, H., Eckert, J., Thompson, R.C.A., 1983. Proliferation and metastaSeS formation of larval *Echinococcus multilocularis*. II. Ultrastructural investigations. Z. ParasitKde 69, 749–763.
Mettrick, D.F., Podesta, R.B., 1974. Ecological and physiological aspects of helminth-host interactions in the mammalian gastrointestinal canal. Adv. Parasitol. 12, 183–277.
Moks, E., Jogisalu, I., Valdmann, H., Saarma, U., 2008. First report of *Echinococcus granulosus* genotype G8 in Eurasia and a reappraisal of the phylogenetic relationships of 'genotypes' G5-G10. Parasitology 135, 647–654.
Monteiro, K.M., de Carvalho, M.O., Zaha, A., Ferreira, H.B., 2010. Proteomic analysis of the *Echinococcus granulosus* metacestode during infection of its intermediate host. Proteomics 10, 1985–1999.
Morales, G.A., Guzman, V.H., Wells, E.A., Angel, D., 1979. Polycystic echinococcosis in Columbia: the larval cestodes in infected rodents. J. Wilde Dis. 15, 421–428.
Moro, P., Schantz, P.M., 2009. Echinococcosis: a review. Int. J. Inf. Dis. 13, 125–133.
Morseth, D.J., 1965. Ultrastructure of developing taeniid embryophores and associated structures. Exp. Parasitol. 16, 207–216.
Morseth, D.J., 1966. Chemical composition of embryophoric blocks of *Taenia hydatigena, Taenia ovis* and *Taenia pisiformis* eggs. Exp. Parasitol. 18, 347–354.
Morseth, D.J., 1967. Fine structure of the hydatid cyst and protoscolex of *Echinococcus granulosus*. J. Parasitol. 53, 312–325.
Nakao, M., Lavikainen, A., Yanagida, T., Ito, A., 2013. Phylogenetic systematics of the genus *Echinococcus* (Cestoda: Taeniidae). Int. J. Parasitol. 43, 1017–1029.
Nicolao, M.C., Denegri, G.M., Cárcamo, J.G., Cumino, A.C., 2014. P-glycoprotein expression and pharmacological modulation in larval stages of *Echinococcus granulosus*. Parasitol. Int. 263, 1–8.
Nieland, M.L., 1968. Electron microscope observations on the egg of *Taenia taeniaeformis*. J. Parasitol. 54, 957–969.
Noya, V., Bay, S., Festari, M.F., García, E.P., Rodriguez, E., Chiale, C., Ganneau, C., Baleux, F., Astrada, S., Bollati-Fogolín, M., Osinaga, E., Freire, T., 2014. Mucin-like peptides from *Echinococcus granulosus* induce antitumor activity. Int. J. Oncol. 43, 775–784.
Ohbayashi, M., Rausch, R.L., Fay, F.H., 1971. On the ecology and distribution of *Echinococcus* spp. (Cestoda: Taeniidae), and characteristics of their development in the intermediate host. II. Comparative studies on the larval *E. multilocularis* Leuckart, 1863, in the intermediate host. Jpn. J. Vet. Res. 19 (Suppl).
Ohman-James, C., 1973. Cytology and cytochemistry of the scolex gland cells in *Diphyllobothrium ditremum* (Creplin, 1825). Z. ParasitKde 42, 77–86.
Ohnishi, K., Kutsumi, H., 1991. Possible formation of calcareous corpuscles by the brood capsule in secondary hepatic metacestodes of *Echinococccus multilocularis*. Parasitol. Res. 77, 600–601.
Pan, D., Das, S., Bera, A.K., Bandyopadhyay, S., Bandyopadhyay, S., De, S., Rana, T., Das, S.K., Suryanaryana, V.V., Deb, J., Bhattacharya, D., 2011. Molecular and

biochemical mining of heat-shock and 14-3-3 proteins in drug induced protoscoleces of *Echinococcus granulosus* and the detection of a candidate gene for anthelmintic resistance. J. Helminthol. 85, 196–203.

Parkinson, J., Wasmuth, J.D., Salinas, G., Bizarro, C.V., Sanford, C., Berriman, M., Ferreira, H.B., Zaha, A., Blaxter, M.L., Maizels, R.M., Fernández, C., 2012. A transcriptomic analysis of *Echinococcus granulosus* larval stages: implications for parasite biology and host adaptation. PLoS Negl. Trop. Dis. 6, e1897.

Pawlowski, I.D., Yap, K.W., Thompson, R.C.A., 1988. Observations on the possible origin, formation and structure of calcareous corpuscles in taeniid cestodes. Parasitol. Res. 74, 293–296.

Pednekar, R.P., Gatner, M.L., Thompson, R.C.A., Traub, R.J., 2009. Molecular and morphological characterisation of *Echinococcus* from food producing animals in India. Vet. Parasitol. 165, 58–65.

Pergrine, A.S., Jenkins, E.J., Barnes, B., Johnson, S., Polley, L., Barker, I.K., De Wolf, B., Gottstein, B., 2012. Alveolar hydatid disease (*Echinococcus multilocularis*) in the liver of a Canadian dog in British Columbia, a newly endemic region. Can. Vet. J. 53, 870–874.

Pospekhova, N.A., Bondarenko, S.K., 2014. Morpho-functional characteristics of the scolex of *Wardium chaunense* (Cestoda: Aploparaksidae) penetrated into host intestine. Parasit. Res. 113, 131–137.

Ranasinghe, S.L., Fischer, K., Zhang, W., Gobert, G.N., McManus, D.P., 2015. Cloning and characterization of two potent Kunitz type protease inhibitors from *Echinococcus granulosus*. PLoS Negl. Trop. Dis. 9 (12), e0004268, 8.

Rausch, R.L., Wilson, J.F., 1973. Rearing of the adult *Echinococcus multilocularis* Leukart, 1863, from sterile from sterile larvae from man. Am. J. Trop. Med. Hyg. 22, 357–360.

Rausch, R.L., D'Alessandro, A., Rausch, V.R., 1981. Characteristics of the larval *Echinococcus vogeli* Rausch and Bernstein, 1972 in the natural intermediate host, the paca, *Cuniculus paca* L. (Rodentia: Dasyproctidae). Am. J. Trop. Med. Hyg. 30, 1043–1052.

Rausch, R.L., 1954. Studies on the helminth fauna of Alaska. XX. The histogenesis of the alveolar larva of *Echinococcus* species. J. Infect. Dis. 94, 178–186.

Rausch, R.L., 1975. Taeniidae. In: Hubbert, W.T., McCulloch, W.F., Schurrenberger, P.R. (Eds.), Diseases Transmitted From Animals to Man. Thomas, Springfield, IL, pp. 678–707.

Rausch, R.L., 1997. *Echinococcus granulosus*: biology and ecology. In: Anderson, F.L., Ouhelli, H., Kachani, M. (Eds.), Compendium of Cystic Echinococcosis in Africa and in Middle Eastern Countries with Special Reference to Morocco. Brigham Young University, Provo, UT, pp. 18–53.

Richards, K.S., Arme, C., 1981. The ultrastructure of the scolex-neck syncytium, neck cells and frontal gland cells of *Caryophyllaeus laticeps* (Caryophyllidea: Cestoda). Parasitology 83, 477–487.

Richards, K.S., Arme, C., Bridges, J.F., 1983. *Echinococcus granulosus equinus*: an ultrastructural study of the laminated layer, including changes on incubating cysts in various media. Parasitology 86, 399–405.

Rodriguez-Caabeiro, F., Casado, N., 1988. Evidence of in vitro germinal layer development in *Echinococcus granulosus* cysts. Parasitol. Res. 74, 558–562.

Rogan, M.T., Richards, K.S., 1987. *Echinococcus granulosus*: changes in the surface ultrastructure during protoscolex formation. Parasitology 94, 359–367.

Romig, T., Dinkel, A., Mackenstedt, U., 2006. The present situation of echinococcosis in Europe. Parasit. Int. 55, S187–S191.

Romig, T., Ebi, D., Wassermann, M., 2015. Taxonomy and molecular epidemiology of *Echinococcus granulosus* sensu lato. Vet. Parasitol. 213, 76–84.

Roneus, O., Christensson, D., Nilsson, N.-G., 1982. The longevity of hydatid cysts in horses. Vet. Parasitol. 11, 149–154.

Saarma, U., Jogisalu, I., Moks, E., Varcasia, A., Lavikainen, A., Oksanen, A., Simsek, S., Andresiuk, V., Denegri, G., Gonzalez, L.M., Ferrer, E., Garate, T., Rinaldi, L., Maravilla, P., 2009. A novel phylogeny for the genus *Echinococcus*, based on nuclear data, challenges relationships based on mitochondrial evidence. Parasitology 136, 317–328.
Sakamoto, T., Sugimura, M., 1970. Studies on echinococcosis. XXIII. Electron microscopical observations on histogenesis of larval *Echinococcus multilocularis*. Jpn. J. Vet. Res. 18, 131–144.
Sakamoto, T., 1981. Electron microscopical observations on the egg of *Echinococcus multilocularis*. Mem. Fac. Agr. Kagoshima. Univ. 17, 165–174.
Sawada, I., 1973. The mode of attachment of the larval tapeworm to the mucosa of the chicken intestine. Jpn. J. Zool. 17, 1–9.
Schantz, P.M., Chai, J., Craig, P.S., Eckert, J., Jenkins, D.J., Macpherson, C.N.L., Thakur, A., 1995. In: Thompson, R.C.A., Lymbery, A.J. (Eds.), Epidemiology and Control of Hydatid Disease. CAB International, Wallingford, pp. 233–331.
Schantz, P.M., 1982. Echinococcosis. In: Steele, J. (Ed.), CRC Handbook Series in Zoonoses, Section C: Parasitic Zoonoses, 1. CRC Press, Boca Raton, FL, pp. 231–277.
Schurer, J.M., Ndao, M., Skinner, S., Irvine, J., Elmore, S.A., Epp, T., Jenkins, E.J., 2013. Parasitic zoonoses: one health surveillance in northern Saskatchewan. PLoS Negl. Trop. Dis. 7, e2141.
Schurer, J.M., Ndao, M., Quewezance, H., Elmore, S.A., Tasha, E., Jenkins, E.J., 2014. People, pets, and parasites: one health surveillance in southern Saskatchewan. Am. J. Trop. Med. Hyg. 90, 1184–1190.
Sharbatkhori, M., Harandi, M.F., Mirhendi, H., Hajialilo, E., Kia, E.B., 2011. Sequence analysis of cox1 and nad1 genes in *Echinococcus granulosus* G3 genotype in camels (*Camelus dromedaries*) from central Iran. Parasitol. Res. 108, 521–527.
Siles-Lucas, M., Nunes, C.P., Zaha, A., Breijo, M., 2000. The 14-3-3 protein is secreted by the adult worm of *Echinococcus granulosus*. Parasite Immunol. 22, 521–528.
Slais, J., Vanek, M., 1980. Tissue reactions to spherical and lobular hydatid cysts of *Echinococcus granulosus* (Batsch, 1786). Folia Parasitol. 27, 135–143.
Slais, J., 1973. Functional morphology of cestode larvae. Adv. Parasitol. 11, 395–480.
Slais, J., 1980. Experimental infection on sheep and pigs with *Echinococcus granulosus* (Batsch, 1786), and the origin of pouching in hydatid cysts. Acta Vet. Acad. Sci. Hung. 28, 375–387.
Smyth, J.D., Barrett, N.J., 1979. *Echinococcus multilocularis*: further observations on strobilar differentiation in vitro. Rev. Iber. Parasitol. 39, 39–53.
Smyth, J.D., Davies, Z., 1974a. In vitro culture of the strobilar stage of *Echinococcus granulosus* (sheep strain): a review of basic problems and results. Int. J. Parasitol. 4, 631–644.
Smyth, J.D., Davies, Z., 1974b. Occurrence of physiological strains of *Echinococcus granulosus* demonstrated by in vitro culture of protoscoleces from sheep and horse hydatid cysts. Int. J. Parasitol. 4, 443–445.
Smyth, J.D., Davies, Z., 1975. In vitro suppression of segmentation in *Echinococcus multilocularis* with morphological transformation of protoscoleces into monozoic adults. Parasitology 71, 125–135.
Smyth, J.D., Heath, D.D., 1970. Pathogenesis of larval cestodes in mammals. Helminthol. Abstr. A 39, 1–23.
Smyth, J.D., Smyth, M.M., 1969. Self insemination in *Echinococcus granulosus* in vivo. J. Helminthol. 43, 383–388.
Smyth, J.D., Howkins, A.B., Barton, M., 1966. Factors controlling the differentiation of the hydatid organism, *Echinococcus granulosus*, into cystic or strobilar stages in vitro. Nature 211, 1374–1377.

Smyth, J.D., Morseth, D.J., Smyth, M.M., 1969. Observations on nuclear secretions in the rostellar gland cells of *Echinococcus granulosus* (Cestoda). Nucleus 12, 47–56.
Smyth, J.D., 1964a. The biology of the hydatid organism. Adv. Parasitol. 2, 169–219.
Smyth, J.D., 1964b. Observations on the scolex of *Echinococcus granulosus*, with special reference to the occurrence and cytochemistry of secretory cells in the rostellum. Parasitology 54, 515–526.
Smyth, J.D., 1967. Studies on tapeworm physiology. XI. In vitro cultivation of *Echinococcus granulosus* from the protoscolex to the strobilate stage. Parasitology 57, 111–133.
Smyth, J.D., 1969. Parasites as biological models. Parasitology 59, 73–91.
Smyth, J.D., 1971. Development of monozoic forms of *Echinococcus granulosus* during in vitro culture. Int. J. Parasitol. 1, 121–124.
Smyth, J.D., 1979. *Echinococcus granulosus* and *E. multilocularis:* in vitro culture of the strobilar stages from protoscoleces. Angew. Parasitol. 20, 137–147.
Smyth, J.D., 1982. The insemination-fertilization problem in cestodes cultured in vitro. In: Meerovitch, E. (Ed.), Aspects of Parasitology. A Festschrift Dedicated to the Fiftieth Anniversary of the Institute of Parasitology of McCill University 1932–1982. McGill University, Montreal, pp. 393–406.
Smyth, J.D., 1987. Changing concepts in the microecology, macroecology and epidemiology of hydatid disease. In: Geerts, S., Kumar, V., Brandt, J. (Eds.), Helminth Zoonoses. Martinus Nijhoff, Dordrecht, pp. 1–11.
Sousa, O.E., Thatcher, V.E., 1969. Observations on the life-cycle of *Echinococcus oligarthrus* (Diesing, 1863) in the Republic of Panama. Ann. Trop. Med. Parasitol. 63, 165–175.
Specian, R.D., Lumsden, R.D., 1981. Histochemical, cytochemical and autoradiographic studies on the rostellum of *Hymenolepsis diminuta.* Z. ParasitKde 64, 335–345.
Spiliotis, M., Brehm, K., 2009. Axenic in vitro cultivation of *Echinococcus multilocularis* metacestode vesicles and the generation of primary cell cultures. Mol. Biol. 470, 245–262.
Spiliotis, M., Lechner, S., Tappe, D., Scheller, C., Krohne, G., Brehm, K., 2008. Transient transfection of *Echinococcus multilocularis* primary cells and complete in vitro regeneration of metacestode vesicles. Int. J. Parasitol. 38, 1025–1039.
Spruance, S.L., 1974. Latent period of 53 years in a case of hydatid cyst disease. Arch. Int. Med. 134, 741–742.
Stadelmann, B., Spiliotis, M., Müller, J., Scholl, S., Müller, N., Gottstein, B., Hemphill, A., 2010. *Echinococcus multilocularis* phosphoglucose isomerase (EmPGI): a glycolytic enzyme involved in metacestode growth and parasite-host cell interactions. Int. J. Parasitol. 40, 1563–1574.
Sulgostowska, T., 1972. The development of organ systems in cestodes. I. A study of histology of *Hymenolepsis diminuta* (Rudolphi, 1819) Hymenolepididae. Acta Parasitol. Pol. 20, 449–462.
Sulgostowska, T., 1974. The development of organ systems in cestodes. II. Histogenesis of the reproductive system in *Hymenolepis diminuta* (Rudolphi, 1819) (Hymenolepididae). Acta Parasitol. Pol. 22, 179–190.
Swiderski, Z., 1982. *Echinococcus granulosus*: embryonic envelope formation. In: Proceedings of 10th International Congress Electron Microscopy, 3, p. 513.
Swiderski, Z., 1983. *Echinococcus granulosus*: hook-muscle systems and cellular organisation of infective oncospheres. Int. J. Parasitol. 13, 289–299.
Thevenet, P.S., Jensen, O., Drut, R., Cerrone, G.E., Grenóvero, M.S., Alvarez, H.M., Targovnik, H.M., Basualdo, J.A., 2005. Viability and infectiousness of eggs of *Echinococcus granulosus* aged under natural conditions of inferior arid climate. Vet. Parasitol. 133, 71–77.
Thompson, R.C.A., Eckert, J., 1982. The production of eggs by *Echinococcus multilocularis* in the laboratory following in vivo and in vitro development. Z. ParasitKde 68, 227–234.

Thompson, R.C.A., Eckert, J., 1983. Observations on *Echinococcus multilocularis* in the definitive host. Z. ParasitKde 69, 335–345.
Thompson, R.C.A., Jenkins, D.J., 2014. *Echinococcus* as a model system. Int. J. Parasitol. 44, 865–877.
Thompson, R.C.A., Lymbery, A.J., 1988. The nature, extent and significance of variation within the genus *Echinococcus*. Adv. Parasitol. 27, 210–263.
Thompson, R.C.A., Lymbery, A.J., 1990. *Echinococcus*: biology and strain variation. Int. J. Parasitol. 20, 457–470.
Thompson, R.C.A., Lymbery, A.J., 1996. Genetic variability in parasites and host-parasite interactions. Parasitology 112, 57–522.
Thompson, R.C.A., Lymbery, A.J., 2013. Lets not forget the thinkers. Trends Parasitol. 29, 581–584.
Thompson, R.C.A., McManus, D.P., 2001. Aetiology: parasites and life-cycles. In: Eckert, J., Gemmell, M.A., Mesin, F.X., Pawlowski, Z.S. (Eds.), WHO/OIE Manual on *Echinococcus* in Humans and Animals: A Zoonosis of Global Concern. World Organisation for Animal Health (OIE), Paris, pp. 1–19.
Thompson, R.C.A., McManus, D.P., 2002. Towards a taxonomic revision of the genus *Echinococcus*. Trends Parasitol. 18, 452–457.
Thompson, R.C.A., Smyth, J.D., 1975. Equine hydatidosis: a review of the current status in Great Britain and the results of an epidemiological survey. Vet. Parasitol. 1, 107–127.
Thompson, R.C.A., Dunsmore, J.D., Hayton, A.R., 1979. *Echinococcus granulosus*: secretory activity of the rostellum of the adult cestode in situ in the dog. Exp. Parasitol. 48, 144–163.
Thompson, R.C.A., Houghton, A., Zaman, V., 1982. A study of the microtriches of adult *Echinococcus granulosus* by scanning electron microscopy. Int. J. Parasitol. 12, 579–583.
Thompson, R.C.A., Kumaratilake, L.M., Eckert, J., 1984. Observations on *Echinococcus granulosus* of cattle origin in Switzerland. Int. J. Parasitol. 14, 283–291.
Thompson, R.C.A., Deplazes, P., Eckert, J., 1990. Uniform strobilar development of *Echinococcus multilocularis* in vitro from protoscolex to immature stages. J. Parasitol. 76, 240–247.
Thompson, R.C.A., Lymbery, A.J., Constantine, C.C., 1995. Variation in *Echinococcus*: towards a taxonomic revision of the genus. Adv. Parasitol. 35, 145–176.
Thompson, R.C.A., Boxell, A.C., Ralston, B.J., Constantine, C.C., Hobbs, R.P., Shury, T., Olson, M.E., 2006a. Molecular and morphological characterization of *Echinococcus* in cervids from North America. Parasitology 132, 439–447.
Thompson, R.C.A., Kapel, C.M., Hobbs, R.P., Deplazes, P., 2006b. Comparative development of *Echinococcus multilocularis* in its definitive hosts. Parasitology 19, 1–8.
Thompson, R.C.A., Jenkins, E.J., Schurer, J.M., Thompson, R.C.A., 2014. *Echinococcus canadensis, E. borealis*, and *E. intermedius*. What's in a name? Trends Parasitol. 31, 23–29.
Thompson, R.C.A., 1976. The development of brood capsules and protoscoleces in secondary hydatid cysts of *Echinococcus granulosus*. Z. ParasitKde 51, 31–36.
Thompson, R.C.A., 1977. Growth, segmentation and maturation of the British horse and sheep strains of *Echinococcus granulosus* in dogs. Int. J. Parasitol. 7, 281–285.
Thompson, R.C.A., 1995. Biology and systematics of *Echinococcus*. In: Thompson, R.C.A., Lymbery, A.J. (Eds.), *Echinococcus* and Hydatid Disease. CAB International, Wallingford, Oxon, UK, pp. 1–50.
Thompson, R.C.A., 2001. Echinococcosis. In: Gillespie, S.H., Pearson, R.D. (Eds.), Principles and Practice of Clinical Parasitology. Wiley, Sussex, pp. 595–612.
Thompson, R.C.A., 2008. The taxonomy, phylogeny and transmission of *Echinococcus*. Exp. Parasitol. 119, 439–446.

Thompson, R.C.A., 2013. Parasite zoonoses and wildlife: one health, spillover and human activity. Int. J. Parasitol. 43, 1079–1088.

Thompson, R.C.A., 2015. Neglected zoonotic helminths: *Hymenoleps nana, Echinococcus canadensis* and *Ancylostoma ceylanicum*. Clin. Microbiol. Inf. 21, 426–432.

Tigre, W., Deresa, B., Haile, A., Gabriel, S., Victor, B., Pelt, J.V., Devleesschauwer, B., Vercruysse, J., Dorny, P., 2016. Molecular characterization of *Echinococcus granulosus* s.l. cysts from cattle, camels, goats and pigs in Ethiopia. Vet. Para. 215, 17–21.

Vanek, M., 1980. A morphological study of hydatids of *Echinococcus granulosus* (Batsch, 1786) from pigs. Fol. Parasitol. 27, 37–46.

Veit, P., Bilger, B., Schad, V., Schäfer, J., Frank, W., Lucius, R., 1995. Influence of environmental factors on the infectivity of *Echinococcus multilocularis* eggs. Parasitology 110, 79–86.

Vogel, H., 1978. Wie wachst der Alveolarechinokokkus? Tropenmed. Parasitol. 29, 1–11.

Wassermann, M., Woldeyes, D., Banchwosen Mechal Gerbi, B.M., Ebi, D., Zeyhle, E., Mackenstedt, U., Petros, B., Tilahun, G., Peter Kern, P., Romig, T., 2016. A novel zoonotic genotype related to *Echinococcus granulosus* sensu stricto from southern Ethiopia. Int. J. Parasitol. pii: S0020-7519(16)30066-2. http://dx.doi.org/10.1016/j.ijpara.2016.04.005 [Epub ahead of print].

Weiss, A.T., Bauer, C., Köhler, K., 2010. Canine alveolar echinococcosis: morphology and inflammatory response. J. Comp. Pathol. 143, 233–238.

Whitfield, P.J., Evans, N.A., 1983. Pathogenesis and asexual multiplication among parasitic platyhelminths. Parasitology 86, 121–160.

Williams, J.F., Colli, C.W., 1970. Primary cystic infection with *Echinococcus granulosus* and *Taenia hydatigena* in *Meriones unguiculatus*. J. Parasitol. 56, 509–513.

Williams, R.J., Sweatman, G.K., 1963. On the transmission, biology and morphology of *Echinococcus granulosus equinus*, a new subspecies of hydatid tapeworm in horses in Great Britain. Parasitology 53, 391–407.

Wilson, J.F., Rausch, R.L., 1980. Alveolar hydatid disease. A review of clinical features of 33 indigenous cases of *Echinococcus multilocularis* infection in Alaskan Eskimos. Am. J. Trop. Med. Hyg. 29, 134–155.

Woolsey, I.D., Jensen, P.M., Deplazes, P., Kapel, C.M., 2015. Establishment and development of *Echinococcus multilocularis* metacestodes in the common vole (*Microtus arvalis*) after oral inoculation with parasite eggs. Parasitol. Int. 64, 571–575.

Yamashita, K., Uchino, J., Sato, N., Furuya, K., Namieno, T., 1997. Establishment of a primary culture of *Echinococcus multilocularis* cells. J. Gastroenterol. 32, 344–350.

Zd'árská, Z., Nebesárová, J., 2003. Ultrastructure of the early rostellum of *Silurotaenia siluri* (Batsch, 1786) (Cestoda: Proteocephalidae). Parasitol. Res. 89, 495–500.

CHAPTER THREE

Phylogenetic Pattern, Evolutionary Processes and Species Delimitation in the Genus *Echinococcus*

A.J. Lymbery
Murdoch University, Murdoch, WA, Australia
E-mail: a.lymbery@murdoch.edu.au

Contents

Abstract

An accurate and stable alpha taxonomy requires a clear conception of what constitutes a species and agreed criteria for delimiting different species. An evolutionary

Advances in Parasitology, Volume 95
ISSN 0065-308X
http://dx.doi.org/10.1016/bs.apar.2016.07.002

or general lineage concept defines a species as a single lineage of organisms with a common evolutionary trajectory, distinguishable from other such lineages. Delimiting evolutionary species is a two-step process. In the first step, phylogenetic reconstruction identifies putative species as groups of organisms that are monophyletic (share a common ancestor) and exclusive (more closely related to each other than to organisms outside the group). The second step is to assess whether members of the group possess genetic exchangeability (where cohesion is maintained by gene flow among populations) or ecological exchangeability (where cohesion is maintained because populations occupy the same ecological niche). Recent taxonomic reviews have recognized nine species within the genus *Echinococcus*. Phylogenetic reconstructions of the relationships between these putative species using mtDNA and nuclear gene sequences show that for the most part these nine species are monophyletic, although there are important incongruences that need to be resolved. Applying the criteria of genetic and ecological exchangeability suggests that seven of the currently recognized species represent evolutionarily distinct lineages. The species status of *Echinococcus canadensis* and *Echinococcus ortleppi* could not be confirmed. Coalescent-based analyses represent a promising approach to species delimitation in these closely related taxa. It seems likely, from a comparison of sister species groups, that speciation in the genus has been driven by geographic isolation, but biogeographic scenarios are largely speculative and require further testing.

1. INTRODUCTION

Taeniid tapeworms (Eucestoda: Cyclophyllidae: Taeniidae) are important parasites of people throughout the world. Although as many as 13 genera have been described in the family, the most recent taxonomic revision recognized only four; *Hydatigera*, *Taenia*, *Versteria*, and *Echinococcus* (Nakao et al., 2013a). The genus *Echinococcus* is a monophyletic group of species characterized by small adult worms and larvae (metacestodes) with extensive asexual reproduction. Definitive hosts are carnivores, usually canids or felids, and infection is acquired by eating herbivorous or omnivorous intermediate hosts. Humans are accidental intermediate hosts, with the infection being known as echinococcosis or hydatid disease. There are three different types of echinococcosis, which result from infection with different species of *Echinococcus* and are named for the structure of the metacestode; cystic, alveolar or polycystic. Cystic and alveolar echinococcosis are major public health issues in many countries throughout the world and are recognized as neglected parasitic zoonoses (Moro and Schantz, 2009; Torgerson, 2013).

Classification and nomenclature within the genus *Echinococcus* have long been controversial topics, but in recent years molecular phylogenetic analyses have promised a resolution to this controversy. In this paper, I will briefly review the taxonomic history and currently accepted taxonomic designations within the genus, attempt to define an appropriate species concept, examine both the phylogenetic and population genetic data that are required to correctly delimit species according to that concept, apply criteria for delimitation to currently described species of *Echinococcus* and, finally, explore the phenotypic consequences of genetic variation among species.

2. SPECIES OF *ECHINOCOCCUS*

Prior to the widespread application of molecular genetic techniques, a total of 16 species and 13 subspecies had been described in the genus based on morphology, but most of these taxa were subsequently invalidated by Rausch (1953), Vogel (1957), Rausch and Nelson (1963) and Schantz et al. (1976), leaving only four valid species: *Echinococcus granulosus* (with the subspecies *E. g. granulosus* and *E. g. canadensis*); *Echinococcus multilocularis* (with the subspecies *E. m. multilocularis* and *E. m. sibiricensis*); *Echinococcus oligarthra* (= *oligarthus*); and *Echinococcus vogeli*. The term 'strain' was used to refer to intraspecific variants of uncertain taxonomic status, including many of the invalidated taxa (Thompson, 1986; Thompson and Lymbery, 1988). Strains were initially described from differences in host occurrence, geographic distribution, morphology and developmental biology, with most strains ascribed to the species *E. granulosus*. Molecular genetic studies, based principally on partial sequencing of mtDNA, clarified the extent of strain variation, leading to a 'genotype' nomenclature of intraspecific variants within *E. granulosus*: G1 (sheep strain); G2 (Tasmanian sheep strain); G3 (buffalo strain); G4 (horse strain); G5 (cattle strain); G6 (camel strain); G7 (pig strain); G8 (American cervid strain); and G10 (European or Fennoscandian cervid strain) (Bowles et al., 1992, 1994; Bowles and McManus, 1993).

By the mid-1990s, it was suggested from phylogenetic analyses of both morphological (Lymbery, 1992) and mtDNA sequence data (Bowles et al., 1995; Lymbery, 1995) that *E. granulosus* was a paraphyletic group, with some strains being more closely related to *E. multilocularis* than to other strains within the complex. Initial taxonomic revisions split *E. granulosus* (sensu lato; s.l) into three species; *Echinococcus equinus* for G4, *Echinococcus ortleppi* for G5 and *E. granulosus* sensu stricto (s.s.) for G1, G2 and G3 (Thompson et al., 1995; Thompson and McManus, 2002). Nakao et al. (2007) subsequently

Table 1 Currently recognized species within the genus *Echinococcus*

Species	Definitive hosts	Intermediate hosts	Distribution
Echinococcus oligarthra[a]	Wild felids	Agouti	Central and South America
Echinococcus vogeli	Bush dog	Paca	Central and South America
Echinococcus granulosus[b]	Domestic dog	Sheep, many other ungulates	Cosmopolitan
Echinococcus felidis	Lion, hyena	Unknown	Africa
Echinococcus equinus	Domestic dog	Horse, other equids	Eurasia, Africa
Echinococcus multilocularis	Foxes	Arvicolid rodents	Central and northern Eurasia, northern North America
Echinococcus shiquicus	Tibetan fox	Pika	Tibetan Plateau
Echinococcus ortleppi	Domestic dog	Cattle	Eurasia, Africa
Echinococcus canadensis	Domestic dog, wolves	Pig, camel, cervids	Eurasia, Africa, South America

Hosts and geographic distribution are indicative. Only those hosts thought to be major contributors to transmission cycles are included. Continental range is shown, but distribution within those continents may be patchy.
[a]Usually referred to as *E. oligarthrus*. Originally described as *Taenia oligarthra*. The spelling of the specific epithet appears to have been changed, incorrectly, from *oligarthra* to *oligarthrus* by Cameron (1926); see Hüttner and Romig (2009) for details. By the Principle of Priority, *E. oligarthra* is the correct name.
[b]This is *Echinococcus granulosus* sensu stricto (s.s.); the name *E. granulosus* sensu lato (s.l.) is often used to refer to those taxa formerly included in the species, i.e. *E. granulosus* s.s., *E. felidis*, *E. equinus*, *E. ortleppi* and *E. canadensis*. Throughout this review, the term *E. granulosus* always refers to *E. granulosus* s.s.
After Nakao, M., Lavikainen, A., Yanagida, T., Ito, A., 2013b. Phylogenetic systematics of the genus *Echinococcus* (Cestoda: Taeniidae). Int. J. Parasitol. 43, 1017–1029.

proposed that the remaining strains (G6, G7, G8 and G10) be afforded species status as *Echinococcus canadensis*. In addition to these taxonomic changes, Xiao et al. (2005) described a new species (*Echinococcus shiquicus*), closely related to *E. multilocularis*, from Tibet, and Hüttner et al. (2008) resurrected the species *Echinococcus felidis*, originally described by Ortlepp (1934, 1937) from South Africa. As a consequence of these revisions and new descriptions, nine valid species were recognized in the most recent taxonomic study of the genus *Echinococcus* (Table 1).

While substantial progress has been made in elucidating the species-level taxonomy of *Echinococcus*, the current species designations are unlikely to be the final word. Genetically differentiated populations within described species may at some stage warrant further taxonomic revision. For example, Lymbery et al. (2015) suggested that *E. canadensis* may consist of three separate species, based on differences in mtDNA sequences, morphology and life history of (apparently) sympatric populations. Substantial genetic diversity has been found among geographically separated populations of *E. multilocularis* (Nakao et al., 2009), *E. vogeli* (Santos et al., 2012) and *E. oligarthra* (Soares et al., 2013); at present there is no basis for taxonomic recognition of these differences, but that may change with further studies. There may also be undiscovered species, particularly in parts of the world where the parasite fauna has been less well studied than in Europe and North America. *Echinococcus shiquicus*, for example, was recently described from the Tibetan Plateau (Xiao et al., 2005). Prior to this study, the adult stage of *E. shiquicus* in Tibetan foxes (*Vulpes ferrilata*) had been regarded as a different morphological form of *E. multilocularis*, while the larval stage in the plateau pika (*Ochotona curzoniae*) was misidentified as *E. granulosus* (Xiao et al., 2006). In Africa, early taxonomists described a number of species of *Echinococcus* which were subsequently invalidated; recent studies, however, have highlighted that the genetic diversity of *Echinococcus* spp. in African wildlife is far from clear and it is possible that some of these invalidated taxa may need to be revisited (Romig et al., 2011; Wassermann et al., 2015).

Resolving controversies about species status and inferring the taxonomic rank of newly discovered variant populations requires a clear conception of what constitutes a species and what criteria can be validly used to delimit species. This has not always been the case with respect to the species-level taxonomy of *Echinococcus*. For example, Thompson et al. (1995) in initially proposing a taxonomic revision in the genus, favoured an evolutionary species concept; a species is a single lineage of ancestral descendant populations which maintains its identity from other such lineages and which has its own evolutionary tendencies and historical fate (Wiley, 1978). Most subsequent taxonomic studies, however, have not identified a species concept or proposed clear criteria for delimiting different species. In many cases, where there have been no obvious differences in morphology between putative species, a genetic yardstick has been used; species status has been recommended if taxa are as genetically different from each other as the difference between well-established species in the same or related genera (e.g., Thompson and McManus, 2002; Nakao et al., 2007; Hüttner et al., 2008). While a genetic yardstick approach may sometimes provide a useful guide to species status, it is essentially an appeal to authority, or at least to past practice. Like all general rules for historical inference it may provide a misleading picture of lineage separation, because of large stochastic variation in evolutionary processes over time and space (Knowles and Carstens, 2007).

Nakao et al. (2010a, 2013b), in the most recent reviews of the genus, proposed the application of a phylogenetic species concept, defining species as the smallest set whose members are descended from a common ancestor. In practice, however, this definition does not lead to workable criteria for delimiting species (i.e., for determining what is the 'smallest set'), and Nakao et al. (2010a, 2013b) variously used a genetic yardstick, the presence (or absence) of fixed morphological and ecological characters, and evidence of lack of gene flow to determine which set of intraspecific variants should be accorded species status.

While deciding on an appropriate species concept may seem to be an arcane academic exercise, in reality it has great practical value if we wish to have an accurate and stable alpha taxonomy. Accurate delimitation of species provides the foundation of our knowledge of parasite life history, distribution and disease processes (Hoberg, 2006). In addition, controversy and instability in species names can have important public health implications. For example, *E. canadensis*, but not *E. granulosus* s.s., is found in Canada (Moro and Schantz, 2009). Government health regulations, however, do not make the distinction between *E. granulosus* s.s. and *E. granulosus* s.l. Officially, Canada is considered endemic for *E. granulosus* and import requirements for dogs into Canada do not require treatment with an anthelmintic. Such treatment could prevent the introduction of *E. granulosus* s.s., a species of far more economic and public health significance than *E. canadensis* (Lymbery et al., 2015).

3. SPECIES CONCEPTS AND SPECIES DELIMITATION

An accurate and stable alpha taxonomy requires agreement about what the term species actually means; without this there can be no objective way of deciding whether one particular proposal for species-level nomenclature is any more valid than another proposal. For such a fundamental unit of biological organization, there has been a surprising amount of debate as to what constitutes a species, with at least 24 different species concepts having been proposed (Mayden, 1997). Most of these concepts, however, differ in their criteria for delimiting species rather than in their theoretical understanding of how species exist. The basic theoretical view of a species in almost all concepts is that species represent independently evolving lineages (de Queiroz, 1999, 2007; Hey, 2006). This general lineage concept of species is essentially an updated version of the evolutionary species concept and I will use the operational definition suggested by Lymbery and Thompson

(2012): a species is a single lineage of organisms with a common evolutionary trajectory, distinguishable from other such lineages.

The evolutionary species concept is applicable to all organisms, regardless of their mode of reproduction or breeding system. Because there are two parts to the concept (single lineage and common evolutionary trajectory), delimiting evolutionary species requires two steps. The first step is phylogenetic reconstruction to determine the pattern of evolutionary relationships among lineages. Putative species identified through phylogenetic reconstruction should be both monophyletic (sharing a common ancestor) and exclusive (more closely related to each other than to any organisms outside the group), as these two properties are not necessarily congruent in a reticulate (as opposed to a purely diverging) genealogy (de Queiroz and Donoghue, 1990; Velasco, 2009). While monophyly and exclusivity are necessary conditions for a group of organisms to have species status, they are not of themselves sufficient, because they do not address the processes responsible for maintaining a cohesive evolutionary trajectory within the group. The second step in the delimitation of evolutionary species is therefore to assess whether members of the group possess genetic exchangeability (where cohesion is maintained by gene flow among populations) or ecological exchangeability (where cohesion is maintained because populations occupy the same ecological niche and selective regime) (Templeton, 1989; Crandall et al., 2000).

In practice, the lack of exchangeability among groups is much more easily inferred when they are sympatric than when they are allopatric. Groups which maintain fixed genetic differences in sympatry can be confidently asserted to lack genetic exchangeability and therefore to be different species. In allopatry, however, lack of exchangeability is usually inferred from the extent of genetic, phenotypic or life history differences among groups, which is a much more subjective exercise. The distinction between sympatric and allopatric populations is not as clear in parasites as in free-living organisms, as parasite populations may be physically separated by different host associations even within the same geographic area (McCoy, 2003; Huyse et al., 2005). When considering the capacity for genetic exchangeability among parasite populations, therefore, a distinction should be made between occurrence in the same geographic region (which I will call *broad* sympatry) and occurrence in the same host associations in a region (*strict* sympatry).

Do the groups which have been conferred species status in the genus *Echinococcus* fit the criteria for evolutionary species? Answering this question requires us to consider both the phylogenetic pattern among groups in the genus and the evolutionary processes responsible for maintaining a cohesive evolutionary trajectory within each group.

4. PHYLOGENETIC PATTERN

The phylogeny of genetic variants within the genus *Echinococcus* has been reconstructed using both mtDNA sequences (e.g., Le et al., 2002; McManus et al., 2002; Obwaller et al., 2004; Lavikainen et al., 2003, 2006; Thompson et al., 2006; Nakao et al., 2007, 2013b,c; Hüttner et al., 2008; Moks et al., 2008) and nuclear DNA sequences (e.g., Lavikainen et al., 2003; Bart et al., 2006; Saarma et al., 2009; Knapp et al., 2011). Figs 1 and 2 show the best current estimates of these phylogenies, from studies using a number of different sequences and a complete (or almost complete) range of major genetic variants. There are a number of common features to these phylogenies; *E. felidis* and *E. granulosus* s.s. are clearly sister species, and *E. ortleppi* is closely related to the different genotypes of *E. canadensis*. However, there are also some major discrepancies. While the mtDNA phylogenies appear quite robust, there are differences between the mtDNA and nuclear DNA phylogenies and among the nuclear DNA phylogenies themselves. These differences principally concern the position of the South American species *E. vogeli* and *E. oligarthra* (basal or nonbasal), relationships of clades in the *E. granulosus* s.l. complex (monophyletic or paraphyletic), relationships between the different genotypes of *E. canadensis* (monophyletic or paraphyletic) and whether *E. multilocularis* and *E. shiquicus* are sister species. With respect to the basal position of *E. vogeli* and *E. oligarthra*, the paraphyletic nature of the *E. granulosus* s.l. complex and the sister species relationship between *E. multilocularis* and *E. shiquicus*, the nuclear DNA phylogenies of Knapp et al. (2011) are in agreement with the mtDNA phylogenies of Nakao et al. (2007, 2013b,c) and Hüttner et al. (2008), while the phylogeny of Saarma et al. (2009) suggests a different topology. With respect to *E. canadensis*, all phylogenies indicate a monophyletic origin, except for that based on exon sequences of nuclear DNA genes by Knapp et al. (2011), which suggests that the G6 genotype is more closely related to *E. ortleppi* than to the other genotypes of *E. canadensis*.

Clearly, further analyses are required, using more nuclear DNA sequences, to completely resolve the pattern of relationships among putative species within the genus. Nakao et al. (2013b) argued that broad agreement between the mtDNA phylogenies and the complete nuclear

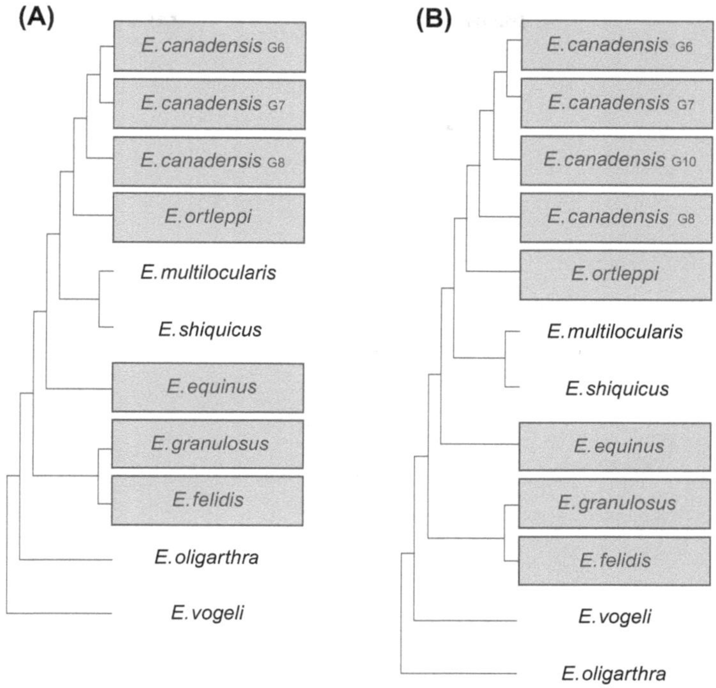

Figure 1 Cladograms of species of *Echinococcus* based on phylogenetic analyses of mtDNA sequences. (A) Analysis of concatenated sequences from 12 protein-coding, two rRNA and 22 tRNA genes (Nakao et al., 2007, 2013c). (B) Analysis of concatenated sequences from three protein-coding and one rRNA gene (Hüttner et al., 2008). *Shaded taxa* are those formerly placed in the species *Echinococcus granulosus* s.l.

phylogeny of Knapp et al. (2011) provided a solid foundation for delimiting species. From these phylogenies (Figs 1 and 2B), it appears that each of the nine currently described species forms a monophyletic and exclusive group. They therefore fulfil the first criterion to be recognized as separate evolutionary species. This is not evidence, however, that they fulfil the second criterion of being independent lineages, i.e., on separate evolutionary trajectories. To determine this, we need to consider the evolutionary processes operating within the groups.

5. EVOLUTIONARY PROCESSES

In contrast to the large number of studies which have aimed at reconstructing the phylogeny of species of *Echinococcus*, the study of population genetic structure has been relatively neglected. This is unfortunate, because

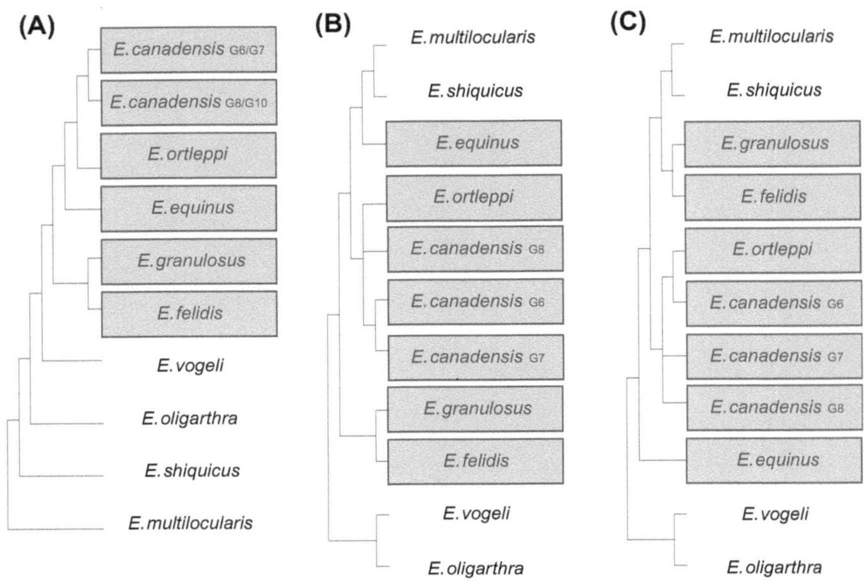

Figure 2 Cladograms of species of *Echinococcus* based on phylogenetic analyses of nuclear DNA sequences. (A) Analysis of concatenated sequences from five protein-coding genes (Saarma et al., 2009). (B) Analysis of concatenated sequences (including both introns and exons) from three protein-coding genes (Knapp et al., 2011). (C) Analysis of concatenated sequences (exons only) from three protein-coding genes (Knapp et al., 2011). *Shaded taxa* are those formerly placed in the species *Echinococcus granulosus* s.l.

analyzing the distribution of genetic variation within and among populations of a species can provide information on evolutionary processes such as gene flow, genetic drift and selection, and on the biological factors, such as mode of reproduction, breeding system, effective population size and dispersal ability, which influence these evolutionary processes.

It has often been suggested that the reproductive biology of species of *Echinococcus*, with a combination of self-fertilization and extensive asexual reproduction, has a profound effect on evolutionary processes, leading to the genetic uniformity of local populations and rapid genetic differentiation among populations subject to different selection pressures (Smyth and Smyth, 1964; McManus and Smyth, 1986; Bryant and Flockhart, 1986; Haag et al., 2008; Nakao et al., 2009, 2010a, 2013b). This is thought to occur because the population genetic consequences of obligate self-fertilization and asexual reproduction are almost complete homozygosity, extensive linkage disequilibrium (nonrandom association of alleles at different loci) and a distribution of genetic diversity between, rather than within family groups, which will lead to spatial structuring of genetic variation if dispersal is limited. The empirical population genetic data, although not extensive, suggest a much more complex picture than this.

5.1 Adult worms reproduce by both self-fertilization and cross-fertilization

Substantial deficiencies of heterozygous genotypes have been reported in populations of *E. granulosus* in Australia (Lymbery and Thompson, 1988; Lymbery et al., 1990, 1997), and *E. granulosus* and *E. ortleppi* in Brazil (Badaraco et al., 2008). These heterozygote deficiencies cannot be accounted for by unobserved population structuring (Wahlund effect), at least for the Australian samples (see Lymbery et al., 1997), and the most likely explanation is predominant self-fertilization. Cross-fertilization must also occur, however, because heterozygous genotypes have been found, not only in *E. granulosus* (Lymbery and Thompson, 1988; Lymbery et al., 1990, 1997; Badaraco et al., 2008) but also in *E. multilocularis* (Nakao et al., 2003; Knapp et al., 2007) and *E. vogeli* (Santos et al., 2012). Furthermore Haag et al. (1999, 2011) found no evidence of heterozygote deficiencies in populations of *E. granulosus* in Brazil in two separate studies.

It appears, then, that both cross-fertilization and self-fertilization occur within species of *Echinococcus*. In organisms with alternating sexual and asexual stages in their life cycle, self-fertilization can be achieved in two ways; ova may be fertilized by sperm of the same individual (autogamy) or by the sperm of clonally identical individuals (geitonogamy). Both self-insemination and cross-insemination have been observed microscopically in *Echinococcus* spp. (Smyth and Smyth, 1969; Kumaratilake et al., 1986; Wang, 1998). The interesting question is: What is the relative frequency of cross-fertilization (mating between genetically different individuals) versus self-fertilization through either autogamy or geitonogamy? At present, the genetic data are too few for any definitive answer; three studies in Australia and one in Brazil gave high rates of self-fertilization, while two studies in Brazil suggest cross-fertilization is the norm. Lymbery et al. (1997) argued that, while high rates of self-fertilization were found in Australian populations of *E. granulosus*, an absence of linkage disequilibrium indicated that this selfing rate is an accident of colonization (i.e., due to low genetic diversity in the founding population, leading to geitonogamy) rather than the natural breeding system of the species. A more recent study, however, has found that genetic diversity within Australian populations of *E. granulosus* is much greater than previously thought (Alvaraz Rojas et al., 2016), suggesting that this interpretation may need to be revised.

Cross-fertilization requires definitive hosts to be multiply infected with genetically different metacestodes and for adult worms derived from these metacestodes to be in contact in the small intestine. Multiple infections are likely to be common, since the distribution of infected definitive hosts is typically overdispersed (e.g., Lahmar et al., 2001; Ziadinov et al., 2010), and worms in infrapopulations are aggregated within the small intestine, which will increase contact rates (Lymbery et al., 1989). The important issue is whether definitive hosts are often concurrently infected with genetically different cysts. A number of studies have used mtDNA markers to demonstrate mixed infections of different species and strains of *Echinococcus* in individual dogs and wolves (Stefanic et al., 2004; Xiao et al., 2006; Zhang et al., 2006; Schurer et al., 2014). More pertinently, nuclear DNA markers have been used to find different genotypes of *E. multilocularis* in individual foxes and *E. granulosus* in individual dogs. Genetically different *E. multilocularis* worms were found in 38% of 13 foxes in Hokkaido, Japan (Nakao et al., 2003), 52% of 25 foxes in France (Knapp et al., 2008) and 35% of 125 foxes throughout Europe (Knapp et al., 2009), while genetically different *E. granulosus* worms were found in 50% of dogs in Argentina (Haag et al., 2011). These data suggest that it is not unusual for definitive hosts to be infected with genetically different worms of the same species.

In summary, the available evidence indicates that species of *Echinococcus* normally have a mixed mating system, with sexual reproduction by both self-fertilization and cross-fertilization. Mixed mating systems are common in both hermaphroditic plants (Goodwillie et al., 2005) and animals (Jarne and Auld, 2006) and present an evolutionary conundrum because theoretical studies predict the evolution of mating systems towards either pure selfing or pure outcrossing (Lande and Schmenske, 1985). The most common explanation for the maintenance of mixed mating systems is the reproductive assurance hypothesis; outcrossing is usually favoured by selection, but selfing can be favoured if mate availability varies, because selfing is better than not mating at all (Holsinger, 1996). For species of *Echinococcus*, the most parsimonious explanation of the available data is that adult worms normally reproduce by mating with other individuals, with high rates of self-fertilization due to geitonogamy rather than autogamy. Geitonogamy will be common where infection rates of definitive hosts are low, so that multiple infections are rare, or where genetic diversity has been reduced, for example, through recent colonization of an area by a small number of founders. Cross-fertilization will be common where infection rates and genetic diversity are high.

5.2 Genetic diversity within populations reflects time since colonization

Initial studies on *E. granulosus* in Australia found much lower levels of genetic diversity at isozyme loci than reported for other species of parasitic helminths or for free-living organisms, consistent with a founder effect from recent colonization (Lymbery and Thompson, 1988; Lymbery et al., 1990; Lymbery, 1995). Subsequently, more comprehensive analyses of mtDNA variation in *E. granulosus* from the Middle East, Europe, China, South Asia, Africa, South America and Australia found relatively low levels of nucleotide diversity, although greater levels of haplotype diversity, again suggesting rapid population expansion from small founder populations (Moro et al., 2009; Nakao et al., 2010b; Casulli et al., 2012; Yanagida et al., 2012; Sharma et al., 2013). Nakao et al. (2013b) suggested that a comparison of haplotype diversities supported a Middle Eastern origin for this species (or at least for the switch from a wildlife to a domestic host cycle), with subsequent spread to other parts of the world through anthropogenic transport of hosts. The results of Alvarez Rojas et al. (2016) indicate that the picture is a little more complicated than this. They found levels of mtDNA nucleotide and haplotype diversity in *E. granulosus* in Australia to be as great as those in the Middle East, perhaps throwing some doubt on the presumed Middle Eastern origin of the species and also indicating that there may have been multiple introductions of the parasite into Australia. Regardless of where the species originated, genetic diversity is likely to reflect both the time since colonization and the number of colonization events from different sources.

There have been far fewer studies on other species, but Knapp et al. (2009) found microsatellite allelic diversity in *E. multilocularis* in Europe to be greatest in historically endemic areas centred on Switzerland and lower in northern and eastern regions where the parasite has only been recently recorded. Nakao et al. (2010b) found low levels of mtDNA diversity in *E. multilocularis* in the Tibetan Plateau and suggested that the species was introduced to this region by a recent range extension of the red fox (*Vulpes vulpes*), although this hypothesis requires confirmation because there does not seem to be independent evidence of such a range extension (Kutschera et al., 2013). However, relatively high levels of mtDNA diversity have been found for *E. shiquicus* in the Tibetan Plateau (Nakao et al., 2010b) and *E. vogeli* in Brazil (Santos et al., 2012). Both *E. shiquicus* and *E. vogeli* are principally maintained in wildlife cycles and have presumably been present in their respective areas for the last 1.5–3.0 million years (based on the chronogram of Knapp et al., 2011).

In summary, the data suggest that genetic diversity is greatest in species that persist in (presumably ancestral) wildlife cycles, while in species that cycle (at least partially) in domesticated hosts, genetic diversity can vary widely and is positively related to the time since colonization of a region and the number of colonization events.

5.3 Gene flow prevents differentiation among populations in the absence of geographic or ecological barriers

The data on this point are rather sparse because not many studies have measured genetic variation at the appropriate scale, i.e., within and among local populations of a species. Those data that are available, however, indicate that most genetic variation occurs within, rather than among, geographically defined populations.

In *E. granulosus*, Lymbery et al. (1997) found only 2.4% of total genetic variance between populations separated by approximately 3500 km in mainland Australia; Haag et al. (1999) found effectively no variance among localities in Southern Brazil, Australia and Europe (Germany, Poland, Spain, Switzerland and Ireland); and Casulli et al. (2012) found 4% of variance among populations from different countries in eastern Europe (Bulgaria, Hungary and Romania). More substantial genetic structuring has been found in some cases. Lymbery et al. (1997) found 5.8% of genetic variance between populations of *E. granulosus* in mainland Australia and the island state of Tasmania, Casulli et al. (2012) found 18.7% of genetic variance between populations in Hungary and Italy, and Wang et al. (2014) reported 9.3% of variance among populations in different regions of south-western China. In a few cases, the majority of genetic variation has been found to be distributed among, rather than within populations. Haag et al. (2004) found 40.8% of genetic variance in *E. granulosus* from different geographical regions (north, central, south) of Argentina, although nested clade analysis indicated that the most likely cause of this differentiation was historical differences in the time and origin of livestock introductions (carrying different parasite genotypes), rather than restricted gene flow among regions. On a more local scale, Haag et al. (2011) compared genetic diversity within and among dogs from different farms in the same rural area of southern Brazil and found the majority of variance (61.9%) among farms.

There have been very few studies on the genetic structure of other species of *Echinococcus*. Phylogenetic analysis of mtDNA sequences identified geographically distinct clades in Europe, Asia and North America (Nakao et al., 2009). Within Europe, Knapp et al. (2009) measured genetic variation at a microsatellite locus in *E. multilocularis* from foxes in Switzerland, France, Germany, Austria, Czech Republic, Slovakia and Poland. Although they did not statistically partition genetic variation at different hierarchical levels, there was no evidence of geographic clustering of genotypes and no relationship between genetic and geographic distance among regions. In contrast, Santos et al. (2012) found substantial structuring (39% of total genetic variance) in nuclear and mitochondrial markers among populations of *E. vogeli* from eastern and western regions of the Brazilian Amazon, separated by approximately 2500 km.

In summary, despite the capacity for self-fertilization and asexual reproduction in species of *Echinococcus*, the available population genetic data suggest that the majority of genetic variation is usually found within, rather than among, local populations with the extent of spatial genetic structuring determined by host vagility.

6. ARE CURRENTLY DESCRIBED SPECIES OF *ECHINOCOCCUS* EVOLUTIONARILY INDEPENDENT?

6.1 *Echinococcus oligarthra* and *Echinococcus vogeli*

These taxa are basally placed in most phylogenetic trees and sister species in the nuclear gene phylogeny of Knapp et al. (2011). Both taxa occur throughout South and Central America, often in the same geographic locality (e.g., in Columbia; D'Alessandro and Rausch, 2008), where they maintain consistent differences in nuclear and mtDNA sequences, as well as in adult morphology and host occurrence; *E. oligarthra* using mainly wild felids as definitive hosts, with a wide intermediate host range, while *E. vogeli* is found principally in bush dogs (*Speothos venaticus*) and pacas (*Cuniculus paca*), although there is some overlap in intermediate host occurrence (D'Alessandro and Rausch, 2008; Nakao et al., 2013b). This is strong evidence for a lack of ecological exchangeability between the taxa, confirming their status as different evolutionary species. If *E. oligarthra* and *E. vogeli* are sister species and/or basally placed within the phylogeny of *Echinococcus*, then logically each must also be a different evolutionary species to all other taxa in the genus.

6.2 *Echinococcus granulosus* and *Echinococcus felidis*

Echinococcus granulosus and *E. felidis* are sister taxa on all mitochondrial and nuclear phylogenies. They occur in the same localities in southern and eastern Africa, with fixed differences in both mtDNA and nuclear DNA sequences, as well as consistent differences in adult morphology and definitive host occurrence; usually canids for *E. granulosus* and lion (*Panthera leo*) or spotted hyena (*Crocuta crocuta*) for *E. felidis* (Hüttner et al., 2008; Kagendo et al., 2014). Separation of host cycles is not complete, however, and eggs of both taxa have been found (using molecular markers) from a single lion faecal sample (Kagendo et al., 2014). This indicates a lack of both genetic and ecological exchangeability between the taxa, and therefore confirms their status as separate evolutionary species. As they are sister taxa, then they must also be evolutionarily independent of all other taxa in the genus. In a recent study, an isolate from a human host in Ethiopia was found to cluster with the *E. granulosus/E. felidis* clade in mtDNA phylogenies, but was genetically distinct from each (Wasserman et al., 2016). Much further work is required to determine whether this is a member of a new species, but the study highlights the incomplete nature of our understanding of genetic diversity within the genus.

6.3 *Echinococcus equinus*

In the phylogenetic tree of Saarma et al. (2009), *E. equinus* clusters with *E. ortleppi* and *E. canadensis*, but in all other phylogenies it forms a separate branch, with no closely related sister taxa. It is distinct genetically and morphologically from all other taxa, and has a strong intermediate host preference for equids (Thompson et al., 1995; Thompson and McManus, 2002). It is found in broad sympatry with *E. granulosus*, and probably also with *E. felidis*, *E. ortleppi*, *E. canadensis* and *E. multilocularis* (Thompson et al., 1995; Romig et al., 2015; Wassermann et al., 2015), although strict sympatry has not been demonstrated in all cases. Nevertheless the consistent morphological and ecological distinctiveness of *E. equinus* over a wide geographic area is evidence of a lack of ecological exchangeability with all other taxa, and therefore of separate species status.

6.4 *Echinococcus multilocularis* and *Echinococcus shiquicus*

In most (but not all) phylogenies, *E multilocularis* and *E. shiquicus* are sister taxa and positioned within the *E. granulosus* s.l. complex. The two taxa occur sympatrically in the Tibetan Plateau, with a partial, but not complete, separation of life cycles. Mixed infections of *E. multilocularis* and *E. shiquicus* have been found in the Tibetan fox (*Vulpes ferrilata*) (Jiang et al., 2012) and the plateau pika (Xiao et al., 2006). Importantly, differences in nuclear DNA sequences, as well as mtDNA sequences, have been found in strict sympatry, providing strong evidence for a lack of genetic exchangeability (Xiao et al., 2005). This is supported by small, but consistent differences in adult morphology and by different metacestode structures (alveolar cysts for *E. multilocularis* and unilocular cysts for *E. shiquicus*, which are maintained even in coinfections of the same intermediate host; Xiao et al., 2006). If *E. multilocularis* and *E. shiquicus* are different evolutionary species, and if they are sister taxa, as indicated in most phylogenetic trees, then they must also be separate evolutionary species from all other taxa in the genus.

6.5 *Echinococcus canadenis*

There is still some doubt about the monophyletic origin of *E. canadensis*. While phylogenies from mtDNA and some nuclear gene sequences show *E. ortleppi* as a sister group to the genotypes of *E. canadensis* (Nakao et al., 2007, 2013b; Saarma et al., 2009), other nuclear phylogenies are unable to resolve the relationship (Knapp et al., 2011). If the genotypes of *E. canadensis* are not monophyletic, they clearly cannot be regarded as a single evolutionary species, but if we accept the mtDNA phylogeny as a true reconstruction of evolutionary history, then they may be.

Nakao et al. (2007; 2013b,c) proposed the unification of the G6, G7, G10 and G8 genotypes into the species *E. canadensis*, while Lymbery et al. (2015) suggested that there were three different evolutionary lineages within this group; G6/G7 as one species, G10 as a second and G8 as a third. Both proposals are consistent with the phylogenetic pattern shown by analyses of mtDNA sequence data, as indeed would be a third proposal; to consider each genotype as a separate species. The challenge is to find evidence that the taxa either have or lack genetic or ecological exchangeability.

The G6 and G7 genotypes are largely allopatric, making it difficult to establish the presence or absence of genetic exchangeability (Lymbery et al., 2015). Although the metacestode of G6 is often found in camels, while that of G7 is often found in pigs, there is an overlap in life cycles, as well as similarity in morphological traits and close genetic similarity in mtDNA and nuclear DNA sequences (Nakao et al., 2007, 2013b,c; Lymbery et al., 2015; Romig et al., 2015). These data are all consistent with ecological exchangeability between the G6 and G7 genotypes (and hence conspecific status), although, as pointed out by Romig et al. (2015), there remains the possibility of biologically important variation between these groups.

The G6 genotype has been found in the same area in Far East Russia as the G10 (and G8) genotypes (Konyaev et al., 2013), although strict sympatry has not been demonstrated. In addition, the genotypes in this region have only been characterized at nonrecombining mtDNA loci, so a lack of genetic exchangeability cannot be inferred. Lack of ecological exchangeability is suggested by differences in intermediate hosts (principally domestic livestock for G6/G7 and cervids for G10; Thompson, 2008), although this separation is not complete, as G6 has been found in reindeer (Konyaev et al., 2013). Morphological differences between protoscoleces and adult worms of pig or camel origin and those of cervid origin (both G8 and G10 genotypes) are also suggestive of a lack of ecological exchangeability (Lymbery et al., 2015), although some morphological traits can be substantially influenced by intermediate host origin (Lymbery, 1998).

The G8 and G10 genotypes have been found in strict sympatry; not only in the same species of intermediate and definitive hosts in the same localities in northern Eurasia and North America (Thompson et al., 2006; Moks et al., 2008; Konyaev et al., 2013; Schurer et al., 2014), but adult worms of both genotypes have been found in the same individual wolf (Schurer et al., 2014). This is not necessarily an indication of a lack of genetic exchangeability, because these worms were only genotyped at mitochondrial (nonrecombining) loci, so it is not clear if gene flow among the genotypes is restricted in sympatry. It could be argued that morphological and ecological (life cycle) differences between these mitochondrial lineages indicate a lack of genetic exchangeability (Lymbery et al., 2015), but confirmation of this would require some knowledge of genetic and environmental effects on these traits.

At present, it is difficult to say whether *E. canadensis*, as currently described, represents a single evolutionary species or several species. The question of genetic exchangeability between the G8 and G10 genotypes could possibly be settled by a population genetic study (using nuclear markers) in sympatry, but strict sympatric occurrence is rare and, even if genetic differences were found at nuclear loci, it may be that the two mitochondrial lineages have only recently come into contact through natural or artificial host movements, leaving insufficient time to detect genetic signatures of introgression (Nakao et al., 2015). The G6, G7 and G10 genotypes are largely allopatric, so genetic exchangeability cannot be tested and arguments for species status rest either on a genetic yardstick or on similarities and differences in life cycles and morphology. These are necessarily subjective judgements. Lymbery et al. (2015) suggested that an integrative or iterative approach (see Padial et al., 2010; Yeates et al., 2011), using molecular, ecological and morphological data together, might provide more objectivity, but unfortunately we know very little about the heritability of morphological and ecological traits in *Echinococcus* spp. The best way to resolve this issue may be to use multilocus genotypic data in a coalescent-based approach to species delimitation (see Section 7).

6.6 *Echinococcus ortleppi*

Echinococcus ortleppi is clearly closely related to the genotypes of *E. canadensis*. If we assume that the genotypes of *E. canadensis* are monophyletic and can be regarded as a single species, then we need to consider whether *E. ortleppi* represents a lineage on a different evolutionary trajectory (of course, if the genotypes of *E. canadensis* are regarded as separate species, then they do not need to form a monophyletic group to the exclusion of *E. ortleppi*). *Echinococcus ortleppi* co-occurs with the G6 and G7 genotypes in Africa, South America and Europe (Thompson and McManus, 2002; Tigre et al., 2016), but it is not clear whether they are strictly sympatric and therefore whether genetic exchangeability can be tested. There are differences in morphology and host occurrence, but, as described for *E. canadensis*, the genetic basis of these differences has not been established. Nakao et al. (2007; 2013b,c) used a genetic yardstick to suggest that *E. ortleppi* and *E. canadensis* are separate species; the genetic distances at mtDNA loci between *E. ortleppi* and the genotypes of *E. canadensis* are similar to the distance between *Taenia saginata* and *Taenia asiatica*, the most closely related known sister species within the family. As previously discussed, while the genetic yardstick may be useful in some situations, it is an arbitrary and unreliable criterion on its own, and more evidence of a lack of ecological exchangeability is required to confidently assign *E. ortleppi* as a separate evolutionary species.

7. A COALESCENT-BASED APPROACH TO SPECIES DELIMITATION

The delimitation of evolutionary species is reasonably straightforward when taxa are sympatric or have well-defined genetic, morphological and ecological differences in allopatry. It is much more problematic, however, for cases such as *E. ortleppi* and the genotypes of *E. canadensis*, where lineage separation appears to be recent or incomplete, so that we cannot recognize fixed diagnostic states or reciprocal monophyly (concordance of all gene trees). In situations such as this, a coalescent-based approach might provide a more objective means of delimitation. Coalescent theory is a mathematical formulation of the genealogical history of neutral alleles in a population and, when considering a number of different populations, the multispecies coalescent can be used to describe a probability distribution for gene trees evolving along the branches of a species tree (Degnan and Rosenberg, 2009). Because coalescent approaches are probabilistic, they allow for gene tree discordance when testing alternative hypotheses of lineage separation (Fujita et al., 2012).

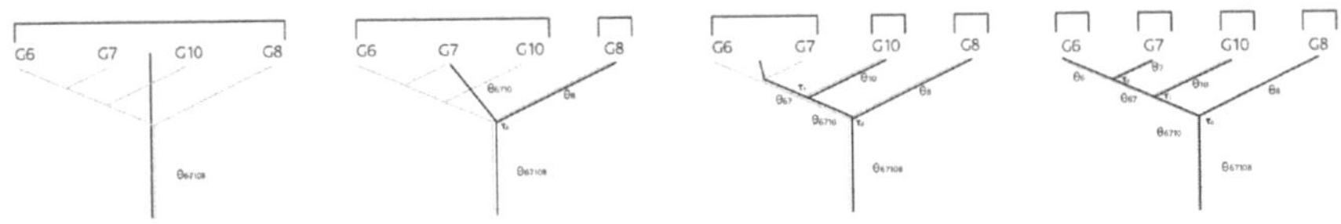

Figure 3 Example of alternative species trees for the genotypes of *Echinococcus canadensis*. The fully resolved phylogeny for all four genotypes, based on analyses of mtDNA by Nakao et al. (2013c) is shown in grey. The *black lines* indicate different species delimitation models that are compatible with the phylogeny, with associated population size (θ) and coalescence time (τ) parameters. The number of parameters varies from two, when all genotypes are regarded as a single species, to 10, when each genotype is regarded as a separate species.

Coalescent theory tells us that the time at which two alleles share a most recent common ancestor (coalescence time) is a function of the demographic history of the population (Emerson et al., 2001). Individuals that belong to the same species will therefore share demographic parameters such as population size (θ, the product of effective population size and mutation rate) and lineage divergence time (τ). This provides the basis for using a collection of gene trees to evaluate the probabilities of different species delimitation models, i.e., of determining which of several alternative species trees best explains the gene tree data (Rannala and Yang, 2003; Knowles and Carstens, 2007). For example, if we consider the genotypes of *E. canadensis*, there are four different species trees that are compatible with the mtDNA phylogeny and these species trees contain different numbers of population demographic parameters (Fig. 3). Bayesian or maximum likelihood approaches can be used to evaluate the fit of parameter estimates from gene trees to those expected from different species trees (Fujita et al., 2012).

Coalescent-based species delimitation requires a number of methodological considerations. First, multilocus data are needed on a number of different individuals of each putative species. Simulation studies suggest that 5–10 individuals, sequenced at each of 5–10 independent loci provide sufficient power for species delimitation (Zhang et al., 2011), although a larger number of individuals may need to be sampled for genetically diverse taxa. Sampling should also encompass the geographical range of the putative species and incorporate any existing information on diversity. Second, multispecies coalescent models assume neutrality, so the loci chosen for study should not be subject to strong natural selection (Degnan and Rosenberg, 2009). Empirical studies of species delimitation with coalescent-based methods have used both mitochondrial and nuclear loci, and coding and noncoding sequences (e.g., Camargo et al., 2012; Harrington and Near, 2012; Niemiller et al., 2012), but loci which are likely to be under active selection pressure may produce misleading signals of lineage divergence. Third, separate gene trees need to be constructed for each locus. Although it has been quite common to concatenate DNA sequences from different loci prior to phylogenetic analysis of species of *Echinococcus* (e.g., Nakao et al., 2007, 2013b,c; Saarma et al., 2009; Knapp et al., 2011; but see Hüttner et al., 2008), this results in a loss of information and will not necessarily yield a more reliable species tree. When gene trees from different loci are discordant, as is expected when species have recently diverged, then concatenation across loci can actually provide misleading inferences about lineage separation (Kubatko and Degnan, 2007). Finally, different analytical methods should be compared for species delimitation. A range of different software packages are available, using different approaches to evaluate the probability of alternative species trees (e.g., Pons et al., 2006; Ence and Carstens, 2010; O'Meara, 2010; Yang and Rannala, 2010).

8. BIOGEOGRAPHY AND SPECIATION

Acceptance of a particular species concept constrains our view of how speciation occurs. If species are regarded as lineages with separate evolutionary trajectories, then speciation must involve the evolution of traits which limit genetic or ecological exchangeability. There is abundant theoretical and empirical evidence that for the majority of free-living organisms, speciation usually occurs as the result of genetic drift or adaptive divergence between allopatric (geographically separated) populations (Turelli et al., 2001; Coyne and Orr, 2004). In parasites, however, there has been a long-held view that sympatric speciation, mediated by host switching leading to ecological isolation within the same geographic region, is relatively more common (Price, 1980; de Meeûs et al., 1998; Kunz, 2002; Huyse et al., 2005).

It has been suggested that the mode of reproduction in *Echinococcus* spp. (high levels of selfing in definitive hosts and asexual reproduction in intermediate hosts) might predispose to speciation through the rapid (essentially instantaneous) production of genetically different forms (Smyth and Smyth, 1964; Kumaratilake and Thompson, 1982; McManus and Smyth, 1986). This seems rather unlikely, given current knowledge of the breeding system and population genetic structure of species of *Echinococcus* (see Section 5.1), but the issue of speciation in the genus has rarely been addressed empirically. Speciation can almost never be directly observed, but must be inferred from the current pattern of evolutionary relationships among different species and from the evolutionary processes operating within and among contemporary populations. The sympatric occurrence of sister species provides an initial indication that speciation may have occurred by host switching. Although there is some disagreement among studies, *E. granulosus* and *E. felidis*, *E. multilocularis* and *E. shiquicus*, *E. canadensis* and *E. ortleppi*, and *E. vogeli* and *E. oligarthra* have all been identified as sister species in at least one phylogeny of the genus, and all of these pairs except *E. canadensis* and *E. ortleppi* have been shown to be broadly sympatric. Current distribution, however, does not necessarily reflect distribution during the time that speciation occurred, and a historical biogeographic perspective is necessary to infer the processes by which speciation has occurred.

For parasitic organisms, biogeographical patterns have to be interpreted within a coevolutionary context and a number of recent studies have attempted such an interpretation for species of *Echinococcus*. If the basal position of the neotropical species *E. oligarthra* and *E. vogeli* in most (but not all) phylogenies is taken as a starting point, then it has been argued that the ancestor of these species invaded South America in carnivorous (canid or felid) definitive hosts with the opening of the Panamanian land bridge around 3 million years ago (Knapp et al., 2011). Using this time for calibration of their nuclear exon phylogeny, Knapp et al. (2011) estimated that the *Echinococcus* clade began to diversify from other taeniids around 5.8 million years ago (Fig. 4). Nakao et al. (2013b) extrapolated from this to suggest that the ancestor of the *Echinococcus* clade may have originated in either North America, where modern canids evolved approximately 10 million years ago or in Asia where modern felids arose approximately 11 million years ago. Genetic diversity among species of *Echinococcus*, compared to that among other taeniids (Knapp et al., 2011), suggests that speciation and global radiation in the genus has occurred recently and rapidly, although that of course assumes constancy in the molecular clock across the different genera.

Geographic comparisons of genetic diversity have been used to suggest that *E. granulosus* evolved in the Middle East, coincident with the domestication of sheep about 10,000–12,000 years ago, and subsequently spread worldwide with livestock movements (Yanagida et al., 2012; Nakao et al., 2013b; Sharma et al., 2013). Huttner et al. (2008) suggested an Asian origin for *E. felidis*, as modern Felidae apparently arose in Asia, prior to spreading into Africa approximately 3 million years ago (Johnson et al.,

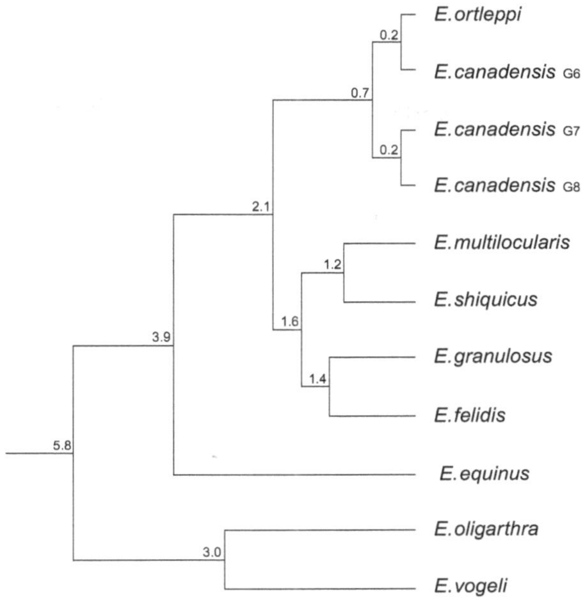

Figure 4 Chronogram of species of *Echinococcus*, reconstructed from an exon dataset of nuclear protein-coding genes (Knapp et al., 2011). Time estimates are based on the assumption that the most recent common ancestor of *Echinococcus vogeli* and *Echinococcus oligarthra* dates from the opening of the Panamanian land bridge between South and North America.

2006; although the demographic history of lions appears to have involved multiple movements between Africa and Asia, see Barnett et al., 2014). As *E. felidis* and *E. granulosus* are sister species, then presumably they shared a common ancestor in Asia prior to this time, and divergence of the lineages occurred in geographic isolation (although Wassermann et al. (2016) have suggested an African origin for *E. granulosus*, which would substantially alter this scenario).

Nakao et al. (2009) suggested an origin for *E. multilocularis* in the region of the land bridge between Asia and North America (Beringia) during the Pleistocene (2.6 million–11,700 years ago), with subsequent spread via foxes across North America, Asia and Europe. The North American, Asian and European clades are hypothesized to have then undergone cycles of isolation in glacial refugia and dispersal from these refugia during interglacial periods (Nakao et al., 2013b). In Europe, for example, Nakao et al. (2009) proposed an introduction of *E. multilocularis* in the late Pleistocene, with contraction and subsequent spread from glacial refugia, such as the Iberian, Italian and

Balkan peninsulas, initially to the genetically diverse Alpine arch region of Switzerland, southern Germany, eastern France and Austria (Knapp et al., 2009, 2015). *Echinococcus shiquicus* is thought to have differentiated from *E. multilocularis* in isolation, presumably during a period of range contraction, with the current sympatric occurrence a result of recent introduction of *E. mulilocularis* into the Tibetan Plateau with red foxes (Nakao et al., 2010b).

These biogeographic scenarios are all plausible with respect to the putative timing of species-splitting events (Fig. 4) and they can be related to the taxon pulse theory, whereby cyclical episodes of isolation and range expansion create opportunities for a complex interplay of cospeciation, host switching and coadaptation (Hoberg and Brooks, 2008, 2010). However, they are all essentially historical narratives, which have not yet been tested empirically. What is required to move these narratives into a more objective realm is to derive from them explicit predictions which can be tested against phylogeographic data. To avoid circularity, it is important that the data used to test the predictions are independent of those used to derive them (Crisp et al., 2011). Until we can get to this stage, hypotheses about the timing and mode of speciation in the genus *Echinococcus* will remain speculative.

9. THE PHENOTYPIC CONSEQUENCES OF SPECIATION

Under an evolutionary (or general lineage) species concept, lineages are recognized as different species when they are on different evolutionary pathways. We therefore expect species to diverge phenotypically, as a result of genetic drift or selective responses to the environment. These phenotypic differences may include traits of clinical or epidemiological importance, so the differentiation of species is of importance to the treatment and control of echinococcosis.

Traits of clinical or epidemiological importance, which may differ among species, include host range, propensity to infect people, cyst structure, cyst growth rate, adult size (related to fecundity) and prepatent period (Lymbery, 1995). The genetic architecture of these traits is not known, but it seems likely that they are multifactorial, with both a genetic component, influenced by a number of polygenes or quantitative trait loci, and an environmental component. In some cases, the genetic contribution to phenotypic differences among species appears quite obvious. The two major contributors to the burden of echinococcosis in humans are *E. granulosus* and *E. multilocularis*. These two species differ markedly in the structure of the metacestode. *Echinococcus granulosus* typically forms single-chambered, unilocular cysts, surrounded by a host-produced fibrous adventitial layer (Thompson, 1995). Growth occurs by concentric enlargement and proliferation of the germinal layer is entirely endogenous (within the laminated layer). The metacestode of *E. multilocularis*, on the other hand, consists of numerous small vesicles, with no limiting adventitial layer; exogenous proliferation of the germinal layer leads to infiltrating growth through host tissue (Thompson, 1995). Although some variation exists in each of these metacestode structures, the extent and consistency of the differences between the two species clearly indicates a substantial genetic component. The infiltrative growth form of *E. multilocularis* is different to that of any other described species of *Echinococcus*, including its sister species *E. shiquicus*, so presumably arose in this lineage.

For many traits of clinical or epidemiological importance, however, it is much more difficult to determine the extent to which differences among species are due to genetic, as opposed to environmental factors. For example, approximately 88% of all documented cases of cystic echinococcosis in people are caused by *E. granulosus*, with about 11% due to infection with the G6 and G7 genotypes of *E. canadensis* and less than 1% to *E. ortleppi* and the G8 and G10 genotypes of *E. canadensis*; there have been no recorded cases of human infection with *E. equinus* or *E. felidis* (Alvarez Rojas et al., 2014). This may be indicative of fundamental differences in the infectivity of these lineages to humans, but is also very likely to be influenced by differences in exposure, which will depend on a complex interplay of geographic range and host occurrence of the parasite, and social behaviour of people who may interact with these hosts.

Determining the relative importance of genetic and environmental factors to phenotypic differences among species or genotypes is challenging, because of the obvious difficulties with experimental approaches which would allow the same genotype to be expressed in different environments. Lymbery (1998) used a quantitative genetic analysis to estimate the contribution of intermediate host origin to variation of hook size and shape in protoscoleces, but this method is really only applicable to traits expressed by protoscoleces within the metacestode. Of much greater value would be the actual identification of the genes involved in traits of clinical or epidemiological importance, so they can be directly compared among species.

Some progress has been made to this end. Siles-Lucas et al. (2001), for example, found differences in the expression profile of the *14-3-3* gene, involved in abnormal cell proliferation, between *E. granulosus* and

E. multilocularis, and suggested that this may partially explain differences in metacestode growth behaviour. The recent publication of whole genome sequences for *E. granulosus* and *E. multilocularis* (Tsai et al., 2013; Zheng et al., 2013) enhances the potential of this approach. As yet, only limited interpretations have been made from the genomic data, but even these open up new avenues for research. For example, differences in regulation profiles among developmental stages of *E. granulosus* suggest candidate genes for regulating larval development and differentiation, and for strobilisation and reproduction in adults (Zheng et al., 2013); and sequence differences in the apomucin gene family between *E. granulosus* and *E. multilocularis* may indicate the genetic basis for differences in the thickness of the laminated layer in the metacestodes of these species (Tsai et al., 2013). The resources provided by these studies will undoubtedly be used much more in the future to identify candidate genes for traits of clinical and epidemiological importance, allowing us to evaluate the extent to which differences in these traits between species are genetically determined.

10. CONCLUSIONS

When I first reviewed this topic in 1995, there were only four recognized species in the genus, with a number of strains of uncertain taxonomic status. My conclusions at that time were that the molecular data which were starting to be collected were not consistent with the prevailing view of phylogeny within the genus, but it was not yet clear how many species existed and how they were related to each other. Thanks to a large number of careful molecular phylogenetic studies in the intervening 20 years, particularly those by Minoru Nakao and his colleagues, we now have a much more robust species-level taxonomy, with nine described species.

There are still many unanswered questions, however. The transmission patterns and zoonotic potential of a number of described species have not yet been fully elucidated. Phylogenies using different nuclear DNA sequences are not fully congruent, or congruent with phylogenies from mtDNA sequences. There is debate about the species status of some intraspecific variants, for example, the G6/7 genotypes in *E. canadensis*, and the recent description of *E. shiquicus* in Tibet suggests that more species may remain to be discovered. Many of these uncertainties, of course, can be resolved with more data, but there is also a need to have a clear idea of what constitutes a species and what criteria are appropriate for delimiting species in the genus. I argue that an evolutionary species concept, which regards a species as a single lineage of organisms with a common evolutionary trajectory, provides an appropriate conceptual framework.

Delimiting species under an evolutionary species concept requires consideration of both the pattern of evolutionary relationships among lineages and the processes responsible for maintaining a cohesive evolutionary trajectory. It is in an understanding of evolutionary processes, such as gene flow and ecological constraints, where we have most to learn. Population genetic studies in species of *Echinococcus* are few and far between, but those that are available indicate that most genetic variation occurs within, rather than between, local populations, with genetic structuring only apparent over larger geographic scales. These studies suggest that, despite asexual reproduction in intermediate hosts, which would tend to favour clumped transmission of clones and effective self-fertilization (through geitonogamy) in definitive hosts, gene flow among clonal groups may be extensive. This interpretation is supported by ecological data, which, while also rather sparse, indicate multiple infections of definitive hosts and an aggregated distribution of adult worms within the small intestine of definitive hosts. These factors would tend to promote cross-fertilization among clones and reduce the potential for geographic structuring of genotypes. Under this scenario, the absence of genetic or ecological exchangeability among clades provides good evidence that they are separate evolutionary species.

Accepting that there is still some uncertainty in the species-level phylogeny, seven of the nine currently described species of *Echinococcus* would appear to lack genetic or ecological exchangeability and therefore be on different evolutionary trajectories. I would be hesitant at this point to regard either *E. canadensis* or *E. ortleppi* as good evolutionary species. Further molecular phylogenetic analyses, comparing gene trees in a hypothesis-testing, coalescent framework, and supported by careful morphological and ecological studies, should resolve this issue.

From a practical perspective, the importance of an evolutionarily accurate alpha taxonomy is that it should provide a deeper understanding of differences among species in traits of medical or epidemiological importance, and a better predictive capability for the treatment and control of echinococcosis. In 1995, I concluded that the difficulty of applying the statistical techniques of quantitative genetics to partition genetic and environmental influences on traits of medical or epidemiological importance, meant that molecular techniques were needed to locate the genes involved. That seemed a long way off at the time. However, important progress has already been made in this field and the recent publication of whole genome maps for *E. granulosus* and *E. multilocularis* heralds exciting times ahead for research on this fascinating and important genus.

ACKNOWLEDGEMENT

Thanks to Andy Thompson for encouraging me to write the review and to him and Thomas Romig for comments on the manuscript. Jessica Morrison and Adam White produced the figures with great care and efficiency.

REFERENCES

Alvarez Rojas, C.A., Romig, T., Lightowlers, M.W., 2014. *Echinococcus granulosus* sensu lato genotypes infecting humans – review of current knowledge. Int. J. Parasitol. 44, 9–18.

Alvarez Rojas, C.A., Ebi, D., Gauci, C.G., Scheerlinck, J.P., Wassermann, M., Jenkins, D.J., Lightowlers, M.W., Romig, T., 2016. Microdiversity of *Echinococcus granulosus* in Australia, Parasitology 143, 1026–1033.

Badaraco, J.L., Ayala, F.J., Bart, J.M., Gottstein, B., Haag, K.L., 2008. Using mitochondrial and nuclear markers to evaluate the degree of genetic cohesion among *Echinococcus* populations. Exp. Parasitol. 119, 453–459.

Barnett, R., Yamaguchi, N., Shapiro, B., Ho, S.Y.W., Barnes, I., Sabin, R., Werdelin, L., Cuisin, J., Larson, G., 2014. Revealing the maternal demographic history of *Panthera leo* using ancient DNA and a spatially explicit genealogical analysis. BMC Evol. Biol. 14, 70.

Bart, J.M., Knapp, J., Gottstein, B., El-Garch, F., Giraudoux, P., Glowatzki, M.L., Berthoud, H., Maillard, S., Piarroux, R., 2006. EmsB, a tandem repeated multi-loci microsatellite, new tool to investigate the genetic diversity of *Echinococcus multilocularis*. Infect. Genet. Evol. 6, 390–400.

Bowles, J., McManus, D.P., 1993. Molecular variation in *Echinococcus*. Acta Trop. 53, 291–305.

Bowles, J., Blair, D., McManus, D.P., 1992. Genetic variants within the genus *Echinococcus* identified by mitochondrial DNA sequencing. Mol. Biochem. Parasitol. 54, 165–173.

Bowles, J., Blair, D., McManus, D.P., 1994. Molecular genetic characterization of the cervid strain ('northern form') of *Echinococcus granulosus*. Parasitology 109, 215–221.

Bowles, J., Blair, D., McManus, D.P., 1995. A molecular phylogeny of the genus *Echinococcus*. Parasitology 110, 317–328.

Bryant, C., Flockhart, H.A., 1986. Biochemical strain variation in parasitic helminths. Adv. Parasitol. 25, 275–319.

Camargo, A., Morando, M., Avila, L.J., Sites Jr., J.W., 2012. Species delimitation with ABC and other coalescent-based methods: a test of accuracy with simulations and an empirical example with lizards of the *Liolaemus darwinii* complex (Squamata: Liolaemidae). Evolution 66, 2834–2849.

Cameron, T.W.M., 1926. Observations on the genus *Echinococcus* Rudolphi, 1801. J. Helminthol. 4, 13–22.

Casulli, A., Interisano, M., Sreter, T., Chitimia, L., Kirkova, Z., La Rosa, G., Pozio, E., 2012. Genetic variability of *Echinococcus granulosus* sensu stricto in Europe inferred by mitochondrial DNA sequences. Infect. Genet. Evol. 12, 377–383.

Coyne, J.A., Orr, H.A., 2004. Speciation. Sinuer, Sunderland, MA.

Crandall, K.A., Bininda-Emonds, O.R.P., Mace, G.M., Wayne, R.K., 2000. Considering evolutionary processes in conservation biology. Trends Ecol. Evol. 15, 290–295.

Crisp, M.D., Trewick, S.A., Cook, L.G., 2011. Hypothesis testing in biogeography. Trends Ecol. Evol. 26, 66–72.
D'Alessandro, A., Rausch, R.L., 2008. New aspects of neotropical polycystic (*Echinococcus vogeli*) and unicystic (*Echinococcus oligarthrus*) echinococcosis. Clin. Microbiol. Rev. 21, 380–401.
de Meeûs, T., Michalakis, Y., Renaud, F., 1998. Santa Rosalia revisited: or why are there so many kinds of parasites in 'the garden of earthly delights'? Parasitol. Today 14, 10–13.
de Queiroz, K., Donoghue, M.J., 1990. Phylogenetic systematics or Nelson's version of cladistics? Cladistics 6, 61–75.
de Queiroz, K., 1999. The general lineage concept of species and the defining properties of the species category. In: Wilson, R.A. (Ed.), Species. MIT Press, pp. 49–89.
de Queiroz, K., 2007. Species concepts and species delimitation. Syst. Biol. 56, 879–886.
Degnan, J.H., Rosenberg, N.A., 2009. Gene tree discordance, phylogenetic inference and the multispecies coalescent. Trends Ecol. Evol. 24, 332–340.
Emerson, B.C., Paradis, E., Thebaud, C., 2001. Revealing the demographic histories of species using DNA sequences. Trends Ecol. Evol. 16, 707–716.
Ence, D., Carstens, B., 2010. SpedeSTEM: a rapid and accurate method for species delimitation. Mol. Ecol. Resour. 24, 473–480.
Fujita, M.K., Leache, A.D., Burbrink, F.T., McGuire, J.A., Moritz, C., 2012. Coalescent-based species delimitation in an integrative taxonomy. Trends Ecol. Evol. 27, 480–488.
Goodwillie, C., Kalisz, S., Eckert, C.G., 2005. The evolutionary enigma of mixed mating systems in plants: occurrence, theoretical explanations, and empirical evidence. Ann. Rev. Ecol. Evol. Syst. 47–79.
Haag, K.L., Araújo, A.M., Gottstein, B., Siles-Lucas, M., Thompson, R.C.A., Zaha, A., 1999. Breeding systems in *Echinococcus granulosus* (Cestoda; Taeniidae): selfing or outcrossing? Parasitology 118, 63–71.
Haag, K.L., Ayala, F.J., Kamenetzky, L., Gutierrez, A.M., Rosenzvit, M., 2004. Livestock trade history, geography, and parasite strains: the mitochondrial genetic structure of *Echinococcus granulosus* in Argentina. J. Parasitol. 90, 234–239.
Haag, K.L., Gottstein, B., Ayala, F.J., 2008. Taeniid history, natural selection and antigenic diversity: evolutionary theory meets helminthology. Trends Parasitol. 24, 96–102.
Haag, K.L., Marin, P.B., Graichen, D.A.S., De La Rue, M.L., 2011. Reappraising the theme of breeding systems in *Echinococcus*: is outcrossing a rare phenomenon? Parasitology 138, 298–302.
Harrington, R.C., Near, T.J., 2012. Phylogenetic and coalescent strategies of species delimitation in snubnose darters (Percidae: *Etheostoma*). Syst. Biol. 61, 63–79.
Hey, J., 2006. On the failure of modern species concepts. Trends Ecol. Evol. 21, 447–450.
Hoberg, E.P., Brooks, D.R., 2008. A macroevolutionary mosaic: episodic host-switching, geographical colonization and diversification in complex host-parasite systems. J. Biogeogr. 35, 1533–1550.
Hoberg, E.P., Brooks, D.R., 2010. Beyond vicariance: integrating taxon pulses, ecological fitting, and oscillation in evolution and historical biogeography. In: Morand, S., Krasnov, B.R. (Eds.), The Biogeography of Host-Parasite Interactions. Oxford University Press.
Hoberg, E.P., 2006. Phylogeny of *Taenia*: species definitions and origins of human parasites. Parasitol. Int. 55, S23–S30.
Holsinger, K.E., 1996. Pollination biology and the evolution of mating systems in flowering plants. Evol. Biol. 29, 107–149.
Hüttner, M., Romig, T., 2009. *Echinococcus* species in African wildlife. Parasitology 136, 1089–1095.
Hüttner, M., Nakao, M., Wassermann, T., Siefert, L., Boomker, J.D.F., Dinkel, A., Sako, Y., Mackenstedt, U., Romig, T., Ito, A., 2008. Genetic characterization and phylogenetic

position of *Echinococcus felidis* Ortlepp, 1937 (Cestoda: Taeniidae) from the African lion. Int. J. Parasitol. 38, 861–868.

Huyse, T., Poulin, R., Théron, A., 2005. Speciation in parasites: a population genetics approach. Trends Parasitol. 21, 469–475.

Jarne, P., Auld, J.R., 2006. Animals mix it up too: the distribution of self-fertilization among hermaphroditic animals. Evolution 60, 1816–1824.

Jiang, W., Liu, N., Zhang, G., Renqing, P., Xie, F., Li, T., Wang, Z., Wang, X., 2012. Specific detection of *Echinococcus* spp. from the Tibetan fox (*Vulpes ferrilata*) and the red fox (*V.vulpes*) using copro-DNA PCR analysis. Parasitol. Res. 111, 1531–1539.

Johnson, W.E., Eizirik, E., Pecon-Slattery, J., Murphy, W.J., Antunes, A., Teeling, E., O'Brien, S.J., 2006. The Late Miocene radiation of modern Felidae: a genetic assessment. Science 311, 73–77.

Kagendo, D., Magambo, J., Agola, E.L., Njenga, S.M., Zeyhle, E., Mulinge, E., Gitonga, P., Mbae, C., Muchiri, E., Wassermann, M., Kern, P., Romig, T., 2014. A survey for *Echinococcus* spp. of carnivores in six wildlife conservation areas in Kenya. Parasitol. Int. 63, 604–611.

Knapp, J., Bart, J.A., Glowatzki, M.L., Ito, A., Gerard, S., Maillard, S., Piarroux, R., Gottstein, B., 2007. Assessment of use of microsatellite polymorphism analysis for improving spatial distribution tracking of *Echinococcus multilocularis*. J. Clin. Microbiol. 45, 2943–2950.

Knapp, J., Guislain, M.H., Bart, J.M., Raoul, F., Gottstein, B., Giraudoux, P., Piarroux, R., 2008. Genetic diversity of *Echinococcus multilocularis* on a local scale. Infect. Genet. Evol. 8, 367–373.

Knapp, J., Bart, J.-M., Giraudoux, P., Glowatzki, M.-L., Breyer, I., Raoul, F., Deplazes, P., Duscher, G., Martinek, K., Dubinsky, P., Guislain, M.-H., Cliquet, F., Romig, T., Malczewski, A., Gottstein, B., Piarroux, R., 2009. Genetic diversity of the cestode *Echinococcus multilocularis* in red foxes at a continental scale in Europe. PLoS Negl. Trop. D 3, e452.

Knapp, J., Nakao, M., Yanagida, T., Okamoto, M., Saarma, U., Lavikainen, A., Ito, A., 2011. Phylogenetic relationships within *Echinococcus* and *Taenia* tapeworms (Cestoda: Taeniidae): an inference from nuclear protein-coding genes. Mol. Phylogenet. Evol. 61, 628–638.

Knapp, J., Gottstein, B., Saarma, U., Millon, L., 2015. Taxonomy, phylogeny and molecular epidemiology of *Echinococcus multiocularis*: from fundamental knowledge to health ecology. Vet. Parasitol. 213, 85–91.

Knowles, L.L., Carstens, B.C., 2007. Delimiting species without monophyletic gene trees. Syst. Biol. 56, 887–895.

Konyaev, S.V., Yanagida, T., Nakao, M., Ingovatova, G.M., Shoykhet, Y.N., Bondarev, A.Y., Odnokurtsev, V.A., Loskutova, K.S., Lukmanova, G.I., Dokuchaev, N.E., Spiridonov, S., Alshinecky, M.V., Sivkova, T.N., Andreyanov, O.N., Abramov, S.A., Krivopalov, A.V., Karpenko, S.V., Lopatina, N.V., Dupal, T.A., Sako, Y., Ito, A., 2013. Genetic diversity of *Echinococcus* spp. in Russia. Parasitology 140, 1637–1647.

Kubatko, L.S., Degnan, J.H., 2007. Inconsistency of phylogenetic estimates from concatenated data under coalescence. Syst. Biol. 56, 17–24.

Kumaratilake, L.M., Thompson, R.C.A., 1982. A review of the taxonomy and speciation of the genus *Echinococcus*. Rudolphi 1801. Z. Parasitenkd. 68, 121–145.

Kumaratilake, L.M., Thompson, R.C.A., Eckert, J., D'Alessandro, A., 1986. Sperm transfer in *Echinococcus* (Cestoda: Taeniidae). Z. Parasitenkd. 72, 265–269.

Kunz, W., 2002. Species concepts versus species criteria. Trends Parasitol. 18, 440.

Kutschera, V.E., Lecomte, N., Janke, A., Selva, N., Sokolov, A.A., Haun, T., Steyer, K., Nowak, C., Hailer, F., 2013. A range-wide synthesis and timeline for phylogeographic events in the red fox (*Vulpes vulpes*). BMC Evol. Biol. 13, 114.

Lahmar, S., Kilani, M., Torgerson, P.R., 2001. Frequency distributions of *Echinococcus granulosus* and other helminths in stray dogs in Tunisia. Ann. Trop. Med. Parasitol. 95, 69–76.

Lande, R., Schmenske, D.W., 1985. The evolution of self-fertilization and inbreeding depression in plants. I. Genetic models. Evolution 39, 24–40.

Lavikainen, A., Lehtinen, M.J., Meri, T., Hirvela-Koski, V., Meri, S., 2003. Molecular genetic characterization of the Fennoscandian cervid strain, a new genotypic group (G10) of *Echinococcus granulosus*. Parasitology 127, 207–215.

Lavikainen, A., Lehtinen, M.J., Laaksonen, S., Ågren, E., Oksanen, A., Meri, S., 2006. Molecular characterization of *Echinococcus* isolates of cervid origin from Finland and Sweden. Parasitology 133, 565–570.

Le, T.H., Pearson, M.S., Blair, D., Dai, N., Zhang, L.H., McManus, D.P., 2002. Complete mitochondrial genomes confirm the distinctiveness of the horse-dog and sheep-dog strains of *Echinococcus granulosus*. Parasitology 124, 97–112.

Lymbery, A.J., 1992. Inbreeding, monophyly and the genetic yardstick: species concepts in parasites. Parasitol. Today 8, 208–211.

Lymbery, A.J., 1995. Genetic diversity, genetic differentiation and speciation. In: Thompson, R.C.A., Lymbery, A.J. (Eds.), *Echinococcus* and Hydatid Disease. CAB International, Wallingford, U.K, pp. 51–87.

Lymbery, A.J., 1998. Combining data from morphological traits and genetic markers to determine transmission cycles in the tapeworm, *Echinococcus granulosus*. Parasitology 117, 185–192.

Lymbery, A.J., Thompson, R.C.A., 1988. Electrophoretic analysis of genetic variation in *Echinococcus granulosus* from domestic hosts in Australia. Int. J. Parasitol. 18, 803–811.

Lymbery, A.J., Thompson, R.C.A., 2012. The molecular epidemiology of parasite infections: tools and applications. Mol. Biochem. Parasitol. 181, 102–116.

Lymbery, A.J., Hobbs, R.P., Thompson, R.C.A., 1989. The dispersion of *Echinococcus granulosus* in the intestine of dogs. J. Parasitol. 75, 562–570.

Lymbery, A.J., Thompson, R.C.A., Hobbs, R.P., 1990. Genetic diversity and genetic differentiation in *Echinococcus granulosus* (Batsch, 1786) from domestic and sylvatic hosts on the mainland of Australia. Parasitology 101, 283–289.

Lymbery, A.J., Constantine, C.C., Thompson, R.C.A., 1997. Self-fertilization without genomic or population structuring in a parasitic tapeworm. Evolution 51, 289–294.

Lymbery, A.J., Jenkins, E.J., Schurer, J.M., Thompson, R.C.A., 2015. *Echinococcus canadensis, E. borealis,* and *E. intermedius*. What's in a name? Trends Parasitol. 31, 23–29.

Mayden, R.L., 1997. A hierarchy of species concepts: the denouement in the saga of the species problem. In: Claridge, M., Darwah, H.A., Wilson, M.R. (Eds.), Species: The Units of Biodiversity. Chapman & Hall, London, pp. 381–424.

McCoy, K.D., 2003. Sympatric speciation in parasites - what is sympatry? Trends Parasitol. 19, 400–404.

McManus, D.P., Smyth, J.D., 1986. Hydatidosis: changing concepts in epidemiology and speciation. Parasitol. Today 2, 163–168.

McManus, D.P., Zhang, L., Castrodale, L.J., Le, T.H., Pearson, M., Blair, D., 2002. Molecular genetic characterization of an unusually severe case of hydatid disease in Alaska caused by the cervid strain of *Echinococcus granulosus*. Am. J. Trop. Med. Hyg. 67, 296–298.

Moks, E., Jõgisalu, I., Valdmann, H., Saarma, U., 2008. First report of *Echinococcus granulosus* G8 in Eurasia and a reappraisal of the phylogenetic relationships of 'genotypes' G5-G10. Parasitology 135, 647–654.

Moro, P., Schantz, P.M., 2009. Echinococcosis: a review. Int. J. Infect. Dis. 13, 125–133.

Moro, P.L., Nakao, M., Ito, A., Schantz, P.M., Cavero, C., Cabrera, L., 2009. Molecular identification of *Echinococcus* isolates from Peru. Parasitol. Int. 58, 184–186.

Nakao, M., Sako, Y., Ito, A., 2003. Isolation of polymorphic microsatellite loci from the tapeworm *Echinococcus multilocularis*. Infect. Genet. Evol. 3, 159–163.
Nakao, M., McManus, D.P., Schantz, P.M., Craig, P.S., Ito, A., 2007. A molecular phylogeny of the genus *Echinococcus* inferred from complete mitochondrial genomes. Parasitology 134, 713–722.
Nakao, M., Xiao, N., Okamoto, M., Yanagida, T., Sako, Y., Ito, A., 2009. Geographic pattern of genetic variation in the fox tapeworm *Echinococcus multilocularis*. Parasitol. Int. 58, 384–389.
Nakao, M., Yanagida, T., Okamoto, M., Knapp, J., Nkouawa, A., Sako, Y., Ito, A., 2010a. State-of-the-art *Echinococcus* and *Taenia*: phylogenetic taxonomy of human-pathogenic tapeworms and its application to molecular diagnosis. Infect. Genet. Evol. 10, 444–452.
Nakao, M., Li, T., Han, X., Ma, X., Xiao, N., Qiu, J., Wang, H., Yanagida, T., Mamuti, W., Wen, H., Moro, P.L., Giraudoux, P., Craig, P.S., Ito, A., 2010b. Genetic polymorphisms of *Echinococcus* tapeworms in China as determined by mitochondrial and nuclear DNA sequences. Int. J. Parasitol. 40, 379–385.
Nakao, M., Lavikainen, A., Iwaki, T., Haukisalmi, V., Konyaev, S., Oku, Y., Okamoto, M., Ito, A., 2013a. Molecular phylogeny of the genus *Taenia* (Cestoda: Taeniidae): proposals for the resurrection of *Hydatigera* Lamarck, 1816 and the creation of the new genus *Versteria*. Int. J. Parasitol. 43, 427–437.
Nakao, M., Lavikainen, A., Yanagida, T., Ito, A., 2013b. Phylogenetic systematics of the genus *Echinococcus* (Cestoda: Taeniidae). Int. J. Parasitol. 43, 1017–1029.
Nakao, M., Yanagida, T., Konyaev, S., Lavikainen, A., Odnokurtsev, V.A., Zaikov, V.A., Ito, A., 2013c. Mitochondrial phylogeny of the genus *Echinococcus* (Cestoda: Taeniidae) with emphasis on relationships among *Echinococcus canadensis* genotypes. Parasitology 140, 1625–1636.
Nakao, M., Lavikainen, A., Hoberg, E., 2015. Is *Echinococcus intermedius* a valid species? Trends Parasitol. 8, 342–343.
Niemiller, M.L., Near, T.J., Fitzpatrick, B.M., 2012. Delimiting species using multilocus data: diagnosing cryptic diversity in the southern cavefish, *Typhlichthys subterraneus* (Teleostei: Amblyopsidae). Evolution 66, 846–866.
O'Meara, B.C., 2010. New heuristic methods for joint species delimitation and species tree inference. Syst. Biol. 59, 59–73.
Obwaller, A., Schneider, R., Walochnik, J., Gollackner, B., Deutz, A., Janitschke, K., Aspock, H., Auer, H., 2004. *Echinococcus granulosus* strain differentiation based on sequence heterogeneity in mitochondrial genes of cytochrome c oxidase-1and NADH dehydrogenase-1. Parasitology 128, 569–575.
Ortlepp, J.R., 1934. *Echinococcus* in dogs from Pretoria and vicinity. Onderstepoort J. Vet. Sci. Anim. Ind. 3, 97–108.
Ortlepp, R.J., 1937. South African helminths, part I. Onderstepoort J. Vet. Sci. Anim. Ind. 9, 311–336.
Padial, J.M., Miralles, A., De la Riva, I., Vences, M., 2010. The integrative future of taxonomy. Front. Zool. 7, 16.
Pons, J., Barraclough, T.G., Gomez-Zurita, J., Cardoso, A., Duran, D.P., Hazell, S., Kamoun, S., Sumlin, W.D., Vogler, A.P., 2006. Sequence based species delimitation for the DNA taxonomy of undescribed insects. Syst. Biol. 55, 595–609.
Price, P.W., 1980. Evolutionary Biology of Parasites. Princeton University Press, Princeton.
Rannala, B., Yang, Z., 2003. Bayes estimation of species divergence times and ancestral population sizes using DNA sequences from multiple loci. Genetics 164, 1645–1656.
Rausch, R.L., 1953. The taxonomic value and variability of certain structures in the cestode genus *Echinococcus* (Rud., 1801) and a review of recognised species. Thapar Commem. Vol. 233–246.

Rausch, R.L., Nelson, G.S., 1963. A review of the genus *Echinococcus* Rudolphi, 1801. Ann. Trop. Med. Parasitol. 57, 127–135.

Romig, T., Omer, R.A., Zeyhle, E., Huettner, M., Dinkel, A., Siefert, L., Elmahdi, I.E., Magambo, J., Ocaido, M., Menezes, C.N., Ahmed, M.E., Mbae, C., Grobusch, M.P., Kern, P., 2011. Echinococcosis in sub-Saharan Africa: emerging complexity. Vet. Parasitol. 181, 43–47.

Romig, T., Ebi, D., Wassermann, M., 2015. Taxonomy and molecular epidemiology of *Echinococcus granulosus* sensu lato. Vet. Parasitol. 213, 76–84.

Saarma, U., Jogisalu, I., Moks, E., Varcasia, A., Lavikainen, A., Oksanen, A., Simsek, S., Andresiuk, V., Denegri, G., Gonzalez, L.M., Ferrer, E., Garate, T., Rinaldi, L., Maravilla, P., 2009. A novel phylogeny for the genus *Echinococcus*, based on nuclear data, challenges relationships based on mitochondrial evidence. Parasitology 136, 317–328.

Santos, G.B., Soares, M.D.P., Brito, E.M.D., Rodrigues, A.L., Siqueira, N.G., Gomes-Gouvea, M.S., Alves, M.M., Carneiro, L.A., Malheiros, A.P., Povoa, M.M., Zaha, A., Haag, K.L., 2012. Mitochondrial and nuclear sequence polymorphisms reveal geographic structuring in Amazonian populations of *Echinococcus vogeli* (Cestoda: Taeniidae). Int. J. Parasitol. 42, 1115–1118.

Schantz, P.M., Colli, C., Cruz-Reyes, A., Prezioso, U., 1976. Sylvatic echinococcosis in Argentina. II. Susceptibility of wild carnivores to *Echinococcus granulosus* (Batsch, 1786) and host-induced morphological variation. Tropenmed. Parasitol. 27, 70–78.

Schurer, J.M., Gesy, K.M., Elkin, B.T., Jenkins, E.J., 2014. *Echinococcus multilocularis* and *Echinococcus canadensis* in wolves from western Canada. Parasitology 141, 159–163.

Sharma, M., Fomda, B.A., Mazta, S., Sehgal, R., Singh, B.B., Malla, N., 2013. Genetic diversity and population genetic structure analysis of *Echinococcus granulosus* sensu stricto complex based on mitochondrial DNA signature. PLoS One e82904.

Siles-Lucas, M., Nunes, C.P., Zaha, A., 2001. Comparative analysis of the 14-3-3 gene and its expression in *Echinococcus granulosus* and *Echinococcus multilocularis* metacestodes. Parasitology 122, 281–287.

Smyth, J.D., Smyth, M.M., 1964. Natural and experimental hosts of *Echinococcus granulosus* and *E. multilocularis*, with comments on the genetics of speciation the genus *Echinococcus*. Parasitology 54, 493–514.

Smyth, J.D., Smyth, M.M., 1969. Self insemination in *Echinococcus granulosus* in vivo. J. Helminthol. 43, 383–388.

Soares, M.C.P., dos Santos Rodrigues, A.L., Moreira Silva, C.A., de Figuerido Brito, E.M., Gomes-Gouvêa, M.S., dos Santos Corrêa, I.R., Pinho, J.R.R., Malheiros, A.P., Nunes, H.M., Pávoa, M.M., 2013. Anatomo-clinical and molecular description of liver neotropical echinococcosis caused by *Echinococcus oligarthrus* in human host. Acta Trop. 125, 110–114.

Stefanic, S., Shaikenov, B.S., Deplazes, P., Dinkel, A., Torgerson, P.R., Mathis, A., 2004. Polymerase chain reaction for detection of patent infections of *Echinococcus granulosus* ("sheep strain") in naturally infected dogs. Parasitol. Res. 92, 347–351.

Templeton, A.R., 1989. The meaning of species and speciation: a genetic perspective. In: Otte, D., Endler, J.A. (Eds.), Speciation and Its Consequences. Sinauer Associates, pp. 3–27.

Thompson, R.C.A., 1986. Biology and systematics of *Echinococcus*. In: Thompson, R.C.A. (Ed.), The Biology of *Echinococcus* and Hydatid Disease. Allen and Unwin, London, pp. 5–43.

Thompson, R.C.A., 1995. Biology and systematics of *Echinococcus*. In: Thompson, R.C.A., Lymbery, A.J. (Eds.), *Echinococcus* and Hydatid Disease. CAB International, Wallingford, UK, pp. 1–50.

Thompson, R.C.A., 2008. The taxonomy, phylogeny and transmission of *Echinococcus*. Exp. Parasitol. 119, 439–446.
Thompson, R.C.A., Lymbery, A.J., 1988. The nature, extent and significance of variation within the genus *Echinococcus*. Adv. Parasitol. 27, 209–258.
Thompson, R.C.A., McManus, D.P., 2002. Towards a taxonomic revision of the genus *Echinococcus*. Trends Parasitol. 18, 452–457.
Thompson, R.C.A., Lymbery, A.J., Constantine, C.C., 1995. Variation in *Echinococcus*: towards a taxonomic revision of the genus. Adv. Parasitol. 35, 145–176.
Thompson, R.C.A., Boxell, A.C., Ralston, B.J., Constantine, C.C., Hobbs, R.P., Shury, T., Olson, M.E., 2006. Molecular and morphological characterization of *Echinococcus* in cervids from North America. Parasitology 132, 439–447.
Tigre, W., Deresa, B., Haile, A., Gabriël, S., Victor, B., Van Pelt, J., Devleesschauwer, B., Vercruysse, J., Dorny, P., 2016. Molecular characterization of *Echinococcus granulosus* s.l. cysts from cattle, camels, goats and pigs in Ethiopia. Vet. Parasitol. 215, 17–21.
Torgerson, P.R., 2013. One world health: socioeconomic burden and parasitic disease control priorities. Vet. Parasitol. 195, 223–232.
Tsai, I.J., Zarowiecki, M., Holroyd, N., Garciarrubio, A., Sanchez-Flores, A., Brooks, K.L., Tracey, A., Bobes, R.J., Fragoso, G., Sciutto, E., Aslett, M., Beasley, H., Bennett, H.M., Cai, J.P., Camicia, F., Clark, R., Cucher, M., De Silva, N., Day, T.A., Deplazes, P., Estrada, K., Fernández, C., Holland, P.W.H., Hou, J.L., Hu, S.N., Huckvale, T., Hung, S.S., Kamenetzky, L., Keane, J.A., Kiss, F., Koziol, U., Lambert, O., Liu, K., Luo, X.N., Luo, Y.F., Macchiaroli, N., Nichol, S., Paps, J., Parkinson, J., Pouchkina-Stantcheva, N., Riddiford, N., Rosenzvit, M., Salinas, G., Wasmuth, J.D., Zamanian, M., Zheng, Y.D., The Taenia solium genome consortium, Cai, X.P., Soberón, X., Olson, P.D., Laclette, J.P., Brehm, K., Berriman, M., 2013. The genomes of four tapeworm species reveal adaptations to parasitism. Nature 496, 57–63.
Turelli, M., Barton, N.H., Coyne, J.A., 2001. Theory and speciation. Trends Ecol. Evol. 16, 330–342.
Velasco, J.D., 2009. When monophyly is not enough: exclusivity as the key to defining a phylogenetic species concept. Biol. Phil. 24, 473–486.
Vogel, H., 1957. Uber den *Echinococcus multilocularis* Suddeutschlands. I. Das Bandwurmstadium von Stammen menschlicher und tierischer Herkunft. Z. Tropenmed. Parasitology 8, 404–454.
Wang, J.H., Wang, N., Hu, D.D., Zhong, X.Q., Wang, S.X., Gu, X.B., Peng, X.R., Yang, G.Y., 2014. Genetic diversity of *Echinococcus granulosus* in southwest China determined by the mitochondrial NADH dehydrogenase subunit 2 gene. Sci. World J. 214, e867839.
Wang, H., 1998. A study on morphology of reproductive organs of *Echinococcus granulosus* by light microscopy, transmission and scanning electron. Endem. Dis. Bull. (China) 3, 31–33.
Wassermann, M., Aschenborn, O., Aschenborn, J., Mackenstedt, U., Romig, T., 2015. A sylvatic lifecycle of *Echinococcus equinus* in the Etosha National Park, Namibia. Int. J. Parasitol. Parasites Wildl. 4, 97–103.
Wiley, E.O., 1978. The evolutionary species concept reconsidered. Syst. Zool. 27, 17–26.
Wassermann, M., Woldeyes, D., Gerbi, B.M., Ebi, D., Zeyhle, E., Mackenstedt, U., Petros, B., Tilahun, G., Kern, P., Romig, T., 2016. A novel zoonotic genotype related to *Echinococcus granulosus* sensu stricto from southern Ethiopia, *Int. J. Parasitol.* in press.
Xiao, N., Qiu, J.M., Nakao, M., Li, T.Y., Yang, W., Chen, X.W., Schantz, P.M., Craig, P.S., Ito, A., 2005. *Echinococcus shiquicus* n. sp., a taeniid cestode from Tibetan fox and plateau pika in China. Int. J. Parasitol. 35, 693–701.
Xiao, N., Nakao, M., Qiu, J., Budke, C.M., Giraudoux, P., Craig, P.S., Ito, A., 2006. Dual infection of animal hosts with different *Echinococcus* species in the eastern Qinhai-Tibet plateau region of China. Am. J. Trop. Med. Hyg. 75, 292–294.

Yanagida, T., Mohammadzadeh, T., Kamhawi, S., Nakao, M., Sadjjadi, S.M., Hijjawi, N., Abdel-Hafez, S.K., Sako, Y., Okamoto, M., Ito, A., 2012. Genetic polymorphisms of *Echinococcus granulosus* sensu stricto in the Middle East. Parasitol. Int. 61, 599–603.

Yang, Z., Rannala, B., 2010. Bayesian species delimitation using multilocus sequence data. Proc. Natl. Acad. Sci. U.S.A. 107, 9264–9269.

Yeates, D.K., Seago, A., Nelson, L., Cameron, S.L., Joseph, L., Trueman, J.W.H., 2011. Integrative taxonomy, or iterative taxonomy? Syst. Entomol. 36, 209–217.

Zhang, Y.L., Bart, J.M., Giraudoux, P., Craig, P., Vuitton, D., Wen, H., 2006. Morphological and molecular characteristics of *Echinococcus multilocularis* and *Echinococcus granulosus* mixed infection in a dog from Xinjiang, China. Vet. Parasitol. 148, 384.

Zhang, C., Zhang, D.X., Zhu, T.Q., Yang, Z.H., 2011. Evaluation of a Bayesian coalescent method of species delimitation. Syst. Biol. 60, 747–761.

Zheng, H., Zhang, W., Zhang, L., Zhang, Z., Li, J., Lu, G., Zhu, Y., Wang, Y., Huang, Y., Liu, J., Kang, H., Chen, J., Wang, L., Chen, A., Yu, S., Gao, Z., Jin, L., Gu, W., Wang, Z., Zhao, L., Shi, B., Wen, H., Lin, R., Jones, M.K., Brejova, B., Vinar, T., Zhao, G., McManus, D.P., Chen, Z., Zhou, Y., Wang, S., 2013. The genome of the hydatid tapeworm *Echinococcus granulosus*. Nat. Genet. 45, 1168–1177.

Ziadinov, I., Deplazes, P., Mathis, A., Mutunova, B., Abdykerimov, K., Nurgaziev, R., Torgerson, P.R., 2010. Frequency distribution of *Echinococcus multilocularis* and other helminths of foxes in Kyrgyzstan. Vet. Parasitol. 171, 286–292.

CHAPTER FOUR

Echinococcus—Host Interactions at Cellular and Molecular Levels

K. Brehm*[,1], **U. Koziol***[,§,1]
*University of Würzburg, Würzburg, Germany
[§]Universidad de la República, Montevideo, Uruguay
[1]Corresponding authors: E-mail: kbrehm@hygiene.uni-wuerzburg.de; ukoziol@gmail.com

Contents

Abstract

The potentially lethal zoonotic diseases alveolar and cystic echinococcosis are caused by the metacestode larval stages of the tapeworms *Echinococcus multilocularis* and *Echinococcus granulosus*, respectively. In both cases, metacestode growth and proliferation occurs within the inner organs of mammalian hosts, which is associated with complex molecular host—parasite interactions that regulate nutrient uptake by the parasite as well as metacestode persistence and development. Using in vitro cultivation systems for parasite larvae, and informed by recently released, comprehensive genome

Advances in Parasitology, Volume 95
ISSN 0065-308X
http://dx.doi.org/10.1016/bs.apar.2016.09.001

and transcriptome data for both parasites, these molecular host–parasite interactions have been subject to significant research during recent years. In this review, we discuss progress in this field, with emphasis on parasite development and proliferation. We review host–parasite interaction mechanisms that occur early during an infection, when the invading oncosphere stage undergoes a metamorphosis towards the metacestode, and outline the decisive role of parasite stem cells during this process. We also discuss special features of metacestode morphology, and how this parasite stage takes up nutrients from the host, utilizing newly evolved or expanded gene families. We comprehensively review mechanisms of host–parasite cross-communication via evolutionarily conserved signalling systems and how the parasite signalling systems might be exploited for the development of novel chemotherapeutics. Finally, we point to an urgent need for the development of functional genomic techniques in this parasite, which will be imperative for hypothesis-driven analyses into *Echinococcus* stem cell biology, developmental mechanisms and immunomodulatory activities, which are all highly relevant for the development of anti-infective measures.

1. INTRODUCTION

Alveolar (AE) and cystic echinococcosis (CE) are zoonotic diseases caused by the metacestode larval stages of the tapeworms *Echinococcus multilocularis* (fox-tapeworm) and *Echinococcus granulosus* (dog-tapeworm), respectively. In both cases, initial infection of the intermediate (mammalian) host occurs by oral uptake of infective eggs that contain the oncosphere larval stage. Upon hatching in the intestine of the host, the oncosphere penetrates the intestinal wall and gains access to the inner organs where it undergoes a metamorphosis towards the metacestode stage which grows as cystic structures either tumour-like, infiltratively (*E. multilocularis*) or as massive cysts (*E. granulosus*). In later stages of the disease, the protoscolex larval stage is produced within the cysts and represents the head structures of the future adult worm, which can develop as a strobilar tapeworm after the protoscoleces have been taken up by the definitive host (Eckert and Deplazes, 2004).

The massive expansion of metacestode tissue within the intermediate host is associated with significant pathologies that, in later stages of the infection, can lead to organ failure and death. Although cure of the disease is principally possible by surgical intervention, the majority of AE cases are detected too late so that complete removal of the parasite tissue is not possible (Brunetti et al., 2010). In these cases, chemotherapy is the only remaining option. However, the currently available anti-*Echinococcus* drugs are associated with significant adverse side effects and act as parasitostatic only (see later discussion), underscoring the urgent need to develop new anti-infectives against the disease. Principally the same problems also apply to CE, although in this case the metacestode is better accessible to surgery and chemotherapy (Brunetti et al., 2010). Worldwide, AE and CE are important diseases with an estimated burden of 600,000 and 1 million disability adjusted life years lost, respectively (Budke et al., 2006; Torgerson et al., 2010). Furthermore, due to the infection of domesticated animals as intermediate hosts, CE also causes massive losses in livestock industry (Cardona and Carmena, 2013). Since echinococcosis infections mostly occur in developing countries, the pharmaceutical industry shows little interest in the development of new therapeutics. AE and CE are, thus, also considered 'neglected (tropical) diseases' by the World Health Organization (Budke et al., 2006; Torgerson et al., 2010).

E. granulosus has a cosmopolitan distribution, with a domestic life cycle that alternates between dogs as definitive hosts, and a wide range of ungulates as intermediate hosts (Moro and Schantz, 2009; Thompson, 1986). Different *E. granulosus* 'strains' show distinct biological characteristics and are better adapted for the infection of different intermediate host species. The important biological differences between *E. granulosus* strains have been known for a long time, and in more recent years these strains have been unambiguously identified with specific genotypes (Alvarez Rojas et al., 2014; Bowles et al., 1992; Nakao et al., 2010; Thompson, 1986). It is currently accepted that *E. granulosus* is actually a species complex, and many of the identified genotypes correspond to independent species (Nakao et al., 2010; Romig et al., 2015). For the sake of simplicity, in this review we will refer to *E. granulosus* in its sensu lato (i.e., comprising all of the genotypes), but a detailed review of this subject is provided by A. Thompson (2017) (chapter: Biology and Systematics of *Echinococcus*). On the other hand, *E. multilocularis* is restricted to endemic regions in the Northern hemisphere, and has a sylvatic cycle in which several wild rodent species serve as intermediate hosts, and foxes represent the definitive hosts (Eckert and Deplazes, 2004; Thompson, 1986; Torgerson et al., 2010). Humans can also be an accidental intermediate host for *E. granulosus* and *E. multilocularis* (thus resulting in the development of CE and AE), although they are a dead end for the life cycle of the parasite. Other species of *Echinococcus* have been described and are currently recognized, including *Echinococcus vogeli* and *Echinococcus oligarthrus*, which may also infect humans (D'Alessandro and Rausch, 2008; Eckert and Deplazes, 2004). These two species are limited in distribution to the neotropical region in Central and South America and are of more limited medical relevance (D'Alessandro and Rausch, 2008;

Eckert and Deplazes, 2004; Moro and Schantz, 2009). Furthermore, very little is known about their basic biology, and they will not be covered in this review. The reader is referred to the comprehensive reviews of Thompson (2017) (chapter: Biology and Systematics of *Echinococcus*), Limbery (2017) (chapter: Phylogenetic Pattern, Evolutionary Processes and Species Delimitation in the Genus *Echinococcus*), Romig et al. (2017) (chapter: Ecology and Life Cycle Patterns of *Echinococcus* Species) and Deplazes et al. (2017) (chapter: Global Distribution of Alveolar and Cystic Echinococcosis) for an in-depth description of the complex taxonomy, ecology and epidemiology of *Echinococcus* species.

As infectious agents, parasitic helminths are insofar 'peculiar', as they share a significant proportion of common genetic heritage with their mammalian hosts, when compared to other pathogens such as bacteria, protozoans or fungi (Brehm et al., 2006). It is expected that this will have implications on the strategies that helminths employ to persist within the organs of the host and, as discussed in depth later, evolutionarily conserved mechanisms that regulate metazoan development do indeed appear to play an important role in helminth—host interactions. This is particularly relevant for the interaction between the host and *Echinococcus* larvae of the early infective and the chronic stages. In the present review we will therefore focus on the oncosphere and the metacestode stages. We will largely avoid discussing immunologically relevant interactions, since this will be covered by the accompanying review article provided by Gottstein et al. (2017) (chapter: Immunology of Alveolar and Cystic Echinococcosis (AE and CE)).

2. THE LARVAL STAGES OF *ECHINOCOCCUS*

2.1 Early interactions of *Echinococcus* larvae and the intermediate hosts

A detailed review of the life cycles of *Echinococcus* spp. is provided by Thompson (2017) (chapter: Biology and Systematics of *Echinococcus*), and we here give only a brief introduction with a focus on what is known about the interactions between the parasite and the host. Intermediate hosts of *Echinococcus* spp. become infected by oral exposure to eggs containing the first larval stage of tapeworms, the oncosphere. The oncosphere is a reduced organism containing few cells (only 54 cells in the oncosphere of *E. granulosus* (Swiderski, 1983)), which is highly specialized for the infection of the intermediate host. Once the oncosphere is accidentally ingested by the intermediate host, it is released from the egg and activated by the action of the digestive secretions of the stomach and intestine (Heath, 1971; Thompson, 1986). It then penetrates through the intestinal epithelium and lamina propia by means of its motile hooks and of the secretions of its penetration glands, resulting in the highly localized lysis of the intestinal epithelia (Heath, 1971). Once in the intestinal submucosa, the oncosphere reaches a blood capillary and is then passively transported by the circulation of the hepatic portal system of the host to the liver (Heath, 1971).

The larvae of *Echinococcus* spp. have a strong organ tropism towards the liver. In the case of *E. multilocularis*, almost all primary infections occur in this organ, whereas in the case of *E. granulosus*, although the great majority of primary infections happen in the liver, other organs (especially the lungs and brain) may also be affected (Eckert and Deplazes, 2004; Thompson, 1986). The reasons for this organ tropism are not completely understood, but it is generally considered that a major factor is the route of infection of the oncosphere via the hepatic portal system, leading directly into the liver. It is clear from experiments of parenteral administration of oncospheres of *E. granulosus* that these larvae can establish an effective infection in a variety of organs, even in the case of oncospheres that were not previously activated (Heath, 1971; Kumaratilake and Thompson, 1981; Thompson, 1986; Williams and Colli, 1970). Therefore most organs are able to provide sufficient conditions (with regards to nutrients, growth factors and immunological responses) for a successful infection under experimental conditions. It has been proposed from experimental infections that the relative size of the oncospheres of *E. granulosus* and the blood capillaries of the hosts may determine whether some of the oncospheres are transported by the intestinal lymphatic capillaries (lacteals) instead of by the portal system, and are thus able to reach other organs besides the liver (Heath, 1971). Although this is a likely factor influencing the relative distribution of larvae in different organs, it fails to explain the difference in tropism between different species of *Echinococcus* in the same host (e.g., *E. multilocularis* and *E. granulosus* in humans), suggesting that specific factors from the liver induce the oncosphere to stop its migration and/or begin post-oncospheral development (see later discussion).

Once in the liver, the oncosphere begins the metamorphosis into the next life stage, the metacestode. The oncosphere possesses a small number of undifferentiated stem cells which are believed to drive the metamorphosis of the larva (Freeman, 1973; Rybicka, 1966) (only 10 undifferentiated cells per oncosphere have been described in detailed ultrastructural reconstructions of *E. granulosus* (Swiderski, 1983)), and most differentiated cells are discarded during this developmental transition. This early phase of development is essential for the success of the infection, and it is at this point that the parasite is most susceptible to the immune response of the host (Thompson, 1986). Unfortunately, this stage is also very difficult to study from a technical point of view in vivo, and few studies have dealt with the early development and metamorphosis of *Echinococcus* larvae in vitro (Harris et al., 1989; Heath and Smyth, 1970; Holcman et al., 1994). Therefore little is known about the interactions between the host and the parasite during the early development of the metacestode. It was shown in classical studies that inoculation with oncospheral antigens induces a strong protection against taeniid cestodes (genera *Echinococcus* and *Taenia*) (Lightowlers et al., 2003). Twenty years ago, elegant experiments from Marshall Lightowlers et al. resulted in the discovery of the oncospheral protein Eg95 of *E. granulosus* (and its homolog Em95 from *E. multilocularis*), which in recombinant form offers a very high level of host protection against these parasites (Heath and Lawrence, 1996; Lightowlers et al., 2003, 1996). Interestingly, this is the only protein specifically known to be expressed in the secretory vesicles of the penetration glands of *E. granulosus*, and it appears to be secreted after activation of the oncosphere (since the distribution of the protein becomes widespread in the rest of the oncospheral tissues) (Jabbar et al., 2011). Clear homologs of Eg95 only occur in taeniid cestodes (including several paralogs in each species) (Lightowlers et al., 2003; Olson et al., 2012). In the case of homologs of Eg95 from the related *Taenia ovis*, the proteins actually reach the surface of the larvae during the middle phase of metamorphosis (Jabbar et al., 2010). This suggests that Eg95 and its homologs have important roles in the early phases of infection, and may be in direct contact with the host. Eg95 is a small secreted protein that is putatively attached to the plasmatic membrane by a glycosylphosphatidylinositol anchor, and most of the polypeptide chain actually comprises a region with similarities to Fibronectin III (FnIII) domains (Lightowlers et al., 2003). A recombinant version of the more distantly related protein TSA-18 of *Taenia saginata* (another host-protective antigen with a FnIII domain) showed activity as an adhesion molecule for mammalian cells in vitro, suggesting that adhesion mediated by this family of proteins might assist taeniid parasites in the invasion of the host (Bonay et al., 2002). Furthermore, evolutionary analysis of Eg95 homologs from different *Echinococcus* species has shown that amino acid residues within the FnIII domain have evolved under positive natural selection, suggesting that they may be of importance for the adaptation to different host species (Haag et al., 2009). However, no further information exists regarding their actual biological function.

2.2 The morphology and development of metacestodes of *Echinococcus* spp.

The metacestodes of *Echinococcus* spp. have unique developmental and morphological characteristics. In more typical cestodes, the metacestode is similar to a 'juvenile' tapeworm, containing the scolex with the attachment organs, but lacking segmentation. In the case of *Echinococcus* spp., the metacestodes consist of fluid-filled vesicles, in which the parasite cells form a thin proliferative layer (the germinal layer) covered by a syncytial tegument that secretes an acellular, carbohydrate-rich external layer (the laminated layer) (Diaz et al., 2011b; Lascano et al., 1975; Morseth, 1967; Sakamoto and Sugimura, 1969; Thompson, 1986). The tegumental syncytium is covered in truncated microtriches (specialized microvilli-like extensions that are typical for cestodes) that greatly increase the surface area (Lascano et al., 1975; Morseth, 1967; Sakamoto and Sugimura, 1969; Smyth and McManus, 1989). The germinal layer contains the nucleated cell bodies of the tegumental cells, which are connected to the superficial tegumental syncytium by thin cytoplasmic strands (Lascano et al., 1975; Morseth, 1967; Sakamoto and Sugimura, 1969, 1970). It also contains other cell types such as glycogen storage cells, muscle cells, nerve cells and undifferentiated stem cells (Koziol et al., 2013, 2014; Lascano et al., 1975; Morseth, 1967; Sakamoto and Sugimura, 1969, 1970). The remaining volume of the vesicles is filled with fluid (hydatid fluid), containing proteins and low-molecular-weight solutes. Within the germinal layer, thickenings appear that invaginate into the vesicle, resulting in the formation of brood capsules (Koziol et al., 2013; Thompson, 1986). In the brood capsules, a new budding process occurs that results in the generation of protoscoleces, the infective form for the definitive host. The protoscolex already resembles the anterior region of the adult form, and remains quiescent with the scolex invaginated within a small posterior body. The life cycle of *Echinococcus* spp. is finally closed when the definitive host ingests the protoscoleces, as it preys on an infected intermediate host. The protoscolex becomes activated and attaches to the intestine, finally developing into the adult stage (Eckert and Deplazes, 2004; Thompson, 1986; chapter: Biology and Systematics of *Echinococcus* by Thompson, 2017).

E. multilocularis and *E. granulosus* differ in the morphology and development of the metacestode stage (Eckert and Deplazes, 2004; Thompson, 1986; chapter: Biology and Systematics of *Echinococcus* by Thompson, 2017). In the case of *E. granulosus*, each oncosphere typically develops into a single vesicle (the hydatid cyst) covered by a massive laminated layer, that can reach several millimetres in thickness. This mode of development is denominated 'unilocular'. Each vesicle can grow to huge dimensions (exceeding 20 cm in diameter). The slow but continuous growth of hydatid cysts within the infected organs can lead to mechanical pressure and to pathological changes associated with compression or obstruction. In contrast, the *E. multilocularis* metacestode has a thin laminated layer and is only unilocular during the very early stages of development. After the first week of infection, new vesicles are generated by exogenous budding of the metacestode, which therefore develops as a multilocular labyrinth of interconnected vesicles (Ohbayashi, 1960; Rausch, 1954; Sakamoto and Sugimura, 1970). This process occurs continuously, and at later stages small protrusions of the metacestode tissue, devoid of laminated layer, have been described to emerge from the periphery of the metacestode mass (Eckert et al., 1983; Mehlhorn et al., 1983). These protrusions infiltrate the host tissues, resulting in the formation of new vesicles not only in the liver but also in neighbouring organs (Eckert et al., 1983; Mehlhorn et al., 1983). The metacestode tissue can even form metastases in distant organs during late stages of infection (Eckert and Deplazes, 2004; Eckert et al., 1983). It is thought that this occurs by infiltration of small vesicles or groups of parasite cells into the blood and lymph vessels, which are then distributed to other organs where they initiate the development of new metacestode tissue (Eckert et al., 1983). Most of the metacestode vesicles in late infections have already ceased to grow, and can even become necrotic in the centre of the metacestode tissue (Thompson, 1986). Only the tissue in the periphery is still active, and continues to grow and infiltrate the organs of the host (Eckert et al., 1983). Growth of *E. multilocularis* is very fast in the natural intermediate hosts (rodents): after a few months the metacestode tissues have taken most of the liver, mature protoscoleces are formed and the host either dies from the infection or is easily preyed on by the definitive hosts (Craig, 2003; Rausch, 1954). In contrast, growth in humans is aberrant, since it is much slower, and usually no protoscoleces are produced (Craig, 2003; Moro and Schantz, 2009). The abnormal growth of *E. multilocularis* in humans, combined with the fact that only a small percentage of humans who are seropositive to *E. multilocularis* antigens (i.e., who have been exposed to *E. multilocularis*) show established infections, suggests that *E. multilocularis* is not well adapted for utilizing humans as intermediate hosts and that many humans are resistant to the disease (Gottstein et al., 2015).

Unlike the case of early developmental stages, great progress has been achieved on the in vitro cultivation of *Echinococcus* metacestodes, greatly facilitating the study of host–parasite interactions through reductionist approaches. In all cases, the successful cultivation of *Echinococcus* spp. larvae requires complex media based on those used for mammalian cell cultures, including the presence of inactivated mammalian serum (Hemphill and Gottstein, 1995; Jura et al., 1996; Rausch and Jentoft, 1957; Spiliotis et al., 2008, 2004). This reflects the fact that, as will be described in more detail later, the biosynthetic capabilities of *Echinococcus* and other tapeworms are very limited, thus requiring the acquisition of many simple substrates from the host. Early classical studies from Smyth et al. (1966) focused on the cultivation of protoscoleces of *E. granulosus*, which under suitable conditions could either develop into the adult stage, or remarkably could also differentiate back into metacestode vesicles. This reveals the true in vivo developmental potential of protoscoleces, since they are able to develop into the adult form if ingested by the definitive host, or can undergo reverse development into vesicles when released into the internal medium of the intermediate host (as may occur accidentally from cyst breakage and protoscolex spillage, resulting in secondary echinococcosis) (Thompson, 1986).

In the case of *E. multilocularis*, because the metacestode stage is able to grow continuously by asexual formation of new vesicles, it can be maintained indefinitely in vivo by serial passage of metacestode tissue from one host to the next (Norman and Kagan, 1961; Spiliotis and Brehm, 2009). The metacestode vesicles can also be successfully cultured in vitro for long periods of time. The first methods developed for the robust culture of metacestode vesicles in vitro are the co-culture systems, in which metacestode vesicles are cultured in media optimized for mammalian cells in the presence of fetal calf serum and mammalian feeder cells (Brehm and Spiliotis, 2008; Hemphill and Gottstein, 1995; Jura et al., 1996; Rausch and Jentoft, 1957; Spiliotis et al., 2008). Optimal growth of the metacestode requires the presence of feeder cells, and different cell lines can serve this function, although the best cells identified so far are primary liver cells and hepatoma cell lines from rodents (Spiliotis et al., 2004). This suggests that specific factors secreted by the liver in vivo are of importance for the organ tropism of *Echinococcus* larvae (Spiliotis et al., 2004). By a series of experiments it was shown that these feeder cells were not only providing soluble factors required by the metacestode vesicles, but also eliminating toxic substances (likely reactive oxygen species) from the cell culture media, and that both processes were necessary for optimal metacestode growth (Brehm and Spiliotis, 2008; Spiliotis and Brehm, 2009; Spiliotis et al., 2004). Based on these experiments, an axenic culture system was developed in which the metacestode vesicles can grow in the absence of host cells, by culturing them in media preconditioned by feeder cells under microaerobic and reducing conditions (Brehm and Spiliotis, 2008; Spiliotis and Brehm, 2009; Spiliotis et al., 2008, 2004). Furthermore, primary cells can be harvested from these axenic vesicles, and under similar conditions these cells are able to completely regenerate new metacestode vesicles in vitro (Spiliotis et al., 2008). The primary cell regeneration system has been proposed as a model for studying the processes that occur during the early formation of new metacestode vesicles in vivo (Nono et al., 2014, 2012; Olson et al., 2012). In particular, the expression of the oncosphere-specific *em95* gene (and of other paralogs of the same gene family) is activated in these cells (Olson et al., 2012). The requirement for serum and soluble host factors in the culture media indicates that a molecular dialog occurs between the metacestode and the host, in which the host cells provide in vitro, and also likely in vivo, signals that promote and regulate the development of the metacestode. These signals must be present in addition to the basic nutrients that are part of the standard culture media.

3. THE INTERFACE BETWEEN *ECHINOCOCCUS* METACESTODES AND THEIR HOSTS

Because *Echinococcus* metacestodes grow in parenteral locations of the intermediate host, especially in the liver, there is a very intimate relationship between both organisms. During the early stages of the infection, the parasite is surrounded by an inflammatory response from the host that includes epithelioid cells, giant multinucleated cells and lymphocytes (Thompson, 1986; Vuitton and Gottstein, 2010; Zhang et al., 2003). In the case of *E. granulosus* infections in humans (and also in some cases in ungulates), the inflammatory response is resolved and the established metacestode is usually surrounded by a host-derived capsule of collagen fibres, called the adventitial layer (Golzari and Sokouti, 2014; Thompson, 1986). This capsule is formed by fibroblasts of the host, and includes small vascular spaces (Golzari and Sokouti, 2014). In the case of *E. multilocularis*, the mass of vesicles is also surrounded by a granulomatous infiltration, which may derive in advanced lesions into a dense fibrosis composed of bundles of collagen that destroys the normal structure of the liver (Guerret et al., 1998; Vuitton et al., 1986). This fibrosis contains abundant vascular structures that are suggestive of vascular neogenesis in the regions surrounding the parasite (Vuitton et al., 1986).

The parasite is in contact to the surrounding host tissues through the acellular laminated layer, which is produced by the tegumental syncytium of the germinal layer. This thick and elastic layer is mainly composed of mucins (glycoproteins with multiple sites of O-glycosylation) (Diaz et al., 2011b). The laminated layer is structurally arranged as an irregular mesh of fibrils made from these mucin chains. The glycans found in the mucins of the laminated layer of *Echinococcus* spp. are based on the same basic structural motifs (cores) as found in other animals, including the mammalian host, but are much richer in non-decorated cores, and also contain particular decorations that are not present in mammalian cells at all (Diaz et al., 2011b, 2009). The formation of the laminated layer begins during early metacestode development (less than 1 week after oncospheral activation for *E. granulosus*, and after approximately 2 weeks in the case of *E. multilocularis*) and coincides with a decrease in the susceptibility of the metacestode to the immune response of the host (Diaz et al., 2011a; Harris et al., 1989; Sakamoto and Sugimura, 1970). Although protoscoleces are not typically found in direct contact with intermediate host cells, in cases of secondary echinococcosis a new laminated layer is also formed during the differentiation of the protoscolex into a metacestode vesicle (Smyth et al., 1966), and this is also thought to be a key event for the successful establishment of the parasite in the intermediate host (Zhang et al., 2005). Therefore besides its structural function, the laminated layer serves as a barrier that prevents the direct contact of the cells of the immune system of the host and the surface of the parasite (Diaz et al., 2011a; Vuitton and Gottstein, 2010). In the case of *E. multilocularis*, it has been described that the buds that extend from the metacestode vesicles and infiltrate the surrounding host tissues are not covered by a laminated layer, so in this species there may always exist direct contact between parts of the tegument of the parasite and the cells of the host (Mehlhorn et al., 1983).

Particles of the laminated layer are shed during the growth of metacestode vesicles, and these particles also make contact with the cells of the immune system (Diaz et al., 2015b). Interestingly, among a wide panel of mammalian innate immune receptors, only the murine Kupffer cell receptor interacted with components of the laminated layer of *Echinococcus* spp. (specifically recognizing the carbohydrate motifs) (Hsu et al., 2013). The Kupffer cells are specialized macrophages of the liver, and because antigen priming in the liver parenchyma through Kupffer cells is tolerogenic, it has been proposed that the carbohydrates of the laminated layer have been evolutionarily optimized for ensuring a tolerogenic response to shed laminated layer particles by the host (Diaz et al., 2015b).

Importantly, the laminated layer does not function as a barrier for macromolecules in vivo (Ahn et al., 2015; Aziz et al., 2011; Coltorti and Varela-Diaz, 1975; Chemale et al., 2003; Diaz et al., 2011b; Hustead and Williams, 1977; Monteiro et al., 2010; Varela-Diaz and Coltorti, 1972; Varela-Diaz et al., 1974), as was clearly demonstrated by the penetration of labelled host proteins (including immunoglobulins) into the laminated layer of *E. granulosus* metacestodes in vitro (Coltorti and Varela-Diaz, 1974). In fact, some host proteins may even be retained in the laminated layer, as has been shown for the complement inhibitor factor H in *E. granulosus* (Ferreira et al., 2000). Factor H is present in the blood plasma of the host, and becomes accumulated in the laminated layer of *E. granulosus*, presumably by sequestration by the mucins of parasite (Diaz et al., 2011a). Because of this, the laminated layer is only a very poor activator of the complement system, in contrast to hydatid cyst fluid and protoscoleces, thus limiting the effect of the complement system on the survival of the parasite (Ferreira et al., 2000). Proteins from the parasite (besides the structural mucins) that are secreted by the germinal layer can also accumulate in the laminated layer (Diaz et al., 2011b; Stadelmann et al., 2010; Vuitton and Gottstein, 2010).

Therefore the main barrier that macromolecules of the host face from the parasite is the tegumental syncytium. Because of the syncytial nature of the tegument (Lascano et al., 1975; Morseth, 1967; Sakamoto and Sugimura, 1969, 1970), there is no possibility of paracellular transport, and solutes must be transported by transcellular pathways. Beneath the tegument, the cells of the germinal layer lack occluding cell junctions, so the syncytial tegument is the only barrier between the host fluids and the intercellular spaces of the parasite, including the hydatid fluid (Lascano et al., 1975).

The hydatid fluid of the vesicles accumulates low-molecular nutrients such as glucose and amino acids at concentrations similar or greater than those found in the serum of the host (Celik et al., 2001; Frayha and Haddad, 1980; Hurd, 1989; Sanchez Franco and Sanchez Acedo, 1971). Hence, the hydatid fluid can be thought of as a large reservoir of nutrients for the parasite. The hydatid fluid also contains proteins secreted by the germinal layer, and proteins from the host, especially plasma proteins such as serum albumin, lipoproteins, haemoglobin and immunoglobulins (Ahn et al., 2015; Aziz et al., 2011; Bernthaler et al., 2009; Coltorti and Varela-Diaz, 1972, 1975; Chemale et al., 2003; Hustead and Williams, 1977; Monteiro et al., 2010; Varela-Diaz and Coltorti, 1972; Varela-Diaz et al., 1974). Therefore proteins of the host are physically able to interact with the cells of the germinal layer and with the proteins of the hydatid fluid of the parasite. The presence of host proteins in the hydatid fluid of *E. granulosus* and in the bladder fluid of the related *Taenia* spp. has been known for a long time from the above-mentioned classical experiments. It has been confirmed more recently by several proteomic analyses and by the identification of particular host proteins by means of specific antibodies (Ahn et al., 2015; Aldridge et al., 2006; Aziz et al., 2011; Bernthaler et al., 2009; Chemale et al., 2003; Monteiro et al., 2010; Navarrete-Perea et al., 2014). However, the abundance of host proteins that has been described from hydatid fluid varies by several orders of magnitude, and importantly, it also seems to vary between different metacestodes even in the same study (Coltorti and Varela-Diaz, 1975; Hustead and Williams, 1977). The mechanism of transport of host proteins into the metacestode vesicles of *Echinococcus* is virtually unknown. In vitro studies have provided some evidence for the non-selective uptake of proteins by metacestodes of *Echinococcus* and *Taenia* from the medium into the fluids of the parasite (Hustead and Williams, 1977). There is more specific evidence of tegumental endocytosis as an uptake mechanism of macromolecules in *Taenia crassiceps* (Ambrosio et al., 1994), and also from classical studies in other tapeworm species (Threadgold and Dunn, 1983, 1984; Threadgold and Hopkins, 1981). However, classical studies supporting the existence of tegumental endocytosis by cestodes have been proposed to be invalidated by experimental artefacts (Conradt and Peters, 1989). Hence, the existence of endocytosis at the tegumental surface of tapeworms is still an unresolved issue requiring urgent investigation. It has even been proposed that in the case of *Echinococcus*, host macromolecules may penetrate in a stochastic fashion, from random events of leakage from tegumental damage and turnover (Coltorti and Varela-Diaz, 1975).

Likewise, there is evidence for the excretion or secretion of parasite proteins towards the host. So far, proteomic analysis of excretion/secretion products (ESPs) of *Echinococcus* spp. has focused on protoscoleces and adult worms (Virginio et al., 2012; Wang et al., 2015), but there is evidence from in vitro cultivation experiments for the secretion of specific parasite macromolecules into the laminated layer and into the surrounding medium (Bernthaler et al., 2009; Gottstein and Hemphill, 2008; Sako et al., 2011, 2007; Stadelmann et al., 2010). Indirect evidence of in vivo secretion of parasite macromolecules into the host medium comes from the immunogenic responses mounted by the hosts to parasite proteins that accumulate in the hydatid fluid, indicating a direct contact between the host cells and the parasite proteins (Diaz et al., 2015a; Silva-Alvarez et al., 2015). The mechanisms of protein transport from the parasite into the intermediate host are not known. They likely involve conventional secretory pathways (exocytosis) (Diaz et al., 2015a; Holcman et al., 1994), but there is also ultrastructural evidence for non-conventional secretion of parasite material (Gottstein and Hemphill, 2008; Lascano et al., 1975).

4. UPTAKE OF NUTRIENTS FROM THE HOST

Because they lack a digestive system, all tapeworms must transport nutrients from the host through the tegumental syncytium (Smyth and McManus, 1989). Tapeworms use glucose from the host as the main or only energy source (Smyth and McManus, 1989; Tielens and van Hellemond, 2006). Transport of glucose in tapeworms is an active process, which depends on extracellular sodium and therefore is likely to be mediated by a glucose/sodium symporter (Pappas, 1975; Smyth and McManus, 1989). No studies have been performed specifically in *Echinococcus* larvae, but it is likely that this is also the case for these species, and genomic and transcriptomic analyses identified glucose/sodium symporters with massive expression in the germinative layer (Tsai et al., 2013). In protoscoleces, glucose is mostly catabolized through fermentative pathways including lactic fermentation and malate dismutation (McManus and Smyth, 1978, 1982). However, the aerobic contribution to glucose catabolism is relatively high as compared with other tapeworms. To the best of our knowledge, no detailed metabolic studies have been performed so far in metacestode vesicles, but transcriptomic analyses indicate a similar scenario as in protoscoleces, with a high level of expression of genes involved in lactic fermentation and malate dismutation, combined with a strong expression of genes related to the tricarboxylic acid cycle and the mitochondrial electron transport chain (Parkinson et al., 2012; Tsai et al., 2013; Zheng et al., 2013). The relatively high concentration of oxygen reported in hydatid fluid (diffusing from the tissues of the host) is congruent with this metabolic scenario (McManus and Smyth, 1978, 1982).

Classical metabolic studies have shown that all parasitic flatworms (including tapeworms) have very limited biosynthetic capacities and must therefore acquire most amino acids, nucleosides and simple lipids from the host as building blocks to generate their own complex macromolecules (Tielens and van Hellemond, 2006). Analyses of the genomes of *Echinococcus* spp. and other related tapeworms have confirmed these classical studies by conclusively demonstrating the lack of key genes involved in de novo synthesis of lipids (such as fatty acids and sterols), purines and most amino acids (Tsai et al., 2013; Zheng et al., 2013). In the case of amino acids, tapeworms do not only lack many biosynthetic pathways that were ancestrally lost in all animals (for histidine, isoleucine, leucine, lysine, methionine, phenylalanine, threonine, tryptophan and valine) but have also lost the ability to synthesize other amino acids such as serine and proline.

Transport systems for amino acids, nucleosides, fatty acids and vitamins have been investigated in great depth in classical studies of model tapeworms, especially *Hymenolepis diminuta* (Pappas, 1975). These studies demonstrated multiple complex systems for the mediated transport of nutrients at the level of their transport activities. A few studies have also been performed in *E. granulosus* metacestode vesicles and protoscoleces, which showed the mediated transport of specific amino acids in vitro (Jeffs and Arme, 1987, 1988) as well as the passive uptake of cholesterol in vitro and in vivo (Bahr et al., 1979). More recently, the amino acid symporter EgAT1 has been described in *E. granulosus* (Camicia et al., 2008). The transcript and protein of this gene are strongly expressed in the tegument of the germinal layer and protoscoleces. Other similar paralogs of this transporter were also found in the genome of *Echinococcus* spp. (Tsai et al., 2013). Therefore it is likely that this family of transporters comprises the genes coding for some of the amino acid transport systems that were described in previous classical studies.

Much research has been conducted on the systems dedicated to the uptake of host lipids in *Echinococcus* and other tapeworms. Tapeworms do not exploit lipids as energy substrates, but use the lipids that they obtain from the host for biosynthetic purposes (Tielens and van Hellemond, 2006). Several lipid-binding proteins have been described in *Echinococcus* and these are among the most highly expressed genes in metacestode larvae (Franchini et al., 2015; Parkinson et al., 2012; Silva-Alvarez et al., 2015; Tsai et al., 2013). The massive energetic budget that the parasite invests into their expression suggests that these genes have essential functions in the uptake and trafficking of host lipids.

One of the most intensively studied proteins of *Echinococcus* spp. is Antigen B (AgB) (Silva-Alvarez et al., 2015). This secreted lipoprotein is massively accumulated in the hydatid fluid of *E. granulosus* and is a major diagnostic antigen for echinococcosis in humans. The native protein exists as a heterogeneous population of lipoprotein particles with average molecular weights of 120–230 kDa, as estimated by different methods

(Obal et al., 2012; Silva-Alvarez et al., 2015). The protein fraction of AgB is composed of multimers of proteins of approximately 8 kDa (AgB8) that are coded by a small family of clustered genes in the genome of *Echinococcus* spp. (Chemale et al., 2001; Olson et al., 2012). Recombinantly produced AgB8 subunits have the ability to bind lipids such as fatty acids in vitro (Chemale et al., 2005). Close homologs of AgB8 genes are found only in taeniid cestodes, but more distant homologs can also be found in other tapeworms, forming the so-called helminth lipid-binding protein (HLBP) family, which includes both secreted and cytoplasmic lipid-binding proteins (Franchini et al., 2015; Saghir et al., 2001; Silva-Alvarez et al., 2015). The complex lipid fraction of native AgB from natural infections has been characterized and shown to comprise up to approximately 50% (w/w) of the lipoprotein (Obal et al., 2012). The lipid fraction was composed of neutral and polar lipids, including fatty acids, phospholipids (especially phosphatidylcholine), triglycerides and sterols. Because the relative abundance of these lipids is very similar between hydatid fluid and purified AgB, it suggests that these are bound in a non-selective manner by AgB8 apolipoproteins.

The biological function of AgB in vivo is not clear. It has been proposed that it could be involved in the uptake of lipids from host lipoproteins and albumin, either in the hydatid fluid (where these host proteins can also be found, as previously discussed) or from the surrounding host tissues (Silva-Alvarez et al., 2015). It is not clear, however, if AgB is actually exported from the metacestode larvae to the host, or how it would be re-imported by the parasite. The strong antibody response mounted by the host against AgB in natural infections indicates that AgB must be released at some point at the host—parasite interface, but it is not known if this occurs continuously or sporadically, or likewise whether it is an active process or the result of hydatid fluid leakage from damaged vesicles (Silva-Alvarez et al., 2015). In intact in vitro cultivated *E. multilocularis* metacestodes, AgB was not detected among ESPs by immunoprecipitation, suggesting that it was not actively exported by the parasite (Bernthaler et al., 2009). Likewise, there is very limited evidence for the excretion or secretion of AgB by protoscoleces in vitro (Virginio et al., 2012). On the other hand, in the case of the *T. saginata* homolog TsM150 it has been shown that the protein is secreted by the parasite in vitro (Lee et al., 2007). Furthermore, TsM150 was localized not only in the parasite tissues but also in lipid droplets of the host granuloma wall that surrounds the parasite in vivo. Finally, in vitro experiments of transport of fluorescent fatty acid analogs indicate that

TsM150 is involved in the transport of fatty acids from the medium into the parasite cyst fluid, as transport is severely impaired by the addition of anti-TsM150 antibodies (Lee et al., 2007). Altogether, the evidence suggests that the *T. saginata* TsM150 protein is involved in the acquisition of lipids from the host by the parasite, perhaps by binding lipids in the exterior of the parasite followed by translocation of the lipoprotein back into the cyst fluid of the parasite. Other functions have also been proposed for AgB, including immune modulation of the host through the delivery of lipid mediators to immune cells (Silva-Alvarez et al., 2015).

Another family of highly expressed lipid-binding proteins are the fatty acid–binding proteins (FABPs) (Alvite and Esteves, 2012; Tsai et al., 2013). Unlike HLBPs, FABPs are intracellular proteins that are highly conserved in all animals, including mammals. FABPs are thought to play key roles in the import, storage and intracellular trafficking of fatty acids in animal cells. Tapeworm FABPs are strongly expressed in the larval tegument, indicating a role in the acquisition of lipids from the host (Alvite et al., 2008). The *E. granulosus* EgFABP1 protein has been characterized in detail biochemically, including its fatty acid binding characteristics and tridimensional structure (Alvite and Esteves, 2012). EgFABP1 has also been detected as an excretory/secretory product despite lacking a signal peptide, and may thus be secreted by non-conventional mechanisms (Aziz et al., 2011; Virginio et al., 2012).

Another protein that has been proposed to be involved in the uptake of lipids by *E. multilocularis* metacestodes, in this case more specifically in the uptake of cholesterol, is a highly conserved homolog of human apolipoprotein A-I–binding protein (AI-BP) denominated EmABP (Bernthaler et al., 2009). This protein contains a signal peptide and was shown to be secreted into the hydatid fluid and into the external medium in vitro, suggesting that it is also secreted to the surrounding host tissues in vivo. As is also the case of human AI-BP, EmABP of the parasite specifically interacts with human apoliprotein A-I (apoA-I), a major constituent of cholesterol-transporting high-density lipoproteins in plasma. Furthermore, apoA-I from the serum of the host was shown to be present in the hydatid fluid of *E. multilocularis* in vitro cultured metacestodes, as well as in the hydatid fluid of *E. granulosus* natural in vivo infections. Based on these results, it was proposed that EmABP may participate in the mechanisms of cholesterol uptake from the host, although the specific roles that it may play, or the mechanisms by which apoA-I and EmABP are transported across the germinal layer, are not yet known.

5. DIVERSE MOLECULAR SIGNALS FROM THE PARASITE TOWARDS THE HOST

Besides the direct contact of the host cells and the laminated layer, it is thought that specific proteins from the parasite are secreted towards the host and participate in the host—parasite interaction. The antigenic effects and the possible roles of *Echinococcus* secreted factors in the immune modulation of the host have been the subject of intense study, but are beyond the scope of this review and are comprehensively covered in this issue by Gottstein et al. (2017) (chapter: Immunology of Alveolar and Cystic Echinococcosis (AE and CE)).

As previously discussed, it has been shown that several proteins are found in the ESPs of in vitro cultured metacestodes, larvae and adults, although the mechanisms of secretion have not been characterized. Some of these proteins have N-terminal signal peptides suggesting that they are secreted by classical secretory pathways (Bernthaler et al., 2009; Sako et al., 2011, 2007; Virginio et al., 2012; Wang et al., 2015). These proteins include *E. multilocularis* proteases of the cathepsin-L and cathepsin-B families, which can degrade host proteins from the serum and extracellular matrix in vitro, and may therefore have a role in the destruction of surrounding tissues and in the uptake of nutrients.

However, a common finding in proteomic studies of hydatid fluid and ESPs in *Echinococcus* (as well as in other parasitic flatworms) is the abundant detection of proteins lacking conventional secretory signals. This is especially prevalent for highly expressed cytoplasmic proteins such as glycolytic enzymes, cytoskeletal proteins, lipid-binding proteins and many others (Aziz et al., 2011; Monteiro et al., 2010; Stadelmann et al., 2010; Virginio et al., 2012; Wang et al., 2015). It is possible that these proteins are released into ESPs from damaged parasite cells or from tegument shedding (Aziz et al., 2011), but it has also been suggested that they could be actively released by non-conventional secretion mechanisms (Lorenzatto et al., 2012). One possible mechanism that has been put forward is the release mediated by extracellular vesicles (exosomes and microvesicles) (Marcilla et al., 2014; Raposo and Stoorvogel, 2013). Extracellular vesicles have already been described as highly abundant in other parasitic flatworms (Marcilla et al., 2014) and in fact, ultrastructural observations of the tegument of *E. granulosus* and *E. multilocularis* metacestodes have already indicated the secretion of membranous structures from the tips of the microtriches into the laminated layer (Gottstein and Hemphill, 2008; Lascano et al., 1975).

It is not uncommon for highly conserved and constitutively expressed proteins to have several mechanistically distinct and independent functions, especially in the case of enzymes involved in sugar metabolism (Huberts and van der Klei, 2010). This phenomenon is denominated 'moonlighting', and has been proposed to explain the abundant presence of enzymes such as enolase and fructose-bisphosphate aldolase among *Echinococcus* ESPs (Lorenzatto et al., 2012). In most cases, no novel function has been characterized for these proteins in *Echinococcus*, although new functions have been found for some of their homologs in other helminths (Figueiredo et al., 2015). One notable secreted protein of *Echinococcus* is EmPGI, the *E. multilocularis* homolog of the glycolytic enzyme phosphoglucose isomerase (Stadelmann et al., 2010). Mammalian phosphoglucose isomerase has known moonlighting functions and can also act as a cytokine, growth factor and inducer of angiogenesis (Watanabe et al., 1996). The *E. multilocularis* EmPGI protein is abundantly found not only in the germinal layer but also in the laminated layer and hydatid fluid, where it is secreted by as yet unknown mechanisms (Stadelmann et al., 2010). Recombinant EmPGI was capable of stimulating the proliferation of *E. multilocularis* primary cells cultured in vitro, and it may act as a growth factor for the parasite. Additionally, it specifically stimulated the proliferation of mammalian endothelial cells but not of other cell types in vitro (Stadelmann et al., 2010). Thus it is possible that EmPGI induces the formation of novel blood vessels around the developing metacestode in vivo (Vuitton et al., 1986), favouring the acquisition of nutrients by the parasite.

6. EVOLUTIONARILY CONSERVED SIGNALLING PATHWAYS AND HOST–PARASITE CROSS-COMMUNICATION

As metazoan organisms, tapeworms and their hosts rely on complex cell–cell communication mechanisms to properly regulate development, homeostasis and physiological functions. Typically, these mechanisms employ peptidic or lipophilic hormones and cytokines that are released by one defined cell type and that act, either over short or long distance, on receptor molecules expressed by target cells or tissues to induce appropriate cellular responses. With the exception of several vertebrate-specific hormones and cytokines that mainly function in regulating adaptive immune responses, such as interleukins or interferons (Kaiser et al., 2004), the vast majority of signalling systems for animal cell–cell communication already evolved long before the Cambrian explosion and are, thus, evolutionarily conserved in deuterostomes and protostomes (Brehm, 2010a, 2010b; Pires-daSilva and Sommer, 2003). These systems include hormones and cytokines such as insulin, epidermal growth factor (EGF) or fibroblast growth factor (FGF) and their cognate, membrane-bound receptor tyrosine kinases (RTKs), the cytokines of the transforming growth factor-β (TGF-β)/ bone morphogenetic protein (BMP) – superfamily and their respective receptor serine/threonine kinases (STKs), lipophilic hormones and their cognate nuclear hormone receptors (NHRs), as well as ligands of the hedgehog (Hh)- and wingless-related (wnt)-cytokine families and respective transmembrane receptors of the patched (PTCH) and frizzled (FRZ) families (Brehm, 2010a, 2010b; Pires-daSilva and Sommer, 2003). Apart from the fact that non-vertebrates share many structurally related signalling systems with mammals, early studies on model systems such as *Drosophila melanogaster* and *Caenorhabditis elegans* already showed that, in principle, mammalian-derived cytokines can functionally activate non-vertebrate receptors of the corresponding signalling pathway and vice versa (Brehm, 2010b). Particularly in the case of helminth parasites which grow in close association with host tissue, such as tapeworm larvae, this has led to the theory of 'host–parasite cross-communication via evolutionarily conserved signalling molecules' (Brehm, 2010b; Vicogne et al., 2004). According to this theory, helminth parasites should, in principle, be able to sense host hormones and cytokines by the respective, evolutionarily conserved signalling systems, utilizing this information, e.g., for organ tropism or for regulating growth and development. Likewise, helminths could also release such hormones and cytokines and, by activating respective receptors on host cells, might modulate immune responses or other physiological functions of the host. During recent years, several evolutionarily conserved signalling pathways have been structurally and functionally analyzed in *Echinococcus* spp. and shall be discussed in the following paragraphs. Since the genomes of *E. multilocularis* and *E. granulosus* proved to be highly similar in gene content and homologies of coding sequences (Tsai et al., 2013; Zheng et al., 2013), most of what we present for the one species will also be applicable to the second species.

6.1 Insulin-signalling pathways in *Echinococcus*

Two members of the insulin receptor family of RTKs, EmIR1 and EmIR2, have been described in *E. multilocularis* and, in almost identical form, are also expressed by *E. granulosus* (Hemer et al., 2014; Konrad et al., 2003) (Table 1). Both receptors display a domain structure that is typical for insulin

Table 1 Membrane-associated *Echinococcus multilocularis* kinase receptors

Protein name	GeneDB accession[b]	Receptor class	Expression in[a] PC	MC	PS	Ad	References
EmIR1	EmuJ_000962900	Insulin receptor TK	+	+	++	+	Konrad et al. (2003)
EmIR2	EmuJ_000981300	Insulin receptor TK	+	+	+	(+)	Hemer et al. (2014)
EmER	EmuJ_000075800	EGF receptor TK	+	+	+	(+)	Spiliotis et al. (2003)
EmERb	EmuJ_000617300	EGF receptor TK	(+)	+	+	(+)	Tsai et al. (2013)
EmERc	EmuJ_000969600	EGF receptor TK	+	+	+	(+)	Tsai et al. (2013)
EmFR1	EmuJ_000833200	FGF receptor TK	+	+	+	+	Tsai et al. (2013)
EmFR2	EmuJ_000196200	FGF receptor TK	(+)	(+)	+	–	Tsai et al. (2013)
EmFR3	EmuJ_000893600	FGF receptor TK	+	(+)	+	(+)	Tsai et al. (2013)
EmTR1	EmuJ_000195400	TGF receptor type I	+	+	+	+	Zavala-Gongora et al. (2006)
EmTR2	EmuJ_000817800	TGF receptor type II	+	+	+	+	Tsai et al. (2013)
EmTR3	EmuJ_000800500	TGF receptor type I	++	+++	++	+	Tsai et al. (2013)
EmTR4	EmuJ_000758400	TGF receptor type I	+	+	++	+	Tsai et al. (2013)
EmVKR	EmuJ_000619800	Venus kinase receptor	+	+	+	(+)	Tsai et al. (2013)

Ad, pre-gravid adult worm; *EGF*, epidermal growth factor; *FGF*, fibroblast growth factor; *MC*, metacestode vesicles without brood capsules; *PC*, primary cells after 2 days of culture; *PS*, low pH/pepsin-activated protoscolex; *TGF*, transforming growth factor; *TK*, tyrosine kinase.

[a]Expression values in fpkm (fragments per kilobase of exon per million fragments mapped) according to Tsai et al. (2013). fpkm <1 depicted as '–', fpkm >1 to 10 as '(+)', fpkm >10 to 50 as '+', fpkm >50 to 100 as '++', fpkm >100 as '+++'.

[b]Accession number as assembled under http://www.genedb.org/Homepage/Emultilocularis.

receptors such as an extracellular ligand-binding domain (LBD) followed by three fibronectin 3 domains, a transmembrane region and an intracellular tyrosine kinase domain. By immunohistochemical analyses, EmIR1 has been found to be mainly located in glycogen-storing cells of the metacestode germinative layer, implying a function in regulating glucose homeostasis (similar to the mammalian insulin receptor). EmIR2, on the other hand, appears to be mainly expressed in proliferative tissue of the parasite larvae and might be specifically expressed in germinative (stem) cells and/or their direct offspring. Similar to the mammalian insulin-like growth factor receptor 1, EmIR2 might have a crucial function in regulating cellular proliferation and differentiation (Hemer et al., 2014).

Apart from insulin-receptor-like RTKs, *E. multilocularis* also expresses two insulin-like ligands, named EmILP1 and EmILP2 (insulin-like peptides 1 and 2) (Hemer et al., 2014; Wang et al., 2014), which, in principle, might act as the cognate ligands for EmIR1 and EmIR2. According to transcriptome data and reverse transcription polymerase chain reaction experiments, however, the parasite ILPs are predominantly expressed in the adult stage, whereas only slight (*emilp*2) or no (*emilp*1) transcription was detected for the metacestode stage or parasite primary cell cultures (Hemer et al., 2014). Furthermore, the *Hymenolepis microstoma* orthologs of both insulin-like peptides show maximal expression in the posterior strobila of the adult worm, and the *Taenia solium* ortholog of Em-ILP1 is specifically expressed in the ovary of the adult proglottids (Wang et al., 2014). The interaction between the parasite-derived insulin receptors and putative ligands was investigated using the yeast two-hybrid system and the LBDs of both parasite receptors were found to strongly interact with human insulin (Hemer et al., 2014; Konrad et al., 2003), indicating that the host hormone might indeed directly bind to the parasite RTKs. For the EmIR1 LBD, no interaction with any parasite ILP was found, whereas the EmIR2 LBD moderately interacted with both EmILP1 and EmILP2 in the yeast two-hybrid system (Hemer et al., 2014). Taking these results in combination with the *emilp* expression patterns, it thus appears that host insulin is the only cognate ligand for EmIR1 that is present in considerable amounts at the site of infection. Although EmIR2 might interact with both host insulin and the parasite ILPs, it is highly likely that host insulin is the major ligand during development within the intermediate host and that the ILPs largely act on EmIR2 for regulating adult development and homeostasis.

If EmIR1 and EmIR2 are able to functionally interact with host insulin at the site of infection, effects of exogenously added host hormone on parasite development would be expected. As demonstrated by Hemer et al. (2014), this is indeed the case. When these authors added physiologically relevant concentrations (10 nM) of human insulin to in vitro cultivated *E. multilocularis* metacestode vesicles, glucose uptake was significantly stimulated which most probably involved an activation of EmIR1 in glycogen-storing cells (Hemer et al., 2014). Furthermore, physiological concentrations of host insulin also stimulated the generation of metacestode vesicles from parasite primary cell cultures (harbouring a high concentration of parasite stem cells (Koziol et al., 2014)), induced the proliferation of germinative cells in mature metacestode vesicles and stimulated the 're-differentiation' of *E. multilocularis* protoscoleces towards the metacestode stage (Hemer et al., 2014). Although it cannot be excluded that these effects are mainly mediated by insulin stimulation of EmIR1, it is at least highly likely that particularly the effects of insulin on stem cell proliferation are the result of a direct stimulation of EmIR2.

In an additional set of experiments, Hemer et al. (2014) also investigated the effects of human insulin on parasite signalling mechanisms that act downstream of EmIR1 and EmIR2. One of the signalling pathways that is often induced in metazoans after stimulation of insulin receptors is the phosphatidyl-inositol-3-kinase (PI3K)/Akt-pathway (Fig. 1), and several

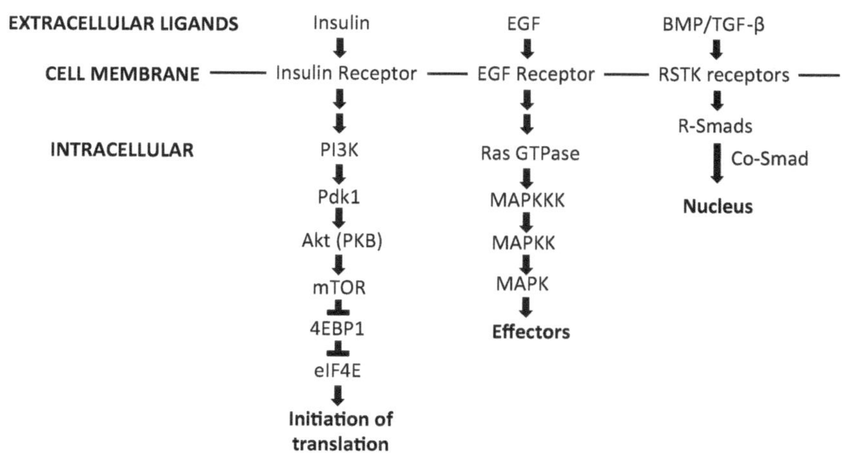

Figure 1 Simplified diagrams of selected conserved signalling pathways. (Left) Insulin/phosphatidyl-inositol-3-kinase/Akt signalling. (Middle) Epidermal growth factor/mitogen-activated protein kinase signalling. (Right) Bone morphogenetic protein/transforming growth factor-β/smad signalling. *Double arrows* indicate steps that are omitted in the diagram.

key components of this pathway such as a catalytic subunit of PI3K, an ortholog to 'mammalian target of rapamycin' (mTOR), a glycogen synthase kinase ortholog, as well as orthologs to protein kinase B (called Akt kinase) and the 'eukaryotic translation initiation factor 4E-binding protein' (4E-BP) are all expressed by the *E. multilocularis* metacestode stage (Hemer et al., 2014). Interestingly, exogenous addition of host insulin to parasite vesicles resulted in a significant phosphorylation, and thus activation, of Em4E-BP and several EmAkt substrates within 5 min, indicating that the parasite also employs downstream insulin signalling components that are responsive to host-derived insulin. Again, this stimulation appeared to mostly involve EmIR1 since in all systems investigated so far the 'bridging' molecules between insulin receptor RTKs and the PI3K/Akt-pathway are so-called insulin receptor substrates which act as molecular scaffolds and which bind to a conserved NPXY-motif in the receptor juxtamembrane domain that is tyrosine phosphorylated upon receptor activation. In EmIR1 such a motif is present, whereas it is absent in EmIR2 (Hemer et al., 2014). Furthermore, and probably the most convincing evidence that an *Echinococcus* insulin receptor is directly stimulated by host insulin, tyrosine phosphorylation of EmIR1 (likely at the NPXY motif) was shown to be drastically induced upon exogenous addition of human insulin to metacestode vesicles (Hemer et al., 2014).

In conclusion, the interaction of EmIR1 with host insulin provides a particularly convincing example for host–helminth cross-communication via evolutionarily conserved signalling molecules. Based on the fact that, at least in yeast two-hybrid investigations, EmIR1 is not able to interact with any of the parasite-derived ILPs, it is even tempting to speculate that this receptor has only been retained by the parasite for interacting with host-derived hormones. Since the highest concentrations of insulin in mammals occur at the junction between the portal vein and the liver parenchyma (Hemer et al., 2014 and references therein), which is also the liver entry site of the oncosphere, the stimulating effects of host insulin on parasite development could also be involved in the pronounced organ-tropism of *E. multilocularis* and, in part, of *E. granulosus* towards the liver. Although it is clear that the route of infection via the portal vein already determines to a great extent the observed organ-tropism, the high insulin concentrations within the liver could serve as a signal to the parasite to 'not move further' and undergo metamorphosis from the oncosphere towards the metacestode stage. In this regard, it would be highly interesting to investigate insulin signalling mechanisms also in other cestodes such as *T. solium*, since in this case the route of infection is similar, but the metamorphosis occurs at different locations.

6.2 EGF/FGF signalling and the mitogen-activated protein kinase cascade in *Echinococcus*

RTKs of the EGF- and FGF-receptor families are activated by binding cognate ligands of the EGF and FGF families and typically signal through downstream situated signalling cascades of the mitogen-activated protein kinase (MAPK) family (Fig. 1). Three major branches of the MAPK family exist in almost all metazoans, namely, the Erk (extracellular signal regulated kinase)-like MAPK, the p38-like MAPK and the JNK (C-jun N-terminal kinase) MAPK, the members of which can be distinguished according to specific sequence motifs at the kinase activation loop (T-E-Y for Erk kinases; T-G-Y for p38 kinases; T-P-Y for JNK) (Kultz, 1998). The activity of each MAPK is regulated by upstream acting, dual-specific MAPK kinases (MKKs) which, in turn, are regulated by upstream acting MKK kinases (MKKKs). The activities of MKKKs are often regulated by small GTP-binding proteins of the Ras family, which act as bridging molecules to the respective RTKs. In some cases, additional MKKK kinases are involved, providing an additional level of regulation between the MAPK module and the RTKs (Sabio and Davis, 2014).

In *E. multilocularis*, a complete Erk-like MAPK module, consisting of the MAPK EmMPK1, the MKK EmMKK2 and the MKKK EmRaf, has been described and the interaction of the respective kinases was verified by yeast two-hybrid analyses and co-immunoprecipitation (Gelmedin et al., 2010; Spiliotis et al., 2006, 2005). Furthermore, at least one small GTP-binding protein with high homology to mammalian Ras, EmRas, has been characterized and found to interact with EmRaf in the yeast two-hybrid system (Spiliotis et al., 2005). Additional MAPK cascade components characterized so far in *E. multilocularis* are a p38-like MAPK, EmMPK2 (Gelmedin et al., 2008), a second MAPK kinase, EmMKK1 (Gelmedin et al., 2010) and the GTP-binding protein EmRal (Spiliotis and Brehm, 2004). However, in none of these latter cases have interacting MAPK cascade components been identified. Unlike EmRas, EmRal did not interact with EmRaf in protein–protein interaction assays (Spiliotis et al., 2005) so that its cognate MKKK is currently unknown. EmMKK1, which structurally is a clear MKK, did not interact with the Erk-like MAPK EmMPK1 or with the p38-like EmMPK2 (Gelmedin et al., 2010) and thus most probably interacts with a yet uncharacterized parasite MAPK. Based on recently released genomic information, the genomes of *E. multilocularis* and *E. granulosus* contain additional genes encoding at least one MAPK (of the JNK family), three additional MKKs as well as numerous MKKK and small GTP-binding protein family members (Tsai et al., 2013). Using yeast two-hybrid analyses and protein—protein interaction assays, it will be a challenging future task to unravel the precise interaction network of MAPK cascade components in these parasites, which is relevant not only for understanding the processing of information that derives from membrane-bound receptors, but also for the design of anti-infectives directed against parasite protein kinases (see later discussion).

Since EGF- and FGF-like ligands are produced by liver parenchyma, particularly during regeneration processes that might follow parasite-induced tissue damage (Ye et al., 2015), these cytokines are, like insulin, possible host-derived growth stimulators of metacestode development. So far, only one potential receptor has been analyzed in more detail: the EGFR-like RTK EmER, which displays a domain structure typical of (mammalian) EGF receptor (EGFR) members including an N-terminal LBD, separated from a transmembrane region by several furin-like domains, and an intracellular tyrosine kinase domain (Spiliotis et al., 2003). Although in the metacestode stage *emer* is surely the most abundantly expressed EGFR-encoding gene, genomic and transcriptomic information indicates that there are at least two additional EGFRs encoded by the *E. multiocularis* genome (Table 1), which both display a typical domain structure and are expressed to a certain level in the parasite larval stages. Structurally, FGF receptors can be clearly distinguished from EGFR and display characteristic immunoglobulin domains in their extracellular regions. Although so far no *Echinococcus* FGF receptor has been investigated in detail, three respective genes are present in the parasite genome and, according to next generation sequencing transcriptomic data, are well expressed in larval stages (Table 1).

Although so far no direct evidence has been obtained that *Echinococcus* EGFRs can interact with host-derived EGF, studies on the parasite's Erk-like MAPK module at least indirectly indicate that this is the case. Using an in vitro cultivation system for the *E. multilocularis* metacestode stage in which host serum is an essential component to support parasite survival and development (Spiliotis et al., 2004), Spiliotis et al. (2006) demonstrated that the parasite's Erk-like MAPK EmMPK1 is specifically phosphorylated, and thus activated, in response to exogenously added host (bovine) serum. This indicates that serum stimulates a membrane-bound *Echinococcus* receptor which subsequently transmits the signal via EmRas, EmRaf and

EmMKK2 to EmMPK1. Interestingly, the phosphorylation state of EmMPK1 was also drastically induced when physiological concentrations of human EGF were added to the culture (Spiliotis et al., 2006), indicating that this host hormone, which regularly occurs in mammalian serum preparations, was also responsible for the serum effect on EmMPK1. Although no experiments were carried out so far to directly measure the interaction between EmER and host EGF, such analyses have previously been made for an EGFR-like RTK, SER, of the related schistosomes (Vicogne et al., 2004). These authors made elegant use of the *Xenopus* oocyte system to clearly show that SER is activated in response to human EGF. In BLASTP searches on the *E. multilocularis* proteome, highest homologies can be detected between *Schistosoma* SER and *Echinococcus* EmER not only in the kinase domain but also in the LBD. It is thus reasonable to assume that EmER is the ortholog of schistosome SER and that the cestode receptor can, like its trematode counterpart, also functionally interact with host-derived EGF, resulting in a subsequent activation of the parasite's Erk-like MAPK cascade.

Altogether, the analyses carried out on the *Echinococcus* RTKs and the corresponding intracellular signalling pathways indicate that, in addition to EmIR1/EmIR2 and insulin, EmER and EGF might be another example in which the *Echinococcus* metacestode makes use of evolutionarily conserved signalling receptors to sense corresponding host hormones for regulating growth and development. This does not exclude, however, that also the remaining two EGFR-like receptors of the parasite could contribute to EmMPK1 activation in response to EGF. Furthermore, it is also possible that the *Echinococcus* FGF receptors might sense host-derived FGF and then transmit signals to the parasite's Erk-like MAPK cascade. Studies to investigate the role of these receptors in *Echinococcus* development are under way.

6.3 TGF-β/BMP signalling in *Echinococcus*

The cytokines of the TGF-β/BMP superfamily signal through membrane-bound STK receptors, of which two sub-forms, type I and type II receptors, exist. In the case of the TGF-β branch of these cytokines (e.g., TGF-β sensu stricto or Activin), the cytokine first binds to the cognate type II receptor which subsequently recruits the corresponding type I receptor, resulting in the phosphorylation and activation of the type I receptor through the constitutive kinase activity of the type II receptor. In case of the BMP subfamily, on the other hand, the cytokine first binds the type I receptor, followed by recruitment of the corresponding type II receptor. Once activated, the type I receptors then phosphorylate, and thus activate, downstream situated DNA-binding proteins of the receptor-regulated R-Smad (Sma/MAD) family (Fig. 1). At least in mammals, the signalling pathways for the TGF-β and the BMP subtypes are strictly separated, leading to an activation of Smad2 and Smad3 (AR-Smads) by TGF-β/Activins and Smad1, Smad5 or Smad8 (BR-Smads) by BMPs. Once phosphorylated, the R-Smads form protein complexes with a so-called Co-Smad (Smad4 in mammals), which are finally translocated into the nucleus to regulate the expression of target genes (Brehm, 2010a, 2010b).

In *E. multilocularis*, the full complement of Smad factors has been described and functionally analyzed (Zavala-Gongora et al., 2003, 2008; Epping and Brehm, 2011) and one study was also carried out on related Smads of *E. granulosus* (Zhang et al., 2014). In total, *E. multilocularis* expresses two R-Smads, EmSmadA and EmSmadC, which structurally belong to the TGF-β/Activin (AR-Smad) signalling branch (Zavala-Gongora et al., 2008, 2003), two R-Smads of the BMP signalling pathway (BR-Smad), EmSmadB and EmSmadE (Epping and Brehm, 2011; Zavala-Gongora et al., 2003), and one single co-Smad, EmSmadD (Zavala-Gongora et al., 2008). Interestingly, the parasite's BR-Smads and the co-Smad EmSmadD comprise the highly conserved MH1 and MH2 domains that are typical for Smad transcription factors, whereas its AR-Smads contain only the MH2 domain which regulates Smad interactions, but have lost the MH1 domain necessary for DNA interactions (Zavala-Gongora et al., 2003, 2008). In yeast two-hybrid experiments, all *Echinococcus* R-Smads were found to associate with the co-Smad EmSmadE and also showed complex interactions with each other (Zavala-Gongora et al., 2008; Epping and Brehm, 2011). Furthermore, all parasite R-Smads were found to functionally interact with human TGF-β/BMP type I receptors (Zavalao-Gongora et al., 2003, 2008; Epping and Brehm, 2011). However, although EmSmadB and EmSmadC strictly interacted with either a human type I BMP receptor or a human type I TGF-β receptor, respectively, as expected from their sequence (Zavala-Gongora et al., 2003, 2008), EmSmadA and EmSmadE interacted with type I receptors of both branches (Epping and Brehm, 2011; Zavala-Gongora et al., 2003), indicating that TGF-β and BMP signalling pathways are not as strictly separated in the parasite as it is in mammals. Finally, biochemical evidence indicated that, at the level of the Smad transcription factors, *E. multilocularis* TGF-β/BMP signalling pathways are cross-interacting with other parasite pathways in a complex way. The co-Smad EmSmadD, for example, was shown to be phosphorylated by EmMPK1 (Zavala-Gongora et al., 2008), indicating that the activation of the Erk-like MAPK cascade (e.g., through EGF) could interfere with the processing of TGF-β/BMP signals. Furthermore, EmSmadA/B/C (but not EmSmadD/E) are capable of interacting with EmSKIP, a member of the SNW/SKIP family of transcriptional co-regulators that usually function in NHR signalling (Gelmedin et al., 2005; Zavala-Gongora et al., 2008) and also direct interactions between R-Smads and nuclear hormone receptors of the parasite were observed (Förster et al., 2011).

So far, only one member of the TGF-β/BMP receptor family has been characterized in *E. multilocularis* (or any other cestode), EmTR1, and it displayed a structure typical of the type I receptor family (Zavala-Gongora et al., 2006). According to genome information, however, genes for at least two additional type I receptors and one single type II receptor are present in the parasite (Tsai et al., 2013; Table 1). In biochemical assays, EmTR1 was able to functionally interact with human type II receptors and phosphorylated EmSmadB, but none of the remaining parasite R-Smads, indicating that this receptor, once activated by the cognate ligand and the type II receptor, signals through EmSmadB/EmSmadD complexes. Most interestingly, the phosphorylation of EmSmadB was strongly induced when human BMP2 was added to the receptor (Zavala-Gongora et al., 2006), providing yet another example for an *Echinococcus* surface signalling receptor that is able to functionally interact with a host-derived cytokine of the respective, evolutionarily conserved signalling cascade.

As yet it is unclear whether the functional interaction between EmTR1 and human BMP2 is of any physiological consequence during the infection. At least in our hands, no significant in vitro effects of exogenously added BMP2 on the proliferation of parasite stem cells or the growth of metacestode vesicles were measurable. Furthermore, it is unclear whether the developing parasite would have access to significant amounts of host BMP2 (or other BMPs) while infecting the liver of the intermediate host. The situation would surely be different in the case of host TGF-β since this cytokine can be found in significant concentrations around parasitic liver lesions, where it is secreted by host immune cells (Wang et al., 2013; Zhang et al., 2008), most probably in association with parasite-induced immunosuppression. Although other than BMP2, host TGF-β did not stimulate the activity of EmTR1 (Zavala-Gongora et al., 2006), it is conceivable that one or more of the additional parasite TGF-β/BMP receptors can functionally interact with host TGF-β. This aspect clearly requires further attention and should involve not only biochemical investigations on interactions between *Echinococcus* TGF-β/BMP receptors and host TGF-β, but also effects of TGF-β on parasite proliferation, differentiation and, e.g., protoscolex production.

6.4 Nuclear hormone receptor signalling in *Echinococcus*

Unlike peptidic hormones and cytokines, the receptors of which are located in the cytoplasmic membrane, lipophilic hormones can pass through the cell surface and usually act on members of the NHR family that are either present in the cytosol (class I NHRs) or the nucleus (class II NHRs) of the target cell. Respective ligands are steroid hormones such as testosterone, androgen or glucocorticoids, as well as thyroid hormone, vitamin A and related compounds (e.g., retinoic acids), vitamin D, fatty acids or cholesterol and oxysterols (Bridgham et al., 2010). Whether *E. multilocularis* or *E. granulosus* larvae alter their gene expression patterns in response to the presence of such ligands in the medium has not yet been clarified. There is, however, circumstantial evidence suggesting an important role of these molecules in parasite biology. First of all, *E. multilocularis* and *E. granulosus* (and other cestodes), like many other parasitic helminths, are not able to synthesize de novo cholesterol and fatty acids and have to take them up from the host (Tsai et al., 2013; Zheng et al., 2013). Corresponding sensor mechanisms, which might involve NHR-like receptors, would thus be helpful for the parasites to ensure development at an appropriate host site. Second, most ligands for NHRs are transported in host serum in close association with albumin, which is one of the most abundant proteins in hydatid fluid. As revealed by studies on in vitro cultivation systems for *Echinococcus* metacestodes (Spiliotis et al., 2004, 2008), the presence of serum in the culture medium is also crucial for parasite development and the utilization of serum-free media usually does not lead to parasite survival for longer than a few days (our own observations), indicating that the complex lipid composition of mammalian serum, which is only poorly reflected in serum-free media, is important for parasite survival. Third, closely related cestodes such as *T. crassiceps* or *T. solium* are apparently capable of producing NHR ligands such as steroid hormones (Valdez et al., 2014) and display certain in vitro responses to host-derived steroids (Ambrosio et al., 2014; Escobedo et al., 2004), which might also apply to the genus *Echinococcus*. This is supported by a report on the identification of a gene encoding a 17b-hydroxysteroid dehydrogenase of *T. solium* which, when recombinantly expressed in mammalian cells, was able to transform 3H-androstenedione into testosterone (Aceves-Ramos et al., 2014). Interestingly, and as outlined by these authors, similar enzymes are also encoded by the *E. multilocularis* genome, indicating that this cestode could also produce steroid hormones. Finally, it has been reported that tamoxifen, a competitive antagonist of the estrogen receptor alpha, has anti-parasitic activities against *E. granulosus* protoscoleces in vitro which, according to these authors, could best be explained by interactions of the drug with an as yet uncharacterized, *Echinococcus* estrogen-responsive NHR (Nicolao et al., 2014). Furthermore, activity of tamoxifen against *E. multilocularis* metacestode vesicles was also demonstrated in a pharmacological screening (Stadelmann et al., 2014).

Only one member of the NHR family, EmNHR1, has so far been functionally characterized in *E. multilocularis* (or other cestodes) to a certain extent (Förster et al., 2011). EmNHR1 displayed a domain structure typical of NHRs and contained a DNA-binding domain with two zinc finger motifs, which is characteristic for the NHR family, as well as a so-called HOLI-domain which is important for ligand binding. On the amino acid sequence level, EmNHR1 displayed significant homologies to *C. elegans* DAF-12, which is a homolog of mammalian Liver X receptors, the Pregnane X receptor and the Vitamin D receptor, and to *Drosophila* HR-96 in both the DNA-binding domain and the LBD (Förster et al., 2011). Interestingly, the DAF-12/HR-96 subfamily of NHRs is considered to be an evolutionarily conserved receptor group that regulates cholesterol homeostasis and longevity in diverse metazoans, and all members of this group, from mammals to invertebrates, directly interact with cholesterol or cholesterol derivatives (Förster et al., 2011). Whether EmNHR1 also interacts with cholesterol or cholesterol-like compounds (e.g., bile acids) has not been biochemically addressed so far. However, yeast two-hybrid analyses clearly showed that the dimerization of the EmNHR1 LBD is greatly stimulated when 5% bovine or human serum is added to the medium, indicating that the cognate ligand for EmNHR1 is a serum component. Furthermore, since hydatid fluid also stimulated the dimerization of the EmNHR1 LBD in these experiments, it can be assumed that the respective ligand is transported into metacestode vesicles, maybe associated with other serum components such as albumin or lipoproteins (Förster et al., 2011).

Upon activation, NHRs are typically translocated into the nucleus where they, either alone or in combination with other transcription factors, bind to specific recognition sites, called hormone-responsive elements (HRE), in the promoter regions of their target genes. Since EmNHR1 contains a DNA-binding domain very similar to those of *C. elegans* DAF-12 and *D. melanogaster* HR-96 (Förster et al., 2011) it can be expected that the cognate HREs are also similar. For DAF-12 and HR-96, these are usually hexameric DR5 direct repeats that are separated by 5 nt (e.g., AGTTCA-n5-AGTGCA for DAF-12 and AGTGCA-n5-GTGTCA for HR-96) and, interestingly, in a bioinformatics approach Förster et al. (2011) found similar sequences in the promoter region of an *Echinococcus* ortholog to the Niemann-Pick disease type C1 gene (NPC1; EmuJ_00107700) which encodes a member of the patched protein family. In *Drosophila*, this protein is required for sterol absorption, controlled by HR-96, and in mammals, the respective gene is under control of the DAF-12/HR-96 ortholog LXR (see Förster et al., 2011). This indicates that, like in *Drosophila* and mammals, the *E. multilocularis* NPC1 gene could be under control of a DAF-12/HR-96 ortholog, which is EmNHR1. Another candidate gene for EmNHR1, which also harbours a potential DAF-12/HR-96-responsive HRE in the promoter region is *emabp*, which encodes an apolipoprotein A1—binding protein (Bernthaler et al., 2009; see Section 4 above). Since ApoA1 is a major component of high-density lipoprotein particles which transport cholesterol in mammalian serum, EmABP might be crucially involved in cholesterol uptake by the parasite. Hence, several lines of evidence indicate that EmNHR1 might indeed be a 'cholesterol sensor' of *Echinococcus* that controls the expression of target genes involved in cholesterol uptake by the parasite. Since EmNHR1 also interacted with EmSmadC in yeast two-hybrid assays (Förster et al., 2011), it is also expected that this regulatory network involves cross-communication between NHR and TGF-β signalling in the parasite, similar to the situation in *C. elegans*.

The total number of NHR-encoding genes significantly varies in different organisms and can comprise, e.g., 48 members in humans, 18 members in *Drosophila* or as many as 270 genes in *C. elegans*, where this receptor family underwent a massive expansion during evolution. In a bioinformatic approach carried out on the first genome assembly versions, Förster et al. (2011) identified a set of 17 NHR-encoding genes in *E. multilocularis* (including *emnhr*1), which broadly overlapped with the NHR gene set of schistosomes. In the full draft version of the *E. multilocularis* genome (Tsai et al., 2013), all these genes are present, although two have been wrongly annotated and lack the conserved LBD (see Table 2). Three of the 17 *Echinococcus* NHRs belong to a lophotrochozoan-specific NHR family, called the 2DBD NHRs, which contain two LBDs instead of just one as typical for NHRs (Table 2). Several of the *Echinococcus* NHR genes are quite highly expressed in the metacestode, including *emnhr*1 and genes encoding orthologs to NHR-48, to the retinoid X receptor family, and two of the

Table 2 *Echinococcus multilocularis* nuclear hormone receptors (NHRs)

Protein name	GeneDB accession[b]	Receptor class	Expression in[a]				References 1	References 2[c]
			PC	MC	PS	Ad		
EmNHR1	EmuJ_001198700	Daf12-like NHR	+	+	+	(+)	Förster et al. (2011)	3395
EmNHR48	EmuJ_000623600	NHR48-like NHR	+	+	+	(+)	Tsai et al. (2013)	3309
EmPPAR	EmuJ_001013700	PPAR-like NHR	(+)	(+)	–	–	Tsai et al. (2013)	2122
EmTHR	EmuJ_000317100	Thyroid hormone NR	–	–	–	(+)	Tsai et al. (2013)	1680
EmRAR	EmuJ_000088800	RAR receptor-like	(+)	(+)	(+)	(+)	Tsai et al. (2013)	1702
EmNR96	EmuJ_001078800	NR96-family NHR	–	–	(+)	–	Tsai et al. (2013)	2891
Em2DBDa	EmuJ_000240200	2DBD NHR	(+)	–	(+)	–	Tsai et al. (2013)	1872
EmRXR	EmuJ_000434500	RXR-like NHR	+	+	++	+	Tsai et al. (2013)	3440
Em2DBDb	EmuJ_000458200	2DBD NHR	+	++	+	(+)	Tsai et al. (2013)	2681
EmFTZF1a	EmuJ_000763500	FTZ-F1-like NHR	+	+	+	–	Tsai et al. (2013)	3400
EmNHR41	EmuJ_000234100	NR41-like NHR	+	+	+	–	Tsai et al. (2013)	1644
EmCOUPTF	EmuJ_000802400	COUP-TF-like	(+)	+++	–	–	Tsai et al. (2013)	2822
Em2DBDc	EmuJ_000379600	2DBD NHR	+	+	+	–	Tsai et al. (2013)	1779
EmNR4A	EmuJ_001032900	NR4A-like NHR	–	–	(+)	–	Tsai et al. (2013)	3425
EmFTZF1b	EmuJ_000814300	FTZ-F1-like NHR	+	+	+	+	Tsai et al. (2013)	3442
EmNRX	EmuJ_000937000	Cestode-spec. NHR	–	–	(+)	–	Tsai et al. (2013)	1798
EmNR1D2	EmuJ_000671100	NR1D2-like NHR	–	–	(+)	(+)	Tsai et al. (2013)	2878

Ad, pre-gravid adult worm; *COUP-TF*, chicken ovalbumin upstream promoter transcription factor; *DBD*, DNA-binding domain; *FTZ*, fushi tarazu; *MC*, metacestode vesicles without brood capsules; *PC*, primary cells after 2 days of culture; *PPAR*, peroxisome proliferator-activated receptor; *PS*, low pH/pepsin-activated protoscolex; *RAR*, retinoic acid receptor.

[a]Expression values in fpkm (fragments per kilobase of exon per million fragments mapped) according to Tsai et al. (2013). fpkm <1 depicted as '–', fpkm >1 to 10 as '(+)', fpkm >10 to 50 as '+', fpkm >50 to 100 as '++', fpkm >100 as '+++'.

[b]Accession number assigned by GeneDB under http://www.genedb.org/Homepage/Emultilocularis.

[c]Contig number of bioinformatics analysis made by Förster et al. (2011).

2DBD family members (Table 2). Interestingly, one of the NHR genes, encoding a member of the chicken ovalbumin upstream promoter transcription factor (COUP TF) II family, is highly expressed only in the metacestode but very lowly in all other developmental stages. This protein displays high amino acid sequence homologies to mammalian COUP TF II, which binds both 9-cis- and all-trans-retinoic acid (vitamin A metabolites) and in the case of the schistosome ortholog RXR1, direct interaction of the flatworm receptor with retinoic acid was already demonstrated (Qiu et al., 2013), indicating that the *Echinococcus* COUP TF II receptor is also able to sense retinoic acid. Whether the protein could be interacting with host-derived retinoic acid has not been investigated so far. However, genomic information (Tsai et al., 2013; Zheng et al., 2013) indicates that *Echinococcus* is able to synthesize retinoic acid on its own since it contains genes for all key enzymes for retinoic acid synthesis from retinol such as retinol dehydrogenases (EmuJ_000422100; EmuJ_000079200) and aldehyde dehydrogenase (EmuJ_000389100). Likeliest candidates for the transport of retinol into *Echinococcus* are the FABPs which have been biochemically investigated to a certain level and which are obviously able to directly interact with *Echinococcus* phospholipid membranes (Porfido et al., 2012).

Apart from EmNHR1, which likely interacts with a cholesterol derivative present in host serum, and COUP TF II, which potentially senses retinoic acid, all other *Echinococcus* NHR are still 'orphan receptors' with no known ligand. Future biochemical and cell biological work which, in the case of NHRs, is much more difficult to perform than studying interactions between receptors and peptidic hormones, will be necessary to unravel the complex interactions between parasite NHR signalling and host- as well as parasite-derived lipophilic ligands. It should be noted in this context that Escobedo et al. (2010, 2004) already claimed the identification of NHRs for estrogen, androgen and progesterone in the related cestode *T. crassiceps*. However, none of these studies provided the actual sequences of the cestode NHRs. Furthermore, in none of the published genomes of *E. multilocularis*, *E. granulosus* or *T. solium* can NHR sequences with significant homologies to mammalian estrogen, androgen or progesterone receptors be found (Tsai et al., 2013; Zheng et al., 2013). It is thus highly likely that any observed in vitro effects of host steroid hormones on taeniid cestodes, if they are physiologically relevant, are mediated by one of the other NHRs or even receptor molecules that are unrelated to the NHR family.

6.5 Other receptors and signalling pathways

By studying insulin signalling mechanisms in the flatworm parasite Schistosoma *mansoni*, Vicogne et al. (2003) identified an unusual receptor composed of an intracellular tyrosine kinase domain with high sequence homologies to insulin receptors, linked to an extracellular venus flytrap (VFT) module, which are typically present in class C G-protein-coupled receptors (GPCR) such as the γ-aminobutyric acid type B receptor. Later, a second member of this protein family, now called VKR (VFT kinases), was identified in schistosomes and a number of additional receptors with VKR domain composition was characterized in insects (Ahier et al., 2009). The VKRs are, thus, an invertebrate-specific family of insulin-like receptors which do not occur in mammals. In insects, the VKRs are activated by binding of natural amino acids or amino acid derivatives to the VFT module (Ahier et al., 2009) and, likewise, at least one *S. mansoni* receptor, SmVKR1, responds to L-arginine, whereas the second one, EmVKR2, is activated by calcium ions (Vanderstraete et al., 2014). Interestingly, one single gene encoding a VKR with both an insulin-like TK domain and an extracellular VFT module is also present on the genome of *E. multilocularis* (as well as *E. granulosus*) and the respective protein, EmVKR (EmuJ_619800), displays higher homologies to SmVKR1 than to SmVKR2. Transcriptome data (Tsai et al., 2013) also indicate that the EmVKR gene is well expressed in larval stages, whereas only slight expression can be found in the adult. Although our own analyses clearly indicated that EmVKR contains an active tyrosine kinase domain, the *Echinococcus* receptor was not activated by L-arginine when expressed in *Xenopus* oocytes (Schubert et al., unpublished). This does not exclude, however, that other amino acids might act on EmVKR. It should be noted in this context that a sensing mechanism for host-derived amino acids could be advantageous for the parasite since as described earlier, genome information indicates that *Echinococcus* has lost a number of amino acid biosynthesis pathways and, thus, has to take these up from the host (Tsai et al., 2013; Zheng et al., 2013).

Tumour necrosis factor alpha (TNF-α) is one of the cytokines that is produced during an immune response against *Echinococcus* (Amiot et al., 1999) and is present in higher concentrations in patients of 'active' cystic echinococcosis when compared to those with 'inactive' lesions (Petrone et al., 2015). In mammals, TNF-α signals through the membrane-bound TNF receptor family which is characterized by extracellular cysteine-rich domains that are a hallmark of the TNF receptor superfamily

(Locksley et al., 2001). In principle, three major sub-families of TNF receptors are expressed in mammalian cells (mostly cells associated with adaptive immune responses) and include the so-called death receptors, which mediate cell death through their cytoplasmic death domain, the non-death receptors, which signal through intracellularly located TNF receptor-associated factors (TRAFs) and the 'decoy receptors', which usually bind apoptosis-inducing TNF ligands (Locksley et al., 2001). Interestingly, the occurrence of a TNF receptor-like molecule, SmTNFR, with clear homologies to the non-death receptors has already been described for the flatworm parasite *S. mansoni* (Oliveira et al., 2009) and on the parasite's genome, these authors also identified genes encoding intracellular signalling molecules associated with TNF receptor signalling, such as components of the JNK cascade and a TRAF ortholog, indicating that schistosomes might process TNF-related signals that derive from the host (since the parasite does not contain a TNF-a-encoding gene). That this might indeed be the case has also been shown by these authors, at least in vitro, since the addition of human TNF-α to cultured *S. mansoni* adults led to substantial transcriptional changes as measured by microarray assays (Oliveira et al., 2009). In *E. multilocularis*, a gene encoding a receptor with high homologies to SmTNFR is also present (EmuJ_000990500) and transcriptome analyses (Tsai et al., 2013) indicate that this gene is highly active in the primary cell system and the metacestode, whereas low expression occurs in protoscoleces and adults (Table 3). Hence, the *Echinococcus* TNFR appears to be mainly produced by larvae that, during an infection, are likely to encounter host TNF. Since the parasite's genome also contains genes for downstream signalling factors such as JNK cascade factors and TRAF (EmuJ_000607600), it is at least possible that *E. multilocularis* (and *E. granulosus*, which contains the same factors) might respond to host TNF-α. Future studies concerning the effects of host TNF-α on *Echinococcus* larvae and on the interaction of TNF-α with the parasite TNFR will be helpful to understand the complex parasite—host interplay associated with anti-*Echinococcus* immune responses.

Members of the superfamily of GPCRs are expressed by all metazoans investigated so far and are usually involved in sensing pheromones, various hormones and neurotransmitters (Venkatakrishnan et al., 2013). Although no functional analyses have so far been carried out on *Echinococcus* GPCRs, the genes for 70 members of this protein family were identified in the *E. multilocularis* genome (also present in *E. granulosus* (Tsai et al., 2013; Zheng et al., 2013)), of which 9 structurally belonged to the amine-responsive GPCRs, 28 to the peptide-responsive GPCR, 4 to a platyhelminth-specific

Table 3 Membrane-associated *Echinococcus multilocularis* non-kinase receptors

Protein name	GeneDB accession[b]	Receptor class	Expression in[a]				References
			PC	MC	PS	Ad	
EmTNFR	EmuJ_000990500	TNFα receptor	+++	+++	(+)	(+)	Tsai et al. (2013)
EmPTCH	EmuJ_000443400	Patched Hh receptor	+	+	+	(+)	Tsai et al. (2013)
EmFZ1	EmuJ_000682100	Frizzled GPCR	+	+	++	(+)	Tsai et al. (2013)
EmFZ5	EmuJ_000996400	Frizzled GPCR	(+)	(+)	(+)	(+)	Tsai et al. (2013)
EmFZ4	EmuJ_000636500	Frizzled GPCR	+	+	+	(+)	Koziol et al. (2016)
EmFZ10	EmuJ_000085700	Frizzled GPCR	(+)	–	+	(+)	Tsai et al. (2013)
EmFZ4b	EmuJ_000438200	Frizzled GPCR	–	–	(+)	(+)	Tsai et al. (2013)

Ad, pre-gravid adult worm; *GPCR*, G-protein-coupled receptor; *MC*, metacestode vesicles without brood capsules; *PC*, primary cells after 2 days of culture; *PS*, low pH/pepsin-activated protoscolex; *TNF*, tumour necrosis factor.

[a]Expression values in fpkm (fragments per kilobase of exon per million fragments mapped) according to Tsai et al. (2013). fpkm <1 depicted as '–', fpkm >1 to 10 as '(+)', fpkm >10 to 50 as '+', fpkm >50 to 100 as '++', fpkm >100 as '+++'.

[b]Accession number assigned by GeneDB under http://www.genedb.org/Homepage/Emultilocularis.

rhodopsin family of GPCR, 16 to the orphan rhodopsin-like family, 3 to the metabotropic glutamate-like GPCR and 5 to the frizzled-like GPCR (Tsai et al., 2013). Although in schistosomes GPCRs have been functionally characterized to some extent and shown to be activated, e.g., by histamine (Hamdan et al., 2002), dopamine (Taman and Ribeiro, 2009) or serotonin (Patocka et al., 2014), it is not expected that the respective ligands activating these receptors necessarily derive from the host since the parasite's genome contains all key enzymes necessary for their production (Protasio et al., 2012). In *E. granulosus*, Camicia et al. (2013) reported effects of exogenously added serotonin on the motility of protoscoleces and on the re-differentiation of protoscoleces towards the metacestode stage, although rather high serotonin concentrations (up to 1 mM) were necessary to provoke these effects. Furthermore, these authors identified serotonergic elements associated with the protoscolex nervous system and identified several key components for serotonin synthesis in the parasite genome, indicating that, like in the case of schistosomes, the parasite on its own is capable of producing the ligand. Hence, although the *Echinococcus* GPCR family might yield attractive target molecules for the development of novel chemotherapeutics (see later discussion), it is completely unclear whether these receptors and their cognate ligands play any role in host–parasite interplay during echinococcosis.

Two additional evolutionarily conserved signalling pathways, hedgehog (Hh) and wingless-related (wnt) signalling, play an important role in proper body axis formation (particularly of the anterior–posterior axis) in free-living flatworms (Petersen and Reddien, 2008; Yazawa et al., 2009). In Hh signalling, a peptidic ligand (e.g., mammalian DHH, IHH or SHH) directly interacts with membrane-bound surface receptors of the PTCH family, leading to PTCH inactivation and accumulation of another membrane-associated protein, smoothened (SMO), which prevents the proteolytic cleavage of the zinc-finger transcription factor Cubitus interruptus (Ci) to the 75 kDa Ci-fragment CiR, which acts as a co-repressor of Hh target genes, allowing their transcription. In the absence of Hh, PTCH prevents the accumulation of SMO, resulting in an accumulation of CiR, which then translocates to the nucleus and prevents the transcription of Hh target genes. In planarians, several WNT-encoding genes, which specify the posterior pole, are regulated by Hh signalling (Yazawa et al., 2009). WNTs are lipid-modified signalling glycoproteins that bind to the extracellular, Cysteine-rich domain of members of the frizzled (FRZ) receptor family (a sub-family of GPCRs), leading, via several additional signalling proteins, to a subsequent accumulation of the transcription factor β-catenin which regulates wnt target genes. In the absence of a signal, β-catenin is degraded within the cytoplasm by a so-called destruction complex, thus preventing the transcription of target genes. Both the Hh and the WNT ligands can bind to multiple receptors of an organism and, although it has not been studied yet whether vertebrate ligands of this family can bind to invertebrate receptors or vice versa, it is at least likely that cross-interactions as observed for insulin- and EGF-like ligands (Hemer et al., 2014; Vicogne et al., 2004; see earlier discussion), are also possible for Hh and WNT ligands. Furthermore, the ligands of both families can form morphogen gradients over moderate distances which, in the case of the *Echinococcus* metacestode, could allow interactions between host Hh/WNT ligands and parasite receptors. Another prerequisite for such interactions would be that WNT and Hh ligands are expressed at the site of infection, e.g., the liver. This seems at least to be the case for mammalian WNT ligands of which several are clearly expressed in the fully developed liver (Zeng et al., 2007). Although Hh signalling plays an important role in liver development, it is known to be significantly downregulated in normal liver parenchyma. However, a significant upregulation of Hh ligand expression has frequently been observed during liver regeneration processes (Omenetti et al., 2011) which, in the case of *Echinococcus* infections, might be active after the parasite, or the periparasitic immune response, has inflicted damage to liver tissue. Hence, the concept of host–parasite cross-communication via evolutionarily conserved signalling molecules could principally also apply to Hh and WNT signalling in echinococcosis.

The analysis of the *E. multilocularis* and *E. granulosus* genomes (Tsai et al., 2013; Zheng et al., 2013) has revealed that all components for functional WNT and Hh signalling are present and expressed in *Echinococcus* and, given the importance of these signalling pathways in development of all metazoans, it is expected that they are also fully functional in these parasites. As already mentioned earlier, a total of five genes encoding a GPCR of the frizzled subfamily have been identified (Table 3) and transcriptomic analyses revealed that at least two of these, encoded by EmuJ_000682100 and EmuJ_000636500, are well expressed in the metacestode (Table 3). As concerning receptors of the Hh pathway, *E. multilocularis*, like *Drosophila* and the majority of metazoans, only contains one single gene (EmuJ_000443400) which is well expressed in all larval stages, and which encodes a protein that displays high similarity to human PTCH1, the major receptor for all three mammalian Hh ligands. In case these ligands (or one of them) are upregulated in the intermediate host's liver upon *Echinococcus* infection, interactions with the parasite PTCH ortholog might occur. As in the case of NHRs, biochemical investigations on ligand–receptor interactions in the Hh and WNT systems are highly complex. It may thus take considerable time till questions on a possible cross-communication between host-derived Hh and WNT factors and respective *Echinococcus* receptors are addressed.

6.6 Conserved parasite-derived ligands

As already discussed, the interactions between *Echinococcus* and its host involving evolutionarily conserved signalling molecules do not have to be confined to effects of host hormones/cytokines on parasite signalling pathways, but could also apply to effects of parasite ligands on host signalling. Two respective ligands, the parasite's insulin-like peptides ILP1 and ILP2 (Table 4), have been introduced in the paragraph on insulin signalling. Based on expression studies and yeast two-hybrid analyses it is, however, rather unlikely that ILP1 and ILP2 affect human insulin signalling pathways. First of all, both ILPs are mainly expressed in the parasite's adult stage where they most probably act on EmIR2 (Hemer et al., 2014). Although it cannot be excluded that they are released into the intestinal lumen by adult *Echinococcus*, it is not expected that they encounter host insulin receptors. Second, at least in the yeast two-hybrid system, neither EmILP1 nor EmILP2 interacted with the LBD of the human insulin receptor (Hemer et al., 2014). Again, this does not exclude that activities on other host RTKs might occur (e.g., the insulin-like growth factor receptor), but overall there is no single piece of evidence to date supporting a role of the parasite ILPs in host parasite interplay during echinococcosis.

Interestingly, although *E. multilocularis* clearly expresses typical members of the FGF receptor family, no canonical FGF ligand has so far been identified by mining genomic information (Tsai et al., 2013; Zheng et al., 2013). On the other hand, the *E. multilocularis* genome analysis revealed the presence of one single gene, EmuJ_000514500, annotated as 'epidermal growth factor' which, according to transcriptome data, is well expressed in all larval stages as well as in the adult (Tsai et al., 2013; Table 4). A highly similar cDNA was previously identified in differential display analyses on trans-spliced *Echinococcus* genes and was found to be induced in the metacestode upon addition of host hepatocytes to the medium (Brehm et al., 2003). Based on the presence of an EGF domain in the extracellular portion, including several highly conserved cysteine residues, this gene was originally named *egfd* (epidermal growth factor domain protein). Apparently, two

Table 4 *Echinococcus multilocularis* secreted ligands

Protein name	GeneDB accession[b]	Ligand class	Expression in[a]				References
			PC	MC	PS	Ad	
EmILP1	EmuJ_000045300	Insulin-like	–	–	(+)	++	Hemer et al. (2014)
EmILP2	EmuJ_000045400	Insulin-like	(+)	(+)	(+)	++	Hemer et al. (2014)
EmEGFD	EmuJ_000514500	EGF-like	+	++	+	+	Brehm et al. (2003)
EmBMP1	EmuJ_000460200	BMP-like	–	–	(+)	+	Tsai et al. (2013)
EmBMP2	EmuJ_000181200	BMP-like	+	+	+	(+)	Tsai et al. (2013)
EmACT	EmuJ_000178100	Activin-like	+	(+)	+	–	Tsai et al. (2013)
EmTIP	EmuJ_000440000	TIP-like	++	+++	+++	+++	Nono et al. (2014)
EmHh	EmuJ_001200300	Hedgehog	+	(+)	+	(+)	Tsai et al. (2013)
EmWNT1	EmuJ_000349900	wnt	++	(+)	(+)	–	Koziol et al. (2016)
EmWNT11A	EmuJ_000907500	wnt	+	+	(+)	+	Koziol et al. (2016)
EmWNT11B	EmuJ_000104000	wnt	(+)	(+)	–	(+)	Koziol et al. (2016)
EmWNT2	EmuJ_000748600	wnt	(+)	(+)	(+)	(+)	Koziol et al. (2016)
EmWNT4	EmuJ_000211300	wnt	(+)	(+)	+	(+)	Koziol et al. (2016)
EmWNT5	EmuJ_000804000	wnt	(+)	(+)	+	(+)	Koziol et al. (2016)

Ad, pre-gravid adult worm; *BMP*, bone morphogenetic protein; *EGF*, epidermal growth factor; *MC*, metacestode vesicles without brood capsules; *PC*, primary cells after 2 days of culture; *PS*, low pH/pepsin-activated protoscolex; *TIP*, T-cell immunomodulatory protein.

[a]Expression values in fpkm (fragments per kilobase of exon per million fragments mapped) according to Tsai et al. (2013). fpkm <1 depicted as '–', fpkm >1 to 10 as '(+)', fpkm >10 to 50 as '+', fpkm >50 to 100 as '++', fpkm >100 as '+++'.

[b]Accession number assigned by GeneDB under http://www.genedb.org/Homepage/Emultilocularis.

different transcripts of *egfd* are present in the parasite, one of which is trans-spliced, whereas the other is conventionally spliced, leading to slight differences in the 5′ region of the transcript when the already published (trans-spliced) version (Brehm et al., 2003) is compared with the annotated version (Tsai et al., 2013). In both versions, the translated protein contains one full EGF domain at the N-terminus (including six highly conserved cysteines), followed by a transmembrane region and a short intracellular domain without homologies (Brehm et al., 2003; Tsai et al., 2013). This closely resembles the structure of mammalian EGF, which is produced as a membrane-bound pro-form that is subsequently processed into mature EGF by proteolytic cleavage (Brehm et al., 2003). It is thus very likely that *egfd* indeed encodes a parasite EGF factor. In this case it is highly interesting that the expression of *egfd* is apparently induced in the presence of host hepatocytes (Brehm et al., 2003). In all in vitro cultivation systems developed for *E. multilocularis* so far, host hepatocytes (and their secretions) were the best growth-stimulating medium compounds (e.g., Spiliotis et al., 2004). It could thus be that secretory products of host hepatocytes stimulate the expression of a parasite EGF factor which, in turn, acts on EmER or the other parasite EGFRs to support growth and development.

On the other hand, it cannot be excluded that EGFD is secreted by the metacestode to act on host EGFRs, although no respective experiments have been undertaken so far. However, at least indirect evidence for a possible secretion of an EGF-like factor by the *E. multilocularis* metacestode has been obtained by Lin et al. (2009)who showed that the Erk-like MAPK cascade is activated in host hepatocytes of infected mice, particularly in the vicinity of parasite lesions. Furthermore, these authors also observed Erk kinase activation of cultured hepatocytes exposed to parasite hydatid fluid (Lin et al., 2009), again indicating that EGF-like substances are present in this compartment. However, care has to be taken in the interpretation of these results since hydatid fluid could also contain host EGF or other host cytokines/hormones that may stimulate the Erk kinase cascade. Further experiments are thus clearly necessary to closer investigate possible interactions between EGFD and host EGFRs.

Surely one of the most interesting cytokine families concerning effects of *Echinococcus* on host cells (and the immune response) is the TGF-β/BMP-like cytokines. Since human BMP2 has already been shown to potentially interact with a parasite receptor of the corresponding class, EmTR1 (Zavala-Gongora et al., 2006), it is likely that parasite cytokines of this family are also able to stimulate host TGF-β/BMP receptors, provided that they are secreted by the parasite. Since receptors for the BMP/TGF-β cytokine superfamily are expressed by many host cells that are encountered by *Echinococcus* during an infection, such as hepatocytes or immune regulatory cells, the parasite would thus have many opportunities to generate a permissive environment within the host. As shown in Table 4, the *E. multilocularis* genome contains three genes for ligands of this superfamily, of which EmuJ_000460200 and EmuJ_000181200 encode proteins with structural motifs characteristic of the BMP sub-family, whereas EmuJ_000178100 apparently encodes a member of the TGF-β/Activin family. At least two of these, EmuJ_000181200 and the TGF-β/Activin-encoding gene, are clearly expressed in the metacestode as well as in primary cell preparations of the parasite (Table 4). In the closely related schistosomes, similar genes have previously been identified and found to be involved in egg development and parasite fertility (Freitas et al., 2007; Liu et al., 2013). It is thus expected that the *Echinococcus* orthologs also have crucial functions in parasite development, e.g., by activating one or more of the parasite's STKs of the BMP/TGF-β receptor family, and preliminary RNA interference experiments carried out in our laboratory indeed indicate that this is the case (Duvoisin and Brehm, unpublished results). However, this does not exclude that these factors may also be released from the parasite or be part of its excretory/secretory compounds that have direct effects on host cells. Of particular interest in this context would be the parasite Activin ortholog since human Activin is known to induce the conversion of $CD4^+$ T-cells into $Foxp3^+$, regulatory T-cells (Treg; Huber et al., 2009), an immunosuppressive cell type that is apparently also induced during alveolar echinococcosis in permissive hosts (Mejri et al., 2011). Interestingly, as reported by Nono et al. (2012), E/S products of the *E. multilocularis* metacestode were, under certain conditions, also able to induce Treg in in vitro assays, indicating that such a cytokine might indeed be released by the parasite. Possible immunomodulatory activities of the *E. multilocularis* TGF-β/Activin ortholog and its interaction with host receptors are, thus, currently subject to intense investigation in our laboratory.

Although not a 'classical' ligand of metazoan cell signalling systems, the mammalian 'T-cell immunomodulatory protein' (TIP) has, at least in one study, been shown to fulfill important criteria for acting as a 'cytokine' (released by cells to affect the behaviour of other cells) (Fiscella et al., 2003). Although both human and murine TIP do not display significant overall homologies to other protein families, they both contain a C-terminal transmembrane domain, an export-directing, N-terminal signal peptide and one (human) or three (murine) FG-GAP repeats that usually mediate cell—cell and cell—matrix interactions (often found in integrins) (Fiscella et al., 2003). Although mostly membrane bound, TIPs obviously also exist in soluble form and are released from the cell after proteolytic cleavage at the transmembrane domain. Interestingly, treatment of primary human and murine T-cells with TIP resulted in elevated release of the cytokines interferon-γ (IFN-γ), TNF-α and IL-10 by these T-cells and, in vivo, TIP exerted protective effects in a murine graft-versus-host-disease model (Fiscella et al., 2003). As shown by Nono et al. (2014), *E. multilocularis* (and *E. granulosus*) also contains a gene encoding a TIP-like protein, EmTIP, which is highly expressed in all developmental stages (Table 4) and contains two FG-GAP repeats in the extracellular region. In primary cell cultures of the parasite, representing the oncosphere-metacestode transition state, EmTIP was present in the E/S fraction in considerable amounts and, when recombinantly expressed in HEK 293 cells, rEmTIP was released and induced the production of IFN-γ (but not IL-10) by murine T-cells (Nono et al., 2014), indicating that it displays at least some of the cytokine activities of the mammalian ortholog. This is relevant insofar as previous studies had already shown that *E. multilocularis* provokes an early, potentially protective Th-1 dominated immune response in intermediate hosts which, in permissive hosts, is gradually subverted into a Th-2 dominated response (see Nono et al., 2014 and references therein). Since IFN-γ is a key cytokine of Th1 immune responses, the effects of EmTIP on T-cells could thus provide an explanation for the early induced Th1 response in echinococcosis (Nono et al., 2014). However, why should a parasite release a cytokine that provokes potentially protective responses in a mammalian host? It turned out that a large proportion of EmTIP (probably membrane bound) is present in parasite intercellular spaces and that the presence of an anti-EmTIP antibody in primary cell cultures prevented parasite development and the formation of mature metacestode vesicles from parasite stem cells (Nono et al., 2014). Hence, it appears likely that EmTIP, maybe as a cell—cell/matrix interaction component, fulfils an important role in early parasite development, leading to a subsequent release of the protein and to the observed effects on host T-cells, i.e., the production of IFN-γ and the possible induction of a Th1 immune response. In this case, the release of the parasite cytokine would actually result in an 'unwanted' host response, which the parasite could simply not escape because it needs the inducing molecule for an even more important process, i.e., larval development.

Finally, evolutionarily conserved ligands of the WNT and Hh pathways should be mentioned since these, as already discussed, could at least theoretically interact with respective host receptors. Genomic analyses carried out by Riddiford and Olson (2011) already indicated that cestodes have a significantly reduced complement of Wnt factors and in agreement with this, only five genes encoding WNT orthologs have so far been annotated in the *E. multilocularis* genome (Tsai et al., 2013; Table 4). Furthermore, only one gene encoding an Hh homologue is present (EmuJ_001200300) which appears typical, however, since the majority of metazoans also contains only one member of this protein family. Transcriptome analyses showed that at least one WNT ligand, EmuJ_000907500, is very well expressed in the metacestode as is the Hh ligand (Table 4). If they are secreted by metacestode vesicles, which could be addressed by investigating parasite E/S products using specific antibodies, they would be quite good candidates as possible ligands that affect host WNT/Hh signalling pathways. Respective investigations are surely worthwhile to undertake.

7. INTEGRATION OF HOST- AND PARASITE-DERIVED SIGNALS INTO CELLULAR RESPONSES

As is clear from the preceding sections, the tissues of *Echinococcus* metacestodes are exposed and respond to a multiplicity of molecular signals from the host and from the parasite itself. However, it is so far unclear which cells from the parasite actually sense these signals, and how these various signals are integrated into specific physiological responses. Because of the importance of conserved signalling pathways in stem cell biology in animals, they constitute a natural focus of research as the possible targets of host-derived growth factors. In free living flatworms, undifferentiated pluripotent stem cells (the so-called neoblasts) are the only proliferative cell population, and are therefore the source of new cells for normal tissue turnover, growth and regeneration, whereas all differentiated cells are post-mitotic (Peter et al., 2004; Reuter and Kreshchenko, 2004; Rink, 2013; Rossi et al., 2008). This is an unusual cellular mechanism for tissue turnover, since in most animals several tissue-specific stem cells exist and at the same time many differentiated cell types are also able to proliferate (Desai et al., 2014; Dor et al., 2004; Yanger and Stanger, 2011). In cestodes, classical studies have described a population of undifferentiated stem cells similar to the neoblasts, usually referred to as the germinative cells (Gustafsson, 1990; Koziol and Castillo, 2011; Reuter and Kreshchenko, 2004).

In particular, in *E. multilocularis* metacestodes classical ultrastructural studies demonstrated the existence of germinative cells in the germinal layer, which proliferate and accumulate during brood capsule and protoscolex development (Sakamoto and Sugimura, 1970). By means of novel molecular markers of cell proliferation and of different cell types in *E. multilocularis*, the undifferentiated germinative cells have been indeed confirmed to be the only proliferating cells in the metacestode, and are the source of new differentiated cells during parasite growth and regeneration (Koziol et al., 2014). Interestingly, although at least one molecular marker is homogeneously expressed in all germinative cells (Fig. 2; Koziol et al., 2015), these cells show extensive heterogeneity in the expression of conserved regulators of pluripotency, suggesting that many sub-populations may exist, with differing self-renewal and differentiation potencies (Koziol et al., 2014). Therefore it is likely that particular sub-populations of germinative cells are already committed to particular paths of differentiation, and that a hierarchy of stem cells and committed progenitors exists as is common in other models.

How host insulin and EGF stimulate the proliferation and differentiation of *Echinococcus* metacestodes at the cellular level remains an open question. As previously described, it appears that the EmIR2 receptor is preferentially expressed in proliferating regions of the metacestode (Hemer et al., 2014) and is therefore likely to be expressed in some of the germinative cells. The expression patterns of other *E. multilocularis* receptors

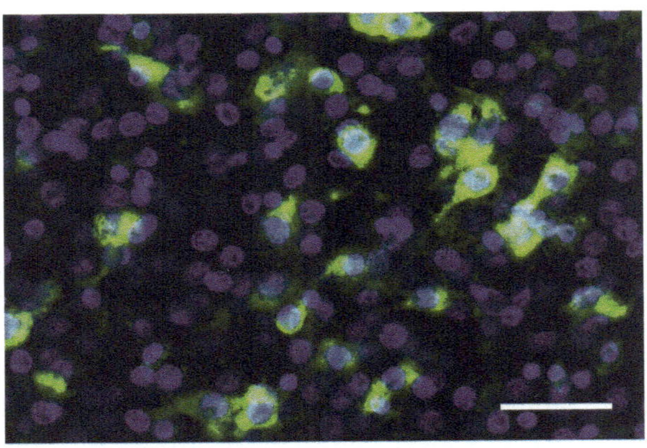

Figure 2 Detection of stem cells in the germinative layer of *Echinococcus multilocularis* by whole-mount in situ hybridization (WMISH) with an antisense probe for the specific marker em-TRIM (Koziol et al., 2015). Green: WMISH signal. Blue: nuclear staining with DAPI. Bar: 20 μm.

are very varied, as some are expressed in virtually all cells of the metacestode, whereas others are expressed in sub-populations of the germinative cells and/or in post-mitotic cells (Koziol and Brehm, unpublished results). Therefore integration of host and parasite signals seems to be a highly complex phenomenon, involving different sub-populations of stem cells and also the response of differentiated cells.

8. TARGETING PARASITE SIGNALLING PATHWAYS FOR CHEMOTHERAPY

The current drugs of choice in anti-echinococcosis chemotherapy are benzimidazoles (BZ; mostly albendazole and mebendazole), the main target molecule of which is β-tubulin, a microtubule component (Martin, 1997). Although β-tubulins are well conserved (more than 90% identical amino acids) between helminths and mammals, the typical helminth β-tubulins show certain amino acid sequence exchanges leading to a higher (100- to 1000-fold) affinity for BZ when compared to mammalian β-tubulin (Lacey, 1990). Since their introduction into echinococcosis chemotherapy, BZs significantly improved the prognosis of patients with AE and CE. However, due to their affinity for host β-tubulin, they are also associated with adverse side effects and, particularly in AE, they act parasitostatic only. As a consequence, they have to be administered for long time periods (sometimes lifelong) and very frequently disease recurrence can be observed upon discontinuation of chemotherapy (Kern, 2010). According to Koziol et al. (2014), the only mitotically active cells of the metacestode are the germinative (stem) cells and it is reasonable to assume that this cell type is not effectively eliminated by BZ during treatment, thus leading to parasite proliferation after chemotherapy discontinuation. At least in the primary cell cultivation system (Spiliotis et al., 2008) which contains up to 85% of germinative cells (Koziol et al., 2014), only limited effects of BZ on parasite cell survival have been observed in our laboratory (Schubert et al., 2014; Hemer and Brehm, unpublished), indicating that the parasite's germinative cell population is largely insensitive to BZ treatment.

The molecular basis for the limited effects of BZ on germinative cells has been discussed by us (Brehm and Koziol, 2014; Koziol and Brehm, 2015) and will only briefly be outlined here. According to genome data (Tsai et al., 2013; Brehm et al., 2000), the parasite contains 10 different β-tubulin-encoding genes, of which three, *tub*-1 to 3, are most abundantly expressed in the metacestode. Two of these, *tub*-1 and *tub*-3, encode β-tubulins with amino acid motifs indicating high affinity to BZ, whereas *tub*-2 codes for a 'mammalian type' β-tubulin isoform which presumably has very limited affinity to BZ (Brehm et al., 2000). Interestingly, transcriptome data on primary cell cultures (Tsai et al., 2013) and our own RT-PCR analyses on metacestode vesicles depleted of germinative cells (Schubert et al., unpublished) indicate that *tub*-2 is specifically expressed in the germinative cells, whereas *tub*-1 and *tub*-3 are the predominant isoforms expressed by differentiated cells. Hence, the parasite's germinative cells are probably insensitive to BZ treatment because they predominantly express a β-tubulin isotype that inherently has low affinity to BZ. For the future development of parasitocidal drugs against echinococcosis it is thus highly important that targets are identified which are also expressed in germinative cells and which either lead to cell killing upon administration or at least inhibit germinative cell proliferation (Brehm and Koziol, 2014).

The fact that *Echinococcus* and mammals use evolutionarily conserved signalling systems for regulating proliferation and differentiation has clear implications for drug design strategies since the underlying pathways are exceptionally well studied, e.g., due to cancer research (Brehm, 2014). Key factors in many of these pathways are kinases which are among the best 'druggable' enzyme families, are biochemically very well investigated and for which a plethora of small molecule compounds has been found to inhibit or to modify enzyme activity (Brehm, 2014). Since many of the respective kinase inhibitors are already in clinical use against different types of cancer, their pharmacokinetic profiles are well known, as are their adverse side effects, providing an 'easy' opportunity to use them either alone or in combination with existing drugs (e.g., albendazole) to treat echinococcosis patients. This might sometimes be problematic since the compounds are originally designed to inhibit the human enzymes so that their affinity for the parasite's enzymes is usually lower, resulting in the need to apply higher drug concentrations to the patient which could result in elevated adverse side effects. On the other hand, existing kinase inhibitors can clearly be used as lead compounds to identify structurally related drugs which have less activity on the human orthologs but higher affinity for the parasite kinases. A few approaches towards using kinases and parasite signalling pathways as drug targets for anti-echinococcosis chemotherapy have been undertaken in the past 10 years and shall be discussed in the following sections.

Imatinib (marketed by Novartis as Gleevec) is an ATP-competitive kinase inhibitor with a relatively benign side effect profile that is in clinical use against a number of different cancers, particularly chronic myelogenous leukaemia (CML) or gastrointestinal stromal tumours (Beckmann and Grevelding, 2010). Imatinib targets kinases of the Abl (Abelson proto oncogene) family such as Abl, which has disordered activity in patients with CML, c-KIT or the platelet-derived growth factor receptor (PDGF-R), but has no significant activity against other tyrosine kinases. First experiments to use imatinib as a drug against flatworm parasites had been carried out by Beckmann and Grevelding (2010) and Beckmann et al. (2011) who demonstrated that schistosomes express several Abl-like kinases that can be inhibited by imatinib. Furthermore, they showed that imatinib, when added to adult schistosomes, affected parasite morphology, pairing and survival (Beckmann and Grevelding, 2010). Encouraged by these experiments, Hemer and Brehm (2012) later also identified related kinases, EmAbl1, EmAbl2 and EmTK6, in *E. multilocularis* where they are expressed in metacestode vesicles and protoscoleces. Furthermore, RT-PCR and transcriptome analyses indicated that the encoding genes are also well expressed in primary cell cultures that are highly enriched in germinative cells (Hemer and Brehm, 2012; Tsai et al., 2013). Interestingly, already at concentrations of 10 μM, which is in the range of imatinib plasma concentrations measured in patients with cancer (6 μM), imatinib had clear anti-parasitic effects on the survival of mature metacestode vesicles and on the formation of vesicles from primary cells in vitro (Hemer and Brehm, 2012). It would thus be worthwhile to also test imatinib in vivo in infected mice either alone or in combination with albendazole. However, care has to be taken in such experiments since serum albumin and the acute-phase factor alpha-1 acidic glycoprotein (AGP) have been shown to negatively affect the activity of imatinib on schistosomes in experimental infections in mice (Beckmann et al., 2014). Since, during infectious processes, AGP levels in mice are up to 8 times higher when compared to humans (see Beckmann et al., 2014), some drugs that might be useful for the treatment of echinococcosis patients might not yield positive results when tested in mice or other rodents. In the *S. mansoni* experiments, the negative effects of AGP could partially be reversed by additional application of the antibiotic erythromycin (Beckmann et al., 2014), which might also be necessary when the anti-*Echinococcus* effects of imatinib are tested in rodents.

A second group of promising compounds with anti-*Echinococcus* activity are the pyridinylimidazoles, which specifically target the p38 MAPK family. In *E. multilocularis*, a member of this kinase family, EmMPK2 (EmuJ_000144900), has been described (Gelmedin et al., 2008) and according to RT-PCR (Gelmedin et al., 2008) and transcriptome analyses (Tsai et al., 2013) is highly expressed in all larval stages (including primary cell cultures) as well as adult worms, indicating that the gene is expressed in both germinative and differentiated cells. Interestingly, the parasite's p38-like MAPK differs biochemically from its human counterpart and appears to be a constitutively active MAPK that is not controlled by upstream MKKs (Gelmedin et al., 2008). When recombinantly expressed in *Escherichia coli*, EmMPK2 showed clear kinase activity which could be inhibited by the pyridinyl imidazole compounds SB202190 and, particularly, ML3403. Furthermore, both compounds had detrimental activities on metacestode vesicles in vitro and prevented the formation of metacestode vesicles from primary cell cultures (Gelmedin et al., 2008). In the case of ML3403 clear anti-parasitic activities were found at concentrations as low as 0.5 μM, which had no effect on the viability of cultured mammalian cells (Gelmedin et al., 2008), indicating that it might also be a potent inhibitor when applied in vivo. Since the primary targets for echinococcosis chemotherapy are the metacestode and the parasite stem cells (as represented by primary cell cultures), no investigations on the effects of pyridinylimidazoles on *E. multilocularis* protoscoleces have been undertaken so far. In the case of *E. granulosus*, however, Lv et al. (2013) measured effects of the inhibitor SB202190 on protoscolex viability in vitro, although quite high drug concentrations (10–80 μM) had been applied. This at least indicates that pyridinylimidazoles might also be effective in the treatment of CE.

An example of a germinative cell-specific drug target has been reported by Schubert et al. (2014) who characterized the *E. multilocularis* cell-cycle regulator EmPlk1 (Polo-like kinase 1). Again, these studies were inspired by previous work in the *Schistosoma* system, where the trematode ortholog, SmPlk1, proved to be a promising target for anti-parasitic chemotherapy (Long et al., 2010). In *E. multilocularis*, the EmPlk1-encoding gene was shown to be specifically expressed in germinative cells, as would be expected since these are the only mitotically active cells of the parasite (Schubert et al., 2014). When recombinantly expressed in the *Xenopus* oocyte system, EmPlk1 displayed clear kinase activity which could be inhibited by the Plk-inhibitor BI2536 and, when added to primary cell cultures, BI 2536 completely prevented the formation of metacestode vesicles from primary cells at concentrations as low as 10 nM (Schubert et al., 2014). Interestingly, when added to mature metacestode vesicles, BI 2536 did not result in structural disintegration but eliminated the parasite's stem cell population and prevented further growth and development (Schubert et al., 2014). Hence, BI 2536 alone might not be an effective drug when given alone, but in combination with drugs that affect the differentiated parasite cells, such as BZ, could potentially prevent the remission of the disease upon discontinuation of chemotherapy.

A number of additional kinase inhibitors had been tested in *Echinococcus* in vitro cultivation systems and either showed limited effects on parasite viability and/or had to be applied at rather high concentrations to observe effects. Although these inhibitors and structurally related compounds might not per se be promising drugs, they are at least informative concerning the identification of druggable parasite pathways. Hemer et al. (2014), for example, had used the insulin receptor inhibitor HNMPA(AM3) on in vitro cultivated parasite larvae. Although the application of concentrations of 25–50 μM were necessary, which is far higher than concentrations that can be achieved in vivo, clear in vitro effects were seen on the formation of metacestode vesicles from primary cells and on the survival of mature metacestode vesicles (Hemer et al., 2014). Since HNMPA(AM3) also clearly inhibited the activation of EmIR1 and downstream insulin signalling components (Hemer et al., 2014), these data at least identify the parasite's insulin signalling pathway as a promising target mechanism. In their screen of Food and Drug Administration–approved drugs, Stadelmann et al. (2014) identified axitinib and sorafenib as being potent killers of germinative cells. Both molecules are multi-kinase inhibitors and probably also affect more than one kinase in the parasite, which could explain the clear effects. In mammals, axitinib is known as an inhibitor of tyrosine kinases of the vascular endothelial growth factor (VEGF)/platelet-derived growth factor (PDGF) receptor family, of which no clear orthologs exist in *Echinococcus* (Tsai et al., 2013). However, FGF receptors are at least distant members of this family and might have acted as targets in the primary cell cultivation system. Sorafenib also acts on VEGF/PDGF receptors but has as another prominent target family the intracellular Raf S/T kinases (MKKKs such as C-Raf and B-Raf), which are orthologs of EmRaf (Spiliotis et al., 2005). The inhibition of EmRaf could indeed lead to effects on parasite stem cells since Gelmedin et al. (2010) already used the Raf inhibitor BAY 34-9006 on metacestode vesicles and, although concentrations of 100 μM of the drug did not result in vesicle disintegration, they were no longer able to proliferate, indicating that the germinative cells were affected. Similar results were also obtained for the MKK inhibitor PD184352, which probably acts on EmMKK2, the target kinase of EmRaf (Gelmedin et al., 2010). The parasite's Erk-like

MAPK cascade module might therefore be a very good target for targeting the stem cells.

Although not directly involving kinases as signal transmitters, the WNT pathway might be another promising target for the development of novel anthelminthics against echinococcosis. As demonstrated by Koziol et al. (2016), the *Echinococcus* metacestode grows within the intermediate host as entirely posteriorized tissue in which WNT ligands are constitutively produced by differentiated muscle cells. Wnt receptors of the Frizzled family are also expressed in the germinative layer (Koziol et al., 2016), including in sub-populations of the germinative stem cells (Koziol and Brehm, unpublished results). An antagonist of WNT signalling, SFRP, the ortholog of which defines the anterior end in planarians, is shut down in metacestode tissue and, later in development, is punctually induced at the sites where protoscoleces are formed, presumably leading to the formation of anterior structures (Koziol et al., 2016). WNT signalling is thus one of the central pathways that regulate metacestode growth and development (Koziol et al., 2016). Interestingly, it has emerged during recent years that WNT signalling also has an important role in cancer development and progression. Like in the case of the *Echinococcus* metacestode, SFRP expression is downregulated in many cancers, usually involving epigenetic silencing of the SFRP promoter (Polakis, 2012a). Likewise, overexpression of β-catenin, the central component and signal transducer of WNT signalling, is associated with many cancer types such as hepatocellular carcinoma, colorectal cancer and lung cancer (Polakis, 2012a). Although there are still many obstacles to effectively target the WNT pathway components by small molecule compounds, significant research efforts are ongoing to identify drugs that either directly or indirectly affect the activities of the intracellular WNT signalling transducers (Polakis, 2012b). It is expected that the outcome of these analyses will also be relevant for the development of anti-echinococcosis drugs.

9. CONCLUSIONS AND OUTLOOK

As is expected in cases when parasites grow for long time periods in close contact to the inner organs of mammals, the molecular interaction mechanisms between *Echinococcus* larvae and their hosts are highly complex. As we have seen, we can basically distinguish between two different evolutionary mechanisms that define these interactions. On the one hand,

E. multilocularis and *E. granulosus* (and probably all cestodes) have undergone a massive reduction of genome size since they separated from the free-living flatworm lineages about 300–500 million years ago, which is clearly an adaptation to the parasitic lifestyle (Tsai et al., 2013; Zheng et al., 2013). This was accompanied by significant gene loss, e.g., for components that define body plan complexity such as *wnt* and *hox* genes (Olson et al., 2012; Tsai et al., 2013). Furthermore, they have lost the capacity to de novo synthesize many important cellular components such as cholesterol, fatty acids and essential amino acids, which, as a consequence, they now have to take up from the host (Tsai et al., 2013; Zheng et al., 2013). To this end, they have newly evolved or expanded gene families that encode factors for lipid binding and uptake such as the AgB subunits or the FABP family (Tsai et al., 2013). In some other cases, such as amino acid transporters, they appear to massively express genes that were already present in the genome before the separation from their free-living ancestors. It can be expected that they also expanded or newly evolved genes to modify the host immune response, although this has still to be investigated in more detail. At least for the cestode-specific AgB family, an additional role in immunomodulation is likely (Rigano et al., 2007), and the same might apply to the Hsp70 family, which has undergone massive expansion in *Echinococcus* (Tsai et al., 2013). Also, the laminated layer as a newly evolved and very prominent *Echinococcus*-specific feature is clearly involved in protecting the larvae from the host immune response (Diaz et al., 2011a).

The second mechanism relies on the close evolutionary relationship between helminths and their mammalian hosts, which share a common ancestor who lived 500–600 million years ago, and who already employed the basic toolkit of animal cell–cell communication mechanisms discussed earlier. This opened the possibility of host–parasite cross-communication via evolutionarily conserved signalling pathways and, in the case of *E. multilocularis*, host insulin and EGF are most likely governing larval development via stimulating parasite insulin and EGF signalling pathways (Spiliotis et al., 2003, 2006; Hemer et al., 2014), whereas some other, as yet undefined factors, likely act on NHR and TGF-β/BMP signalling. Based on the fact that *Echinococcus* larvae also express ligands with high structural similarity to host hormones and cytokines, it is expected that this type of communication is not unidirectional, but bidirectional, i.e., parasite hormones/cytokines are likely to influence corresponding host signalling systems to alter physiological or immunological responses. EmTIP is one telling example of a parasite 'cytokine' that elicits immunomodulatory effects on host cells (Nono et al., 2014).

Concerning proliferation, differentiation and development, the (undifferentiated) germinative cells may be considered the parasite's most important cell type, as they are the only parasite cells capable of division, giving rise to all differentiated cells (Koziol et al., 2014). It is likely that the germinative cells are responsive to parasite-derived cytokines (e.g., WNT factors; Koziol et al., 2016), but based on the stimulating effects of host insulin and EGF on parasite proliferation, they most likely are also responsive to host hormones. How these different signals from parasite and host are integrated into proper stem cell maintenance, proliferation and differentiation, is clearly one of the major challenges for future research. Other open questions for which the molecular and cellular basis have not yet been fully uncovered are, for example, how host macromolecules are transported into the hydatid fluid compartment or how these nutrients are made available to the parasite, either from within the cyst compartment or from outside through the syncytium.

Using the available in vitro cultivation systems for *Echinococcus* larvae (Spiliotis et al., 2004, 2008), and aided by comprehensive genome and transcriptome data (Tsai et al., 2013; Zheng et al., 2013), these questions can now principally be addressed. However, what is still lacking are robust methods for functional genomic analyses to assign functions to all parasite genes with homology to known factors, but particularly to the many genes with no homologies (Tsai et al., 2013). Some first steps have been made in developing RNA-interference protocols for *E. multilocularis* primary cells (Spiliotis et al., 2010) and protoscoleces (Mizukami et al., 2010), which, at least in the case of primary cells, may yield phenotypes that are also maintained in newly regenerated metacestode vesicles (Duvoisin and Brehm, unpublished). Reliable methods for gene knock-in or even targeted knock-out are, however, missing so far. Likewise, the community has not yet achieved the generation of *E. multilocularis* cell lines, which would be an important resource for functional biochemical analyses and for distributing living *E. multilocularis* cell material to a larger research community. It will be interesting to see in the near future whether the recently reported cell line deriving from the *E. granulosus* germinal layer (Albani et al., 2013) can serve these purposes. The urgent need to establish functional genomic methodology not only for *Echinococcus*, but also for a wide variety of parasitic flatworm model systems, has been recognized and respective techniques including virus-based gene knock-in or CRISPR-Cas-mediated gene knock-out, which in other model organisms are well established, should also be operative in flatworms (Hoffmann et al., 2014). Concerted efforts to firmly establish these techniques in *Echinococcus* and schistosomes are ongoing (http://www.sanger.ac.uk/science/projects/fugi) and will hopefully lead to a drastically improved methodological repertoire by which questions of *Echinococcus* development and host–parasite interaction can be addressed in the near future.

ACKNOWLEDGMENTS

The authors wish to thank all colleagues and PhD students working at the Institute of Hygiene and Microbiology (University of Würzburg) and our many collaboration partners who, over the last years, have contributed to the results discussed in this review. We thank the Deutsche Forschungsgemeinschaft (DFG; grants BR 2045/4-1 and IRTG1522 to KB), the Wellcome Trust (grant 107475/Z/15/Z to KB), the 'Echinococcosis Consultant Laboratory' (Robert Koch Institute) and the Wellhöfer Foundation for continuous support of the work discussed herein. UK was supported by a grant of the German Excellence Initiative to the Graduate School of Life Sciences (GSLS), University of Würzburg.

REFERENCES

Aceves-Ramos, A., de la Torre, P., Hinojosa, L., Ponce, A., Garcia-Villegas, R., Laclette, J.P., Bobes, R.J., Romano, M.C., 2014. Cloning, characterization and functional expression of *Taenia solium* 17 beta-hydroxysteroid dehydrogenase. Gen. Comp. Endocrinol. 203, 186–192.

Ahier, A., Rondard, P., Gouignard, N., Khayath, N., Huang, S., Trolet, J., Donoghue, D.J., Gauthier, M., Pin, J.P., Dissous, C., 2009. A new family of receptor tyrosine kinases with a venus flytrap binding domain in insects and other invertebrates activated by aminoacids. PLoS One 4, e5651.

Ahn, C.S., Han, X., Bae, Y.A., Ma, X., Kim, J.T., Cai, H., Yang, H.J., Kang, I., Wang, H., Kong, Y., 2015. Alteration of immunoproteome profile of *Echinococcus granulosus* hydatid fluid with progression of cystic echinococcosis. Parasit. Vectors 8, 10.

Albani, C.M., Cumino, A.C., Elissondo, M.C., Denegri, G.M., 2013. Development of a cell line from *Echinococcus granulosus* germinal layer. Acta Trop. 128, 124–129.

Aldridge Jr., J.R., Jennette, M.A., Kuhn, R.E., 2006. Uptake and secretion of host proteins by *Taenia crassiceps* metacestodes. J. Parasitol. 92, 1101–1102.

Alvarez Rojas, C.A., Romig, T., Lightowlers, M.W., 2014. *Echinococcus granulosus* sensu lato genotypes infecting humans–review of current knowledge. Int. J. Parasitol. 44, 9–18.

Alvite, G., Canclini, L., Corvo, I., Esteves, A., 2008. Two novel *Mesocestoides vogae* fatty acid binding proteins–functional and evolutionary implications. FEBS J. 275, 107–116.

Alvite, G., Esteves, A., 2012. Lipid binding proteins from parasitic platyhelminthes. Front. Physiol. 3, 363.

Ambrosio, J., Landa, A., Merchant, M.T., Laclette, J.P., 1994. Protein uptake by cysticerci of *Taenia crassiceps*. Arch. Med. Res. 25, 325–330.

Ambrosio, J.R., Ostoa-Saloma, P., Palacios-Arreola, M.I., Ruiz-Rosado, A., Sanchez-Orellana, P.L., Reynoso-Ducoing, O., Nava-Castro, K.E., Martinez-Velazquez, N., Escobedo, G., Ibarra-Coronado, E.G., et al., 2014. Oestradiol and progesterone differentially alter cytoskeletal protein expression and flame cell morphology in *Taenia crassiceps*. Int. J. Parasitol. 44, 687–696.

Amiot, F., Vuong, P., Defontaines, M., Pater, C., Dautry, F., Liance, M., 1999. Secondary alveolar echinococcosis in lymphotoxin-alpha and tumour necrosis factor-alpha deficient mice: exacerbation of *Echinococcus multilocularis* larval growth is associated with cellular changes in the periparasitic granuloma. Parasite Immunol. 21, 475–483.

Aziz, A., Zhang, W., Li, J., Loukas, A., McManus, D.P., Mulvenna, J., 2011. Proteomic characterisation of *Echinococcus granulosus* hydatid cyst fluid from sheep, cattle and humans. J. Proteomics 74, 1560–1572.

Bahr, J.M., Frayha, G.J., Hajjar, J.-J., 1979. Mechanism of cholesterol absorption by the hydatid cysts of *Echinococcus granulosus* (Cestoda). Comp. Biochem. Physiol.Part A Physiol. 62, 485–489.

Beckmann, S., Grevelding, C.G., 2010. Imatinib has a fatal impact on morphology, pairing stability and survival of adult *Schistosoma mansoni* in vitro. Int. J. Parasitol. 40, 521–526.

Beckmann, S., Hahnel, S., Cailliau, K., Vanderstraete, M., Browaeys, E., Dissous, C., Grevelding, C.G., 2011. Characterization of the Src/Abl hybrid kinase SmTK6 of *Schistosoma mansoni*. J. Biol. Chem. 286, 42325–42336.

Beckmann, S., Long, T., Scheld, C., Geyer, R., Caffrey, C.R., Grevelding, C.G., 2014. Serum albumin and alpha-1 acid glycoprotein impede the killing of *Schistosoma mansoni* by the tyrosine kinase inhibitor Imatinib. Int. J. Parasitol. Drugs Drug Resist. 4, 287–295.

Bernthaler, P., Epping, K., Schmitz, G., Deplazes, P., Brehm, K., 2009. Molecular characterization of EmABP, an apolipoprotein A-I binding protein secreted by the *Echinococcus multilocularis* metacestode. Infect. Immun. 77, 5564–5571.

Bonay, P., Gonzalez, L.M., Benitez, L., Foster, M., Harrison, L.J., Parkhouse, R.M., Garate, T., 2002. Genomic and functional characterisation of a secreted antigen of *Taenia saginata* oncospheres. Mol. Biochem. Parasitol. 121, 269–273.

Bowles, J., Blair, D., McManus, D.P., 1992. Genetic variants within the genus *Echinococcus* identified by mitochondrial DNA sequencing. Mol. Biochem. Parasitol. 54, 165–173.

Brehm, K., 2014. Protein kinases as drug targets in the treatment of alveolar echinococcosis. In: Doerig, C., Späth, G., Wiese, M. (Eds.), Protein Phosphorylation in Parasites, vol. 5. Wiley Blackwell, pp. 357–373.

Brehm, K., 2010a. *Echinococcus multilocularis* as an experimental model in stem cell research and molecular host-parasite interaction. Parasitology 137, 537–555.

Brehm, K., 2010b. The role of evolutionarily conserved signalling systems in *Echinococcus multilocularis* development and host-parasite interaction. Med. Microbiol. Immunol. 199, 247–259.

Brehm, K., Koziol, U., 2014. On the importance of targeting parasite stem cells in anti-echinococcosis drug development. Parasite 21, 72.

Brehm, K., Kronthaler, K., Jura, H., Frosch, M., 2000. Cloning and characterization of beta-tubulin genes from *Echinococcus multilocularis*. Mol. Biochem. Parasitol. 107, 297–302.

Brehm, K., Spiliotis, M., Zavala-Gongora, R., Konrad, C., Frosch, M., 2006. The molecular mechanisms of larval cestode development: first steps into an unknown world. Parasitol. Int. 55, S15–S21.

Brehm, K., Spiliotis, M., 2008. Recent advances in the in vitro cultivation and genetic manipulation of *Echinococcus multilocularis* metacestodes and germinal cells. Exp. Parasitol. 119, 506–515.

Brehm, K., Wolf, M., Beland, H., Kroner, A., Frosch, M., 2003. Analysis of differential gene expression in *Echinococcus multilocularis* larval stages by means of spliced leader differential display. Int. J. Parasitol. 33, 1145–1159.

Bridgham, J.T., Eick, G.N., Larroux, C., Deshpande, K., Harms, M.J., Gauthier, M.E., Ortlund, E.A., Degnan, B.M., Thornton, J.W., 2010. Protein evolution by molecular tinkering: diversification of the nuclear receptor superfamily from a ligand-dependent ancestor. PLoS Biol. 8.

Brunetti, E., Kern, P., Vuitton, D.A., 2010. Expert consensus for the diagnosis and treatment of cystic and alveolar echinococcosis. Acta Trop. 114, 1–16.

Budke, C.M., Deplazes, P., Torgerson, P.R., 2006. Global socioeconomic impact of cystic echinococcosis. Emerg. Inf. Dis. 12, 296–303.

Camicia, F., Herz, M., Prada, L.C., Kamenetzky, L., Simonetta, S.H., Cucher, M.A., Bianchi, J.I., Fernandez, C., Brehm, K., Rosenzvit, M.C., 2013. The nervous and prenervous roles of serotonin in *Echinococcus* spp. Int. J. Parasitol. 43, 647–659.

Camicia, F., Paredes, R., Chalar, C., Galanti, N., Kamenetzky, L., Gutierrez, A., Rosenzvit, M.C., 2008. Sequencing, bioinformatic characterization and expression pattern of a putative amino acid transporter from the parasitic cestode *Echinococcus granulosus*. Gene 411, 1–9.

Cardona, G.A., Carmena, D., 2013. A review of the global prevalence, molecular epidemiology and economics of cystic echinococccosis in production animals. Vet. Parasitol. 192, 10–32.

Celik, C., Amanvermez, R., Ozkan, K., 2001. Free amino acid concentration in hydatid cyst fluids from fertile and infertile human and animal *Echinococcus granulosus*. Parasite 8, 343–348.

Coltorti, E.A., Varela-Diaz, V.M., 1972. IgG levels and host specificity in hydatid cyst fluid. J. Parasitol. 58, 753–756.

Coltorti, E.A., Varela-Diaz, V.M., 1974. *Echinococcus granulosus*: penetration of macromolecules and their localization in the parasite membrane of cysts. Exp. Parasitol. 35, 225–231.

Coltorti, E.A., Varela-Diaz, V.M., 1975. Penetration of host IgG molecules into hydatid cysts. Z. Parasitenkd. 48, 47–51.

Conradt, U., Peters, W., 1989. Investigations on the occurrence of pinocytosis in the tegument of *Schistocephalus solidus*. Parasitol. Res. 75, 630–635.

Craig, P., 2003. Echinococcus multilocularis. Curr. Opin. Infect. Dis. 16, 437–444.

Chemale, G., Ferreira, H.B., Barrett, J., Brophy, P.M., Zaha, A., 2005. *Echinococcus granulosus* antigen B hydrophobic ligand binding properties. Biochim. Biophys. Acta 1747, 189–194.

Chemale, G., Haag, K.L., Ferreira, H.B., Zaha, A., 2001. *Echinococcus granulosus* antigen B is encoded by a gene family. Mol. Biochem. Parasitol. 116, 233–237.

Chemale, G., van Rossum, A.J., Jefferies, J.R., Barrett, J., Brophy, P.M., Ferreira, H.B., Zaha, A., 2003. Proteomic analysis of the larval stage of the parasite *Echinococcus granulosus*: causative agent of cystic hydatid disease. Proteomics 3, 1633–1636.

D'Alessandro, A., Rausch, R.L., 2008. New aspects of neotropical polycystic (*Echinococcus vogeli*) and unicystic (*Echinococcus oligarthrus*) echinococcosis. Clin. Microbiol. Rev. 21, 380–401.

Deplazes, P., Rinaldi, L., Alvarez Rojas, C.A., Torgerson, P., Harandi, M.F., Romig, T., Antolova, D., Schurer, J., Lahmar, S., Cringoli, G., Magambo, J., Thompson, A., Jenkins, E., 2017. Global distribution of alveolar and cystic Echinococcosis. In: Thompson, A. (Ed.), *Echinococcus* and Echinococcosis, Part A, vol. 95, pp. 315–494.

Desai, T.J., Brownfield, D.G., Krasnow, M.A., 2014. Alveolar progenitor and stem cells in lung development, renewal and cancer. Nature 507, 190–194.

Diaz, A., Casaravilla, C., Allen, J.E., Sim, R.B., Ferreira, A.M., 2011a. Understanding the laminated layer of larval *Echinococcus* II: immunology. Trends Parasitol. 27, 264–273.

Diaz, A., Casaravilla, C., Barrios, A.A., Ferreira, A.M., 2015a. Parasite molecules and host responses in cystic echinococcosis. Parasite Immunol. 38.

Diaz, A., Casaravilla, C., Irigoin, F., Lin, G., Previato, J.O., Ferreira, F., 2011b. Understanding the laminated layer of larval *Echinococcus* I: structure. Trends Parasitol. 27, 204–213.

Diaz, A., Fernandez, C., Pittini, A., Seoane, P.I., Allen, J.E., Casaravilla, C., 2015b. The laminated layer: recent advances and insights into *Echinococcus* biology and evolution. Exp. Parasitol. 158, 23–30.

Diaz, A., Fontana, E.C., Todeschini, A.R., Soule, S., Gonzalez, H., Casaravilla, C., Portela, M., Mohana-Borges, R., Mendonca-Previato, L., Previato, J.O., et al., 2009. The major surface carbohydrates of the *Echinococcus granulosus* cyst: mucin-type O-glycans decorated by novel galactose-based structures. Biochemistry 48, 11678–11691.

Dor, Y., Brown, J., Martinez, O.I., Melton, D.A., 2004. Adult pancreatic beta-cells are formed by self-duplication rather than stem-cell differentiation. Nature 429, 41–46.

Eckert, J., Deplazes, P., 2004. Biological, epidemiological, and clinical aspects of echinococcosis, a zoonosis of increasing concern. Clin. Microbiol. Rev. 17, 107–135.

Eckert, J., Thompson, R.C., Mehlhorn, H., 1983. Proliferation and metastases formation of larval *Echinococcus multilocularis*. I. Animal model, macroscopical and histological findings. Z. Parasitenkd. 69, 737–748.

Epping, K., Brehm, K., 2011. *Echinococcus multilocularis*: molecular characterization of EmSmadE, a novel BR-Smad involved in TGF-beta and BMP signaling. Exp. Parasitol. 129, 85–94.

Escobedo, G., Camacho-Arroyo, I., Hernandez-Hernandez, O.T., Ostoa-Saloma, P., Garcia-Varela, M., Morales-Montor, J., 2010. Progesterone induces scolex evagination of the human parasite *Taenia solium*: evolutionary implications to the host-parasite relationship. J. Biomed. Biotechnol. 2010, 591079.

Escobedo, G., Larralde, C., Chavarria, A., Cerbon, M.A., Morales-Montor, J., 2004. Molecular mechanisms involved in the differential effects of sex steroids on the reproduction and infectivity of *Taenia crassiceps*. J. Parasitol. 90, 1235–1244.

Ferreira, A.M., Irigoin, F., Breijo, M., Sim, R.B., Diaz, A., 2000. How *Echinococcus granulosus* deals with complement. Parasitol. Today 16, 168–172.

Figueiredo, B.C., Da'dara, A.A., Oliveira, S.C., Skelly, P.J., 2015. Schistosomes enhance plasminogen activation: the role of tegumental enolase. PLoS Pathog. 11, e1005335.

Fiscella, M., Perry, J.W., Teng, B., Bloom, M., Zhang, C., Leung, K., Pukac, L., Florence, K., Concepcion, A., Liu, B., et al., 2003. TIP, a T-cell factor identified using high-throughput screening increases survival in a graft-versus-host disease model. Nat. Biotechnol. 21, 302–307.

Förster, S., Gunthel, D., Kiss, F., Brehm, K., 2011. Molecular characterisation of a serum-responsive, DAF-12-like nuclear hormone receptor of the fox-tapeworm *Echinococcus multilocularis*. J. Cell Biochem. 112, 1630–1642.

Franchini, G.R., Porfido, J.L., Ibanez Shimabukuro, M., Rey Burusco, M.F., Belgamo, J.A., Smith, B.O., Kennedy, M.W., Corsico, B., 2015. The unusual lipid binding proteins of parasitic helminths and their potential roles in parasitism and as therapeutic targets. Prostaglandins Leukot. Essent. Fatty Acids 93, 31–36.

Frayha, G.J., Haddad, R., 1980. Comparative chemical composition of protoscolices and hydatid cyst fluid of *Echinococcus granulosus* (Cestoda). Int. J. Parasitol. 10, 359–364.

Freeman, R.S., 1973. Ontogeny of cestodes and its bearing on their phylogeny and systematics. Adv. Parasitol. 11, 481–557.

Freitas, T.C., Jung, E., Pearce, E.J., 2007. TGF-beta signaling controls embryo development in the parasitic flatworm *Schistosoma mansoni*. PLoS Pathog. 3, e52.

Gelmedin, V., Caballero-Gamiz, R., Brehm, K., 2008. Characterization and inhibition of a p38-like mitogen-activated protein kinase (MAPK) from *Echinococcus multilocularis*: antiparasitic activities of p38 MAPK inhibitors. Biochem. Pharmacol. 76, 1068–1081.

Gelmedin, V., Spiliotis, M., Brehm, K., 2010. Molecular characterisation of MEK1/2- and MKK3/6-like mitogen-activated protein kinase kinases (MAPKK) from the fox tapeworm *Echinococcus multilocularis*. Int. J. Parasitol. 40, 555–567.

Gelmedin, V., Zavala-Gongora, R., Fernandez, C., Brehm, K., 2005. *Echinococcus multilocularis*: cloning and characterization of a member of the SNW/SKIP family of transcriptional coregulators. Exp. Parasitol. 111, 115–120.

Golzari, S.E., Sokouti, M., 2014. Pericyst: the outermost layer of hydatid cyst. World J. Gastroenterol. 20, 1377–1378.

Gottstein, B., Hemphill, A., 2008. *Echinococcus multilocularis*: the parasite-host interplay. Exp. Parasitol. 119, 447–452.

Gottstein, B., Soboslay, P., Ortona, E., Wang, J., Siracusano, A., Vuitton, D.A., 2017. Immunology of alveolar and cystic Echinococcosis (AE and CE). In: Thompson, A. (Ed.), *Echinococcus* and Echinococcosis, Part B, vol. 96, pp. 1–54.

Gottstein, B., Wang, J., Boubaker, G., Marinova, I., Spiliotis, M., Muller, N., Hemphill, A., 2015. Susceptibility versus resistance in alveolar echinococcosis (larval infection with *Echinococcus multilocularis*). Vet. Parasitol. 213, 103–109.

Guerret, S., Vuitton, D.A., Liance, M., Pater, C., Carbillet, J.P., 1998. *Echinococcus multilocularis*: relationship between susceptibility/resistance and liver fibrogenesis in experimental mice. Parasitol. Res. 84, 657–667.

Gustafsson, M.K.S., 1990. The cells of a cestode. *Diphyllobothrium dendriticum* as a model in cell biology. Acta Acad. Abo. Ser. B 50, 13–44.

Haag, K.L., Gottstein, B., Ayala, F.J., 2009. The EG95 antigen of *Echinococcus* spp. contains positively selected amino acids, which may influence host specificity and vaccine efficacy. PLoS One 4, e5362.

Hamdan, F.F., Abramovitz, M., Mousa, A., Xie, J., Durocher, Y., Ribeiro, P., 2002. A novel *Schistosoma mansoni* G protein-coupled receptor is responsive to histamine. Mol. Biochem. Parasitol. 119, 75–86.

Harris, A., Heath, D.D., Lawrence, S.B., Shaw, R.J., 1989. *Echinococcus granulosus*: ultrastructure of epithelial changes during the first 8 days of metacestode development in vitro. Int. J. Parasitol. 19, 621–629.

Heath, D.D., 1971. The migration of oncospheres of *Taenia pisiformis*, *T. serialis* and *Echinococcus granulosus* within the intermediate host. Int. J. Parasitol. 1, 145–152.

Heath, D.D., Lawrence, S.B., 1996. Antigenic polypeptides of *Echinococcus granulosus* oncospheres and definition of protective molecules. Parasite Immunol. 18, 347–357.

Heath, D.D., Smyth, J.D., 1970. In vitro cultivation of *Echinococcus granulosus*, *Taenia hydatigena*, *T. ovis*, *T. pisiformis* and *T. serialis* from oncosphere to cystic larva. Parasitology 61, 329–343.

Hemer, S., Brehm, K., 2012. In vitro efficacy of the anticancer drug imatinib on *Echinococcus multilocularis* larva. Int. J. Antimicrob. Agents 40, 458–462.

Hemer, S., Konrad, C., Spiliotis, M., Koziol, U., Schaack, D., Förster, S., Gelmedin, V., Stadelmann, B., Dandekar, T., Hemphill, A., et al., 2014. Host insulin stimulates *Echinococcus multilocularis* insulin signalling pathways and larval development. BMC Biol. 12, 5.

Hemphill, A., Gottstein, B., 1995. Immunology and morphology studies on the proliferation of in vitro cultivated *Echinococcus multilocularis* metacestodes. Parasitol. Res. 81, 605–614.

Hoffmann, K.F., Brindley, P.J., Berriman, M., 2014. Medicine. Halting harmful helminths. Science 346, 168–169.

Holcman, B., Heath, D.D., Shaw, R.J., 1994. Ultrastructure of oncosphere and early stages of metacestode development of *Echinococcus granulosus*. Int. J. Parasitol. 24, 623–635.

Hsu, T.L., Lin, G., Koizumi, A., Brehm, K., Hada, N., Chuang, P.K., Wong, C.H., Hsieh, S.L., Diaz, A., 2013. The surface carbohydrates of the *Echinococcus granulosus* larva interact selectively with the rodent Kupffer cell receptor. Mol. Biochem. Parasitol. 192, 55–59.

Huber, S., Stahl, F.R., Schrader, J., Luth, S., Presser, K., Carambia, A., Flavell, R.A., Werner, S., Blessing, M., Herkel, J., et al., 2009. Activin a promotes the TGF-beta-induced conversion of $CD4^{+}CD25^{-}$ T cells into $Foxp3^{+}$ induced regulatory T cells. J. Immunol. 182, 4633–4640.

Huberts, D.H., van der Klei, I.J., 2010. Moonlighting proteins: an intriguing mode of multitasking. Biochim. Biophys. Acta 1803, 520–525.

Hurd, H., 1989. *Echinococcus granulosus*: a comparison of free amino acid concentration in hydatid fluid from primary and secondary cysts and host plasma. Parasitology 98 (Pt 1), 135–143.

Hustead, S.T., Williams, J.F., 1977. Permeability studies on taenid metacestodes: I. Uptake of proteins by larval stages of *Taenia taeniaeformis*, *T. crassiceps*, and *Echinococcus granulosus*. J. Parasitol. 63, 314–321.

Jabbar, A., Jenkins, D.J., Crawford, S., Walduck, A.K., Gauci, C.G., Lightowlers, M.W., 2011. Oncospheral penetration glands are the source of the EG95 vaccine antigen against cystic hydatid disease. Parasitology 138, 89–99.

Jabbar, A., Kyngdon, C.T., Gauci, C.G., Walduck, A.K., McCowan, C., Jones, M.K., Beveridge, I., Lightowlers, M.W., 2010. Localisation of three host-protective oncospheral antigens of *Taenia ovis*. Int. J. Parasitol. 40, 579–589.

Jeffs, S.A., Arme, C., 1987. *Echinococcus granulosus*: specificity of amino acid transport systems in protoscoleces. Parasitology 95 (Pt 1), 71–78.

Jeffs, S.A., Arme, C., 1988. *Echinococcus granulosus* (Cestoda): uptake of L-amino acids by secondary hydatid cysts. Parasitology 96 (Pt 1), 145–156.

Jura, H., Bader, A., Hartmann, M., Maschek, H., Frosch, M., 1996. Hepatic tissue culture model for study of host-parasite interactions in alveolar echinococcosis. Infect. Immun. 64, 3484–3490.

Kaiser, P., Rothwell, L., Avery, S., Balu, S., 2004. Evolution of the interleukins. Dev. Comp. Immunol. 28, 375–394.

Kern, P., 2010. Clinical features and treatment of alveolar echinococcosis. Curr. Opin. Infect. Dis. 23, 505–512.

Konrad, C., Kroner, A., Spiliotis, M., Zavala-Gongora, R., Brehm, K., 2003. Identification and molecular characterisation of a gene encoding a member of the insulin receptor family in *Echinococcus multilocularis*. Int. J. Parasitol. 33, 301–312.

Koziol, U., Brehm, K., 2015. Recent advances in *Echinococcus* genomics and stem cell research. Vet. Parasitol. 213, 92–102.

Koziol, U., Castillo, E., 2011. Cell proliferation and differentiation in cestodes. In: Esteves, A. (Ed.), Research in Helminths. Transworld Research Network, Kerala, India, pp. 121–138.

Koziol, U., Jarero, F., Olson, P.D., Brehm, K., 2016. Comparative analysis of Wnt expression identifies a highly conserved developmental transition in flatworms. BMC Biol. 14, 10.

Koziol, U., Krohne, G., Brehm, K., 2013. Anatomy and development of the larval nervous system in *Echinococcus multilocularis*. Front. Zool. 10, 24.

Koziol, U., Radio, S., Smircich, P., Zarowiecki, M., Fernandez, C., Brehm, K., 2015. A novel terminal-repeat retrotransposon in miniature (TRIM) is massively expressed in *Echinococcus multilocularis* stem cells. Genome Biol. Evol. 7, 2136–2153.

Koziol, U., Rauschendorfer, T., Zanon Rodriguez, L., Krohne, G., Brehm, K., 2014. The unique stem cell system of the immortal larva of the human parasite *Echinococcus multilocularis*. Evodevo 5, 10.

Kultz, D., 1998. Phylogenetic and functional classification of mitogen- and stress-activated protein kinases. J. Mol. Evol. 46, 571–588.

Kumaratilake, L.M., Thompson, R.C., 1981. Maintenance of the life cycle of *Echinococcus granulosus* in the laboratory following in vivo and in vitro development. Z. Parasitenkd. 65, 103–106.

Lacey, E., 1990. Mode of action of benzimidazoles. Parasitol. Today 6, 112–115.

Lascano, E.F., Coltorti, E.A., Varela-Diaz, V.M., 1975. Fine structure of the germinal membrane of *Echinococcus granulosus* cysts. J. Parasitol. 61, 853–860.

Lee, E.G., Kim, S.H., Bae, Y.A., Chung, J.Y., Suh, M., Na, B.K., Kim, T.S., Kang, I., Ma, L., Kong, Y., 2007. A hydrophobic ligand-binding protein of the *Taenia solium* metacestode mediates uptake of the host lipid: implication for the maintenance of parasitic cellular homeostasis. Proteomics 7, 4016–4030.

Lightowlers, M.W., Gauci, C.G., Chow, C., Drew, D.R., Gauci, S.M., Heath, D.D., Jackson, D.C., Dadley-Moore, D.L., Read, A.J., 2003. Molecular and genetic characterisation of the host-protective oncosphere antigens of taeniid cestode parasites. Int. J. Parasitol. 33, 1207–1217.

Lightowlers, M.W., Lawrence, S.B., Gauci, C.G., Young, J., Ralston, M.J., Maas, D., Health, D.D., 1996. Vaccination against hydatidosis using a defined recombinant antigen. Parasite Immunol. 18, 457–462.

Lin, R.Y., Wang, J.H., Lu, X.M., Zhou, X.T., Mantion, G., Wen, H., Vuitton, D.A., Richert, L., 2009. Components of the mitogen-activated protein kinase cascade are activated in hepatic cells by *Echinococcus multilocularis* metacestode. World J. Gastroenterol. 15, 2116–2124.

Liu, R., Zhao, Q.P., Ye, Q., Xiong, T., Tang, C.L., Dong, H.F., Jiang, M.S., 2013. Cloning and characterization of a bone morphogenetic protein homologue of *Schistosoma japonicum*. Exp. Parasitol. 135, 64–71.

Locksley, R.M., Killeen, N., Lenardo, M.J., 2001. The TNF and TNF receptor superfamilies: integrating mammalian biology. Cell 104, 487–501.

Long, T., Cailliau, K., Beckmann, S., Browaeys, E., Trolet, J., Grevelding, C.G., Dissous, C., 2010. *Schistosoma mansoni* Polo-like kinase 1: a mitotic kinase with key functions in parasite reproduction. Int. J. Parasitol. 40, 1075–1086.

Lorenzatto, K.R., Monteiro, K.M., Paredes, R., Paludo, G.P., da Fonseca, M.M., Galanti, N., Zaha, A., Ferreira, H.B., 2012. Fructose-bisphosphate aldolase and enolase from *Echinococcus granulosus*: genes, expression patterns and protein interactions of two potential moonlighting proteins. Gene 506, 76–84.

Lymbery, A.J., 2017. Phylogenetic pattern, evolutionary processes and species delimitation in the genus *Echinococcus*. In: Thompson, A. (Ed.), *Echinococcus* and Echinococcosis, Part A, vol. 95, pp. 111–146.

Lv, H., Li, S., Zhang, J., Liang, W., Mu, X., Jiang, Y., 2013. In vitro effects of SB202190 on *Echinococcus granulosus*. Korean J. Parasitol. 51, 255–258.

Marcilla, A., Martin-Jaular, L., Trelis, M., de Menezes-Neto, A., Osuna, A., Bernal, D., Fernandez-Becerra, C., Almeida, I.C., Del Portillo, H.A., 2014. Extracellular vesicles in parasitic diseases. J. Extracell. Vesicles 3, 25040.

Martin, R.J., 1997. Modes of action of anthelmintic drugs. Vet. J. 154, 11–34.

McManus, D.P., Smyth, J.D., 1978. Differences in the chemical composition and carbohydrate metabolism of *Echinococcus granulosus* (horse and sheep strains) and *E. multilocularis*. Parasitology 77, 103–109.

McManus, D.P., Smyth, J.D., 1982. Intermediary carbohydrate metabolism in protoscoleces of *Echinococcus granulosus* (horse and sheep strains) and *E. multilocularis*. Parasitology 84, 351–366.

Mehlhorn, H., Eckert, J., Thompson, R.C., 1983. Proliferation and metastases formation of larval *Echinococcus multilocularis*. II. Ultrastructural investigations. Z. Parasitenkd. 69, 749–763.

Mejri, N., Muller, N., Hemphill, A., Gottstein, B., 2011. Intraperitoneal *Echinococcus multilocularis* infection in mice modulates peritoneal $CD4^+$ and $CD8^+$ regulatory T cell development. Parasitol. Int. 60, 45–53.

Mizukami, C., Spiliotis, M., Gottstein, B., Yagi, K., Katakura, K., Oku, Y., 2010. Gene silencing in *Echinococcus multilocularis* protoscoleces using RNA interference. Parasitol. Int. 59, 647–652.

Monteiro, K.M., de Carvalho, M.O., Zaha, A., Ferreira, H.B., 2010. Proteomic analysis of the *Echinococcus granulosus* metacestode during infection of its intermediate host. Proteomics 10, 1985–1999.

Moro, P., Schantz, P.M., 2009. Echinococcosis: a review. Int. J. Infect. Dis. 13, 125–133.
Morseth, D.J., 1967. Fine structure of the hydatid cyst and protoscolex of *Echinococcus granulosus*. J. Parasitol. 53, 312–325.
Nakao, M., Yanagida, T., Okamoto, M., Knapp, J., Nkouawa, A., Sako, Y., Ito, A., 2010. State-of-the-art *Echinococcus* and *Taenia*: phylogenetic taxonomy of human-pathogenic tapeworms and its application to molecular diagnosis. Infect. Genet. Evol. 10, 444–452.
Navarrete-Perea, J., Moguel, B., Mendoza-Hernandez, G., Fragoso, G., Sciutto, E., Bobes, R.J., Laclette, J.P., 2014. Identification and quantification of host proteins in the vesicular fluid of porcine *Taenia solium* cysticerci. Exp. Parasitol. 143, 11–17.
Nicolao, M.C., Elissondo, M.C., Denegri, G.M., Goya, A.B., Cumino, A.C., 2014. In vitro and in vivo effects of tamoxifen against larval stage *Echinococcus granulosus*. Antimicrob. Agents Chemother. 58, 5146–5154.
Nono, J.K., Lutz, M.B., Brehm, K., 2014. EmTIP, a T-Cell immunomodulatory protein secreted by the tapeworm *Echinococcus multilocularis* is important for early metacestode development. PLoS Negl. Trop. Dis. 8, e2632.
Nono, J.K., Pletinckx, K., Lutz, M.B., Brehm, K., 2012. Excretory/secretory-products of *Echinococcus multilocularis* larvae induce apoptosis and tolerogenic properties in dendritic cells in vitro. PLoS Negl. Trop. Dis. 6, e1516.
Norman, L., Kagan, I.G., 1961. The maintenance of *Echinococcus multilocularis* in gerbils (*Meriones unguiculatus*) by intraperitoneal inoculation. J. Parasitol. 47, 870–874.
Obal, G., Ramos, A.L., Silva, V., Lima, A., Batthyany, C., Bessio, M.I., Ferreira, F., Salinas, G., Ferreira, A.M., 2012. Characterisation of the native lipid moiety of *Echinococcus granulosus* antigen B. PLoS Negl. Trop. Dis. 6, e1642.
Ohbayashi, M., 1960. Studies on echinococcosis. X. Histological observations on experimental cases of multilocular echinococcosis. Jpn. J. Vet. Res. 8, 134–160.
Oliveira, K.C., Carvalho, M.L., Venancio, T.M., Miyasato, P.A., Kawano, T., DeMarco, R., Verjovski-Almeida, S., 2009. Identification of the *Schistosoma mansoni* TNF-alpha receptor gene and the effect of human TNF-alpha on the parasite gene expression profile. PLoS Negl. Trop. Dis. 3, e556.
Olson, P.D., Zarowiecki, M., Kiss, F., Brehm, K., 2012. Cestode genomics – progress and prospects for advancing basic and applied aspects of flatworm biology. Parasite Immunol. 34, 130–150.
Omenetti, A., Choi, S., Michelotti, G., Diehl, A.M., 2011. Hedgehog signaling in the liver. J. Hepatol. 54, 366–373.
Pappas, P.W., 1975. Membrane transport in helminth parasites: a review. Exp. Parasitol. 37, 469–530.
Parkinson, J., Wasmuth, J.D., Salinas, G., Bizarro, C.V., Sanford, C., Berriman, M., Ferreira, H.B., Zaha, A., Blaxter, M.L., Maizels, R.M., et al., 2012. A transcriptomic analysis of *Echinococcus granulosus* larval stages: implications for parasite biology and host adaptation. PLoS Negl. Trop. Dis. 6, e1897.
Patocka, N., Sharma, N., Rashid, M., Ribeiro, P., 2014. Serotonin signaling in *Schistosoma mansoni*: a serotonin-activated G protein-coupled receptor controls parasite movement. PLoS Pathog. 10, e1003878.
Peter, R., Gschwentner, R., Schürmann, W., Rieger, R.M., Ladurner, P., 2004. The significance of stem cells in free-living flatworms: one common source for all cells in the adult. J. Appl. Biomed. 2, 21–35.
Petersen, C.P., Reddien, P.W., 2008. Smed-betacatenin-1 is required for anteroposterior blastema polarity in planarian regeneration. Science 319, 327–330.
Petrone, L., Vanini, V., Petruccioli, E., Ettore, G.M., Schinina, V., Busi Rizzi, E., Ludovisi, A., Corpolongo, A., Ippolito, G., Pozio, E., Teggi, A., Goletti, D., 2015. Polyfunctional specific response to *Echinococcus granulosus* associates to the biological activity of the cyst. PLoS Negl. Trop. Dis. 9, e0004209.

Pires-daSilva, A., Sommer, R.J., 2003. The evolution of signalling pathways in animal development. Nat. Rev. Genet. 4, 39—49.
Polakis, P., 2012a. Wnt signalling in cancer. Cold Spring Harb. Perspect. Biol. 4, 5.
Polakis, P., 2012b. Drugging Wnt signalling in cancer. EMBO J. 31, 2737—2746.
Porfido, J.L., Alvite, G., Silva, V., Kennedy, M.W., Esteves, A., Corsico, B., 2012. Direct interaction between EgFABP1, a fatty acid binding protein from *Echinococcus granulosus*, and phospholipid membranes. PLoS Negl. Trop. Dis. 6, e1893.
Protasio, A.V., Tsai, I.J., Babbage, A., Nichol, S., Hunt, M., Aslett, M.A., De Silva, N., Velarde, G.S., Anderson, T.J., Clark, R.C., et al., 2012. A systematically improved high quality genome and transcriptome of the human blood fluke *Schistosoma mansoni*. PLoS Negl. Trop. Dis. 6, e1455.
Qiu, C., Fu, Z., Shi, Y., Hong, Y., Liu, S., Lin, J., 2013. A retinoid X receptor (RXR1) homolog from *Schistosoma japonicum*: its ligand-binding domain may bind to 9-cis-retinoic acid. Mol. Biochem. Parasitol. 188, 40—50.
Raposo, G., Stoorvogel, W., 2013. Extracellular vesicles: exosomes, microvesicles, and friends. J. Cell Biol. 200, 373—383.
Rausch, R., 1954. Studies on the helminth fauna of Alaska. XX. The histogenesis of the alveolar larva of *Echinococcus* species. J. Infect. Dis. 94, 178—186.
Rausch, R., Jentoft, V.L., 1957. Studies on the helminth fauna of Alaska. XXXI. Observations on the propagation of the larval *Echinococcus multilocularis* Leuckart, 1863, in vitro. J. Parasitol. 43, 1—8.
Reuter, M., Kreshchenko, N., 2004. Flatworm asexual multiplication implicates stem cells and regeneration. Can. J. Zool. 82, 334—356.
Rigano, R., Buttari, B., Profumo, E., Ortona, E., Delunardo, F., Margutti, P., Mattei, V., Teggi, A., Sorice, M., Siracusano, A., 2007. *Echinococcus granulosus* antigen B impairs human dendritic cell differentiation and polarizes immature dendritic cell maturation towards a Th2 response. Infect. Immun. 75, 1667—1678.
Riddiford, N., Olson, P.D., 2011. Wnt gene loss in flatworms. Dev. Genes Evol. 221, 187—197.
Rink, J.C., 2013. Stem cell systems and regeneration in planaria. Dev. Genes Evol. 223, 67—84.
Romig, T., Ebi, D., Wassermann, M., 2015. Taxonomy and molecular epidemiology of *Echinococcus granulosus* sensu lato. Vet. Parasitol. 213, 76—84.
Romig, T., Deplazes, P., Jenkins, D., Giraudoux, P., Massolo, A., Craig, P.S., Wassermann, M., Takahashi, K., de la Rue, M., 2017. Ecology and life cycle patterns of *Echinococcus* species. In: Thompson, A. (Ed.), *Echinococcus* and Echinococcosis, Part A, vol. 95, pp. 213—314.
Rossi, L., Salvetti, A., Batistoni, R., Deri, P., Gremigni, V., 2008. Planarians, a tale of stem cells. Cell Mol. Life Sci. 65, 16—23.
Rybicka, K., 1966. Embryogenesis in cestodes. Adv. Parasitol. 4, 107—186.
Sabio, G., Davis, R.J., 2014. TNF and MAP kinase signalling pathways. Semin. Immunol. 26, 237—245.
Saghir, N., Conde, P.J., Brophy, P.M., Barrett, J., 2001. Biochemical characterisation of a hydrophobic ligand binding protein from the tapeworm *Hymenolepis diminuta*. Int. J. Parasitol. 31, 653—660.
Sakamoto, H., Sugimura, M., 1969. Studies on echinococcosis. XXI. Electron microscopical observations on general structure of larval tissue of multilocular *Echinococcus*. Jpn J. Vet. Res. 17, 67—80.
Sakamoto, T., Sugimura, M., 1970. Studies on echinococcosis. XXIII. Electron microscopical observations on histogenesis of larval *Echinococcus multilocularis*. Jpn J. Vet. Res. 18, 131—144.
Sako, Y., Nakaya, K., Ito, A., 2011. *Echinococcus multilocularis*: identification and functional characterization of cathepsin B-like peptidases from metacestode. Exp. Parasitol. 127, 693—701.

Sako, Y., Yamasaki, H., Nakaya, K., Nakao, M., Ito, A., 2007. Cloning and characterization of cathepsin L-like peptidases of *Echinococcus multilocularis* metacestodes. Mol. Biochem. Parasitol. 154, 181–189.

Sanchez Franco, A., Sanchez Acedo, C., 1971. Estudio de algunas propiedades físicas y componentes químicos del líquido y pared germinativa de quistes hidáticos de diversas especies y de diferente localización. Rev. Iber. Parasitol. 31, 347–366.

Schubert, A., Koziol, U., Calilliau, K., Vanderstraete, M., Dissous, C., Brehm, K., 2014. Targeting *Echinococcus multilocularis* stem cells by inhibition of the Polo-like kinase EmPlk1. PLoS Negl. Trop. Dis. 8, e2870.

Silva-Alvarez, V., Folle, A.M., Ramos, A.L., Zamarreno, F., Costabel, M.D., Garcia-Zepeda, E., Salinas, G., Corsico, B., Ferreira, A.M., 2015. *Echinococcus granulosus* antigen B: a hydrophobic ligand binding protein at the host-parasite interface. Prostaglandins Leukot. Essent. Fatty Acids 93, 17–23.

Smyth, J.D., Howkins, A.B., Barton, M., 1966. Factors controlling the differentiation of the hydatid organism, *Echinococcus granulosus*, into cystic or strobilar stages in vitro. Nature 211, 1374–1377.

Smyth, J.D., McManus, D.P., 1989. The Physiology and Biochemistry of Cestodes. Cambridge University Press.

Spiliotis, M., Brehm, K., 2004. *Echinococcus multilocularis*: identification and molecular characterization of a Ral-like small GTP-binding protein. Exp. Parasitol. 107, 163–172.

Spiliotis, M., Brehm, K., 2009. Axenic in vitro cultivation of *Echinococcus multilocularis* metacestode vesicles and the generation of primary cell cultures. Methods Mol. Biol. 470, 245–262.

Spiliotis, M., Konrad, C., Gelmedin, V., Tappe, D., Bruckner, S., Mosch, H.U., Brehm, K., 2006. Characterisation of EmMPK1, an ERK-like MAP kinase from *Echinococcus multilocularis* which is activated in response to human epidermal growth factor. Int. J. Parasitol. 36, 1097–1112.

Spiliotis, M., Kroner, A., Brehm, K., 2003. Identification, molecular characterization and expression of the gene encoding the epidermal growth factor receptor orthologue from the fox-tapeworm *Echinococcus multilocularis*. Gene 323, 57–65.

Spiliotis, M., Lechner, S., Tappe, D., Scheller, C., Krohne, G., Brehm, K., 2008. Transient transfection of *Echinococcus multilocularis* primary cells and complete in vitro regeneration of metacestode vesicles. Int. J. Parasitol. 38, 1025–1039.

Spiliotis, M., Mizukami, C., Oku, Y., Kiss, F., Brehm, K., Gottstein, B., 2010. *Echinococcus multilocularis* primary cells: improved isolation, small-scale cultivation and RNA interference. Mol. Biochem. Parasitol. 174, 83–87.

Spiliotis, M., Tappe, D., Bruckner, S., Mosch, H.U., Brehm, K., 2005. Molecular cloning and characterization of Ras- and Raf-homologues from the fox-tapeworm *Echinococcus multilocularis*. Mol. Biochem. Parasitol. 139, 225–237.

Spiliotis, M., Tappe, D., Sesterhenn, L., Brehm, K., 2004. Long-term in vitro cultivation of *Echinococcus multilocularis* metacestodes under axenic conditions. Parasitol. Res. 92, 430–432.

Stadelmann, B., Aeschbacher, D., Huber, C., Spiliotis, M., Muller, J., Hemphill, A., 2014. Profound activity of the anti-cancer drug bortezomib against *Echinococcus multilocularis* metacestodes identifies the proteasome as a novel drug target for cestodes. PLoS Negl. Trop. Dis. 8, e3352.

Stadelmann, B., Spiliotis, M., Muller, J., Scholl, S., Muller, N., Gottstein, B., Hemphill, A., 2010. *Echinococcus multilocularis* phosphoglucose isomerase (EmPGI): a glycolytic enzyme involved in metacestode growth and parasite-host cell interactions. Int. J. Parasitol. 40, 1563–1574.

Swiderski, Z., 1983. *Echinococcus granulosus*: hook-muscle systems and cellular organisation of infective oncospheres. Int. J. Parasitol. 13, 289–299.

Taman, A., Ribeiro, P., 2009. Investigation of a dopamine receptor in *Schistosoma mansoni*: functional studies and immunolocalization. Mol. Biochem. Parasitol. 168, 24–33.
Thompson, R.C.A., 1986. Biology and systematics of *Echinococcus*. In: Thompson, R.C.A. (Ed.), The Biology of *Echinococcus* and Hydatid Disease. George Allen & Unwin, pp. 5–43.
Thompson, R.C.A., 2017. Biology and systematics of *Echinococcus*. In: Thompson, A. (Ed.), *Echinococcus* and Echinococcosis, Part A, vol. 95, pp. 65–110.
Threadgold, L.T., Dunn, J., 1983. *Taenia crassiceps*: regional variations in ultrastructure and evidence of endocytosis in the cysticercus' tegument. Exp. Parasitol. 55, 121–131.
Threadgold, L.T., Dunn, J., 1984. *Taenia crassiceps*: basic mechanisms of endocytosis in the cysticercus. Exp. Parasitol. 58, 263–269.
Threadgold, L.T., Hopkins, C.A., 1981. *Schistocephalus solidus* and *Ligula intestinalis*: pinocytosis by the tegument. Exp. Parasitol. 51, 444–456.
Tielens, A.G.M., van Hellemond, J.J., 2006. Unusual aspects of metabolism in flatworm parasites. In: Maule, A.G., Marks, N.J. (Eds.), Parasitic Flatworms. Molecular Biology, Biochemistry, Immunology and Physiology. Oxfordshire: CABI Publishing, pp. 387–407.
Torgerson, P.R., Keller, K., Magnotta, M., Ragland, N., 2010. The global burden of alveolar echinococcosis. PLoS Negl. Trop. Dis. 4, e722.
Tsai, I.J., Zarowiecki, M., Holroyd, N., Garciarrubio, A., Sanchez-Flores, A., Brooks, K.L., Tracey, A., Bobes, R.J., Fragoso, G., Sciutto, E., et al., 2013. The genomes of four tapeworm species reveal adaptations to parasitism. Nature 496, 57–63.
Valdez, R.A., Jimenez, P., Fernandez Presas, A.M., Aguilar, L., Willms, K., Romano, M.C., 2014. *Taenia solium* tapeworms synthesize corticosteroids and sex steroids in vitro. Gen. Comp. Endocrinol. 205, 62–67.
Vanderstraete, M., Gouignard, N., Cailliau, K., Morel, M., Hahnel, S., Leutner, S., Beckmann, S., Grevelding, C.G., Dissous, C., 2014. Venus kinase receptors control reproduction in the platyhelminth parasite *Schistosoma mansoni*. PLoS Pathog. 10, e1004138.
Varela-Diaz, V.M., Coltorti, E.A., 1972. Further evidence of the passage of host immunoglobulins into hydatid cysts. J. Parasitol. 58, 1015–1016.
Varela-Diaz, V.M., Coltorti, E.A., Ricardes, M.I., Guisantes, J.A., Yarzabal, L.A., 1974. The immunoelectrophoretic characterization of sheep hydatid cyst fluid antigens. Am. J. Trop. Med. Hyg. 23, 1092–1096.
Venkatakrishnan, A.J., Deupi, X., Lebon, G., Tate, C.G., Schertler, G.F., Babu, M.M., 2013. Molecular signatures of G-protein-coupled receptors. Nature 494, 185–194.
Vicogne, J., Cailliau, K., Tulasne, D., Browaeys, E., Yan, Y.T., Fafeur, V., Vilain, J.P., Legrand, D., Trolet, J., Dissous, C., 2004. Conservation of epidermal growth factor receptor function in the human parasitic helminth *Schistosoma mansoni*. J. Biol. Chem. 279, 37407–37414.
Vicogne, J., Pin, J.P., Lardans, V., Capron, M., Noel, C., Dissous, C., 2003. An unusual receptor tyrosine kinase of *Schistosoma mansoni* contains a Venus Flytrap module. Mol. Biochem. Parasitol. 126, 51–62.
Virginio, V.G., Monteiro, K.M., Drumond, F., de Carvalho, M.O., Vargas, D.M., Zaha, A., Ferreira, H.B., 2012. Excretory/secretory products from in vitro-cultured *Echinococcus granulosus* protoscoleces. Mol. Biochem. Parasitol. 183, 15–22.
Vuitton, D.A., Gottstein, B., 2010. *Echinococcus multilocularis* and its intermediate host: a model of parasite-host interplay. J. Biomed. Biotechnol. 2010, 923193.
Vuitton, D.A., Guerret-Stocker, S., Carbillet, J.P., Mantion, G., Miguet, J.P., Grimaud, J.A., 1986. Collagen immunotyping of the hepatic fibrosis in human alveolar echinococcosis. Z. Parasitenkd. 72, 97–104.
Wang, J., Zhang, C., Wei, X., Blagosklonov, O., Lv, G., Lu, X., Mantion, G., Vuitton, D.A., Wen, H., Lin, R., 2013. TGF-beta and TGF-beta/Smad signaling in the interactions between *Echinococcus multilocularis* and its hosts. PLoS One 8, e55379.

Wang, S., Luo, X., Zhang, S., Yin, C., Dou, Y., Cai, X., 2014. Identification of putative insulin-like peptides and components of insulin signaling pathways in parasitic platyhelminths by the use of genome-wide screening. FEBS J. 281, 877–893.

Wang, Y., Xiao, D., Shen, Y., Han, X., Zhao, F., Li, X., Wu, W., Zhou, H., Zhang, J., Cao, J., 2015. Proteomic analysis of the excretory/secretory products and antigenic proteins of *Echinococcus granulosus* adult worms from infected dogs. BMC Vet. Res. 11, 119.

Watanabe, H., Takehana, K., Date, M., Shinozaki, T., Raz, A., 1996. Tumor cell autocrine motility factor is the neuroleukin/phosphohexose isomerase polypeptide. Cancer Res. 56, 2960–2963.

Williams, J.F., Colli, C.W., 1970. Primary cystic infection with *Echinococcus granulosus* and *Taenia hydatigena* in *Meriones unguiculatus*. J. Parasitol. 56, 509–513.

Yanger, K., Stanger, B.Z., 2011. Facultative stem cells in liver and pancreas: fact and fancy. Dev. Dyn. 240, 521–529.

Yazawa, S., Umesono, Y., Hayashi, T., Tarui, H., Agata, K., 2009. Planarian Hedgehog/Patched establishes anterior-posterior polarity by regulating Wnt signaling. Proc. Natl. Acad. Sci. U.S.A. 106, 22329–22334.

Ye, J.S., Su, X.S., Stoltz, J.F., de Isla, N., Zhang, L., 2015. Signalling pathways involved in the process of mesenchymal stem cells differentiating into hepatocytes. Cell Prolif. 48, 157–165.

Zavala-Gongora, R., Derrer, B., Gelmedin, V., Knaus, P., Brehm, K., 2008. Molecular characterisation of a second structurally unusual AR-Smad without an MH1 domain and a Smad4 orthologue from *Echinococcus multilocularis*. Int. J. Parasitol. 38, 161–176.

Zavala-Gongora, R., Kroner, A., Bernthaler, P., Knaus, P., Brehm, K., 2006. A member of the transforming growth factor-beta receptor family from *Echinococcus multilocularis* is activated by human bone morphogenetic protein 2. Mol. Biochem. Parasitol. 146, 265–271.

Zavala-Gongora, R., Kroner, A., Wittek, B., Knaus, P., Brehm, K., 2003. Identification and characterisation of two distinct Smad proteins from the fox-tapeworm *Echinococcus multilocularis*. Int. J. Parasitol. 33, 1665–1677.

Zeng, G., Awan, F., Otruba, W., Muller, P., Apte, U., Tan, X., Gandhi, C., Demetris, A.J., Monga, S.P., 2007. Wnt'er in liver: expression of Wnt and frizzled genes in mouse. Hepatology 45, 195–204.

Zhang, S., Hue, S., Sene, D., Penfornis, A., Bresson-Hadni, S., Kantelip, B., Caillat-Zucman, S., Vuitton, D.A., 2008. Expression of major histocompatibility complex class I chain-related molecule A, NKG2D, and transforming growth factor-beta in the liver of humans with alveolar echinococcosis: new actors in the tolerance to parasites? J. Infect. Dis. 197, 1341–1349.

Zhang, W., Li, J., McManus, D.P., 2003. Concepts in immunology and diagnosis of hydatid disease. Clin. Microbiol. Rev. 16, 18–36.

Zhang, W.B., Jones, M.K., Li, J., McManus, D.P., 2005. *Echinococcus granulosus*: pre-culture of protoscoleces in vitro significantly increases development and viability of secondary hydatid cysts in mice. Exp. Parasitol. 110, 88–90.

Zhang, C., Wang, L., Wang, H., Pu, H., Yang, L., Li, J., Wang, J., Lü, G., Lu, X., Zhang, W., Vuitton, D.A., Wen, H., Lin, R., 2014. Identification and characterization of functional Smad8 and Smad4 homologues from *Echinococcus granulosus*. Parasitol. Res. 113, 3745–3757.

Zheng, H., Zhang, W., Zhang, L., Zhang, Z., Li, J., Lu, G., Zhu, Y., Wang, Y., Huang, Y., Liu, J., et al., 2013. The genome of the hydatid tapeworm *Echinococcus granulosus*. Nat. Genet. 45, 1168–1175.

CHAPTER FIVE

Ecology and Life Cycle Patterns of *Echinococcus* Species

T. Romig*[,1], P. Deplazes[§], D. Jenkins[¶], P. Giraudoux[||], A. Massolo[#], P.S. Craig, M. Wassermann*, K. Takahashi[§§], M. de la Rue[¶¶]**
*University of Hohenheim, Stuttgart, Germany
[§]University of Zürich, Zurich, Switzerland
[¶]Charles Sturt University, Wagga Wagga, NSW, Australia
[||]University of Franche-Comté and Institut Universitaire de France, Besancon, France
[#]University of Calgary, Calgary, Alberta, Canada
**University of Salford, Greater Manchester, United Kingdom
[§§]Rakuno Gakuen University, Ebetsu, Japan
[¶¶]University of Santa Maria, Santa Maria RS, Brazil
[1]Corresponding author: E-mail: Thomas.Romig@uni-hohenheim.de

Contents

Advances in Parasitology, Volume 95
ISSN 0065-308X
http://dx.doi.org/10.1016/bs.apar.2016.11.002

Abstract

The genus *Echinococcus* is composed of eight generally recognized species and one genotypic cluster (*Echinococcus canadensis* cluster) that may in future be resolved into one to three species. For each species, we review existing information on transmission routes and life cycles in different geographical contexts and – where available – include basic biological information of parasites and hosts (e.g., susceptibility of host species). While some *Echinococcus* spp. are transmitted in life cycles that involve predominantly domestic animals (e.g., dog – livestock cycles), others are wildlife parasites that do or do not interact with domestic transmission. In many cases, life cycle patterns of the same parasite species differ according to geography. Simple life cycles contrast with transmission patterns that are highly complex, involving multihost systems that may include both domestic and wild mammals. Wildlife transmission may be primary or secondary, i.e., resulting from spillovers from domestic animals. For most of the species and regions, existing information does not yet permit a conclusive description of transmission systems. Such data, however, would be highly relevant, e.g., for anticipation of geographical changes of the presence and frequency of these parasites in a warming world, or for initiating evidence-based control strategies.

1. INTRODUCTION

Like all taeniid cestodes, species of *Echinococcus* exploit predator–prey systems for the maintenance of their life cycles. Definitive hosts are exclusively members of the order Carnivora, mainly of the dog family (Canidae), to a lesser degree cats (Felidae) and hyenas (Hyaenidae). Intermediate hosts of *Echinococcus* spp. cover a much wider range. Considering the large number of species where metacestodes can develop – as distantly related to each other as marsupials, rodents and ruminants – it seems that the parasite is principally able to adapt to any prey species of a particular definitive host, provided it is a mammal. This intrinsic capacity is also emphasized by the long list of 'dead-end hosts' for the various species that do not play a role in the life cycles, but where metacestodes can develop even to fertile condition.

This general feature of the genus *Echinococcus*, however, is very differently expressed when looking at the species level. Here, we find marked differences in intermediate host specificity, ranging, e.g., from *Echinococcus granulosus* sensu stricto (s.s.), whose metacestode can reach fertility in a wide range of hosts, to *Echinococcus equinus*, whose metacestodes seem (almost) exclusively to develop in members of the horse family (Equidae). Incidentally, the degree of intermediate host specificity emerges to be reasonably well correlated with the potential of the various *Echinococcus* species to infect humans. *E. granulosus* s.s., with low specificity, has the highest impact on public health in terms of case numbers, and cysts of that species can grow rapidly to large size and usually produce proficient numbers of protoscoleces in human patients. In contrast, humans seem to be partially or completely refractory to infection with other *Echinococcus* spp., an assessment that is based on low numbers of cases despite local parasite abundance in animal hosts (chapter: Global Distribution of Alveolar and Cystic Echinococcosis by Deplazes et al., 2016) or on the aberrant growth patterns of the metacestode in humans (most obvious in the case of *Echinococcus multilocularis*) (chapter: The Echinococcoses: Diagnosis, Clinical Management and Burden of Disease by Kern et al., 2016).

All species of *Echinococcus* (except *Echinococcus shiquicus*) are geographically spread over vast areas, and some agents of cystic echinococcosis (CE) can be considered as cosmopolitan. Apart from species with predominantly domestic life cycles (that are based on very few globally distributed host species), the parasites can exploit different predator–prey systems in different geographic regions, or even in different habitats of the same area. Life cycles in a given location may be simple (the extreme case of *E. multilocularis* on Svalbard demonstrates that even one species of definitive and intermediate host can be sufficient for persisting transmission), but most often involve multihost systems, where several species of definitive hosts – that also interact with each other – prey on various species of intermediate hosts. Such cycles can be highly complex and are often little understood because host species may have different capacities for parasite development, different population dynamics over seasons or prolonged periods, or differ in their accessibility or attractiveness for the predators. Even the presence of prey species, that are unsuitable as hosts, or other food resources for opportunistic carnivores, may influence the life cycles by causing dilution effects.

Describing a life cycle aims at quantification of the impact of different species on parasite transmission and abundance. The complexity of multi-host systems renders this a challenging task, particularly of species with sylvatic transmission. Even regarding *E. multilocularis*, one of the best studied species, there are few areas where the available biological and ecological information would allow an estimate of the relative impact of the various host species involved. Precise data on susceptibility or biotic potential (to produce eggs or protoscoleces) are lacking for many obviously important host species. The extent of differences in host species adaptation or pathogenicity between parasite isolates of different geographical origin is largely unclear, which means that data from experimental infection are not necessarily valid away from the study area. Lack of data on population densities and predation rates prevents the assessment of certain host species' impact on transmission. Data from numerous studies are difficult to interpret because metacestode fertility is not reported or the age structure of examined host animals is not taken into account in prevalence surveys. Unexplained differences of biological variables (e.g., fertility rates of cysts) are apparent when looking at identical pairs of parasites and livestock species in different regions (see Section 7.1). This may be explained not only by different age at slaughter and different types of husbandry (exposure to dogs), but also by differences in the breed of a certain livestock species. The latter has never been analysed, although there are indications that those may have a drastic influence on susceptibility: in Australia, angora goats were found to be highly susceptible to experimental infection with *E. granulosus* s.s., while feral goats were refractory (Jenkins and Macpherson, 2003).

Data on the reproductive potential of definitive host species and their contribution to the life cycles are particularly difficult to interpret. Prevalence estimates are subject to seasonal variation of metacestode availability (caused, e.g., by annual fluctuations of rodent populations, or seasonal livestock slaughtering practices), information that is often not provided in connection with published data. Susceptibility of carnivore species (concerning establishment rate, time needed for worm maturation and length of egg excretion period) is certainly subject to variation, but is only known for few host/parasite species pairs. Where available, data are often derived from few experimental animals and parasite isolates from single geographical locations, whereas data from the field, e.g., on worm burdens, may be biased by different (and unknown) exposure to infection. The apparent lack of basic biological data for definitive hosts of most *Echinococcus* spp. with cystic metacestodes is largely due to the fact that the various cryptic species had not been discriminated in older infection experiments. In general, there are very few quantitative data on carnivore infection with species other than *E. multilocularis* and *E. granulosus* s.s. (Carmena and Cardona, 2013).

The following sections are therefore in large parts summaries of our gaps of knowledge, and the very unequal length of the sections reflects the different availability of biological and epidemiological information for the various species. Lack of data for species with cystic metacestodes is largely due to the recent split of *E. granulosus* (sensu lato, s.l.). Other species, however, are grossly understudied, e.g., those infecting South American wildlife. The following summaries of epidemiological data are therefore not only meant to provide an update of current knowledge, but also to serve as an encouragement to study the basic biological aspects of these parasites.

In the following species accounts, the term 'domestic life cycle' is used, when domestic dogs and livestock are principal host species, even though wild animals may contribute marginally to transmission. 'Sylvatic life cycle' means the reverse situation, where the transmission is principally maintained by wild mammals without essential involvement of domestic species; life cycles maintained by feral mammals (e.g., dingoes or wild pigs in Australia) would also be sylvatic. Host species that can be infected with the metacestode, but that are clearly not part of the life cycle, are referred to as 'dead-end hosts', while other terms such as 'accidental' or 'aberrant' host are avoided. A dead-end host may not take part in the life cycle because the parasite does not reach fertility or because it is outside the prey range of definitive hosts (e.g., due to size, or in the case of zoo animals).

Unless stated otherwise, species names and distribution of host mammals are reported as described in the IUCN Red List (IUCN, 2016).

2. *ECHINOCOCCUS OLIGARTHRA* (DIESING, 1863)

Echinococcus oligarthra is known to occur from southern Argentina through tropical South and Central America into northern Mexico (D'Alessandro and Rausch, 2008). In view of the extreme diversity of habitats, very little is known about the respective transmission systems.

Most available data indicate life cycles that include different species of wild cats and large rodents (Fig. 1). However, host records are geographically widely scattered and do not allow conclusions on the local transmission patterns. The one exception is an extensive study, conducted on the eastern plains of Colombia, that covered relevant sample sizes of definitive and intermediate host species and that provided a basis to understand the epidemiology of the parasite in that area (D'Alessandro et al., 1981).

Adult worms of *E. oligarthra* seem to be specifically adapted to members of the cat family (Felidae) because fully developed adult worms have so far only been recovered from felids. Experimental infections of domestic dogs and two species of procyonids (*Procyon lotor* and *Nasua narica*) were unsuccessful (Sousa and Thatcher, 1969), and few adult worms recovered from a domestic dog that was naturally exposed to infection (used for hunting of pacas) were described as 'underdeveloped' (D'Alessandro et al., 1981). Six out of the ten wild felid species that occur in South and Central America were found naturally infected: jaguar (*Panthera onca*) in Panama; **puma**

Figure 1 Life cycle of *Echinococcus oligarthra* between various species of wild felids and large rodents. Rabbits and opossums may also contribute as intermediate hosts (omitted). Humans are the only dead-end hosts known so far.

(*Puma concolor*) in Brazil, Colombia, Panama and Costa Rica; **jaguarundi** (*Herpailurus yagouaroundi*) in Colombia and Panama; **ocelot** (*Leopardus pardalis*) in Colombia; and **pampas cat** (*Leopardus colocolo*) and **Geoffroy's cat** (*Leopardus geoffroyi*) in Argentina (Rausch and D'Alessandro, 2002). The only North American record was from a **bobcat** (*Lynx rufus*) in Tamaulipas, northern Mexico (Salinas-Lopez et al., 1996). House cats were experimentally susceptible (Sousa and Thatcher, 1969). The low number of examined specimens per species and locality does not allow prevalence estimates in definitive hosts. An exception is the report of 7 infected out of 46 examined Geoffroy's cats from the south of Argentina (Schantz and Colli, 1973).

Metacestodes of *E. oligarthra* are only known from a surprisingly small number of potential host species, which is clearly due to the lack of field studies. 3 of 325 **pacas** (*Cuniculus paca*) and 2 of 1168 **spiny rats** (*Proechimys semispinosus, Proechimys* cf. *guyannensis, Proechimys* sp.) were found infected during the survey in Colombia, where 1316 animals belonging to >30 species of other rodents, 8 species of primates and some 30 species of marsupials, lagomorphs, insectivores, edentates and ungulates were negative (D'Alessandro et al., 1981; Rausch, 1995). 2 of 74 spiny rats (*P. semispinosus*) and 3 of 39 **agoutis** (*Dasyprocta punctata*) had metacestodes in Panama; other infected agouti species are reported from Colombia (*Dasyprocta fuliginosa*), Venezuela (*Dasyprocta leporina*) and Brazil (*D. leporina, Dasyprocta* sp.) (D'Alessandro et al., 1981; Rausch and D'Alessandro, 2002). Apart from rodents, metacestodes were also found in **eastern cottontails** (*Sylvilagus floridanus*) in Venezuela (Meléndez et al., 1984) and in a **southern opossum** (*Didelphis marsupialis*) in Colombia (Thatcher, 1972); the latter species, however, is not considered to be important for the life cycle (D'Alessandro and Rausch, 2008). The host spectrum which has been reported as particularly relevant for transmission (species of agoutis, pacas and spiny rats) is clearly not representative for the entire distribution area of *E. oligarthra*, as no members of these three rodent genera occur in the southern and northern part of the confirmed range of the parasite. There must be other small mammal species involved in the transmission, which have escaped recognition so far. This is supported by experimental studies showing infectivity of *E. oligarthra* to a wider host range, including **Panama climbing rats** (*Tylomys panamensis*), **hispid cotton rats** (*Sigmodon hispidus*) and **Mongolian gerbils** (*Meriones unguiculatus*) (in addition to agoutis and spiny rats), while laboratory mice, laboratory rats and golden hamsters were refractory to oral inoculation with eggs from a naturally infected puma (Sousa and Thatcher, 1969).

Recently, considerable genetic distances were found between isolates of human origin from Panama and Brazil (Do Carmo Perreira Soares et al., 2013). Further data are needed to conclude on possible relevance of this observation for life cycles and taxonomy.

3. *ECHINOCOCCUS VOGELI* RAUSCH AND BERNSTEIN, 1972

The geographic range of *Echinococcus vogeli* is assumed to be largely identical to that of its only verified natural definitive host, the **bush dog** (*Speothos venaticus*), which stretches from Panama through northern South America and the Amazon basin (east of the Andes) into Paraguay and north-eastern Argentina. The limited set of host records suggests that the parasite is transmitted between bush dogs and large rodent species (Fig. 2). While all records of *E. vogeli* in animals are confined to the bush dog's distribution range, there are some reports on suspected human cases from outside,

Figure 2 Life cycle of *Echinococcus vogeli* between bush dogs and various species of large rodents. Domestic dogs may contribute to the life cycle to some extent, as well as armadillos as intermediate hosts (omitted). Humans are dead-end hosts.

e.g., Nicaragua, Costa Rica and Chile (D'Alessandro and Rausch, 2008). It is unclear whether these cases are due to other *Echinococcus* spp., whether the infections were acquired abroad, or whether the host range of *E. vogeli* is wider than reported.

Bush dogs are small canids that are described as elusive and difficult to locate, which lends some uncertainty to published data on their presence, absence and frequency. They are highly carnivorous with small and medium-sized mammals (including agoutis and pacas) as main prey. Although described to be rare, bush dogs appear to be ecologically adaptable, occurring in a wide range of habitats from primary rainforest to riverine vegetation in ranchland and other landscapes altered by humans; there seems to be some preference for the vicinity of watercourses (IUCN, 2016). There are no data on prevalence of *E. vogeli* in bush dogs, and records are confined to only two reports. One animal, caught in Ecuador in 1970, was found infected at veterinary examination in the Los Angeles zoo — the recovered worms provided the basis for the species description (Rausch and Bernstein, 1972). The second infected animal was obtained from a local hunter in the eastern Amazon basin of Brazil in 2007 (Do Carmo Perreira Soares et al., 2014). The involvement of other wild canids cannot be excluded for lack of surveys, but of all indigenous canids, the bush dog appears to be the only species that regularly preys on the large rodents thought to be the principal intermediate hosts. A survey in Colombia of 50 **crab-eating foxes** (*Cerdocyon thous*), a canid that occurs sympatrically with bush dogs, but specializes on arthropods and small vertebrates as prey, yielded no infected animal (D'Alessandro et al., 1981).

Only two species of rodents are known as intermediate hosts: **pacas** (*C. paca*) were found infected in Colombia, Peru, Brazil, Suriname, Bolivia, and Argentina, and **agoutis** (*D. leporina, Dasyprocta* sp.) in Venezuela and Brazil (D'Alessandro et al., 1981; Mayor et al., 2015, Rausch and D'Alessandro, 2002; Santos et al., 2012, Vizcaychipi et al., 2013). Indications for a life cycle between bush dog and paca exist only for the plains of eastern Colombia, where pacas coexist with bush dogs in gallery forests along watercourses. 22% of 325 pacas caught in this habitat were identified as carriers of *E. vogeli* metacestodes. Prevalence was increasing with age, reaching 29% in adults (D'Alessandro et al., 1981; D'Alessandro and Rausch, 2008). The absence of metacestodes in young animals indicates that infection is a rare event (probably due to the low population density of bush dogs), and it was concluded that relatively long-lived rodents are required to maintain the life cycle: pacas may live up to 12 years in the wild (D'Alessandro et al., 1981). In addition to pacas, five specimens of **nine-banded armadillos** (*Dasypus novemcinctus*) were recently found to carry metacestodes of *E. vogeli* in the eastern Amazon basin of Brazil (Santos et al., 2012). Experimental infections indicate that the restricted intermediate host range in the wild may have ecological instead of physiological reasons: **coypus** (*Myocastor coypus*) were susceptible to oral infection with *E. vogeli* eggs, and a large range of small rodent species produced fertile metacestodes after intraperitoneal inoculation (Rausch and D'Alessandro, 1999).

The rather substantial number of human cases of polycystic echinococcosis caused by *E. vogeli* (D'Alessandro and Rausch, 2008) was attributed to the involvement of domestic dogs in the life cycle. In most of their distribution area, pacas are traditionally hunted by people for their meat, and their viscera are reportedly fed to dogs giving them infection (Mayor et al., 2015, Rausch, 1995). This hypothesis is based on the discovery of one infected hunter's dog (out of 15 examined) from Colombia (D'Alessandro et al., 1981), the proven susceptibility of domestic dogs (in contrast to cats) to experimental infection (Rausch and D'Alessandro, 2002) and reports about dog contacts given by patients (Rausch and D'Alessandro, 2002). In an interview study with seven patients from Brazil, all reported contact to dogs that had been fed with viscera of pacas (Meneghelli et al., 1990).

Pathogenicity of *E. vogeli* for nonhuman primates was exemplified at the Los Angeles zoo, where an infected bush dog was temporarily housed near an enclosure with juvenile apes. Over a consecutive period of ~10 years, seven orangutans, three chimpanzees, five gorillas and three gibbons died of *E. vogeli* echinococcosis or were euthanized at a terminal stage of disease (Howard and Gendron, 1980; O'Grady et al., 1982; Rausch, 1995).

In conclusion, the life cycle is assumed to include bush dogs and pacas as the principal hosts, with domestic dogs becoming infected as a spillover from the wildlife cycle due to human behaviour. However, this presumed life cycle rests largely on very few observations from widely separated parts of the continent, including only three records of adult *E. vogeli*, and the presence of human cases outside the range of bush dogs and pacas remains to be explained. A transmission between domestic dogs and armadillos has been hypothetically proposed for these areas (Santos et al., 2012), but there are no specific data available to support this.

4. *ECHINOCOCCUS MULTILOCULARIS* LEUCKART, 1863

4.1 General life cycle

Despite the large geographical distribution range that includes most of the temperate and cold zones of the northern hemisphere, no subspecific taxonomic units are presently recognized within *E. multilocularis*, and previously erected subspecies (e.g., *E. multilocularis sibiricensis*) as well as proposed species (*Echinococcus russicensis*) are considered to fall within the current concept of *E. multilocularis* as a species (Nakao et al., 2013a; Knapp et al., 2015). Intraspecific genetic structuring of *E. multilocularis* that initially appeared to represent Asian, European and North American lineages (Nakao et al., 2009) requires more investigations, since isolates of the 'North American' lineage were also found in northeastern Russia (Konyaev et al., 2013), the 'Asian' lineage occurs in the European part of Russia (Konyaev et al., 2013) and the 'European' lineage is now known from various locations in western Canada (Gesy et al., 2013). It is not yet clear if this complex geographical pattern reflects a global polymorphism of the parasite or has been influenced by anthropogenic movements of host animals.

E. multilocularis is adapted to circulate between wild and domestic canids as definitive hosts and small mammals as intermediate hosts (Fig. 3). Small mammals combine restricted space for the metacestode growth with a short life span, so that morphological differences to metacestodes of other *Echinococcus* spp. (compact, vesicular growth with high density of protoscoleces) and the short time needed for the development of protoscoleces can be interpreted as an evolutionary response to those conditions. In the largest part of the endemic area *E. multilocularis* life cycles are based on rodents, predominantly voles (Arvicolinae), and various species of canids that prey on them. Based on field data and/or experimental infections, other families of Carnivora are assumed to be partially (some Felidae) or completely (e.g., Mustelidae, Ursidae) refractory to infection. However, sporadic reports indicate an incomplete physiological barrier at least in some species. Examples are the recovery of immature worms after experimental infection of American black bears (*Ursus americanus*) (Rausch and Richards, 1971) and recent reports of very small numbers of gravid worms found in pine martens (*Martes martes*) and stone martens (*Martes foina*) in the European part of Russia (Andreyanov, 2011; Andreyanov, pers. comm.). In general, 'good' definitive hosts for *E. multilocularis* combine competency to infection and to development of fertile

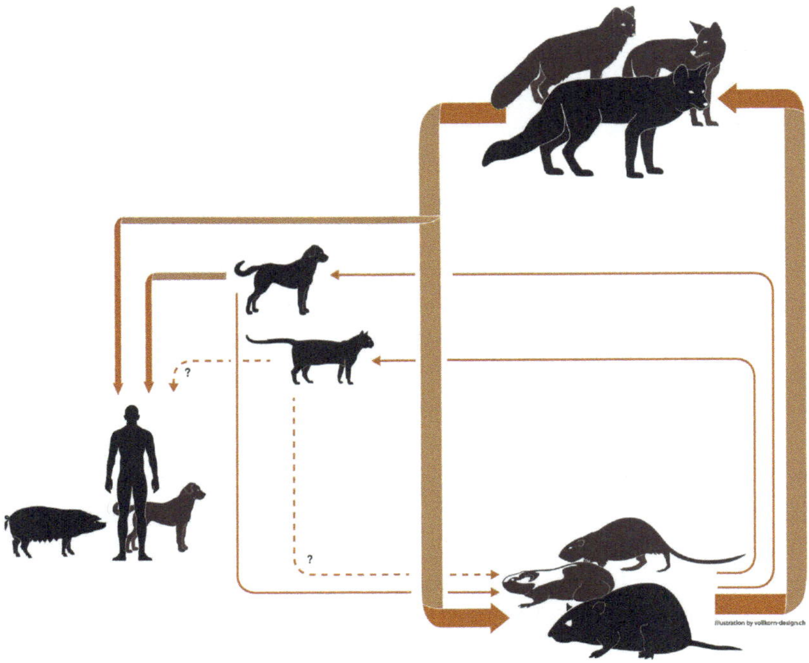

Figure 3 Life cycle of *Echinococcus multilocularis* between wild canids (with the red fox as the most widespread definitive host) and rodents, with small arvicolines as the most important intermediate hosts. Domestic dogs contribute to the life cycle in some areas and are infection sources for humans, while the role of cats is minor and not fully understood. Numerous species are dead-end hosts, either due to partial resistance to infection (abortive or aberrant metacestode development – e.g., humans, pigs) or because they are not available as prey for definitive hosts (e.g., dogs, zoo animals).

adult parasites, having a diet that includes a large proportion of susceptible intermediate host species, and having relatively high population densities compared to other sympatric host species. Moreover, their territorial and dispersal behaviour make them good 'spreaders' as they mark throughout their territory with faeces, and the juveniles usually disperse from their natal sites thus spreading the parasite beyond their original home range. Furthermore, there is evidence that certain final host species specifically mark the sites where they predate on the intermediate hosts and thereby facilitate parasite transmission (Giraudoux et al., 2002; Hegglin et al., 2015).

Across the extensive geographical range of *E. multilocularis*, which covers ecosystems as diverse as arctic tundra, high-altitude grasslands, agricultural landscapes and even cities and towns, the parasite is transmitted in diverse life cycles that involve numerous definitive and intermediate host species. Although life cycles and host contributions are insufficiently known outside Europe, Japan and parts of the arctic regions, there seems to be a general pattern that surprisingly few species are key hosts in any particular region, whereas a large number of others contribute to a marginal degree or are dead-end hosts. This can be due to generally lower population densities (affecting encounter rates with the definitive host); predation behaviour; unattractiveness as prey for definitive hosts due to particular habitat requirements, body size or defence behaviour (e.g., beavers); lower physiological suitability for the establishment, development and persistence of the parasite (insufficiently known for most species, however); and even hibernation patterns and definitive host defaecation behaviour. Despite the data summarized below on the involvement of different host species, far too little information is available on the role of individual species in multihost transmission systems to estimate the quantitative impact of any one species.

There is certainly an evolutionary pressure on the parasite to adapt to the most intensive predator–prey system in a given region, optimizing biological factors such as infectivity and development characteristics to the regionally most frequent hosts. It is therefore not surprising that establishment/development of *E. multilocularis* isolates from distant geographical region can be different after experimental inoculation into the same host species (e.g., Bartel et al., 1992; Rausch and Richards, 1971). However, very few data are available on regional differences in host–parasite adaptation (often anecdotal), and experimental studies on the parasite's development in 'natural' host species are urgently needed to estimate the extent of such mechanisms.

Differences of infectivity and/or pathogenicity to humans may also occur, as large numbers of alveolar echinococcosis (AE) patients in Eurasia contrast with only sporadic cases in North America (with the exception of St. Lawrence Island) (chapter: Global Distribution of Alveolar and Cystic Echinococcosis by Deplazes et al., 2016), and human AE cases are rarely reported from the northernmost regions of Russia (Konyaev et al., 2012). Unfortunately, the worldwide genetic structure of the parasite is insufficiently known. In the absence of systematic studies on correlations between genetic markers of the parasite (including nuclear sequences), geographical distribution and host infectivity, the extent of genetically fixed parasite adaptations to different hosts remains unclear. This is also true for the 'Mongolian' genotype, the most divergent genotype so far discovered within *E. multilocularis*. It had initially been described as a separate species *E. russicensis* (Tang et al., 2007), but host records available so far do not indicate any differences in the host spectrum (Konyaev et al., 2013).

In addition to geographical differences of natural habitats and host species assemblages, human activities have a major impact on the presence and abundance of *E. multilocularis*. Agricultural practices, deforestation, urbanization, hunting pressure and even wildlife conservation change host species assemblages and abundance (Hegglin et al., 2015). Translocations of wildlife species (accidental or by purpose, with or without the parasite) and of domestic animals (particularly dogs) create new host populations and/or introduce *E. multilocularis* (or particular genotypes) into new areas (Davidson et al., 2012). Some canid species are very adaptable and show a great resilience to human activities and interventions (Comte et al., 2014; Hegglin et al., 2015). The red fox (*Vulpes vulpes*), as the major definitive host in the largest part of the parasite's range, has recently established populations living in close contact to humans even in urban areas (Deplazes et al., 2004). This adds additional complexity to the life cycle, and different transmission patterns may exist on small spatial scales in landscapes fragmented by human activities (Liccioli et al., 2015a).

Acknowledging this diversity, typical life cycles of *E. multilocularis* are described separately for major geographical regions.

4.2 Temperate Europe

4.2.1 Life cycles

The presently recognized range of *E. multilocularis* includes all of central Europe, extending westwards to northwestern France; northwards to Denmark and southern Sweden; southwards to the southern Alps, Serbia and Bulgaria; and eastwards to western Ukraine, southern Belarus and western Russia (details see chapter: Global Distribution of Alveolar and Cystic Echinococcosis by Deplazes et al., 2016). Before the 1990s, *E. multilocularis* was considered to be associated with rural areas, based on the habitat requirements of its hosts and the origin of human AE cases. In the last two decades, the additional presence of *E. multilocularis* in and near cities and towns has been documented (Deplazes et al., 2004; Liccioli et al., 2015a).

The life cycle most typical for rural areas in central Europe involves the red fox (*V. vulpes*) and the common vole (*Microtus arvalis*) as the principal definitive and intermediate hosts, although other host species may play a role in different landscapes or habitats. This host–parasite system seems to be most frequent in landscapes fragmented by traditional agricultural land use, which is typical for most of central Europe (Romig et al., 2006). This anthropogenic landscape consists of a mosaic of forests, meadows, pastures, fields and human settlements, supporting high population densities of foxes due to sufficient shelter and stable food sources. At the same time, a sufficient proportion of grassland (as meadows or pastures) is needed as habitat for common voles. It is in this type of landscape that – throughout Europe – *E. multilocularis* is highly abundant, e.g., in the mountainous ranges from central France and Belgium through central/southern Germany and Switzerland into the Carpathians (EFSA, 2015; chapter: Global Distribution of Alveolar and Cystic Echinococcosis by Deplazes et al., 2016). Depending on the environment, the relative importance of different intermediate host species may vary: the montane water vole (*Arvicola scherman*) and common vole (*M. arvalis*) were shown to be crucial for *E. multilocularis* transmission in grassland-dominated parts of eastern France (Raoul et al., 2015); muskrats (*Ondatra zibethicus*) may be important in wetland habitats (Romig et al., 1999; Umhang et al., 2013), while field voles (*Microtus agrestis*) and water voles (*Arvicola amphibius*) probably maintain the life cycle in southern Sweden which is outside the range of the common vole (Miller et al., 2016). In some areas, common voles and water voles show pronounced population cycles, whose amplitudes depend on landscape parameters such as the proportion of grassland (Delattre et al., 1999; Giraudoux et al., 1997, 2002, 2003; Raoul et al., 2001). High population density of rodents has been linked to high prevalence of *E. multilocularis* in definitive hosts, but high abundance of the parasite in foxes in areas without pronounced rodent cycles has been reported (Guislain et al., 2008; Raoul et al., 2015). Rodent population outbreaks leading to extremely high density of voles and, consequently, a high predation rate by the definitive hosts favour transmission in certain periods, but are not a necessary condition for stable *E. multilocularis* life cycles (Raoul et al., 2010, 2015).

Towards the end of the 20th century, red foxes established populations in European human settlements including inner cities. Vaccination against rabies and changing attitude towards wildlife (feeding) are factors that contributed to this development. The absence of larger wild canids, low hunting pressure, and positive attitudes towards foxes presumably have modified the antipredator response of foxes ('landscape of fear') and promoted their tolerance to disturbing factors such as traffic and human presence, which in turn facilitated the colonization of residential areas (Hegglin et al., 2015). Genetic studies showed that these 'urban' fox populations can be self-sustaining, showing reduced gene flow to and from surrounding rural populations (Wandeler et al., 2003). In Zurich (Switzerland) with >10 foxes per km^2, population densities can be far higher than in surrounding rural areas due to anthropogenic food sources (Contesse et al., 2004). In the same city, but also in Geneva and Nancy (France), prevalence of *E. multilocularis* in urban foxes was shown to be lower towards the city centre (Deplazes et al., 2004; Hegglin et al., 2007; Reperant et al., 2009; Robardet et al., 2008). This was explained by decreased availability of suitable intermediate hosts in highly urbanized areas, which depend on extensively managed grassland that becomes increasingly rare towards the city centre (*M. arvalis, A. scherman*), or which are not easily accessible as prey for foxes due to low density, burrowing habits or size (*A. scherman, Myodes glareolus, O. zibethicus*). The availability of anthropogenic food for foxes may also lead to a decreased predation on rodents (Hegglin et al., 2007; Robardet et al., 2011). For this reason, the abundance of *E. multilocularis* is particularly high in the periphery of these urban areas, where high population densities of foxes and suitable rodent habitats cooccur and the fox diet is less affected by alternative anthropogenic food (Hegglin et al., 2007). Nonetheless, emphasis should be placed on the structure of predator—prey communities within different urban areas, which can show completely different ecological and landscape characteristics (habitat diversity, habitat fragmentation, connectivity) and may generate transmission patterns that differ from the one described for Switzerland and Nancy (Liccioli et al., 2014, 2015a). Common voles are thought to be important intermediate hosts also in European urban areas, but only few data on the relative contribution of the different species are available. Unlike water voles, which are frequently found in small, fragmented patches of habitat inside cities (e.g., gardens, parks or cemeteries), common voles prefer larger stretches of meadows, which are in cities usually restricted to large parks and the periphery. Yet, data from France suggest that common voles are also important intermediate hosts in urbanized areas (Robardet et al., 2011), while the role of water voles and muskrats (although frequently infected in city parks – Stieger et al., 2002; Romig, unpublished) is less clear. Interestingly, although *A. scherman* (formerly *Arvicola terrestris*) is the predominant vole species in Zurich (Switzerland) and *M. arvalis* has been found only occasionally on small patches, the latter species was found in the stomachs of foxes as frequently as *A. scherman* (Hegglin et al., 2007). This is in accordance with French rural studies from the Jura mountains and the Ardennes where foxes predated less on *A. scherman* when the *M. arvalis* density increased and where *M. arvalis* was an important prey even when it occurred in very low densities (Raoul et al., 2010, 2015). Domestic dogs have long been considered to contribute little to transmission in urban areas due to the generally low *E. multilocularis* prevalence level. However, this may be compensated by their extremely large numbers in urban and periurban areas. It has been estimated that, under urban conditions, dogs may contribute 6.8–18.9% of the total egg output of all definitive hosts combined (Deplazes et al., 2011; Hegglin and Deplazes, 2013). Generally, it is uncertain whether 'urban' life cycles of *E. multilocularis* can persist independently from the transmission in the surrounding rural areas.

4.2.2 Definitive hosts in temperate Europe

Red foxes (*V. vulpes*) are distributed throughout Europe (with the exception of the high Arctic and high-altitude alpine areas), although at extremely variable population densities depending on food availability (Lindström, 1989; Bino et al., 2010). Throughout temperate Europe, red foxes are considered the principal definitive hosts of *E. multilocularis*. This assessment is based on (usually) high population densities, high susceptibility to infection, high worm burden and high prevalence of infection compared to other sympatric definitive host species (EFSA, 2015; Oksanen et al., 2016). In addition, the defecation behaviour of red foxes – territorial defecation at food-rich sites across the home range, i.e., vole habitats – is assumed to favour transmission to vole intermediate hosts (Stieger et al., 2002; Giraudoux et al., 2002; Raoul et al., 2015). *E. multilocularis* prevalence in red foxes can be extremely high: surveys reporting >50% are known from Switzerland, France, Germany, the Netherlands and Lithuania (EFSA, 2015), a level that is not known from any other carnivore in that region. There are no areas in Europe where *E. multilocularis* was found in other definitive hosts while being absent from red foxes. Yet, the geographical range of the red fox is far more extensive than that of the parasite, so the endemicity status of an area is likely to be determined by intermediate host populations (e.g., Guerra et al., 2014). However, variations in fox population densities within the distribution range of the species should also be considered in some areas. Fox density is also one of the factors affecting abundance of the parasite: the historical drastic increase of fox populations during the 1990s was accompanied by an increase in the prevalence of *E. multilocularis* both in foxes and intermediate hosts and was followed by an increase in the number of human AE cases (Schweiger et al., 2007). Population densities of foxes are highly variable across Europe.

Even within one country (the United Kingdom) it was shown to range between 0.02 and 30 foxes per 1 km^2, depending on abundance of food (Macdonald and Reynolds, 2008). Red foxes are highly susceptible for *E. multilocularis*: in an infection experiment, worm burdens at the onset of patency were higher in red foxes than in raccoon dogs and domestic dogs, although foxes eliminated the worms faster in comparison with dogs (Kapel et al., 2006). Field studies show that the *E. mutlilocularis* worm burden in foxes is typically aggregated (most animals carrying low numbers and few animals being massively infected). This phenomenon is probably caused by a combination of factors, such as old infections with residual worm burdens, variable numbers of protoscoleces in intermediate hosts and individual susceptibility of foxes for intestinal infections. Juveniles (<1 year old) have been reported with higher parasite burdens as compared with adults; 85% of the total worm burden in an urban fox population of Zurich was present in juveniles (e.g., Hofer et al., 2000; Robardet et al., 2008). This has been hypothetically explained by acquired intestinal immunity after repeated infections. However, this view is challenged by the observation of higher worm burden in adult foxes in other studies (Bruzinskaite-Schmidhalter et al., 2012; Ziadinov et al., 2010).

Raccoon dogs (*Nyctereutes procyonoides*) in Europe are allochthonous canids that originate from eastern Asia. Since the control of rabies in the 1990s, there has been a westward expansion of a feral population from eastern Europe, and today they form stable populations from western Russia and the Baltic States through Poland into eastern Germany. Further west, their presence becomes sporadic. Raccoon dogs are opportunistic feeders, whose diet mainly consists of insects, plants, amphibians and small mammals, but is subject to drastic variations according to landscape and season (Sutor et al., 2010). Raccoon dogs consumed a far smaller proportion of small mammals compared to red foxes from the same area in Estonia (28.6% vs 53.4%) (Süld et al., 2014). However, mammals consumed by raccoon dogs in agricultural landscapes of Germany consisted mainly of voles (*Microtus* spp. and *Arvicola* spp.) (Sutor et al., 2010). Raccoon dogs are experimentally highly susceptible to infection with *E. multilocularis,* their biotic potential (calculated number of parasite eggs excreted over the life span of the worm population) being equal to that of red foxes (Kapel et al., 2006). Comparative prevalence data between raccoon dogs and red foxes exist from northern Brandenburg, Germany, where population densities (hunting bag) are similar between the two species, and prevalence levels in raccoon dogs are at a similar level than those reported from red foxes in earlier studies (Schwarz et al., 2011; Tackmann et al., 1998). In contrast, raccoon dogs showed far lower prevalences than red foxes in Lithuania and Estonia (Bruzinskaite-Schmidhalter et al., 2012; Laurimaa et al., 2015, 2016). Raccoon dogs may be less suited as transmitters to voles because they show decreased activity during a 'hibernation' period in winter periods of high snow cover, which reduces their contribution to the life cycle in colder regions (Kauhala et al., 2007). Raccoon dogs tend to defaecate in latrines, which may reduce their potential to spread parasite eggs to vole habitats (Kauhala and Salonen, 2012). Data from Japan suggest a minor role for transmission of this parasite at least in that part of their original range (see Section 4.3.2).

Golden jackals (*Canis aureus*) were largely restricted to southeastern Europe, a region with unclear endemicity status for *E. multilocularis*. In recent years, however, a northward expansion of their distribution can be observed, probably favoured by the population decline of wolves. They are now regularly present in endemic areas such as northern Hungary, Slovenia, eastern Austria and even Estonia, with vagrants reaching as far as Denmark. Golden jackals are omnivorous animals that feed readily on voles and other rodents; where the common vole occurs, it is an important prey species for both red fox and golden jackal, which have a wide overlap in feeding habits (Lanszki et al., 2006; Lalosevic et al., 2016). Jackals infected with *E. multilocularis* have been reported from Hungary and northern Serbia with prevalence estimates in the range of those of red foxes in the same area; published worm burdens appear to be on the lower side compared to foxes, but numbers of infected jackals ($n = 5$) are too small to draw conclusions on susceptibility (Szell et al., 2013). There are no data from experimental infections.

Wolves (*Canis lupus*) have been identified in Latvia and Slovakia as hosts of *E. multilocularis* (Bagrade et al., 2009; Martinek et al., 2001); although due to their preference for large herbivores as prey, wolves were until recently not considered to be playing an important role in the parasite life cycle. Data from North America suggest that wolves may be frequent hosts of *E. multilocularis* (Section 4.4.2), although their low population density compared to red foxes indicates a minor contribution to the life cycle. However, large home ranges and long-distance dispersal of subadult wolves may play a role in the dispersal of the parasite. Also, their range of prey species includes large rodents (muskrats, nutria and beaver), which are known to be frequently infected with *E. multilocularis* metacestodes. There are no experimental data, but the conspecificity of wolves with domestic dogs suggests a high susceptibility. In addition, it has to be considered that wolves can have a strong impact on the distribution and abundance of smaller canid species, which could significantly affect the transmission dynamics of *E. multilocularis* (Hegglin et al., 2015). Studies in North America show that wolves can reduce densities of coyotes, and coyotes are known to affect red fox populations negatively by agonistic behaviour, competition or by direct predation (Berger and Conner, 2008; Fedriani et al., 2000).

Domestic dogs have been recorded as definitive hosts in six European countries (EFSA, 2015). Extremely low prevalence estimates from the general dog population (Deplazes et al., 2011) contrast with high prevalence in unrestrained dogs from rural high-endemicity regions; 7% of 86 dogs were found infected in a hyperendemic focus of Switzerland with exceptionally high prevalence levels in rodents (Gottstein et al., 1996, 2001). Dogs are highly susceptible to experimental infection (Kapel et al., 2006), so low prevalence is clearly a function of low exposure due to limited predation on rodents, possibly in conjunction with occasional anthelmintic treatments. Despite low prevalence, the large number of dogs (compared to foxes) has led to estimates, that they may contribute up to 19% of the *E. multilocularis* egg production in urban or periurban areas of Central Europe (Hegglin and Deplazes, 2013). Dogs are a source of concern for nonendemic areas, as infected companion animals may carry the parasite across country borders (Höjgård et al., 2012).

Domestic cats have been reported as hosts of *E. multilocularis* worms in six European countries, with prevalences comparable to those in dogs (lit. in Deplazes, 2015). Experimental infections showed, however, that worms become only established in some of the animals, worm burdens (although highly variable) are mostly low at the onset of patency, and there is uncertainty about the infectivity of eggs produced in these worms (Jenkins and Romig, 2000, 2003; Kapel et al., 2006; Rausch and Richards, 1971; Thompson et al., 2003). Given those data, the contribution of cats to the life cycle is deemed to be very small. A certain role of domestic cats as transmitters for human AE can, however, not be excluded because massive infections of cats are known to occur occasionally (Vogel, 1957), and in one experiment *E. multilocularis* eggs of a St. Lawrence Island isolate, obtained from an experimentally infected cat, produced metacestodes in a brown lemming (Rausch and Richards, 1971).

Wild cats (*Felis silvestris*). Recently, 6 of 101 examined wild cat carcasses from western Germany, Luxembourg and the French Ardennes had immature *E. multilocularis* worms (Steeb et al., 2012; cited in Deplazes, 2015; Umhang et al., 2015). In Europe these are rare animals with fragmented distribution and extremely low population densities. Even if they might be better suited as hosts than domestic cats (no data exist, but see their role in Asia), their contribution to the life cycle is negligible.

4.2.3 Intermediate hosts in temperate Europe

Common voles (*M. arvalis*) are widespread in Europe and can be the most abundant mammals in some regions. The species does not occur in the British Isles (except for the Orkney Islands), Scandinavia (except Denmark) and most countries/regions bordering the Mediterranean Sea. Where and when common voles are present, even at lower densities, they are a preferred prey of foxes and constitute a large part of fox diets (Guislain et al., 2008; Raoul et al., 2010). The generally very low prevalence of *E. multilocularis* in common voles (as in most intermediate hosts – chapter: Global Distribution of Alveolar and Cystic Echinococcosis by Deplazes et al., 2016) on a large scale is compensated by the frequent predation events and the fact that prevalence hotspots of 10–21% can exist locally (Giraudoux et al., 2002; Gottstein et al., 2001). In urban contexts seasonal prevalence rates in *M. arvalis* were found to range up to 65% in single fields in the periurban area of Zurich (Hegglin, pers. comm.). Common voles are most abundant in agricultural landscapes and prefer permanent grassland that is kept short (e.g., meadows and pastures). They exhibit a tendency to cyclic changes of population density which may in the extreme lead to population outbreaks followed by (almost) complete crashes. These cycles are most pronounced in regions, where suitable vole habitats are a dominating landscape feature, e.g., agriculturally managed grassland, or legume crops (Delattre et al., 1996; Jareno et al., 2015). However, in most of the highly endemic regions of Europe (e.g., agricultural areas of Switzerland, southern Germany, northern France and the High Tatra region), habitats of common voles are widespread but fragmented, and common voles seem to maintain medium to high population densities without extreme cycles in such landscapes. Their importance for the *E. multilocularis* life cycle is supported by a comparative study at the southern limit of the *E. multilocularis* range in Switzerland, which showed a correlation with the distribution of common voles, but not with the distribution of six other vole species, including *M. glareolus* and *A. amphibius* (Guerra et al., 2014). Data from France and Switzerland suggest that common voles may also be crucial for the life cycle in urbanized areas (Hegglin et al., 2007; Hegglin, pers. comm.; Robardet et al., 2011). Experimental inoculations of *E. multilocularis* eggs show high susceptibility of common voles. Metacestode development is rapid with production of protoscoleces commencing at around 8 weeks p.i. (Merli, pers. comm.; Woolsey et al., 2015a; Zeyhle, 1976). In 44 naturally infected *M. arvalis*, protoscolex numbers ranged between 235 and 370,800 (Hegglin, pers. comm.).

A number of other *Microtus* species were found infected with *E. multilocularis* in Europe, but their importance for the persistence of the parasite is unknown. The **field vole** (*M. agrestis*) is widespread and frequent in Europe (including Great Britain and Scandinavia), occurring in moister and more densely vegetated habitats than the common vole. It is highly susceptible to experimental infection with *E. multilocularis* eggs (Woolsey et al., 2015b), but field records of infected animals are only known from southern Sweden (Miller et al., 2016). In this low endemic country, however, this species may be important for the life cycle in the absence of common voles. The **European pine vole** (*Microtus subterraneus*) was only found infected in eastern France (Delattre et al., 1990), while no records of *E. multilocularis* in **arctic voles** (*Microtus oeconomus*) or **East European voles** (*Microtus levis*) exist from temperate parts of Europe; their suitability as hosts, however, is well established from other regions (Section 4.5.3).

Water voles (*Arvicola* spp.) are frequently infected with *E. multilocularis* and are thought to be important for the life cycle in parts of their range, e.g., eastern France, and in some environments, e.g., urban/suburban areas. The species formerly known under the name *A. terrestris* was recently split into two species, although, according to ongoing research, this is not undisputed (Quéré, pers. comm.). The **European water vole** (*A. amphibius*) has a wide European distribution and a preference for aquatic habitats, while the **montane water vole** (*A. scherman*) is largely restricted to grassland environments of elevated regions ranging from northern Spain through central Europe into the Carpathian mountains and has fossorial habits. Both species can also be present in grassy areas within open forests and orchards and may undergo cyclic population changes; they can be pest species during population outbreaks due to damage of grassland and tree roots. In former studies on *E. multilocularis* infection, the two species were not distinguished. Based on distribution and environmental data, most records, e.g., from eastern France, Germany and Switzerland are likely to refer to *A. scherman*. In urban areas of Switzerland, the prevalence of *E. multilocularis* can exceed 10%, reaching 61% in single fields (Burlet et al., 2011). In fragmented urban environments this species is likely to be important for the life cycle (and for the transmission to domestic dogs), since it is less dependent on the presence of coherent stretches of grassland than *M. arvalis*. Due to its fossorial habits it is not as readily available as fox prey *Microtus* spp., but fox predation on *A. scherman* can be substantial where they are prone to cyclic population outbreaks, e.g., in the French Jura mountains (Weber and Aubry, 1993; Raoul et al., 2010). The presence of *A. amphibius* was not associated with the presence of *E. multilocularis* in southern Switzerland. *A. amphibius* was recently found infected in southern Sweden, but its contribution to the life cycle in addition to *M. agrestis* (and in the absence of *M. arvalis*) is still unclear (Miller et al., 2016). Spatial association of fox infection and the presence of water bodies (Staubach et al., 2001; Tolnai et al., 2013) may partially be explained by the presence of aquatic rodents such as *A. amphibius* as intermediate hosts, but could also be attributed to a better survival of the parasite eggs under such humid conditions (Veit et al., 1995). However, there are no data on parasite establishment and development after experimental infection of *Arvicola* sp., and various field studies reported a (very) low proportion of metacestodes that contained protoscoleces among infected *Arvicola* spp. (e.g., 2/19, 2/31, 26/81 and 3/8) (Hofer et al., 2000; Reperant et al., 2009; Stieger et al., 2002; Miller et al., 2016). However, protoscolex numbers as high as 244,000 have been recorded in *Arvicola* spp (Stieger et al., 2002). The range of the southwestern European species *Arvicola sapidus* overlaps with the range of *E. multilocularis* in parts of France, but there are no data on infection.

Bank voles (*M. glareolus*) are frequent throughout Europe with the exception of some regions bordering the Mediterranean Sea. There are numerous records of natural infection with *E. multilocularis* (e.g., from Belgium, France, Switzerland, Germany and the Czech Republic). Yet, the role of bank voles in the life cycle remains unclear because their susceptibility to experimental infection appears to be lower than that of *Microtus* spp (Woolsey et al., 2016); however, once established, the metacestode may produce large numbers of protoscoleces (108,000 were reported by Stieger et al., 2002). The presence of bank voles was not associated with the presence of *E. multilocularis* in southern Switzerland (Guerra et al., 2014), and they are generally less consumed by foxes (Guislain et al., 2008). Prevalence of *E. multilocularis* metacestodes in bank voles is usually low even where other voles are frequently infected (Stieger et al., 2002; Hanosset et al., 2008), although areas of >10% infected bank voles have been reported (Reperant et al., 2009). Fertile metacestodes in *M. glareolus* were recently reported from a wildlife park in France, and an epidemiological role of bank voles was discussed for this rather specific environment (Umhang et al., 2016). Two other *Myodes* species, the **grey-sided vole** (*Myodes rufocanus*) and the **northern red-backed vole** (*Myodes rutilus*), are present in the extreme Northeast of temperate Europe. Both species are frequently infected in parts of Asia and North America (Sections 4.3.3 and 4.5.3), but there are no records from Europe. The general presence of *E. multilocularis* in parts of their European range (northwestern Russia) is also not known.

European snow voles (*Chionomys nivalis*) are specialists for rocky habitats at high elevations and occur in montane to alpine areas of Europe and western Asia. They were found infected with fertile metacestodes of *E. multilocularis* in Bulgaria, which are indeed the only convincing records of *E. multilocularis* in that country (Genov et al., 1980) and in Romania (lit. in Sikó Barabási et al., 2011). In the Alps this vole species is widely distributed in areas where the parasite has never been recorded (e.g., Guerra et al., 2014), which suggests that it plays only a marginal role in the life cycle.

Muskrats (*O. zibethicus*) are, like voles, members of the Arvicolinae and share their high susceptibility as hosts for *E. multilocularis* with the smaller species. As an introduced species originating from North America, it is now found in high population densities in aquatic habitats from western Europe through northern Asia to Japan. *E. multilocularis* is frequently recorded from muskrats of western and central Europe (EFSA, 2015; Oksanen et al., 2016). Prevalence is typically far higher than in voles from the same area and can exceed 30%, e.g., in southern Germany (Romig et al., 1999). To which extent muskrats contribute to the life cycle is unclear. Red foxes do occasionally prey on muskrats (when found outside of the water) and scavenge on carcasses, e.g., of trapped animals that were improperly disposed of or purposefully used as fox baits by hunters. These may be an infection source for foxes and other canids with home ranges that include wetlands. Long-distance migration of muskrats may also contribute to the dispersal of the parasite. They can play a role in life cycles in urban areas, as 15% of 46 muskrats were found infected at a recreational lake within the city of Stuttgart, Germany (Romig, unpublished).

Species of the Murinae (mice and rats) appear to contribute little to the life cycle of *E. multilocularis* in Europe. No infected animals were found in areas of high prevalence in voles (e.g., Stieger et al., 2002). Published infection records of *Apodemus* species in Europe are limited to **striped field mice** (*Apodemus agrarius*) from Belarus (Merkusheva, 1958; cited in Rausch,

1995), Romania (Sikó Barabási et al., 2011) and southwestern Russia (Kirillova and Kirillov, 2008). The susceptibility of *Apodemus* spp. seems to vary greatly according to species, and the consumption by foxes is low (Guislain et al., 2008; Miller et al., 2016). No infected *Apodemus* were found in areas of high prevalence in voles (Stieger et al., 2002; Reperant et al., 2009)

One of three examined **house mice** (*Mus musculus*) was found infected in central France (Petavy et al., 1990). This had initially sparked some discussion on a possible domestic life cycle involving cats and house mice (Petavy et al., 2000), but the (partial) resistance of cats to infection makes the existence of such a cycle unlikely (Deplazes, 2015). Moreover, laboratory strains of mice (*M. musculus*), although rather variable in their susceptibility, show a reduced establishment rate and/or delayed metacestode development after oral infection compared to species of Arvicolinae and some other rodents (Matsumoto et al., 2010). In Europe, only one **brown rat** (*Rattus norvegicus*) was ever recorded with small, nonfertile lesions in eastern France (Umhang et al., 2016; Umhang, pers. comm.). Experimental oral inoculations of laboratory rats were only successful under immunosuppressive drug treatment, while intraperitoneal inoculation of metacestode material usually leads to normal development of the parasite. Complete resistance even of T cell-deficient athymic nude rats to oral inoculation suggests that the resistance mechanisms of *R. norvegicus,* which act during the first stage of oncosphere invasion, are independent of T cell activity (Armua-Fernandez et al., 2016).

Several large rodent species are known to be susceptible for infection with *E. multilocularis* metacestodes. Their contribution to the life cycle is certainly marginal due to their body size, restricted occurrence and comparatively low population densities. They may, however, play a role in long-distance dispersal (and accidental translocation) of the parasite. **Nutria**, or **coypu** (*Myocastor coypus*), is an introduced species originating from South America that has established localized populations in aquatic habitats of climatically favourable areas throughout Europe. In France and southwestern Germany, nutria is frequently infected with fertile metacestodes (Umhang et al., 2013). **European beavers** (*Castor fiber*) were found naturally infected in southern Germany, Switzerland and Austria, while records from Serbia and the United Kingdom concerned animals that had been imported from Germany for conservation purposes (restocking) (Barlow et al., 2011; Cirovic et al., 2012). There are no reliable prevalence data, but recently 1 of 11 wild-caught beavers in Bavaria (Germany) had a metacestode lesion (Campbell-Palmer et al., 2015). In parts of Europe, there are fragmented populations of introduced **American beavers** (*Castor canadensis*). The largest European population, in Finland, exists outside the known range of the parasite, and no infected animals of that species are recorded from Europe (nor from North America, incidentally). Marmots (*Marmota* spp.) are known to be suitable hosts, mainly based on data from Asia (see Section 4.3.3). Little information exists for the **European marmot** (*Marmota marmota*) apart from three positive records from the French Alps; it is worth to note, that one of the animals found infected had been killed by a domestic dog (Callait, 2003).

Lagomorphs are known as hosts for *E. multilocularis* in parts of Asia, but there is only one verified record from Europe: a fertile metacestode was recently found in the liver of a **European hare** (*Lepus europaeus*) in Switzerland (Chaignat et al., 2015). There are no data on prevalence, but the fact that this parasite had not been reported earlier from this frequently hunted species argues against a significant part in the life cycle.

No verifiable records of *E. multilocularis* in Insectivora (shrews and moles) are known from Europe, in contrast to other parts of the world (Sections 4.3.3 and 4.5.3).

4.2.4 Dead-end hosts in temperate Europe

There is a rather long list of mammals, which can harbour metacestodes, but which are clearly not part of the life cycle. These are either species where metacestodes can establish, but do not reach fertility, or species which are suitable for the development of fertile metacestodes but which, for different reasons, cannot transmit the parasite to definitive hosts.

The first category includes **wild boars** (*Sus scrofa*) and **domestic pigs**, where died-out liver lesions have been frequently reported from various European countries, and where the incomplete metacestode development has been experimentally documented (Deplazes et al., 2005; Sydler et al., 1998). Similar lesions were found after experimental infection of domestic ruminants and equids (Ohbayashi et al., 1971), but all macroscopically diagnosed AE infections of ruminants in Europe were, at molecular investigation, shown to be cases of cystic echinococcoses with aberrant growth patterns (Casulli, pers. comm.). In contrast to wild canids, there are numerous reports of lethal metacestode growth in the liver of domestic dogs (with development of protoscoleces) (Deplazes and Eckert, 2001; Corsini et al., 2015).

The second category mainly consists of 'exotic' animal species that are kept in zoos or wildlife parks, whose infection is usually attributed to contaminated food from the local environment. This includes numerous species of primates (from lemurs to apes) (Deplazes and Eckert, 2001), but also systematically diverse taxa as hyraxes (Rietschel, pers. comm.) and macropod marsupials (Peters et al., 2010).

4.3 Temperate Asia

4.3.1 Life Cycles

The distribution of *E. multilocularis* across the Asian landmass (south of the tundra region) is only known in sketches, and rather well researched regions alternate with regions that are either devoid of data, or from where only old records exist that are difficult to verify. The parasite is clearly widespread in Russia from the Ural mountains to the Far East. In fragmented patches of favourable environment (e.g., mountains, wetlands) it ranges southwards into steppe, semidesert and desert regions of central Asia and is highly frequent on the eastern part of the Tibetan plateau. Whether all these areas of confirmed endemicity are part of a continuum in the distribution of the parasite, or represent isolated endemicity regions, or are just ecologically linked to each other (e.g., via long-distance migration of host animals) is not clear. The vast ecological differences which obviously exist between, e.g., subarctic taiga in Russia, semidesert in Kazakhstan, high-altitude grassland in western China and broad-leaved forests in Japan result in a diversity of life cycles that involve a large number of different hosts species (especially on the side of intermediate hosts). Even in regions that are comparatively well studied (e.g., western China, Kazakhstan and Kyrgyzstan), transmission patterns are still incompletely known, mainly due to the lack of ecological and even taxonomic knowledge about the large number of potential rodent hosts and the predator—prey communities existing in any one area.

Boreal forest and steppe zones. Almost the entire Asian part of the Russian Federation can be considered as an endemic region for *E. multilocularis*. Based on the frequency of human AE, transmission is particularly intense in parts of southern Siberia (e.g., Altai, Omsk and Tomsk), central Yakutia and parts of the Far East (Bessonov, 1998). Despite numerous host records, there is only sketchy information on transmission patterns in the temperate zones of Russia. Red foxes are generally considered the principal definitive hosts; wolf and raccoon dog may regionally contribute to the life cycle (Bessonov, 1998; Konyaev et al., 2012). High prevalence estimates (up to 80%) in red foxes are reported from Omsk and Novosibirsk areas and from the Tyva steppe bordering Mongolia, while in steppe areas of northern Kazakhstan red foxes (*V. vulpes*) and corsac foxes (*Vulpes corsac*) both act as definitve hosts at similar prevalence levels (22–30%) (lit. in Shaikenov, 2006). Various voles (*M. agrestis, Microtus obscurus, M. oeconomus, Microtus gregalis, Microtus socialis, Lagurus lagurus, M. rutilus, M. rufocanus, Alticola strelzowi, Alticola olchonensis*) are listed as intermediate hosts in the forest and steppe zones of Russia, and there are also older records from hamsters (*Cricetulus barabensis*), squirrels (*Sciurus vulgaris*) and pikas (*Ochotona dauurica, Ochotona pallasi*) (Abuladze, 1964; Rausch, 1995; Bessonov, 1998; Shaikenov, 2006; Konyaev et al., 2013). Wetlands were considered as transmission foci, where muskrats play a key role as hosts (e.g., in Buryatia – Masur and Fomina, 2012). In southern Siberia and northern Kazakhstan, bobak marmots (*Marmota bobak*) also contribute to the life cycle (Shaikenov, 2006), while *Apodemus* species (*Apodemus sylvaticus, A. agrarius*) are considered unimportant due to extremely low prevalence of the parasite (Shaikenov and Torgerson, 2002). A transmission involving domestic dogs, voles and house mice was proposed to exist around villages in central Yakutia (Yarotsky and Martinenko, 1986; cited in Bessonov, 1998).

Central Asian arid zones. In the more arid regions towards western and southern Kazakhstan, the distribution of *E. mutlilocularis* is described as patchy (Shaikenov, 2006). Based on extensive rodent surveys in western Kazakhstan, the parasite is either absent in deserts and semideserts, or the prevalence in potential host species is low. However, spots of high endemicity can occur in areas of higher ground moisture, which supports meadow vegetation, particularly in the valleys of the Ural and Emba rivers. Both red and corsac foxes are known as definitive hosts in these environments, while the most important intermediate host is the great gerbil (*Rhombomys opimus*). A number of other rodent species were also found naturally infected with fertile metacestodes in this region, but prevalence values are very low, and their contribution to the life cycle is unknown. These include gerbils (*Meriones libycus*), jerboas (*Allactaga elater, Pygeretmus pumilio*), dwarf hamsters (*Cricetulus migratorius*) and ground squirrels (*Spermophilus pygmaeus*). Foci of high endemicity exist near central Asian lakes (e.g., south of Lake Balkhash). This has been linked to the presence of muskrats (with prevalence up to 12%), although water voles (*A. amphibius*) are also present in such habitats. Muskrats are reported to make up 60% of fox diets in parts of Kazakhstan (Shaikenov, 2006;

Shaikenov and Torgerson, 2002). Near Lake Balkhash muskrats may also account for >30% of the diet of wild cats (*F. silvestris*), and high prevalence and worm burden of *E. multilocularis* was reported in wild cats from that area. This observation is in disagreement with other studies on cat infection and was hypothetically linked to muskrats as the source of infection (Bondareva, 1966; cited in Shaikenov and Torgerson, 2002). Muskrats are also assumed to be the source of domestic dog infection in lake regions of eastern Kazakhstan. Dogs of muskrat hunters were infected at prevalence of 33–35% during the winter hunting season (Bondareva, 1972; cited in Shaikenov and Torgerson, 2002).

Altai, Tien Shan and Pamir. A specific transmission pattern is reported from higher-altitude areas of the Altai, Tien Shan and Pamir mountains (from southern Kyrgyzstan and southern Kazakhstan into northern Xinjiang), which possibly extends into Mongolia. Red foxes are the principal definitive hosts, while domestic dogs may play some role in and around villages; in one location of central Kyrgyzstan, 64% of red foxes and 18% of dogs were infected (Ziadinov et al., 2008; 2010). A typical and frequent rodent species (and probably an important host) in productive alpine grassland habitats is the Zaisan mole vole (*Ellobius tancrei*); the species benefits from human farming activities, and an infected specimen caught near a village suggests a transmission between dogs and mole voles (Afonso et al., 2015). Along streams and other wetlands, and also in grassland, *Microtus* species (*M. gregalis, M. oeconomus* and *M. obscurus*) are certainly also part of the life cycle. The Siberian zokor (*Myospalax myospalax*) was found rather frequently infected with fertile metacestodes in the Tarbagatai and Kalbin mountains of East Kazakhstan (Shaikenov and Torgerson, 2002). Marmots (*Marmota baibacina, Marmota caudata*) are apparently important intermediate hosts in alpine habitats. Marmots make up a significant proportion of the summer diet of foxes (Shaikenov, 2006) and are frequently infected with fertile metacestodes (up to 9% prevalence) (Shaikenov and Torgerson, 2002). There is an older host record for *Ochotona* sp. from Kyrgyzstan (Dzhumadilov, 1966; cited in Rausch, 1995).

Southwestern Asia. Only fragmentary data on *E. multilocularis* life cycles exist for southwestern Asia, from Uzbekistan through Iran into the Caucasus and Turkey. For 'Uzbekistan and Turkmenistan', infections of red foxes, corsac foxes and golden jackals (*C. aureus*) are mentioned (Shaikenov, 2006), and in the valley of the lower Amu Darya river (Uzbekistan), muskrats, great gerbils and midday gerbils (*Meriones meridianus*) were infected at low prevalences (Kairov, 1976; cited in Shaikenov, 2006). In the Razavi Khorasan province of northeastern Iran, *E. multilocularis* worms were detected in 5 of 10 golden jackals and 1 examined wolf, and PCR of taeniid eggs in faeces was positive for 6.5% of 77 dogs, all of 3 red foxes and the 1 examined hyena (*Hyaena hyaena*) (Beiromvand et al., 2011). Locally, the Transcaspian vole (*Microtus transcaspicus*) may be important for transmission, as >40% of animals trapped near a village were infected. Additional records were from *M. musculus, Apodemus witherbyi, Ochotona rufescens* and *Crocidura gmelini* (Beiromvand et al., 2013), but fertility of the metacestodes was not determined (Razmjou, pers. comm.). While parasitological confirmation for the suitability of hyenas as hosts are warranted, these results clearly indicate a high transmission intensity and complex multihosts system in areas of higher altitude in northeastern Iran. A similar pattern may exist in the northwest of Iran and further north towards the Caucasus. Older prevalence figures from red foxes (10–23%) and golden jackals (16%) from Ardebil and Azerbaijan provinces prove the endemicity status of northwestern Iran (Mobedi and Sadighian, 1971; Mobedi et al., 1973; Zariffard and Massoud, 1998; cited in Beiromvand et al., 2011), while some older records (often difficult to access and verify) are available for the southern Caucasus, i.e., of wolf and red fox infection in Azerbaijan and Georgia (Elchuev, 1989; Kurashvili, 1966; cited in chapter: Global Distribution of Alveolar and Cystic Echinococcosis by Deplazes et al., 2016), of a jerboa (*A. elater*) in Azerbaijan (Mamedov, 1964; cited in Rausch, 1995) and of three vole species (*M. socialis, M. obscurus, Chionomys roberti*) in Georgia (Kurashvili, 1961; and Matsaberidze, 1966; cited in Rausch, 1995). A life cycle including red foxes and social voles (*M. socialis*) has been mentioned for eastern Georgia (Kurashvili, 1966; cited in Rausch and Richards, 1971). In the northern Caucasus (Ingushetia) 22% of red foxes were found infected, as well as various species of rodents (*M. obscurus, Apodemus uralensis, O. zibethicus, Spermophilus* sp. and *R. norvegicus*) (Pliyeva and Uspenskii, 2006). There was a complete lack of data on animal infection from the Asian part of Turkey, despite the large number of human cases particularly in the eastern part of the country (Altintas, 2003). Only very recently, *E. multilocularis* was confirmed morphologically and molecularly from 1 out of 10 red foxes in Erzurum province (Avcioglu et al., 2016).

Tibetan Plateau. A high-endemicity focus of *E. multilocularis* is known from the high-altitude grassland regions in the eastern parts of the Tibetan plateau (from western Sichuan into Qinghai and Xizang provinces), where the prevalence of AE in the human population is higher than anywhere else in the world (Li et al., 2010). The life cycle is unusual because the main definitive hosts are Tibetan foxes (*Vulpes ferrilata*) rather than red foxes, which are also present. Prevalence of *E. multilocularis* in both fox species seems to be on a similar level: in a survey covering various counties, close to 60% of both Tibetan and red foxes were infected (Qiu et al., 1995), and in Shiqu county (western Sichuan), *E. multilocularis* eggs were found in 42 of 120 faeces of Tibetan foxes (in almost half of them as coinfection with *E. shiquicus*) and in two 2 of 6 faeces of red foxes (Jiang et al., 2012). However, Tibetan foxes outnumber red foxes by far, at least in this part of their range. Plateau pikas (*Ochotona curzoniae*) occur at high population densities, are frequently infected and are preyed upon by both fox species. However, a number of vole species (*Phaiomys leucurus, Neodon irene, Lasiopodomys fuscus, Microtus limnophilus*) and hamsters (*Cricetulus kamensis*) can also reach high population densities and are susceptible to infection (Giraudoux et al., 2013). Approximately 7% of Tibetan hares (*Lepus oiostolus*) in that area harboured metacestodes, most of them fertile, and could also contribute to the life cycle (Xiao et al., 2004). Due to the limited amount of research done in that area, the relative contribution of the various hosts to transmission is not yet clear. Wolves may also be involved as additional definitive hosts, as they are known to prey on pikas (Schaller, 1998). More important, however, should be domestic dogs, at least locally. They frequently prey on pika and rodents, are often found infected and play certainly a key role in transmission of AE to humans (Budke et al., 2005a, 2005b). In the environment of villages more infectious dog faeces than fox faeces were collected (Vaniscotte et al., 2011). A semidomestic life cycle involving dogs and small mammals has been proposed, running in parallel to the sylvatic cycle that exists away from human settlements. However, it is not clear, if such a life cycle would persist without the input of foxes (Giraudoux et al., 2013).

Gansu and Ningxia. In this area of central-western China the landscape is characterized by a mosaic of grassland and farmland created by deforestation of the original subalpine forests. The region is highly endemic for human AE in farmland (Craig et al., 1992, 2000; Yang et al., 2006), but the life cycle of *E. multilocularis* seems to be unusual. Wild canids are virtually absent in the South Ningxia endemicity area, partly due to secondary poisoning with rodenticides in the 1990s (Yang et al., 2012). The only definitive host seems to be the domestic dog. *M. limnophilus* voles are only present in foci of southwestern Gansu, while the small mammal fauna in most of the Gansu/Ningxia region is dominated by the long-tailed dwarf hamster (*Cricetulus longicaudatus*), the Chinese zokor (*Eospalax fontanierii*) and, in the central and eastern part of the region, the Daurian pika (*O. dauurica*). In South Gansu, dogs, zokors and pika were found infected with *E. multilocularis* at considerable levels (5%, 1% and 2%, respectively) (Zhao et al., 2009), and a life cycle may be maintained in the region without the input of wild canids (Giraudoux et al., 2013). One infection source for the dogs may be zokors. These large burrowing rodents can be a pest species in cultivated land, and carcasses of trapped animals are being fed to dogs. Likewise, the viscera of ground squirrels (*Spermophilus dauricus, S. alashanicus*), caught for human consumption, are also used as food for dogs (Giraudoux et al., 2013).

Mongolia and northern China. In the eastern part of Mongolia and Inner Mongolia, the life cycle is maintained by red foxes and corsac foxes as definitive hosts and Brandt's voles (*Lasiopodomys brandtii*) as the most important intermediate host. The extent to which other susceptible small mammals (e.g., *O. dauurica, Meriones unguiculatus*) occurring in the region are involved is uncertain (Tang et al., 2004, 2006; Giraudoux et al., 2013). Old records of metacestodes in ground squirrels (*Spermophilus undulatus*) and gerbils (*M. unguiculatus*) from Buryatia suggest that this transmission system extends north into Russia (Abuladze, 1964). There are no data on the life cycle in the northwestern Chinese province of Heilongjiang, which is known to be endemic for human AE (Vuitton et al., 2003).

Japan. *E. multilocularis* in Japan is restricted to Hokkaido, where it had been accidentally introduced on at least two different occasions in the early and mid-20th century. During the 1980s it gradually spread throughout the island (Ito et al., 2003; Takahashi et al., 2005). The life cycle, now established, differs from transmission patterns in all other regions by its almost exclusive reliance on *Myodes* species as intermediate hosts and is therefore not associated with open grassland such as meadows and pastures where any species of voles are absent. The principal definitive host is the abundant red fox with prevalence of >35%, while raccoon dogs and domestic dogs are only sporadically infected (1% prevalence) (Takahashi et al., 2005); in an area of 57% prevalence in foxes, 3 of 13 raccoon dogs carried *E. multilocularis* worms (only 1 with a patent infection) (Yimam et al., 2002). The grey-sided vole (*M. rufocanus*) is the most common small mammal throughout Hokkaido (Kaneko et al., 1998); the predominant fox prey (Kondo et al., 1986) is highly susceptible to the parasite (Yagi et al., 1984, 1985) and, consequently, is the most important intermediate host species. Two other voles, *M. rutilus* and *Myodes rex*, are also naturally infected, but are less abundant (Ohbayashi, 1996; Takahashi and Nakata,

1995). Typical habitat for the voles (and, therefore, transmission) is the dense undergrowth in woodland formed by dwarf bamboo species (*Sasa* spp.). Bamboo stems and leaves provide food throughout the year and prevent compacting of snow in winter. These conditions facilitate high population densities of *M. rufocanus* in areas of woodland and shrubland, which are also the preferred environment for fox breeding dens, and it has been demonstrated that the life cycle takes place in the immediate vicinity of fox dens (Takahashi et al., 1989; Takahashi and Uraguchi, 1996). Population fluctuations of *M. rufocanus* have been shown to affect infection rates in foxes (Saitoh and Takahashi, 1998).

Among the Murinae, *Apodemus argenteus* and *M. musculus* were confirmed as intermediate hosts (Yagi et al., 1984, 1985; Takahashi et al., 1986). Interestingly, metacestodes were also found on two occasions in brown rats (*R. norvegicus*); one metacestode was fertile (the other containing immature protoscoleces), although brown rats are usually considered refractory to infection (Okamoto et al., 1992; Armua-Fernandez et al., 2016; Iwaki et al., 1993). Two other *Apodemus* species present on Hokkaido, *Apodemus speciosus* and *Apodemus peninsulae*, have never been found infected. A few positive cases have been detected among shrews of two species (*Sorex unguiculatus, Sorex caecutiens*), but their role for transmission is not clear (Takahashi and Uraguchi, 1994).

Foxes on Hokkaido have also established synanthropic populations, e.g., in the outskirts and even central parts of Sapporo (Uraguchi and Takahashi, 1999). Positive foxes found in such areas suggest the presence of a transmission of *E. multilocularis* in urban habitats (Tsukada et al., 2000).

There are case reports about infected dogs on Honshu, the Japanese main island (Yamamoto et al., 2006; Nonaka et al., 2009). However, there is no indication of an established life cycle, although — in addition to foxes and several *Myodes* species — a grassland-adapted vole (*Microtus montebelli*) that was found experimentally highly susceptible to *E. multilocularis* (Ohbayashi et al., 1971) exists on Honshu.

4.3.2 Definitive Hosts in Temperate Asia

Red foxes (*V. vulpes*) are the most widespread of wild canids in temperate Asia. Except for the Tibetan plateau (and possibly some of the arid parts of central and western Asia), they are the most abundant (wild) definitive hosts and are a key species for the life cycles. However, their geographical spread extends southwards far outside the known range of *E. multilocularis*, which highlights the fact that fox presence/absence is not the only parameter to consider to explain transmission. Fox population density variations yet unknown on appropriate spatial scale and resolution, the distribution of suitable intermediate host species assemblages and/or the ecological conditions for parasite egg survival should be considered together. For further details on red fox biology and susceptibility see Section 4.2.2.

Corsac foxes (*V. corsac*) inhabit grassy steppe to semidesert habitats and occur in the drier parts of central Asia from the Caspian Sea to northeastern China. Throughout their range, they live in sympatry with the much larger red foxes and have largely the same diet; in Mongolia, 10–30% of faeces of both species contained rodent remains (mainly gerbils, hamsters and jerboas) (Murdoch et al., 2010). Competition between the two fox species is not well understood, but red foxes are known to kill corsac foxes at encounters, and a decline in corsac fox numbers in Kazakhstan was attributed to rising red fox populations (Heptner et al., 1998). Corsac foxes are less tolerant of high altitude than red foxes and are absent from mountainous areas with snow cover >15 cm. No data on susceptibility of corsac foxes to *E. multilocularis* exist, but similar prevalence levels as red foxes in the same areas suggest a similar host competence (Shaikenov and Torgerson, 2002).

Tibetan foxes (*V. ferrilata*) are geographically restricted to the alpine grassland habitats of the Tibetan plateau, usually occurring at altitudes above 4000–4200 m. Their range almost coincides with that of the plateau pika (*O. curzoniae*), which constitutes an important prey species (IUCN, 2016; Tsukada et al., 2014). Unlike the larger red foxes, they are strictly carnivorous. This specialization, and possibly a better adaptation to cold climate, may be the key factor that makes them competitive against the sympatric red foxes (and wolves) in areas of high density of pikas, but the interaction among the canids of the Tibetan highlands is not well known. In a study in Qinghai (during summer), remains of small mammals were found in 80% of Tibetan fox faeces, but only in 29% of red fox faeces in the same area; unlike Tibetan foxes, red foxes there frequently preyed on insects (Tsukada et al., 2014). Despite this, prevalences of *E. multilocularis* in Tibetan and red foxes seem to be similar.

The natural range of **raccoon dogs** (*N. procyonoides*) is in East Asia, stretching from Vietnam through eastern China and Korea to southeastern Russia, including Japan and Sakhalin. Their diet is highly diversified, with a large proportion of fruits and insects. Yet, they readily feed on rodents, particularly voles and gerbils. Raccoon dogs were found to be highly susceptible to infection with *E. multilocularis* in Europe (Kapel et al., 2006), but worms recovered from a raccoon dog in Japan showed smaller size and lower numbers of eggs compared to worms from foxes (Yagi et al., 1988). Prevalence figures in their natural range only exist for Hokkaido, where they are far lower than those of red foxes (Takahashi et al., 2005).

Wolves (*C. lupus*) are present at varying abundance throughout the Asian range of *E. multilocularis*. Records of *E. multilocularis* in wolves are sporadic but widespread in Asia, including northeastern Iran (Beiromvand et al., 2011), the Russian and Chinese parts of the Altai region (Wang et al., 1989; Konyaev et al., 2012), the Far East of Russia (Tranbenkova, 1992; and Yudin, 2012; cited in chapter: Global Distribution of Alveolar and Cystic Echinococcosis by Deplazes et al., 2016) and Mongolia (Ito et al., 2013). There are no data on prevalences, and their quantitative contribution to the life cycle is unknown.

Dholes (*Cuon alpinus*) may still be present in low numbers in the western Chinese part of the parasite's range. However, they are now so rare (due to human–wildlife conflict) and close to extinction, that they are certainly irrelevant to the life cycle. There are no infection records, and their prey preference of larger mammals also argues against a significant role as host.

The range of **golden jackals** (*C. aureus*) overlaps with the *E. multilocularis* endemicity area only in southwestern Asia, from Turkey and the southern Caucasus to Turkmenistan. The diet of golden jackals is rather opportunistic, with a preference for small to medium-sized animals. They will frequently prey on rodents including gerbils and muskrats (Heptner et al., 1998) and are confirmed as hosts of *E. multilocularis* in Iran and Turkmenistan (Beiromvand et al., 2011; Shaikenov, 2006). Prevalence data are available for northeastern Iran, where 50% of 10 golden jackals were positive at necropsy (Beiromvand et al., 2011). There are no data on the relative contribution of jackals and the sympatric red foxes to transmission.

Throughout the Asian range of the parasite, there is evidence that **domestic dogs** participate to some degree in the maintenance of the life cycle, from Yakutia to Japan, Kyrgyzstan and Iran. Typically, unrestrained dogs in rural villages acquire infection by feeding on commensal rodents, or hunt small mammals in the vicinity of villages. Sometimes, independent life cycles running in parallel to the sylvatic transmission have been proposed, but there is no evidence from anywhere that this 'domestic' transmission would be able to maintain a life cycle in the absence of wild canids (with the possible exception of southern Ningxia – see Section 4.3.1). Studies on the distribution of dog versus fox faeces around Tibetan villages showed that dogs were mainly responsible for contamination of the immediate human environment, but defecated more rarely in habitats of high rodent densities. Thus, even in areas of very high parasite prevalence in dogs their contribution to the life cycle may be limited to the close vicinity of settlements (Vaniscotte et al., 2011).

In contrast to the European, forest-adapted lineage of **wild cats** (*F. silvestris*), the Asian subspecies inhabit drier, more open habitats. An important – if geographically restricted – definitive host role was proposed for wild cats in Kazakhstan, where they heavily prey on muskrats. Three of eleven necropsied cats near Lake Balkhash had up to 200,000 worms (Bondareva, 1966; cited in Shaikenov and Torgerson, 2002). However, wild cats were not found infected in other surveys in Kazakhstan, so these observations need to be verified. The role of **domestic cats** in Asia seems to resemble the situation in Europe. In Japan, there are some records of cat infection, but infectivity of excreted eggs could not be determined (Nonaka et al., 2008). Infection of **Lynx** (*Lynx lynx*) has been reported from the Altai (Pomamarev et al., 2011; cited in chapter: Global Distribution of Alveolar and Cystic Echinococcosis by Deplazes et al., 2016).

4.3.3 Intermediate Hosts in Temperate Asia

Tundra voles (*M. oeconomus*) and **narrow-headed voles** (*M. gregalis*) are known as important hosts in the arctic regions and also have a key role where they occur further south, e.g., in patches of wetlands in steppe regions and in high-altitude grasslands of central Asian mountain ranges. The **lacustrine vole** (*M. limnophilus*) – closely similar to *M. oeconomus* – replaces the latter in high-altitude grassland of the eastern Tibetan plateau, ranging into semidesert areas of Mongolia. The **Altai vole** (*M. obscurus*), formerly included in *M. arvalis*, ranges from western Russia to the Altai region, in a steppe and grassland belt between the boreal forest and the drier parts of central Asia. Information on its role in the life cycle is contradictory: rarely mentioned as a host, it is reported as frequent fox prey in steppe and mountains of Kazakhstan, and prevalences (2.6–2.9%) are among the highest of any intermediate host (Shaikenov and Torgerson, 2002). **Zaisan mole voles** (*E. tancrei*) are widely spread in elevated areas of central Asia from Altai to Pamir through Tian Shan and everywhere observed with pronounced population fluctuations. Its size and ecology is quite similar to montane water voles in Europe, and it may play an important role in transmission (including transmission to domestic dogs) (Afonso et al., 2015; Giraudoux et al., 2013). **Brandt's vole** (*L. brandtii*), a grassland species undergoing cyclic population changes, is considered an ecological key species in its distribution range from Inner Mongolia to southern Siberia. It is probably the most important intermediate host in this region, as it is frequent prey of foxes, and shows high prevalences of *E. multilocularis* (e.g., in Inner Mongolia) (Giraudoux et al., 2013; Tang et al., 2004). Among the very large number of potentially susceptible rodents in temperate Asia, other vole species are either sporadically mentioned as hosts (not always with confirmation of metacestode fertility) without further information on their impact on the life cycle, or have only local importance due to their limited range. Such records exist for additional *Microtus* species (*M. socialis, M. agrestis, M. transcaspicus*) in Georgia, southern Siberia and Iran, **Robert's snow vole** (*C. roberti*) in Georgia, the **flat-headed vole** (*A. strelzowi*) and the **Olkhon mountain vole** (*A. olchonensis*) from southern Siberia, the **steppe vole** (*L. lagurus*) from southern Siberia and Kazakhstan, **Blyth's vole** (*P. leucurus*) and **Irene's mountain vole** (*N. irene*) from the Tibetan plateau (Abuladze, 1964; Bessonov, 1998; Giraudoux et al., 2013; Konyaev et al., 2013; Rausch, 1995; Shaikenov, 2006). *Myodes* species (*M. rufocanus, M. rutilus* and *M. rex*) are key intermediate hosts in Hokkaido, but their role elsewhere is less clear (Ohbayashi, 1996; Takahashi and Nakata, 1995; Yagi et al., 1984, 1985). As species inhabiting forest undercover or rocky areas, they are not a preferred prey of foxes worldwide, although prevalences of *E. mutlilocularis* can be extremely high (e.g., in *M. rutilus* on Kamchatka – Rausch, 1995). **Water voles** (*A. amphibius*), important intermediate hosts in Europe, were only found infected at low to extremely low prevalences in southern Siberia and Kazakhstan (Shaikenov and Torgerson, 2002).

Muskrats (*O. zibethicus*) are an invasive species in Asia and are widely distributed from the Arctic Circle to isolated wetlands within semidesert areas of central Asia. Prevalences of *E. multilocularis* in muskrats are typically high, and a number of authors proposed specific transmission systems maintained by muskrats, and locally a role as infection source for hunting dogs (lit. in Rausch, 1995; Masur and Fomina, 2012; Shaikenov and Torgerson, 2002). Like in North America and Europe, the impact of this species is in need of evaluation, specifically concerning predation rate by wild canids and dogs. It may play a key role in persistence of the life cycle in wetland habitats.

Two species of mole rats (Spalacidae), the **Siberian zokor** (*M. myospalax*) and the **Chinese zokor** (*E. fontanierii*), were recorded with fertile metacestodes in the Altai range and in Gansu/Ningxia, respectively (Shaikenov and Torgerson, 2002; Zhao et al., 2009). They are large, strictly burrowing rodents which are probably rare prey of canid predators. However, some role in transmission was proposed in southern Ningxia and Gansu, where carcasses of zokors trapped as crop pests by farmers are being fed to dogs (Giraudoux et al., 2013).

There are a number of *E. multilocularis* records from **dwarf hamsters** (*C. barabensis* in southern Siberia, *C. migratorius* in Kazakhstan, *C. kamensis* on the Tibetan plateau) (Giraudoux et al., 2013; Rausch, 1995; Shaikenov and Torgerson, 2002). Dwarf hamsters can reach high densities in agricultural land and areas of dry grass cover, where their number can be similar to arvicolines. They are also often synanthropic (occupying the 'house mouse niche') and may therefore be important locally for a 'domestic' transmission to dogs. There is only one record of a nonfertile infection in a **black-bellied hamster** (*Cricetus cricetus*) from western Siberia (Petrov, 1958; cited in Rausch, 1995).

The role of gerbils for transmission is in need of assessment and probably differs in different regions. Most gerbil species occupy the driest parts of central Asia, where transmission of *E. mutlilocularis* appears to be rather tenuous (Shaikenov, 2006). Even in the most frequently infected species, the **great gerbil** (*R. opimus*), prevalence is generally extremely low except in the most humid parts of its range or in microhabitats of increased soil moisture (Shaikenov and Torgerson, 2002). The role of other susceptible species is either marginal (*M. libycus* in Kazakhstan) or not yet assessed (*M. unguiculatus* in Inner Mongolia, Mongolia and Buryatia; *M. meridianus* in Uzbekistan). Experimental infections of *M. unguiculatus* gave ambiguous results. Mongolian gerbils — widely used as laboratory animals — are highly susceptible to intraperitoneal inoculation with metacestode tissue. However, experimental oral infection (using *E. multilocularis* eggs derived from a Japanese laboratory strain) showed extremely low establishment rate of oncospheres compared to various laboratory strains of *M. musculus* and cotton rats (*S. hispidus*), while subsequent development of metacestodes was normal and resulted in large numbers of protoscoleces (Matsumoto et al., 2010).

Various *Apodemus* species (*A. sylvaticus, A. agrarius, A. uralensis, A. witherbyi, A. argenteus*) are confirmed as hosts in Iran, Russia, Kazakhstan and Japan (Beiromvand et al., 2013; Shaikenov and Torgerson, 2002). There are no reports that indicate an important part in the life cycle from anywhere, but their impact needs individual assessment according to region and species, as predation rate by foxes, prevalence and susceptibility seem to vary considerably. Other Murinae recorded as hosts are **house mice** (*M. musculus*) from widely spaced areas (often in association with proposed, but unconfirmed, synanthropic dog–rodent cycles) and **brown rats** (*R. norvegicus*) from Russia and Japan.

Central Asia is home to seven species of marmots, of which one steppe species (*M. bobak*) and two high-altitude grassland species (*M. baibacina, M. caudata*) are confirmed as good hosts of *E. multilocularis*. Both in steppe and highlands, predation by foxes is apparently common, and the high prevalence figures reported from various areas of Kazakhstan indicate an important intermediate host role, at least locally. Interestingly, *Marmota himalayana,* which is locally abundant on the Tibetan plateau, has not been confirmed as intermediate hosts yet. The role of ground squirrels is more difficult to assess: from the large number of Asian species, only four are confirmed as intermediate hosts (*S. alashanicus, S. dauricus, S. pygmaeus, S. undulatus*). *E. multilocularis* prevalences reported from ground squirrels are extremely low, and experimental infections of different species resulted in poor metacestode development (Rausch and Richards, 1971; Zhou et al., 1998). Only 3 of >1000 **red squirrels** (*S. vulgaris*) were infected in southern Yakutia and can be excluded from the list of important hosts (Tavrovskii et al., 1971; cited in Rausch, 1995).

Members of the lagomorph order may play a more significant role for transmission in Asia than on other continents. The **Tibetan hare** (*L. oiostolus*), a leporid, is frequently infected on the Tibetan plateau. Other lagomorphs of the Ochotonidae family, especially **plateau pikas** (*O. curzoniae*) seem to be of some importance. Plateau pikas occur at high population densities, are frequently infected and are heavily preyed on by both Tibetan and red foxes (Giraudoux et al., 2013; Zhang et al., 2015). The contribution of plateau pikas, however, may vary according to locations: a recent survey of 80 animals in Qinghai found no infection with *E. multilocularis* (known to be present in dogs of the study area), while prevalence of *E. shiquicus* was 24% (Fan et al., 2016). The role of other *Ochotona* species found infected in other parts of Asia is less clear: the Daurian pika (*O. dauurica*) in Gansu, Ningxia, Inner Mongolia and southern Siberia; Pallas' pika (*O. pallasi*) in southern Siberia, an undetermined species in Kyrgyzstan; and the Afghan pika (*O. rufescens*) in northern Iran.

Two species of **shrews** were occasionally found with *E. multilocularis* metacestodes on Hokkaido, Japan (*S. caecutiens, S. unguiculatus*). Although one of the four recorded cases (in *S. unguiculatus*) was fertile (Takahashi and Uraguchi, 1994), shrews probably play no significant role in the life cycle as they are rarely preyed on by foxes. There are no further details about an infection found in *C. gmelini* in northern Iran (Beiromvand et al., 2013).

4.3.4 Dead-End Hosts in Temperate Asia

Spontaneously died-out ('calcified') lesions are frequently reported from the livers of **domestic pigs** in Hokkaido (Sakui et al., 1984; Ishige, 1984), which is in accordance with observations from Europe. Several verified records of inactive *E. multilocularis* metacestodes in **horses** are also known from Japan (Miyauchi et al., 1984; Takahashi and Mori, 2001). Only larvae in a 'regressive' state of development could be found in a horse experimentally infected with eggs of an isolate derived from St. Lawrence Island (Ohbayashi et al., 1971). Infection of ruminants (cattle, yak, sheep) have been frequently reported from the Tibetan plateau (Qiu et al., 1989), but most likely refer to inactive lesions or are due to misidentification of CE with atypical morphology (Heath et al., 2005). Experimental infections of ruminants have not resulted in the development of fertile metacestodes (Ohbayashi et al., 1971; Rausch and Fay, 2002).

As in Europe, a range of primate species (lemurs, monkeys and apes) kept in Japanese zoos have succumbed to alveolar echinococcosis, probably due to contaminated food or soil from the environment (Ohbayashi, 1996; Sato et al., 2005; Yamano et al., 2014).

4.4 Temperate North America

4.4.1 Life cycles

Geographic range and epidemiology of *E. multilocularis* in North America are still insufficiently known. In the temperate parts, the parasite is only recorded from the north central region, encompassing southern Canada (from British Columbia to Manitoba) and the northern United States (from Montana and Wyoming in the West to Ohio in the East and central Indiana in the South). In this region, characterized by prairie and boreal forest, the parasite is maintained in a sylvatic cycle that involves only few host species: red fox (*V. vulpes*) and coyote (*Canis latrans*) as definitive hosts, and meadow vole (*Microtus pennsylvanicus*), southern red-backed vole (*Myodes gapperi*) and deer mouse (*Peromyscus maniculatus*) as intermediate hosts, with differential importance depending on the habitat characteristics and on the overall prey—predator community (Liccioli et al., 2014). The role of muskrats is largely unclear due to the lack of recent data. Based on findings of *E. multilocularis* in two cats and one house mouse (*M. musculus*) on a farm in North Dakota, a domestic cycle had initially been proposed, but these records are now considered to be the result of a spillover from the wildlife cycle for the same reasons as in Europe (see Section 4.2.2) (Leiby and Kritsky, 1972). The life cycle in temperate North America is not fully understood due to the complex interaction between coyotes and red foxes (interspecific competition) and the different proportion of rodents (both competent and not competent species; Liccioli et al., 2014) in their respective diets (which also varies according to region). Adding complexity, wolves (*C. lupus*) were recently found regularly infected in western Canada (Schurer et al., 2014), but their role in the life cycle needs further evaluation, e.g., by considering the relative population densities of the different canids and their interactions.

In the city of Calgary (Alberta, Canada), where coyotes are much more common (800–900 coyotes within the urban area) than foxes, almost 30% of coyotes were found infected with *E. multilocularis* (Catalano et al., 2012). With regard to the life cycle, there seems to be no fundamental difference to the sylvatic cycle in rural areas, except for the almost complete exclusion of foxes from the cycle and the possible role of domestic dogs. In fact, the close proximity of the wildlife cycle to human habitations favours the transmission to pet dogs and potentially to humans: in Calgary, 2 out of 218 dogs were found infected, both known to prey on rodents in an endemic area of the city (Massolo et al., 2014; Massolo, unpublished). A significant role of the local domestic dog population in parasite transmission cannot be ruled out because, as in Europe, despite the low prevalence detected (<1%), their large numbers (>125,000 dogs in the city of Calgary) might give them a significant role (in terms of shed eggs) in the life cycle of the parasite in this urban setting (see also Section 4.2.2).

4.4.2 Definitive hosts in temperate North America

Red foxes (*V. vulpes*) are present throughout North America in various subspecies. Until recently, the largest part of the range was considered to be populated by foxes of European origin, which had been introduced from colonial times into the previously red fox-free eastern United States and had supposedly spread over the continent replacing the indigenous subspecies except for remnant populations (Kamler and Ballard, 2002). Based on molecular evidence, this view was later challenged, and the range extensions, population increases and presence in synanthropic environments, which were observed in recent years are now considered due to adaptations by fox populations of native or mixed origin (Statham et al., 2012). The northernmost (and only Canadian) records of *E. multilocularis* in red foxes come from British Columbia (Gesy et al., 2013), while red fox infections were recorded from almost all endemic states of the United States (lit. in Massolo et al., 2014). In some studies from the Dakotas, prevalence estimates in red foxes exceeded 70% (Hildreth et al., 2000; Rausch and Richards, 1971). The role of red foxes in the life cycle, compared to that of coyotes, is difficult to assess due to spatial and temporal heterogeneity of existing prevalence data and probably shows great variation according to landscape and habitat type. When both species were examined in the same area and period, either foxes or coyotes exhibited higher prevalences (lit. in Massolo et al., 2014), and no data on host population densities are available. In parts of North America, declining fox populations are attributed to increasing numbers of coyotes and may result in decreasing parasite abundance due to differences in the prey range between both predators (Melotti et al., 2015).

Coyotes (*C. latrans*) are food generalists with a wide range of prey species. In contrast to foxes, the proportion of rodents in their diet is partially reduced in favour of larger prey (e.g., hares and deer). However, prey preference seems to vary across the coyote's vast range that includes almost all of North America south of the tundra zone. Eastern populations (mainly outside the range of *E. multilocularis*) prefer larger prey, possibly due to hybridizations with dogs and wolves ('eastern coyotes', 'coydogs'), which is reflected by far higher prevalence of taeniid species that are transmitted by deer and lagomorphs compared to red foxes from the same area (Melotti et al., 2015). In the rest of their range, small prey remains can be found in up to 85% of coyote diet (Lukasik and Alexander, 2012) and can seasonally represent the main food item with up to 65% of frequency in coyote scats (Liccioli et al., 2015b). Moreover, up to 80% of the small mammals preyed by coyotes are competent hosts for *E. multilocularis* (Liccioli et al., 2015b).

As coyotes also seem to suppress fox populations (and even prey on them — Liccioli et al., 2015b), the presence and increase of eastern coyotes has been tentatively linked to a decrease (and possible absence) of *E. multilocularis* in eastern North America (Melotti et al., 2015). In the western United States and Canada, however, coyotes seem to be an important component of the parasite's life cycle; reported prevalences of *E. mutlilocularis* were >40% in South Dakota and >60% in Alberta (Catalano et al., 2012; Hildreth et al., 2000). Coyotes have recently formed resident populations in synanthropic environments, where they maintain a life cycle in the immediate environment of humans, which, in turn, exposes domestic dogs to infection (Catalano et al., 2012; Massolo et al., 2014).

Wolves (*C. lupus*) were only recently identified as frequent hosts for *E. multilocularis* in Canada (13% of 93 animals — Schurer et al., 2014). Surprisingly, timber wolves from southern Canada were more frequently infected than wolves from the Northwest Territories, although the southern populations were thought to feed largely on ungulates. It is still unclear whether wolves contribute significantly to the life cycle, or whether their infection depends on the parasite cycle maintained by foxes and coyotes.

There is a single record of *E. multilocularis* from a **grey fox** (*Urocyon cinereoargenteus*), a foxlike canid with southeastern distribution that is only distantly related to *Vulpes* spp (Vande Vusse et al., 1978). It occurs only in a part of the parasite's range, and its susceptibility and possible contribution to the life cycle are unknown. During a recent survey in Michigan, 45 grey foxes were negative, but the parasite was also absent in red foxes and very rare in coyotes from that area (Melotti et al., 2015). **Black bear** (*U. americanus*) and **lynx** (*Lynx canadensis*) developed worms after experimental infection, but there were apparently no eggs present 31 days p.i. (Rausch and Richards, 1971).

Apart from the two aforementioned infected animals in Calgary, there are no reports of **domestic dogs** as definitive hosts in North America with the exception of St. Lawrence Island (see Section 4.5.1). There are only two reports of *E. multilocularis* in **domestic cats**, from Saskatchewan (Wobeser, 1971) and North Dakota (Leiby and Kritsky, 1972). Most cats contained few worms, but worms from some infections were described as gravid. Yet, cats appear to less suitable hosts than canids and probably play a marginal role in the life cycle (see Section 4.2.2).

4.4.3 Intermediate hosts in temperate North America

The **meadow vole** (*M. pennsylvanicus*) is the only *Microtus* species that has ever been recorded as host for *E. multilocularis* in temperate North America. It is a frequent and widespread species that prefers wetlands and moist habitats overgrown by vegetation (similar to the European *M. agrestis*). Reported prevalences range up to 6% (lit. in Massolo et al., 2014), and it might be the most important intermediate host in temperate North America. Meadow voles were highly susceptible to experimental infection with isolates from St. Lawrence Island and from Germany, with 100% establishment rate and subsequent production of fertile metacestodes largely replacing the liver tissue (Ohbayashi et al., 1971). There are at least five additional *Microtus* species in the parasite's North American range whose status as hosts is unconfirmed. Meadow voles were shown to be the most frequent (and preferred) prey species of coyotes in Calgary, Alberta (Liccioli et al., 2015b). At the same location, the **southern red-backed vole** (*M. gapperi*), ranging widely in southern Canada and the northern United States, was for the first time identified as a suitable intermediate host (containing fertile metacestodes). It prefers forested areas, occurs near human habitations, is a preferred prey of coyotes and may therefore play a role in the urban transmission of *E. multilocularis* in North America (Liccioli et al., 2013). Experimental infections were successful in two arvicolines, which have not yet been found naturally infected: in **sagebrush voles** (*Lagurus curtatus*), whose distribution in western North America partly overlaps with the recognized range of *E. multilocularis*, and in the **Amargosa vole** (*Microtus californicus*), whose range in California is clearly outside (Ohbayashi et al., 1971).

The **deer mouse** (*P. maniculatus*) is a member of the New World mice (Neotominae) which are approximate American equivalents to the unrelated Murinae, which were originally restricted to the Old World. Deer mice are the most widespread of all North American rodents and exist as different ecotypes, utilizing virtually every type of habitat within their range. The large number of *E. multilocularis* records (with prevalence up to 6%) correlates with the large numbers of animals caught in most surveys (lit. in Massolo et al., 2014). Yet, there is some doubt about the role of deer mice in the life cycle, as metacestodes grow as rather large, thin-walled cysts that contain relatively few protoscoleces compared to voles (Rausch and Richards, 1971) and experimental infection (using eggs of a St. Lawrence Island-derived parasite isolate) produced metacestodes only in 2 of 12 animals (Ohbayashi et al., 1971). In addition, coyotes were found to select against deer mice as prey in favour of voles (Liccioli et al., 2015b). Despite this, deer mice are likely to contribute to the life cycle to some extent, and the ongoing population decline of the deer mouse in favour of the related **white-footed mouse** (*Peromyscus leucopus*), whose suitability as host is unknown, was recently discussed as one of various possible factors for the observed decrease of *E. multilocularis* abundance in Michigan (Melotti et al., 2015). There is one *E. multilocularis* record, from Wyoming, from another neotomine rodent, the **bushy-tailed woodrat** (*Neotoma cinerea*), but the metacestode was nonfertile and the host status of this species is therefore unconfirmed (Kritsky et al., 1977).

Muskrats (*O. zibethicus*) are highly susceptible and frequent intermediate hosts in Europe and Asia (see Sections 4.2.3 and 4.3.3), where they were introduced early in the 20th century. In contrast, there are only four records of infected animals from their native range in North America, despite large numbers of examined animals: no infection was found in 12,142 animals in North Dakota in the 1960s (Rausch and Richards, 1971), 2 out of 192 were infected in southern Montana in 1976/77 (Eastman and Worley, 1979) and a further 2 of 657 muskrats from the same location were found positive in a survey of 1980 (Feigley and Worley, 1988). The latter surveys, conducted in an area with known fox infection, yielded no metacestodes in almost 800 individuals of various other small mammal species, and the authors proposed an important intermediate host role for muskrats. Muskrats are known to be frequently consumed by foxes in North America (Rausch and Fay, 2002). The low infection prevalence in North American muskrats may be linked to the parasite 'strain', as only one of three muskrats established metacestodes after experimental infection with a North Dakota isolate, or to different susceptibility of muskrats subspecies (Rausch and Richards, 1971). Future studies on the role of muskrats are clearly warranted.

No **beavers** (*C. canadensis*) were ever found infected in North America.

One natural infection in a **house mouse** (*M. musculus*) is known from North Dakota. The metacestode was fertile, which had led to the suggestion of a 'domestic' life cycle involving cats and commensal mice (Leiby et al., 1970; Leiby and Kritsky, 1972) (but see Sections 4.2.2 and 4.4.2). The house mouse is an introduced species in the Americas that is largely restricted to buildings and cultivated fields.

4.4.4 Dead-end hosts in temperate North America

A total of three **domestic dogs** suffering from severe hepatic alveolar echinococcosis were recently identified at distant locations in Canada (lit. In Massolo et al., 2014). This has no consequence for the life cycle, but the case reports may be an indicator for the spread of the parasite (or a consequence of heightened awareness for the disease).

4.5 Circumpolar arctic and subarctic regions

4.5.1 Life Cycles

In the holarctic tundra zone, the natural life cycle depends on the arctic fox (*Vulpes lagopus*) and various species of voles and lemmings; red foxes, wolves and domestic dogs can serve as additional definitive hosts in some areas. The ecological conditions and the biogeography of host species are not uniform in this vast region, which is reflected in the drastic differences of the species communities and abundance of rodents. Rather than the widespread definitive hosts, the local rodent fauna seems to determine the presence and frequency of *E. multilocularis* in the arctic tundra, which ranges from extremely frequent (in areas of high and stable vole populations) to complete absence (e.g., on the west coast of Greenland due to the absence of any species of rodent). Detailed epidemiological data are only available for few places. The life cycle is known in detail for St. Lawrence Island in the Bering Strait, where prevalence in all hosts is subject to strong seasonal fluctuation. At the end of the warm season, during which foxes heavily prey on rodents (particularly the tundra vole *M. oeconomus*), fox infection can be 100% (Rausch et al., 1990a). With increasing snow cover, foxes shift to scavenging of carcasses of larger mammals, and infection declines accordingly towards spring. Conversely, prevalence in the overwintering adult voles is highest at the beginning of the breeding season in spring, when it can reach >80% and declines during summer due to dilution by the new generations which are yet to be infected (Rausch et al., 1990b). There are strong regional variations in the life cycle. In areas with high and stable populations of voles, prevalence in foxes is highest (e.g., on St. Lawrence Island). Where voles are absent and lemmings undergo strong cyclic fluctuations of population density, the parasite is rare despite the presence of foxes (e.g., in parts of the Alaskan coast) (Rausch and Fay, 1988). In various parts of the Russian arctic tundra (Nenetsia, northern Yakutia and Chukotka), prevalence data from arctic foxes vary widely, while different species of lemmings (*Lemmus sibiricus, Dicrostonyx torquatus*) and voles (*M. oeconomus, M. gregalis*) are listed as intermediate hosts (Bessonov, 1998). There is no clear information, whether these differences are due to season or differences in rodent ecology, neither are there data on the infectivity of the local parasite strain to different rodent species. The *E. multilocularis* life cycle in the tundra zone is not completely separated from areas further south, as arctic foxes are known to migrate southwards to the taiga zone especially in periods of low rodent densities, carrying the parasite into areas with higher diversity of potential hosts (e.g., red foxes). Wolves may also play a role in the North–South dispersal of the parasite due to their large home ranges and dispersal distances.

An exceptional life cycle exists on the Norwegian arctic island Spitsbergen in the Svalbard archipelago. Due to the absence of indigenous rodents, there was initially no basis for *E. multilocularis* transmission despite arctic foxes being common. After the accidental introduction of the East European vole (*M. levis*) that must have taken place before the 1960s (probably together with cattle forage imported from western Russia), a viable vole population became established on strips of vegetation below bird cliffs along a few kilometres of coastline. From 1999 onwards, *E. multilocularis* was recorded at extremely high prevalences both in voles and foxes (Henttonen et al., 2001; Fuglei et al., 2008). Genetic analysis of the parasite isolates indicates that the parasite had been introduced by infected arctic foxes migrating over sea ice rather than together with the voles from their original range in the St. Petersburg area of Russia (Knapp et al., 2012). This life cycle is remarkable because it involves only two host species and because it has been persisting for decades in an extremely restricted area.

A synanthropic life cycle was reported in the 1970s from St. Lawrence Island, where cases of human AE infection clustered in two villages of Yupik people. A cyclic transmission was found involving domestic dogs, which lived partly unrestrained in the villages feeding on tundra voles that occurred as commensals around and even inside the traditionally built houses. The commensal voles showed infection prevalences approaching 30% in the overwintering population (Rausch et al., 1990b). It is unclear whether this cycle was self-sustaining or depended on spillover from infected wildlife in the surrounding area. The village transmission was subsequently controlled by anthelmintic treatment of dogs, replacement of dogs by snowmobiles, and probably other alterations of people's lifestyle.

4.5.2 Definitive Hosts in Arctic and Subarctic Regions

Arctic foxes (*V. lagopus*) are common predators and scavengers in the tundra zones of Eurasia and North America. Rodents form a large part of their diet in summer, while snow cover prevents access to this food source in the colder seasons. This causes significant seasonal variation of *E. multilocularis* prevalence, and the grossly divergent prevalence figures from arctic foxes in different regions are therefore difficult to interpret without more detailed information. In winter, arctic foxes are able to migrate several thousands of kilometres over land or sea ice. They have even been observed to follow polar bears as far as the North Pole, scavenging on the remains of bear kills (Feldhamer et al., 2003). They are also known to migrate on and off arctic islands in large numbers, spreading *E. multilocularis* and other parasites over vast areas (Rausch and Fay, 2002). Southward migrations in times of food shortage can link the arctic life cycle to those in subarctic and temperate zones. Arctic foxes are highly susceptible to infection with *E. multilocularis*, exemplified by worm burdens in excess of 100,000 being 'not uncommon' (Rausch and Fay, 2002). Estimated maximum life span of the worms was approximately 7 months, based on an experimental infection of two arctic foxes with 100,000 protoscoleces each, but only few worms were left after that period (with no or few eggs in the uterus) (Rausch and Fay, 2002).

The distribution of **red foxes** (*V. vulpes*) extends far north into the arctic on the mainland and overlaps widely with that of the arctic fox. Where prevalence data of both species are available, red foxes seem to be less frequently infected, e.g., in Nenetsia, northwestern Russia (Peklo, 2014) and in Yakutia (Odnokurtsev and Sadalishchev, 2012; cited in Davidson et al., 2016). Reasons for this are not known, but data from experimental infections indicate that red foxes may be partially resistant to arctic variants of *E. multilocularis* (Rausch and Richards, 1971). Global warming and a continental increase of red fox populations in Europe led to a northward spread of red foxes and to the competitive exclusion of the arctic foxes in those areas (Hamel et al., 2013), which may create more complex life cycles at the limit of the two fox species' distributions.

Wolves (*C. lupus*) were found infected at a remarkably high prevalence level (8.2%) in the Northwest Territories of Canada (Schurer et al., 2014), which corresponds to 'about 10%' infected wolves on the Taimyr Peninsula in northern Siberia (Savelev, 1972; cited in Rausch, 1995). Tundra wolves are known to prey more heavily on small mammals than wolf populations further south and those in coastal areas (lit. in Schurer et al., 2014; Schurer et al., 2016) and may contribute to the arctic fox—vole life cycle to a certain degree. They may be the most important definitive hosts in subarctic and boreal regions of North America (and possibly other regions of similar latitude), bridging the 'boreal gap' between the arctic fox-based life cycle in the far north and the red fox- or coyote-based life cycles in the temperate latitudes due to their large home ranges and long dispersal distances (Davidson et al., 2016; Rausch, 1995) (see also Section 4.2.2).

Domestic dogs may participate in the life cycle in or near human settlements, but data on their role in the far north are only available from St. Lawrence Island (see previous discussion).

4.5.3 Intermediate Hosts in Arctic and Subarctic Regions

Tundra voles (*M. oeconomus*) are ecologically adaptable rodents that occur in a variety of habitats (including forests and wetlands) from Scandinavia and northwestern Europe through most of northern Asia into Alaska and western Canada. They are not present in all tundra regions, but where they occur in dense and stable populations they seem to be a key host for *E. multilocularis*. In Alaska, human AE cases were not found north of the vole's distribution limit (Rausch and Fay, 2002). Experimental infections proved high susceptibility for a parasite isolate from St. Lawrence Island, while oncospheres derived from a North Dakota isolate showed a decreased establishment rate and the metacestodes grew less vigorously (Ohbayashi et al., 1971; Rausch and Richards, 1971). **Narrow-headed voles** (*M. gregalis*) are adapted to open areas and have a patchy distribution in parts of the Russian tundra, but also in steppe regions of central Asia. They were frequently recorded as hosts of *E. multilocularis* throughout their distribution range (lit. In Rausch, 1995) and are considered important for the life cycle (Bessonov, 1998). In addition, **Middendorffs vole** (*Microtus middendorffii*), a specialist for waterlogged tundra with a distribution in Russia east of the Ural mountains, has been mentioned as host (Rausch, 1995). Interestingly, numerous attempts of experimental infections (with parasites deriving from St. Lawrence Island and North Dakota) of the **singing vole** (*Microtus miurus*), distributed from Alaska into northwestern Canada, have always been unsuccessful (Ohbayashi et al., 1971; Rausch and Richards, 1971).

Northern red-backed voles (*M. rutilus*), present throughout the northern parts of Eurasia and America, are most numerous in the subarctic taiga and only venture further north in places with sufficient cover (e.g., rocks). They were frequently found infected with prevalences exceeding 40% locally (lit. in Rausch, 1995). In infection experiments, they showed high susceptibility to parasites of St. Lawrence Island and German origin, but low establishment rates were achieved using material from North Dakota (Ohbayashi et al., 1971; Rausch and Richards, 1971). Yet, their contribution to the life cycle is probably limited because they are more difficult to catch for foxes due to their avoidance of open habitats (Fay and Stephenson, 1989).

Various species of lemmings appear to take part in the life cycle, but published data are partly contradictory. The **brown lemming** (*L. sibiricus*) is abundant in the Eurasian tundra from Nenetsia in the European part to Yakutia; further east, and in North America, it is replaced by the closely related *Lemmus trimucronatus*. Brown lemmings exhibit pronounced 3–4 year population cycles, which is thought to be unfavourable for stable transmission of *E. multilocularis* (Rausch, 1995). Yet, and in contrast to North America, brown lemmings seem to be important intermediate hosts in the Eurasian tundra, and infection records are scattered throughout the species' range (lit. in Bessonov, 1998). Reported prevalences differ drastically, e.g., in Nenetsia (European Russia) between 0.8% and 75% (lit. in Peklo, 2014), and 0.4% at the north coast of Alaska (Holt et al., 2005). The experimental susceptibility of brown lemmings was limited to arctic isolates of *E. multilocularis* (Rausch and Richards, 1971). The **Norway lemming** (*L. lemmus*), known for its population outbreaks every two to three decades, occurs largely outside the range of *E. multilocularis*, but could be experimentally infected (Ohbayashi et al., 1971; Rausch and Richards, 1971). There is no information on the intermediate host role of two additonal *Lemmus* species in the Russian subarctic and on Wrangel Island. **Collared lemmings** (*Dicrostonyx* spp.), occur in eight currently recognized species throughout the holarctic tundra, usually in the drier or elevated parts. Their capacity to develop fertile metacestode seems to vary among the different species (Ohbayashi et al., 1971; Rausch and Richards, 1971). They are supposed to be unsuitable hosts in Alaska (Rausch, 1995), but considered important for the life cycle in the far north of Russia (Bessonov, 1998). 1.2% of 81 *D. torquatus* were found infected in Komi Republic (Russia), while no *Dicrostonyx groenlandicus* was positive among 17 animals at the north coast of Alaska (Holt et al., 2005) and among 72 animals in northern Canada (Gesy et al., 2014).

Infection records from **arctic ground squirrels** (*Urocitellus parryii*) from St. Lawrence Island are doubtful and experimental infection failed (Rausch, 1995; Ohbayashi et al., 1971). An endemic **shrew** species (*Sorex jacksoni*) was often found naturally infected, but its contribution to the life cycle appears insignificant as the capacity to produce protoscoleces seems to be restricted (Ohbayashi et al., 1971).

5. *ECHINOCOCCUS SHIQUICUS* XIAO ET AL., 2005

E. shiquicus seems to be geographically restricted to the Qinghai-Tibet plateau region of China. It has only recently been described as a sister species of *E. multilocularis* (with which it occurs sympatrically) based on differences in adult and larval morphology and gene sequence data (Xiao et al., 2005). Its only known wild definitive host is the **Tibetan fox** (*V. ferrilata*); it has never been found in **red foxes** (*V. vulpes*), which can be sympatric with Tibetan foxes locally. The adult stage of *E. shiquicus* has also been detected by PCR in **domestic dogs**, but the potential role of dogs as definitive hosts for this parasite remains to be elucidated (Boufana et al., 2013a). Only the **plateau pika** (*O. curzoniae*) is known as intermediate host so far, while various other species of small mammals in this region were only found infected with *E. multilocularis* (see Section 4.3.1). From the limited data available, it seems that *E. shiquicus* is restricted to the predator–prey system between Tibetan foxes and plateau pikas (Fig. 4), and its geographical range may be defined by the overlapping distribution of these species, which are

Figure 4 Life cycle of *Echinococcus shiquicus* between Tibetan foxes and plateau pikas. A potential contribution of domestic dogs is unconfirmed. No human cases are known.

endemic to the Tibetan highlands. In contrast, *E. multilocularis,* which is also present in this region, utilizes a clearly wider range of hosts in the same region (including Tibetan foxes, which can have coinfections with both *Echinococcus* species – Jiang et al., 2012). Should *E. shiquicus* have a strict specificity for its intermediate host, this could also account for the fact that no human case of infection with *E. shiquicus* has so far been reported (Xiao et al., 2006a). Considerable sequence variability of mitochondrial genes was recently reported, based on 19 cyst isolates from pika, which indicates long regional evolution without bottleneck events (Fan et al., 2016).

6. *ECHINOCOCCUS EQUINUS* (WILLIAMS AND SWEATMAN, 1963)

E. equinus has formerly been included in *E. granulosus* as the 'horse strain'. An initial characterization of its transmission between **horses** and **domestic dogs** in the United Kingdom and Ireland and its morphological and epidemiological distinction from the 'sheep strain' (*E. granulosus* s.s.) (Smyth, 1977; Kumaratilake et al., 1986) were followed by reports of horse infection from numerous countries worldwide (Thompson and McManus, 2002). However, it is now known that other agents of CE (e.g., *E. granulosus* s.s., *Echinococcus ortleppi*) can also develop metacestodes in species of the Equidae (Boufana et al., 2014; Obwaller et al., 2004; Varcasia et al., 2008), so any records without molecular data are questionable. Today, *E. equinus* is thus confirmed to be present in various European countries, Turkey, Kyrgyzstan, northern Africa and Namibia (Aboelhadid et al., 2013; Boufana et al., 2014; Romig et al., 2015; Simsek et al., 2015; Wassermann et al., 2015; Ziadinov et al., 2008). It probably also occurs in Jordan (Al-Qaoud et al., 2003), and earlier records based on morphology leave little doubt that the parasite is (or was) also present in New Zealand (Kumaratilake et al., 1986).

An exemplary domestic life cycle was described in the 1970s in Britain, where feeding of raw horse meat and offal was considered the most economical way to maintain large packs of hunting dogs, which, after infection, spread the parasite over vast areas of horse pasture. The habit of feeding offal uncooked had apparently started after World War II, when fuel for cooking became expensive, which subsequently led to a steep increase of CE prevalence in horses (Smyth, 1977; Thompson and Smyth, 1975). Apart from the 'classical' horse–dog cycle, a transmission from **donkeys** to stray dogs has been reported for Tunisia, Egypt and Turkey (through offal from abattoirs, or via donkey carcasses left in the environment) (Aboelhadid et al., 2013; Boufana et al., 2014; Simsek et al., 2015) (Fig. 5). Fertile cysts have also been found in a mule in Turkey (Simsek and Cevik, 2014).

An exclusively sylvatic transmission of *E. equinus* was described from the Etosha National Park and other areas of Namibia where **lions** (*Panthera leo*) and **black-backed jackals** (*Canis mesomelas*) act as definitive hosts, and **plains zebras** (*Equus quagga*) are carriers of the metacestodes (Wassermann et al., 2015; Aschenborn, unpublished). The involvement of **mountain zebras** (*Equus zebra*) is unconfirmed but likely because infected lions were found in northwestern Namibia outside the plains zebras' range (Aschenborn, unpublished). The parasite clearly exploits the predator–prey system between lions and zebras, while jackals are likely to acquire infection through scavenging (Fig. 5). Older reports on high prevalences of CE in zebras from northern South Africa indicate that this life cycle is – or has been – widely present in southern Africa, although successful experimental infections of lions with cysts of zebra origin had initially led

Figure 5 Domestic (left) and sylvatic (right) life cycles of *Echinococcus equinus*. Dead-end hosts are omitted (no human cases are known).

to the conclusion that the parasite involved is '*Echinococcus granulosus felidis*' (Young, 1975) (see Section 7). It is not known, whether the lion–zebra transmission is a primary wildlife cycle, or whether the parasite had been introduced at some time into southern African wildlife with dogs or horses, e.g., from Europe. No obvious genetic structuring was found between isolates from Namibian wildlife and isolates from European horses (the United Kingdom, Italy and Germany), which argues against a long-standing separation between domestic and sylvatic transmission systems (Wassermann et al., 2015). High cyst fertility has been reported from *E. equinus* cysts of horse and donkey origin (Boufana et al., 2014; Varcasia et al., 2008). In Namibian zebras, a large number of calcified liver lesions was present in addition to fertile cysts (Wassermann et al., 2015); it is unclear, if this is due to different sampling strategies between the studies, or if zebras respond differently to infection than domestic equids.

At the metacestode stage, *E. equinus* is apparently an almost exclusive parasite of Equidae, although the infection of a **red ruffed lemur** (*Varecia* *rubra*) with multiple abdominal cysts of unrecorded fertility status was recently reported from a UK zoo (Boufana et al., 2012), and earlier experimental infections had produced a few infertile cysts in the lungs of **sheep** (Williams and Sweatman, 1963).

7. *ECHINOCOCCUS GRANULOSUS* (BATSCH, 1786)

7.1 General Life Cycle

Following a revision of the various species that had been described earlier as agents of CE, the name *E. granulosus* was widely used in the second half of the 20th century to cover all taxa of *Echinococcus* that produced cystic metacestodes (Rausch and Nelson, 1963; Romig et al., 2015). To accommodate for the high diversity regarding host range, morphology, developmental biology and geography, a number of 'strains' were erected (Thompson and McManus, 2001). Some are now regarded as separate species, while three of them, the 'sheep', 'Tasmanian sheep' and 'buffalo' strains (G1–3), and several other variants that are genetically closely related, now constitute the species *E. granulosus* s. s. (Nakao et al., 2013a; Romig et al., 2015). Transmitted predominantly in domestic life cycles involving dogs and livestock (Fig. 6), *E. granulosus* s.s. has the widest geographical distribution of all *Echinococcus* species, and, with 88% of 1661 human CE cases characterized to species level worldwide, it has by far the largest impact on public health (Alvarez Rojas et al., 2014). The high number of human cases certainly reflects the wide distribution and high frequency in dogs and livestock, but an apparently low specificity at the intermediate host level may also contribute to an enhanced infectivity or pathogenicity for humans compared to other *Echinococcus* spp. causing CE.

Epidemiological data suggest that this species is particularly well adapted to **sheep** as intermediate hosts, which is reflected in high prevalence (Cardona and Carmena, 2013; chapter: Global Distribution of Alveolar and Cystic Echinococcosis by Deplazes et al., 2016) and high cyst fertility rates. Based on 14 molecular surveys, cyst fertility in sheep was 21–83% (South America), 37–67% (Africa), 50–100% (Asia) and 40–100% (Europe) (lit. in Cardona and Carmena, 2013). Both prevalence and cyst fertility in sheep were shown to be closely correlated with age: 80% of protoscoleces were present in sheep aged 4 years and older in Kyrgyzstan, although such old animals only accounted for 28% of slaughtered sheep (Torgerson et al., 2003, 2009). In addition, almost all other livestock species

Figure 6 Life cycle of *Echinococcus granulosus* sensu stricto with domestic dogs and sheep as the most important hosts. Numerous other species can contribute as competent intermediate hosts (here represented by cattle, dromedary and pig). A number of wild mammal species can locally contribute (here represented by dingo and wallaby), but a (secondary) sylvatic life cycle that is clearly independent from domestic transmission is only documented in Australia. Humans are frequently infected; the numerous other dead-end hosts are omitted.

(goats, cattle, yak, camels, alpacas, pigs, donkeys) are known to develop fertile cysts of *E. granulosus* s.s. and thereby contribute to transmission, but are usually considered to be less important for the life cycle due to lower prevalence, cyst fertility or availability to dogs. However, published data on prevalence and fertility rates are conflicting. **Goats** seem to be less frequently infected than sheep in some surveys (but not in others), and cyst fertility in nine studies ranged from 0% to 100% (Cardona and Carmena, 2013). **Cattle** are frequently found infected. Cyst fertility is low according to most studies (<20%), but can reach 75% (Latif et al., 2010). Low cyst fertility rates are reported from young **Yak** on the Tibetan plateau, but fertility seems strongly associated with old age, and the relative contribution of yak to the life cycle (in comparison to sheep and goats) is not sufficiently explored (Yang et al., 2009a). Cysts from **domestic water buffalo** show high fertility rates in Asia (Latif et al., 2010; Pednekar et al., 2009), but not in Europe (Beyhan and Umur, 2011; Capuano et al., 2006). **Camels** are frequent host of *E. granulosus* s.s., and fertility rates range from 0% (Hailemariam et al., 2012) to 95% (Latif et al., 2010); most studies refer to the dromedary, but for some data from Asia this is not clear. Four cysts recovered from **Alpacas** in Peru belonged to *E. granulosus* s.s. and were fertile (Sanchez et al., 2012). Frequent infections of *E. granulosus* s.s. are known from **domestic pigs** in Europe, Asia and South America, with fertility rates ranging from 0% to 100% (lit. in Cardona and Carmena, 2013; Mbaya et al., 2014; Sanchez et al., 2012; Tigre et al., 2016). Equids can also be infected with *E. granulosus* s.s.; while only infertile cysts are known from **horses** (Utuk and Simsek, 2013; Varcasia et al., 2008), all of 13 cysts derived from **donkeys** in Tunisia were fertile (Boufana et al., 2014). In an older study, experimental infection of two horses with eggs of sheep/dog origin (presumably *E. granulosus* s.s.) had failed (Williams and Sweatman, 1963). The wide range of prevalence and cyst fertility figures among different studies makes it difficult to estimate the quantitative contribution of different livestock species to transmission of *E. granulosus* s.s. (as well as some other agents of CE). Possible reasons for the apparently contradictory values include the wide range of sample sizes among the various studies, different livestock breeds, the usual lack of age stratification in prevalence surveys of livestock and biased selection of cysts for molecular examination regarding size, location and condition. An additional factor may be the variance of pathogenicity among the large number of recognized genotypic variants within *E. granulosus* s.s. (including the 'Tasmanian sheep' and 'buffalo' strains described in the 1990s) (Romig et al., 2015; Thompson and McManus, 2001). At present, there is insufficient evidence for differences in host susceptibility, pathogenicity or maturation periods between any of these variants, although some observations suggest that they exist. As examples, a more rapid development of worms in dogs was reported from *E. granulosus* of Tasmanian origin compared to parasites from mainland Australia (Kumaratilake et al., 1983), and on the Tibetan plateau cysts in sheep, goats and yaks only produce protoscoleces in very old hosts (>10 years) (Yang et al., 2009a). This means that the same species of the parasite may be differently adapted to a given host species in one area compared to another. Apart from the susceptibility to infection and the potential to develop fertile cysts, there are other factors that determine the relative contribution of different livestock species in a given region. Of principal importance is the accessibility of cysts to scavenging by dogs, which — to give an example — may be high in case of sheep slaughtered at home without qualified supervision and low in case of cattle that are often sold alive and slaughtered under better hygienic conditions (Addy et al., 2012).

Typically, transmission in domestic settings involves **dogs** as definitive hosts and livestock as intermediate hosts. Infection of dogs occurs by purposeful feeding of contaminated offal after home slaughter, improper management of abattoirs and slaughterhouses (where roaming dogs have access to condemned offal), or by stray or semistray dogs scavenging on livestock carcasses left on the pasture. However, there are numerous reports of wild animals interacting with the life cycle as definitive or intermediate hosts, and at least one secondary sylvatic cycle has become established (in Australia), that originated from domestic transmission, but is now maintained independently from livestock. In the following, different transmission patterns are addressed according to region.

The apparently low host specificity for development of the *E. granulosus* s.s. metacestode is reflected in the long list of accidental intermediate hosts that do not play a part in the life cycles, ranging from cyst development in rodents (Yang et al., 2009b) to abdominal CE in domestic cats (Deplazes, 2015).

7.2 Europe, Mediterranean Region and Middle East

Based on published polymorphism of mitochondrial genes, the region with the highest intraspecific diversity of *E. granulosus* s.s. stretches from Iran to Turkey and Jordan (Casulli et al., 2012; Yanagida et al., 2012; Romig et al., 2015). As this region includes the initial domestication centres of the major species of livestock (sheep, goats and cattle), it is plausible to hypothesize that in the same region the life cycle of this parasite had been gradually transformed from an ancestral sylvatic to a domestic transmission. According to this hypothesis, the genetic variability that is observed there reflects an ancient polymorphism inherited from the ancestral wildlife cycle(s) that was only partially spread over the rest of the world in the wake of livestock migrations and translocations.

Concerning domestic transmission, the incidence of human infection with *E. granulosus* s.s. coincides with regions of extensive and traditional sheep farming in the Middle East, northern Africa and southern Europe (Alvarez Rojas et al., 2014; Seimenis, 2003; chapter: Global Distribution of Alveolar and Cystic Echinococcosis by Deplazes et al., 2016), a fact that highlights the importance of the sheep–dog cycle for the abundance and infection pressure of the parasite. In most settings, sheep are considered the epidemiologically most relevant intermediate hosts (vs. cattle or water buffalo) because access of dogs to sheep offal (or carcasses) is usually far more likely than to offal from larger livestock that is slaughtered in better supervised facilities (e.g., Cringoli et al., 2007). Detailed epidemiological data on domestic parasite transmission — apart from the general principles highlighted previously — are scarce even for Europe. Especially data on dog infection are fragmentary and are mostly associated with control programmes (e.g., Buishi et al., 2005; Jimenez et al., 2002). In a recent review, large numbers of free roaming dogs (e.g., sheep dogs, semistray and stray dogs) are considered important risk factors (Otero-Abad and Torgerson, 2013). As old sheep have been shown to harbour the majority of the *E. granulosus* protoscolex burden in their cysts (Torgerson et al., 2009), home slaughtering of old sheep for consumption on the farm and scavenging of carcasses of old animals on the pastures are major source of infection for dogs. The age-related risk of fertile infections has to be considered in control programmes.

Today, there is no clear evidence for the existence of a true sylvatic life cycle of *E. granulosus* s.s. in Europe, that is maintained without the involvement of domestic dogs and livestock. As *E. granulosus* s.s. seems to be particularly well adapted to sheep, the ancestral wildlife cycle may have included wild sheep (and possibly wild goats, cervids and wild boar) as intermediate hosts and wolves, golden jackal and possibly striped hyenas as definitive hosts. All these species are known to be susceptible to *E. granulosus* s.s. and develop fertile cysts and gravid worms, respectively (Beiromvand et al., 2011; Breyer et al., 2004; Eslami et al., 2016; Gori et al., 2015; Onac et al., 2013; Simsek and Eroksuz, 2009; Sobrino et al., 2006). However, given the small numbers remaining of some the wild host species, it is likely that these infection records indicate spillover events from the ubiquitous domestic life cycle in this region. Nevertheless, sylvatic transmission may persist or even increase, e.g., in Romania and/or Spain between **wolves** (*C. lupus*), **wild boar** (*S. scrofa*) and **red deer** (*Cervus elaphus*) (Onac et al., 2013; Rojo-Vazquez et al., 2011), possibly also in parts of Iran between wolves and **wild sheep** (*Ovis orientalis*), **goitered gazelles** (*Gazella subgutturosa*) or wild boar (Dalimi et al., 2002, 2006; Eslami et al., 2016; Sarkari et al., 2015), although the species of *Echinococcus* infecting Iranian wildlife is not always clear. South of the Mediterranean Sea, **African golden wolves** (*Canis anthus*) (the species was recently separated from the golden jackal of Asia) and **red foxes** (*Vulpes vulpes atlantica*) were found infected with *E. granulosus* s.s. in northern Tunisia (Lahmar et al., 2009, 2014), where infertile cysts of *E. granulosus* s.s. were found in wild boar (Boufana et al., 2014), and where a case of fertile cysts in an **addax antelope** (*Addax nasomaculatus*) was recently reported (Boufana et al., 2015). The role of these animals in the life cycle requires further investigations, but a link to the widespread domestic transmission in the country is likely. Where wild animal species interact with the domestic life cycle, they may represent an important component of a regional transmission pattern. This is discussed, e.g., for the role of wild boars in southern Europe, which are likely sources for infection of hunting dogs and thus contribute to the spread of the parasite into domestic environments, or which provide a host reservoir that may jeopardize the success of control programmes (Rojo-Vazquez et al., 2011). Wolves and **golden jackals** (*C. aureus*) in Bulgaria were found to carry *E. granulosus* s.s. worms of the same haplotype that circulates in livestock, and infections are probably due to scavenging or preying on livestock (Breyer et al., 2004). The same applies most likely to records of *E. granulosus* s.s. from wolves of northern Italy (Liguria), whose concurrent infection with *Taenia ovis* indicates domestic sheep as an infection source (Gori et al., 2015). **Red foxes** (*V. vulpes*), generally considered as less suitable hosts for *E. granulosus* s.l., were occasionally found infected, with gravid worms, in areas of Britain where the sheep–dog cycle was prevalent. This was attributed to foxes scavenging on sheep carcasses, but was not assumed to play an important role for transmission (Thompson and Smyth, 1975).

7.3 Asia

Epidemiology and transmission biology of *E. granulosus* s.s. in domestic life cycles are reasonably well known for China (Craig, 2004; McManus, 2010; Zhang et al., 2015), Kazakhstan (Torgerson et al., 2003; Abdybekova et al., 2015) and Kyrgyzstan (Torgerson et al., 2006; Ziadinov et al., 2008). There are very few molecularly confirmed records from wild mammals, but there are a substantial number of older records from wild central Asian herbivores that cannot be allocated now to any particular species of *Echinococcus* (Rausch, 1995). Primarily, transmission occurs between **dogs** (prevalence range 10–30%) and livestock, especially **sheep** (prevalence range 10% to >70%) within traditional pastoral systems including settled, seminomadic and nomadic communities (Craig, 2004; Torgerson et al., 2006; Wang et al., 2008). Home slaughter and unregulated slaughter sites, access of dogs to raw offal, dog type (farm dog) and free roaming of dogs are – as in other parts of the world – key risk factors for maintenance of transmission

(Shaikenov et al., 2003; Huang et al., 2008; Otero-Abad and Torgerson, 2013).

Genotypic confirmation of *E. granulosus* s.s. exists for dogs in western China (Xinjiang, Qinghai, Sichuan) (Bart et al., 2006; Xiao et al., 2006b; Ma et al., 2012; Boufana et al., 2013b), Kazakhstan and Kyrgyzstan (Trachsel et al., 2007; Ziadinov et al., 2008). Regarding livestock in western China, the parasite has been confirmed in **sheep**, **goats**, **cattle** and **yak** (Liu et al., 2013; Ma et al., 2012, 2015; Xiao et al., 2003; Yang et al., 2005; Zhang et al., 1998). For wild herbivores, there is a single report of *E. granulosus* s.s. from a naturally infected **ground squirrel** (*S. dauricus*) trapped in Ningxia Hui Autonomous Region, although a role in transmission was considered unlikely (Yang et al., 2009b). Reports of '*E. granulosus*' infections in **plateau pika** (*O. curzoniae*) in Qinghai were most probably due to infection with *E. shiquicus* (Xiao et al., 2005). Analysis of genetic polymorphisms in western China revealed haplotypic variation in the *cox1* gene between isolates (human, sheep and yak hosts) from Xinjiang (northwest China) versus Qinghai/Sichuan (Tibetan Plateau), but overall the haplotypic network displayed one common haplotype for 53.6% of all the isolates (Nakao et al., 2010). This and other studies suggested that *E. granulosus* s.s. on the Tibetan Plateau, which is responsible for some of the highest human CE prevalences recorded globally (Li et al., 2010), does not represent a genotypic cluster different from isolates in other parts of northwest China, nor in other areas of the world (Ma et al., 2008; Nakao et al., 2010). This suggests that the transmission and zoonotic risk for *E. granulosus* s.s. in the China/Central Asia region is probably more associated with human behaviour, dog ownership and husbandry practices, than with genetic variants of the parasite (Craig et al., 2015).

Numerous reports of CE exist from the Russian Federation, but the distribution and host range of different *Echinococcus* species are not always clear. Comparatively few isolates of *E. granulosus* s.s. have been genetically confirmed, and the species seems to occur predominantly in the European part of Russia and southwestern Siberia, infecting sheep (Bessonov, 2002; Konyaev et al., 2013).

7.4 Sub-Saharan Africa

E. granulosus s.s. has a curiously patchy distribution south of the Sahara, being absent from large regions with apparently suitable conditions (in terms of host animals and human lifestyle). Based on the (incomplete) data that are available, this parasite is either absent or occurs only sporadically (focally?)

in the Sahel and the savannah zones stretching from Mauritania through the northern part of West Africa into Sudan. It becomes abundant as an agent of animal and human disease in East Africa, including most of Ethiopia, the southern part of South Sudan, Kenya, Uganda and northern Tanzania. The extent of this focus further south is unclear, but it is rare or possibly absent in large parts of Zambia and Namibia, again to resurface in the Republic of South Africa (in the form of rather frequent human cases) (Cardona and Carmena, 2013; Romig et al., 2011; chapter: Global Distribution of Alveolar and Cystic Echinococcosis by Deplazes et al., 2016). Most details on the life cycle are known from Ethiopia and Kenya, where the globally known transmission pattern involving **dogs** and **sheep** is widespread. **Goats**, **cattle** and **camels** can contribute to the life cycles to different degrees: camels are frequently infected, but are only kept in relevant numbers in the arid regions of northern Kenya and southern/eastern Ethiopia, and are far less frequently slaughtered than small livestock. Goats show lower prevalence than sheep in most studies (chapter: Global Distribution of Alveolar and Cystic Echinococcosis by Deplazes et al., 2016). In cattle, cysts of *E. granulosus* s.s. are often infertile (although this seems to differ among regions – Tigre et al., 2016). Also, even in remote areas cattle are often transported to distant abattoirs for slaughter and therefore contribute little to the infection of local dogs (Addy et al., 2012).

The large number of dogs kept by traditional pastoralists in eastern Africa has been linked to the frequency of human CE cases, but this does not necessarily correlate with the abundance of the parasite in animal hosts: in the Turkana region of Kenya, the human incidence is one of the highest in the world, but prevalence in livestock is low compared to other endemic regions (Macpherson and Wachira, 1997; chapter: Global Distribution of Alveolar and Cystic Echinococcosis by Deplazes et al., 2016). Due to the inadequate amount of data on molecularly confirmed *E. granulosus* s.s. infections, no further conclusions can be drawn on the relative contribution of different livestock species to transmission.

This also applies to the contribution of wild mammals to the transmission of *E. granulosus* s.s. Numerous species of wild mammals have been recorded as definitive and intermediate hosts of *Echinococcus* spp. in sub-Saharan Africa (Hüttner and Romig, 2009; Macpherson and Wachira, 1997; chapter: Global Distribution of Alveolar and Cystic Echinococcosis by Deplazes et al., 2016), but few data exist on the specific identity of the parasites. Based on eggs recovered from faeces collected in the environment, *E. granulosus* s.s. was recently found to be widespread in **lions** (*P. leo*) and **spotted hyenas**

(*Crocuta crocuta*) of four conservation areas of Kenya (Kagendo et al., 2014) and was identified in a **wild dog** (*Lycaon pictus*) in northern Namibia (Aschenborn, pers., comm.). As some of those areas are encroached by pastoralists and used as a part-time grazing area for livestock, spillover from the domestic life cycle to wild mammals is likely (by wild carnivores preying or scavenging on livestock). Unfortunately, almost no recent data on the presence of cysts in potential wild intermediate hosts are available from areas where *E. granulosus* s.s. is endemic; exceptions are the recovery of only four cysts of this parasite from 354 **western white-bearded wildebeest** (*Connochaetes mearnsii*) that had drowned in the Mara River during their annual migration in southern Kenya (Kagendo et al., 2014), and of one cyst of unknown fertility status found in a **warthog** (*Phacochoerus africanus*) in the Queen Elizabeth National Park in Uganda (Hüttner et al., 2009). These data are insufficient to draw any conclusion on the existence of a sylvatic transmission of *E. granulosus* s.s. in African wildlife.

As a matter of curiosity, humans reportedly play an active intermediate host role in parts of northern East Africa, where traditional customs of some groups of nomadic pastoralists facilitate the access of dogs and other scavengers to human corpses (Macpherson, 1983). Despite the high prevalence of CE and the high rate of cyst fertility in people of that region (which are predominantly infected with *E. granulosus* s.s.), quantitative considerations make it unlikely that humans are an important component for the life cycle (Macpherson, 1983; Romig et al., 2011).

An unusual genotype related to *E. granulosus* s.s., but considerably distant from the previously known infraspecific variants, was recently characterized on the basis of a single isolate from a human patient in the South Omo region of Ethiopia (Wassermann et al., 2016). Despite the large number of *Echinococcus* isolates that have been examined in this region, this taxon has never been found in any species of livestock or in human CE patients. Nothing is known about the life cycle, but circumstantial evidence (the patient belongs to a group of pastoralists that live in close vicinity to wildlife conservation areas) makes the involvement of wild mammals not unlikely.

7.5 South America

Despite the high abundance of CE in livestock, there are few studies on molecular identification of the causative parasites. Data available indicate that the **sheep–dog** cycle is the principal transmission system for *E. granulosus* s.s.: it is the only *Echinococcus* species that has ever been identified in sheep in the Americas, and it combines high prevalence (particularly west of the Andes) with high cyst fertility. **Goats** are less frequently infected and cyst fertility is low, which is also true for **cattle** (Cardona and Carmena, 2013; chapter: Global Distribution of Alveolar and Cystic Echinococcosis by Deplazes et al., 2016). **Pig** infection with *E. granulosus* s.s. (in addition to *Echinococcus canadensis* G6/7) seems to be widespread in Mexico, Peru and Argentina, with considerable cyst fertility rates (25–100%) (Kamenetzky et al., 2002; Moro et al., 2009; Sanchez et al., 2012; Villalobos et al., 2007). There is one genotyping study on *Echinococcus* isolates from **alpacas**, where fertile cysts of *E. granulosus* s.s. were recovered in Peru (Sanchez et al., 2012). There is no further information available on the role of alpacas or other New World camelids in the transmission of this or any other species of *Echinococcus*.

A spillover into wildlife was described from the Neuquén province of Argentina in the 1970s, where substantial prevalences of adult worms in **culpeos** (*Lycalopex culpaeus*) and fertile metacestodes in (introduced) **European hares** (*L. europaeus*) were interpreted as a secondary transmission system independent from the domestic life cycle (Schantz et al., 1972). The causative *Echinococcus* sp. had then not been identified, and recent attempts to find *Echinococcus* in European hares elsewhere in South America failed, so that a sylvatic transmission of this parasite may be transient, rare or localized and is considered unimportant (Scioscia et al., 2013). However, unidentified *Echinococcus* had been reported in various – mainly older – studies from *Lycalopex* spp. (lit. in Scioscia et al., 2013). The susceptibility of different *Lycalopex* species to experimental infection with sheep-derived material varied, but at least one species (*L. culpaeus*) appears to be a suitable definitive host that may contribute to transmission by killing sheep or scavenging on carcasses (Schantz et al., 1976).

7.6 Australia and New Zealand

E. granulosus s.s. is the only species occurring in Australia and New Zealand, having been introduced through importation of domestic livestock and dogs during European settlement about 200 years ago, possibly from North Africa and Spain (Jenkins, 2005). A recent molecular survey found a considerably high polymorphism of *E. granulosus* s.s. in Australia, apparently introduced from the regions of livestock origin (Alvarez Rojas et al., 2016). The parasite spreads rapidly in the domestic situation, quickly becoming a major public health issue and being perpetuated in the classical dog–sheep cycle (Gemmell, 1990); prevalence in sheep between 1920 and 1960 in Australia was estimated to range between 16% and 32% (Kumaratilake and

Thompson, 1982). Following improved farmer education, the development of commercial dry dog food, wide availability of cheap deworming tablets, and concerted control programmes, domestic transmission gradually declined, and CE was declared eradicated in New Zealand in 2002 (chapter: Echinococcosis: Control and Prevention by Craig et al., 2016). However, in Australia infection in rural dogs still occurs where owners keep large numbers of dogs and feed them with offal (often only intermittently) from domestic animals and/or wildlife (Jenkins et al., 2006, 2014). Parasite transmission into wildlife populations of mainland Australia occurred through dingo predation of infected sheep and domestic sheep dogs defecating on pasture grazed by macropodid marsupials (mainly wallabies and kangaroos). This host switch is linked to the agricultural practice of transhumant grazing, where large numbers of domestic livestock and dogs spent several months each year in remote high-altitude areas of the Great Dividing Range of eastern Australia (Jenkins, 2005; King, 1959). This switch from domestic to sylvatic transmission is supported by genetic data, which show no differentiation between isolates from domestic and wild animals, and identical variants (haplotypes) were recovered from both livestock and marsupials (Alvarez Rojas, Ebi et al., 2016).

Crucial for the switch from domestic to sylvatic transmission was the presence of a highly susceptible canid, the **dingo** (*Canis lupus dingo*), and large populations of various species of native macropodid marsupials. They are highly susceptible to infection and are the preferred prey of dingoes. Currently, the main, ongoing source for the perpetuation of transmission of *E. granulosus* s.s. in Australia is the extensive wildlife reservoir (dingo – macropodid) through 'spillover' of infection into domestic animals. Infection in **sheep** occurs most commonly on farms located adjacent to parks and forests (Grainger and Jenkins, 1996) and in **cattle** grazed on bush pasture (Durie and Riek, 1952; Fotheringham et al., unpublished data). The key in both these situations is the presence of wild dogs in the vicinity (dingoes and dingo/domestic dog hybrids), which commonly contain worm burdens well above the level that is typical for domestic dogs. In rural domestic dogs, worm burdens of 100 or less are common, and over 1000 worms would be unusual (Gemmell et al., 1986), while in dingoes and their hybrids average infections of several thousand to tens of thousands of worms is the norm (Jenkins and Morris, 1991, 2003; Jenkins et al., 2008). There is possibly a greater biological susceptibility of wild dogs to infection, but an important contributing factor is likely to be the increased exposure to the parasite through predation of infected intermediate hosts. In contrast, **red foxes** (*V. vulpes*) that had been introduced into Australia during European settlement and soon had become a major agricultural pest (Rolls, 1984) have also been shown to act as definitive host for *E. granulosus* s.s. in a range of different locations in Australia, probably by scavenging (e.g., Gemmell, 1959; Jenkins and Morris, 2003). The prevalence among fox populations is usually low with correspondingly low worm burdens of <50 worms, indicating that foxes in rural environments are of minor importance for the life cycle.

The importance of different macropodid species for transmission varies from region to region, reflecting their geographically differing role as prey for wild dogs. In southeastern Australia, **swamp wallabies** (*Wallabia bicolor*) are the favoured prey species of wild dogs, and *Echinococcus* infection in the lungs of these species is common and often massive. In addition, **eastern grey kangaroos** (*Macropus giganteus*) and **red-necked wallabies** (*Macropus rufogriseus*) are also important in southeastern Australia, while in northern Queensland the **black-striped wallaby** (*Macropus dorsalis*) is frequently infected (Banks, 1984). Other species involved in Queensland are **nailtail wallaby** (*Onychogalea fraenata*), various **rock wallabies** (*Petrogale godmani, Petrogale mareeba, Petrogale persephone, Petrogale penicillata*), **whiptail wallaby** (*Macropus parryi*) and **pademelons** (*Thylogale stigmatica*) (lit. in Jenkins and Macpherson, 2003; Barnes et al., 2008; Beveridge et al., 1989), and **western grey kangaroos** (*Macropus fuliginosus*) are frequently infected in western Australia near Perth (Thompson et al., 1988). **Common wombats** (*Vombatus ursinus*) are known as intermediate hosts in Victoria, but are rarely infected and do not appear to be important for the life cycle. Infection in macropodids occurs almost exclusively in the lungs (Thompson et al., 1988; Jenkins and Morris, 2003; Barnes et al., 2007a,b; Barnes et al., 2008) either as multiple cysts or as single cysts that may become large. Pulmonary infection leads to major respiratory impairment (Barnes et al., 2011) eventually causing death (Johnson et al., 1998; Barnes et al., 2007b). Infected animals are rendered more susceptible to predation by wild dogs, a situation similar to that reported in the United States between infected moose and wolves (Mech, 1966; Joly and Messier, 2004). In addition, cysts in macropods, compared to sheep, grow and become fertile more rapidly. Cysts in Australian sheep take 15 months or longer to become fertile (Harris et al., 1980; Gemmell et al., 1986; Barnes et al., 2011), whereas in experimental infection in **Tammar wallabies** (*Macropus eugenii*) the cysts were fertile at 9 months postinfection (Barnes et al., 2007b, 2011). For endangered macropods that are reduced to small populations with restricted distribution, mortality due to CE emerged as a conservation issue, e.g., in the case of the **brush-tailed rock wallaby** (*P. penicillata*) where a substantial proportion of the surviving population was found infected (Barnes et al., 2008).

Cysts are found commonly in **feral pigs**, but cyst fertility is highly variable in different regions (Jenkins and Morris, 2003; Lidetul and Hutchinson, 2007). Since wild dogs mainly prey on piglets (which are least likely to contain fertile cysts), feral pigs are not considered to be important for transmission. However, they may contribute in a small way through dead pigs or offal left by hunters being scavenged by wild dogs and foxes. Infection has never been reported in **feral horses**, **donkeys** or **camels**, and only few cases are known from **feral goats** harvested for human consumption (Jenkins unpublished data). Wild **rabbits** (*Oryctolagus cuniculus*) have been shown to be susceptible to experimental infection with *E. granulosus* (Jenkins and Thompson, 1995), but it is not known if cysts in rabbits would ever reach fertility. Many thousands wild-caught rabbits have been examined for a range of reasons and *Echinococcus* infection has never been reported.

The crucial role of the dingo for the sylvatic cycle is highlighted by the fact that the parasite was never able to switch from the domestic transmission to wildlife hosts in Tasmania, where dingoes have always been absent. None of the various species of marsupial carnivores (Dasyuridae), including **quolls** (*Dasyurus* spp.) and **Tasmanian devils** (*Sarcophilus harrisii*) have ever been found infected, and experimental infections had always failed (Jenkins, 2005). As a hypothesis, this nonsusceptibility of dasyuromorphs also included the extinct **thylacine** (*Thylacinus cynocephalus*) – the top-order predator in Tasmania at the time of European settlement – which prevented the establishment of a sylvatic transmission in Tasmania (Jenkins, 2006). Likewise, the parasite had never spilled over to indigenous animals in New Zealand, as no potential wild hosts were present on the islands (Sweatman and Williams, 1962).

In some parts of eastern Australia wild dogs have now established themselves in urban environments (Allen et al., 2013). These animals supplement traditional prey with scavenging, often having small home ranges of <1 km^2 to over 100 km^2 (Allen et al., 2013). Importantly, there is also a high prevalence of *E. granulosus* among these 'urban' wild dogs (Brown and Copeman, 2003; Jenkins et al., 2008) and some individuals may contain worm burdens in excess of 100,000 worms (Jenkins et al.,

2008). The impact of these populations on public health is in need of close surveillance. Likewise, foxes are known to enter urban areas such as popular picnic sites, which may give them some importance as a source for human infection, despite the modest level of prevalence and worm burden found in most fox studies.

8. *ECHINOCOCCUS FELIDIS* ORTLEPP, 1937

Echinococcus felidis was long known as the 'lion strain' of *E. granulosus* and could only recently be confirmed as an independent species (Hüttner et al., 2008). Originally described from worms found in a South African lion, molecularly diagnosed eggs in faeces of **lions** (*P. leo*) confirmed the actual presence of this parasite in Uganda (Queen Elizabeth National Park) (Hüttner et al., 2009), Kenya (six conservation areas across the country) (Kagendo et al., 2014), Zambia (Kafue National Park) (Banda, pers. comm.) and Namibia (Zambezi region) (Aschenborn, pers. comm.); in addition to lions, eggs were also detected in faeces of **spotted hyenas** (*C. crocuta*), but not in any other large carnivore (Hüttner et al., 2008, 2009; Kagendo et al., 2014; Aschenborn, unpublished). The life cycle was (and partly still is) speculative due to the lack of diagnostic morphological criteria for the metacestode stage. Lions prey on a large number of African herbivore species, and many of these are known to develop *Echinococcus* cysts (Hüttner and Romig, 2009; Macpherson and Wachira, 1997; chapter: Global Distribution of Alveolar and Cystic Echinococcosis by Deplazes et al., 2016). However, during recent molecular surveys a wide spectrum of *Echinococcus* species was found in African herbivores, including *E. equinus*, *E. granulosus* s.s., *E. ortleppi* and *E. canadensis* G6/7 (Obwaller et al., 2004; Hüttner et al., 2008, 2009; Kagendo et al., 2014; Aschenborn, unpublished), so that the intermediate host range of *E. felidis* cannot be based on previous lists of wild mammal species as host of CE. There are only six genetically verified records of *E. felidis* metacestodes, all from **warthogs** (*P. africanus*), one from Uganda and five from Namibia, and only from locations where *E. felidis* was also found in lions (Hüttner et al., 2009; Aschenborn, unpublished). Despite being widespread in eastern Africa, cysts of this species were never recorded from humans, nor from any species of livestock in this region, even where pastoralists graze their animals near or in conservation areas (Hüttner et al., 2009).

Despite the paucity of data (particularly from wild herbivores), the available information indicates that warthogs play a pivotal role in the life cycle of this parasite (Fig. 7). The existence of an *Echinococcus* sp. life cycle including lions, warthogs and other wild Suidae had been proposed earlier for the Central African Republic, where lions, warthogs and **red river hogs** (*Potamochoerus porcus*) were found infected, while other herbivores were not (Graber and Thal, 1980). Experimental infections of **domestic dogs** with cysts from those animals failed to produce worms (Graber and Thal, 1980). Even though it is uncertain whether the parasite used in that study was indeed *E. felidis*, the data presently available indicate a rather narrow host range for this parasite (which is unexpected, given the close phylogenetic relationship with *E. granulosus* s.s.). In any case, as all verified records of *E. felidis* are from African wild mammals, there is no reasonable doubt that this species is a primary wildlife parasite, probably restricted to sub-Saharan Africa.

Figure 7 Life cycle of *Echinococcus felidis* between lions, spotted hyenas and warthogs. No other hosts (including dead-end hosts) are known, but the contribution of other wild suids as intermediate hosts is likely. No human cases are known.

9. *ECHINOCOCCUS ORTLEPPI* LOPEZ-NEYRA AND SOLER PLANAS, 1943

The species was described from adult worms of **dogs** that had been fed with cysts from South African **cattle**. It was initially characterized as the 'cattle strain' of *E. granulosus* and finally recognized as a separate species (Nakao et al., 2013a; Romig et al., 2015; Thompson and McManus, 2002). In contrast to *E. granulosus* s.s., it is well adapted to cattle as intermediate hosts, producing cysts with a high fertility rate predominantly in the lungs (Balbinotti et al., 2012; Grenouillet et al., 2014; Kamenetzky et al., 2002; Mbaya et al., 2014; Monteiro et al., 2016; Pednekar et al., 2009; Tigre et al., 2016). It shows distinct morphological features of the adult worms, and the development time in dogs before the onset of egg production (35 days or less) is – as far as known – shorter than in other *Echinococcus* species causing CE (Thompson et al., 1984). *E. ortleppi* occurs worldwide in domestic life cycles between cattle and dogs (Fig. 8). Compared to

Figure 8 Life cycle of *Echinococcus ortleppi* between domestic dogs (as the only confirmed definitive hosts) and cattle. Several other intermediate and dead-end hosts are known (omitted). Human cases are known, but rare.

E. granulosus s.s., this species is usually far less frequent even in cattle-raising regions. This apparent paradox has been tentatively explained by the fact that even in traditional pastoralist societies (e.g., in sub-Saharan Africa), cattle are a valuable asset which is mostly sold alive to be transported to distant slaughterhouses. In contrast to cysts developing in sheep and goats (that are often slaughtered at home), cysts in cattle are therefore less frequently available for local dogs, which creates a barrier for cattle infection with *E. ortleppi* (Addy et al., 2012; de la Rue, unpublished). This is in contrast to infection with *E. granulosus* s.s., that is acquired by cattle as spillover from the concurrently running sheep—dog cycle. Even where standards of cattle-slaughterhouses permit access to cysts for stray dogs, such slaughterhouses are often in urban environments far away from the cattle-raising areas (Wachira et al., 1993). Sharp increases of *E. ortleppi* infections in cattle in Rio Grande do Sul (Brazil) in recent years (de la Rue et al., 2006; Balbinotti et al., 2012; Urach Monteiro et al., 2016) are tentatively explained by the recently improved availability of electricity that enables rural farmers to run deep freezers. This allows now the home slaughter of cattle, whereas previously only sheep were slaughtered at home for lack of meat storage facilities (de la Rue, unpublished). In western Zambia, the frequent infection of cattle is explained by the common presence of unsupervised cattle slaughter slabs in villages (Banda, pers. comm.). Older data from the Republic of South Africa, where cyst fertility in cattle was ~90% (Verster, 1962), indicate a wide spread of *E. ortleppi*, but no recent confirmation is available.

Although *E. ortleppi* is the most important species of *Echinococcus* that produces fertile cysts in cattle, there is no strict intermediate host specificity. **Goats** and **pigs** are known to develop fertile cysts also (Mbaya et al., 2014; Pednekar et al., 2009), and metacestodes have been reported from a number of other mammals: an unreported **zebra** species and **oryx antelopes** (*Oryx gazella*) were found naturally infected in Namibia (Aschenborn, pers. comm., Obwaller et al., 2004). There is also an interesting observation on unusually rapid development of worms in a dog that had been fed protoscoleces from a cyst of a **giraffe** (*Giraffa camelopardalis*) from southwestern Africa (Tscherner, 1978; cited in Thompson et al., 1984). Given the fact that *E. ortleppi* is widespread in cattle in southern Africa (Aschenborn, Addy, pers. comm.), this might indicate that giraffes are suitable hosts for *E. ortleppi*; however, *E. canadensis* G6/7 is also widespread in the region (see Section 10.2), and the development of this species in dogs seems to be only marginally slower (Eckert et al., 1989, 1993).

Concerning captive dead-end hosts, there are case reports of fatal echinococcoses (with sterile cysts) in a **red-shanked douc langur** (*Pygathrix nemaeus*) kept in a primate breeding centre in Vietnam (Plesker et al., 2009) and in a **Philippine spotted deer** (*Rusa alfredi*) in a UK zoo (Boufana et al., 2012).

Human infections are known from all over the world, but seem to be extremely rare (Alvarez Rojas et al., 2014). It is unclear if this is caused by low exposure due to the sporadic occurrence in animal hosts or by partial resistance to infection.

Phylogenetically, *E. ortleppi* is the sister taxon to the *E. canadensis* cluster. Genetic differences between isolates from Europe, Africa and South America seem to be minor (Addy et al., 2016), but the taxonomy of this clade is not yet fully resolved (chapter: Phylogenetic Pattern, Evolutionary Processes and Species Delimitation in the Genus *Echinococcus* by Lymbery, 2016) and further biological, morphological and molecular data are warranted for the members of this clade, including further information of life cycles and host specificities.

10. *ECHINOCOCCUS CANADENSIS* CLUSTER

10.1 Introduction

This species complex currently comprises the genotypes G6/7 (camel and pig strain), G8 ('American' cervid strain) and G10 ('Fennoscandian' cervid strain) (Nakao et al., 2013a; Romig et al., 2015). The three genotypic groups seem genetically sufficiently distant from each other to warrant recognition as species, and G6/7 (with predominantly domestic transmission and worldwide distribution) seems to be epidemiologically distinct from G8 and G10, which are mainly present in wildlife cycles and are restricted to the north of Eurasia and America. However, as G10 is more closely related to G6/7 than to G8 (at least based on the mitochondrial genome), a division into two species is taxonomically not possible (Nakao et al., 2013a,b). The erection of three species has been proposed, with the name *Echinococcus intermedius* Lopez-Neyra and Soler Planas, 1943, for G6/7; *Echinococcus borealis* (Sweatman and Williams, 1963) for G8; and *E. canadensis* (Webster and Cameron, 1961) for G10 (Lymbery et al., 2015a). Yet, some controversy still exists due to lack of nuclear genomic data (Lymbery et al., 2015b; Nakao et al., 2015), and in this chapter a conservative approach is chosen by using the names of genotypic clusters instead of binomial species names.

10.2 The Camel and Pig Strain (G6/7)

The genotypic cluster G6/7 includes the 'camel strain' and the 'pig strain' that have been described morphologically and based on their transmission patterns in dog–camel and dog–domestic pig cycles (Eckert et al., 1989, 1993; Thompson et al., 1995) (Figs. 9 and 10). Based on short sequences of the mitochondrial cox1 and nad1 genes, they were characterized as genotypes G6 and G7 (Bowles et al., 1992; Bowles and McManus, 1993). Consecutive studies using longer sequences including entire mt genome data concluded that both genotypes differ only in few base pair exchanges and might be better regarded as geographical microvariants of the same taxon (Romig et al., 2015). The difficulty to distinguish between both genotypes is emphasized by cases, where isolates carry markers for both strains in different genes, e.g., numerous isolates from pigs in Corsica

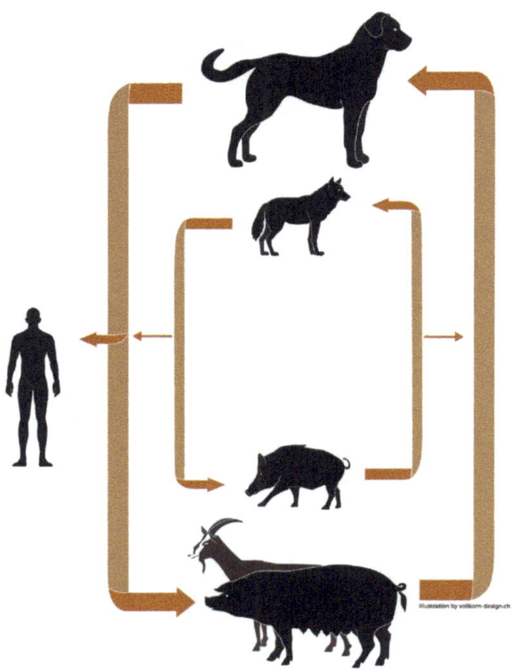

Figure 9 Life cycle of *Echinococcus canadensis* G6/7 in pig-raising regions of western Eurasia and South/Central America. Principal hosts are domestic dogs and pigs, but other livestock (e.g., goats) contribute in some areas. Wild mammals are also involved in Eurasia (here represented by wolf and wild boar), but the interaction between domestic and sylvatic transmission is not fully understood. Dead-end hosts are known (omitted). Human cases occur at moderate numbers.

Figure 10 Life cycle of *Echinococcus canadensis* G6/7 in camel-raising regions (from eastern and northern Africa through the Middle East to central Asia). Principal hosts are domestic dogs and dromedaries, but other livestock can contribute. The involvement of wild mammals is little understood (omitted). Human cases occur at moderate numbers; other dead-end hosts are omitted.

(Addy, pers. comm.). Yet, looking at global host records, isolates with G7 sequences are reported from pigs in areas where isolates conforming to G6 are present in other hosts (e.g., Soriano et al., 2010). Whether this is due to host adaptation of genetic microvariants or reflects historical migration or introduction routes remains to be investigated.

Typically, G6/7 is transmitted in domestic life cycles with **dogs** as definitive hosts and a rather wide range of livestock species as potential intermediate hosts. Topologies of phylogenetic trees (e.g., based on the complete mt genome), where cervid-transmitted genotypes are in basal position within the *E. canadensis* clade, suggest that those domestic life cycles derived from an ancestral transmission system involving cervids and wild canids (Nakao et al., 2013b); the recent discovery of G6 being present in cervids (reindeer) along with G8 and G10 in eastern Russia adds support to this view (Konyaev et al., 2013; Nakao et al., 2013b). Various wild mammals are known to harbour G6/7, but there is no evidence yet, that sylvatic life cycles are maintained independently from domestic transmission in the areas concerned.

In central to eastern Europe, southern Asia and America south of the United States, **dog–pig** transmission systems are known to be associated with smallholder pig farming and unsupervised home slaughter (Fig. 9). This has been emphasized, e.g., in a study in Lithuania, where pigs from family farms showed prevalence of >13%, while pigs from industrialized farms were much less frequently infected (Bruzinskaite et al., 2009). Life cycles involving **dogs** and **camels** are known from the Sahel zone south of the Sahara, the Sudan, the arid parts of northern East Africa, most of the Middle East and parts of central Asia and western China (chapter: Global Distribution of Alveolar and Cystic Echinococcosis by Deplazes et al., 2016) (Fig. 10). Both in pigs and camels, fertility rates of G6/7 are very high: in European and South American studies, cyst fertility in pigs was 19–100% (Bruzinskaite et al., 2009; Daniel Mwambete et al., 2004; Soriano et al., 2010; Sanchez et al., 2012; Varcasia et al., 2006) and ranged from 45% to 100% in camels in Mauritania, Sudan and Ethiopia (Bardonnet et al., 2002; Ibrahim et al., 2011; Omer et al., 2010; Tigre et al., 2016). The high fertility of camel cysts can be explained by the fact that camels slaughtered at abattoirs are usually >10 years of age (Elmahdi et al., 2004), but G6/7 cysts in pigs seem to become fertile in a very short time: the majority of cysts from Lithuanian pigs aged 12–26 months were fertile, and some fertile cysts were even found in pigs aged less than 1 year (Bruzinskaite et al., 2009).

Domestic transmission of G6/7 can also persist where neither pigs nor camels are kept as livestock. Examples are parts of southern Europe and eastern Africa, where **goats** were identified as important hosts, while **sheep** are rarely infected (Addy et al., 2012; Varcasia et al., 2007). However, the role of sheep may differ according to region: in Darfur (western Sudan), sheep were frequently infected with G6/7 (12% prevalence) and cysts had a high fertility rate (19%), which is in contrast to other parts of Sudan where sheep are rarely infected and cysts are mainly sterile (Elmahdi et al., 2004; Omer et al., 2010). Such regional differences are known for other host–parasite systems; it is unclear if this is caused by the livestock breed, or by an adaptive trait of the local parasite variant. The role of **cattle** in the domestic life cycle of G6/7 needs further study. There are few data on cattle infection with this parasite. In Sudan, fertility of G6/7 cysts in cattle ranged from 50% to 100%, although prevalence was much lower than in camels (Ibrahim et al., 2011; Omer et al., 2010). In southern Brazil, fertile cysts in cattle are known, but rare (Urach Monteiro et al., 2016). Different transmission systems may run in parallel, as has been reported from Neuquén province in Argentina, where a dog–pig cycle was reported for periurban areas, while a dog–goat cycle is present where transhumant models of livestock production are practised (Soriano et al., 2010).

There are numerous reports about the involvement of wildlife in the transmission of G6/7. **Wolves** (*C. lupus*) have been found infected (and shedding eggs) in northern Portugal, where a domestic life cycle between dogs and pigs exists (Guerra et al., 2013). This may be an indication for a sylvatic transmission because **wild boars** (*S. scrofa*) were found infected with G6/7 in neighbouring Spain (Daniel Mwambete et al., 2004), but scavenging on domestic animal carcasses cannot be excluded. Wild boar infection was reported from Corsica (Umhang et al., 2014), where domestic pigs are kept free-ranging and home slaughter is common, maintaining a domestic cycle. The infection of wild boar in Corsica is possibly a spillover from the domestic cycle, although involvement of **red foxes** (*V. vulpes*) (by scavenging on boar carcasses) might also contribute. In contrast, wild boars were found infected with fertile cysts of G7 in Romania where wolves are frequent and have the potential to maintain sylvatic transmission (Onac et al., 2013). The same uncertainty about the life cycle applies to records of G6/7 from wild boar in Ukraine (Kedra et al., 2000) and eastern Germany, where wolf populations have established themselves in recent years (Dinkel et al., 2006). New data from southern Africa show that a wide range of wild mammals are suitable hosts for G6/7 and might be important for transmission, including **lions** (*P. leo*) and **oryx antelopes** (*O. gazella*); data available so far do not allow conclusions yet, whether the parasite in this region is transmitted in sylvatic cycles, or whether these records are a result of spillover from domestic transmission (Aschenborn, pers. comm.).

In parts of the endemicity area, transmission of G6/7 may be highly complex. In central Asia, infections have been molecularly confirmed in rural domestic dogs (Bart et al., 2006; Trachsel et al., 2007; Ziadinov et al., 2008; Van Kesteren et al., 2013). It is presumed that the transmission cycles involve domestic livestock. The G6 genotype has currently been molecularly confirmed in **Asian camels** (*Camelus bactrianus*), goats and sheep in western China (Zhang et al., 1998; Liu et al., 2013; Ma et al., 2015). In some regions of central Asia and western China pastoral communities exist in areas where dogs, livestock, wild ungulates and wild carnivores coexist. Infection of wolves was reported from the Altai region of western

Mongolia and Russia (Ito et al., 2013; Konyaev et al., 2013). Sympatric transmission of *E. granulosus* s.s, *E. canadensis, E. equinus* and *E. multilocularis* has been shown to occur in southern Kyrgyzstan (Ziadinov et al., 2008; Abdybekova et al., 2015) and a similar situation probably exists in western China (Craig, 2004; Wang et al., 2008), with likelihood of transmission within domestic and semidomestic cycles, as well as occurrence of wildlife cycles and potential spillovers. Recently, a G6 isolate was recovered from **domestic reindeer** in Yakutia (Russia), a region where the northern cervid strains (see Section 10.3) are also present (Konyaev et al., 2013; Nakao et al., 2013b). Although the fertility status of the cyst was not reported, this increases the potential host range of the G6/7 cluster and adds complexity to the genotypic diversity of the northern sylvatic and semidomestic transmission system (see Section 10.3)

10.3 The Northern Cervid Strains (G8 and G10)

The geographic distribution and host records of G8 and G10 correlate well with the 'northern form' of *E. granulosus* that was considered to have a primary wildlife transmission between **wolves** (*C. lupus*) and wild cervids in the northern hemisphere (Rausch, 2003) (Fig. 11). The genetic distance between the two strains — and the proposal to separate them into species — suggests phenotypic differences, e.g., in ecological adaptations, host susceptibility and pathogenicity. However, to date relatively few isolates have been identified to genotype level and the results do not yet allow definite conclusions. G8 has been reported from southern Canada [wolf, **moose** (*Alces alces*) and **wapiti or elk** (*Cervus canadensis*)], northern Canada [wolf, **muskox** (*Ovibos moschatus*)], the northwestern United States (moose), the northeastern United States (moose), eastern Europe (moose) and Yakutia (wolf and moose) (Konyaev et al., 2013; Lichtenwalner et al., 2014; Moks et al., 2008; Schurer et al., 2013, 2014; Thompson et al., 2006). G10 is known from southern Canada (wolf, moose, wapiti), northern Canada (wolf, **dog**, **caribou**), the western United States (moose), northern Fennoscandia (wolf, **domesticated reindeer** and moose), Eastern Europe (wolf, moose), Yakutia (wolf, moose and domesticated reindeer) and Mongolia (wolf) (Ito et al., 2013; Konyaev et al., 2013; Lavikainen et al., 2003, 2006; Moks et al., 2008; Oksanen and Lavikainen, 2015; Schurer et al., 2013, 2014; Thompson et al., 2006). The record of a sterile cyst from muskox indicates that G8/G10 is not limited to cervids as intermediate hosts, but the genotypic identity of cysts recorded from **mountain goat** (*Oreamnos americanus*) in Alaska (Rausch and Williamson, 1959), **bison**

Figure 11 Life cycle of *Echinococcus canadensis* G8 and G10, based on wild mammals (here represented by wolf and moose), in some areas interlinked with (or replaced by) semidomestic transmission patterns (here represented by domestic dog and semidomesticated reindeer). Human cases are relatively few and often described as benign.

(*Bison bison*) and **domestic pigs** in Alberta (Sweatman and Williams, 1963) and other deer species (Schurer et al., 2013) is now difficult to interpret. The same is true for earlier reports of *E. granulosus* s.l. from **coyotes** (*C. latrans*) in Canada (Alberta, Manitoba and Ontario), as recent surveys failed to detect the parasite (Davidson et al., 2016). Human cases are comparatively rare and were frequently considered to constitute a benign disease (Wilson et al., 1968), although a case of aggressive CE caused by G8 has been reported from Alaska (Castrodale et al., 2002; McManus et al., 2002). As far as known, G8 and G10 have a largely sympatric distribution and occur in the same hosts, a situation that even includes a member of the G6/7 cluster that was found in reindeer in Yakutia. Genetic studies of the nuclear genome are urgently required to find out whether all these parasites of the *E. canadensis* cluster carrying different mt genotypic markers (G610) can recombine or are reproductively isolated (Konyaev et al., 2013). It has been speculated that one or both genotypes (G8 and G10) might have been introduced with domesticated reindeer from Eurasia into northern America in historic times (Rausch, 2003; Thompson et al., 2006).

Three distinct transmission cycles have been proposed (Rausch, 2003; Oksanen and Lavikainen, 2015), namely sylvatic (wolf – wild cervid), semisynanthropic (dog – wild cervid) and domestic (herding dog – domesticated reindeer).

Sylvatic transmission depends on the wolf as definitive host and occurs throughout the circumpolar north [extending south to Washington (United States), Estonia and Mongolia] (Davidson et al., 2016; Oksanen and Lavikainen, 2015). Intermediate host seems to be predominantly moose, which is frequently infected and harbours cysts with high fertility rates, while the roles of wild reindeer (caribou), muskoxen and wapiti (elk) are less clear. In the former two, only single records (with sterile cysts) exist, while cysts in wapiti seem to be often sterile or produce few protoscoleces (Schurer et al., 2013). In North America, G8 and G10 are common where wolves and large herbivores coexist (Jenkins et al., 2013). Stable presence of suitable herbivores in a certain population density seems to be required for transmission, which might explain the absence of the life cycle in the High Arctic islands of North America despite the presence of wolves (Jenkins et al., 2013). The essential role of wolves was highlighted by the absence of transmission in Newfoundland (where wolves were extirpated) despite the abundance of potential intermediate hosts (Jenkins et al., 2013), but recently moose were found to be commonly infected with *E. canadensis* G8 in Maine (northeastern United States), where wolves are also absent and where eastern coyotes or domestic dogs may be responsible for transmission (Lichtenwalner et al., 2014). Increasing density in moose populations was associated with increase in prevalence, and infection of moose with *Echinococcus* cysts was shown to increase vulnerability to predation by wolves (Joly and Messier, 2004). The possibility of a future range extension of the sylvatic transmission of G8/G10 is being discussed in the context of a warming climate in the north (Davidson et al., 2011).

A semisynanthropic cycle was proposed, where free-ranging dogs of indigenous communities are getting infected through offal from hunted cervids in the context of subsistence hunting (Rausch, 2003; Oksanen and Lavikainen, 2015). In northern Canada, 6% of such dogs were found to shed eggs of G10 (Himsworth et al., 2010) and may thus form 'bridging hosts' between the wildlife cycle and human infection (Rausch, 2003).

A domestic cycle occurs (or had occurred) between herding dogs and domesticated reindeer. This was described for northern Fennoscandia, where the parasite (assumed to conform to G10) had been abundant in herded reindeer well into the second half of the 20th century. This transmission system was accompanied by high incidence of human CE (reviewed in Oksanen and Lavikainen, 2015). The consecutive breakdown of the cycle was explained by a combination of control efforts (dog treatment, controlled slaughter of reindeer) and the replacement of working dogs by snowmobiles, therefore removing the definitive host population (Oksanen and Lavikainen, 2015). Recently, this life cycle seems to recover in a modified form, where the increasing wolf population in eastern Finland is assumed to take the role of definitive hosts from dogs and provide an infection source for semidomesticated reindeer (Hirvelä-Koski et al., 2003; Lavikainen et al., 2003, 2006). Little is known about domestic life cycles in northern Asia. Indigenous reindeer herders and hunters are reported with high seroprevalence rates, which have been linked to the habit of feeding dogs with reindeer offal and the close contact of people with their dogs (Konyaev et al., 2013).

11. CONCLUSION

Rapid and reliable molecular methods for detection and differentiation of *Echinococcus* species have recently boosted the available information on ecology and geographical distribution of these parasites, including their zoonotic potential. However, while transmission patterns are reasonably well understood for some *Echinococcus* species in some regions, our gaps of knowledge are still extensive from a global perspective. This is exemplified by the very recent recognition of sub-Saharan Africa as a diversity centre for these parasites and the emerging recognition of *Echinococcus* spp. spread in North America, which suggests that current data on the presence and ecology of *Echinococcus* taxa are to some extent a reflection of research efforts rather than the true situation. The limitations of our ecological knowledge are highlighted in the preceding species accounts and may serve as an encouragement for research on basic biological and ecological aspects of *Echinococcus*. Better understanding of transmission patterns is not an academic exercise, but has an impact on a number of practical issues, including the design of cost-effective programmes for control and prevention of human and livestock disease in resource-poor countries, development of risk-based monitoring schemes, prevention of accidental introduction into nonendemic areas, wildlife conservation issues and studies on the role of parasites in ecosystems of a rapidly changing world.

REFERENCES

Abdybekova, A., Sultanov, A., Karatayev, B., Zhumabayeva, A., Shapiyeva, Z., Yeshmuratov, T., Toksanbayev, D., Shalkeev, R., Torgerson, P.R., 2015. Epidemiology of echinococcosis in Kazakhstan: an update. J. Helminthol. 89, 647–650.

Aboelhadid, S.M., El-Dakhly, K.M., Yanai, T., Fukushi, H., Hassanin, K.M., 2013. Molecular characterization of *Echinococcus granulosus* in Egyptian donkeys. Vet. Parasitol. 193, 292–296.

Abuladze, K.I., 1964. TAENIATA of animals and man and diseases cause by them. In: Skrjabin, A.K.I. (Ed.), Essentials of Cestodology, vol. IV. Israel Program for Scientific Translations, 1970, Jerusalem.

Addy, F., Alakonya, A., Wamae, N., Magambo, J., Mbae, C., Mulinge, E., Zeyhle, E., Wassermann, M., Kern, P., Romig, T., 2012. Prevalence and diversity of cystic echinococcosis in livestock in Maasailand, Kenya. Parasitol. Res. 111, 2289–2294.

Addy, F., Wassermann, M., Banda, F., Mbaya, H., Aschenborn, J., Aschenborn, O., Koskei, P., Umhang, G., de la Rue, M., Elmahdi, I.E., Mackenstedt, U., Kern, P., Romig, T., 2016. Genetic polymorphism and population structure of *Echinococcus ortleppi* Lopes-Neyra and Soler Planas, 1943. Parasitology. http://dx.doi.org/10.1017/S0031182016001840 [Epub ahead of print].

Afonso, E., Knapp, J., Tete, N., Umhang, G., Rieffel, D., van Kesteren, F., Ziadinov, I., Craig, P.S., Torgerson, P.R., Giraudoux, P., 2015. *Echinococcus multilocularis* in Kyrgyzstan: similarity in the Asian EmsB genotypic profiles from village populations of Eastern mole voles (*Ellobius tancrei*) and dogs in the Alay valley. J. Helminthol. 89, 664–670.

Al-Qaoud, K.M., Abdel-Hafez, S.K., Craig, P.S., 2003. Canine echinococcosis in northern Jordan: increased prevalence and dominance of sheep/dog strain. Parasitol. Res. 90, 187–191.

Allen, B.J., Goullet, M., Allen, L.R., Lisle, A., Leung, L.K.P., 2013. Dingoes at the doorstep: preliminary data on the ecology of dingoes in urban areas. Landsc. Urban Plan. 119, 131–135.

Altintas, N., 2003. Past to present: echinococcosis in Turkey. Acta Trop. 85, 105–112.

Alvarez Rojas, C., Romig, T., Lightowlers, M.W., 2014. *Echinococcus granulosus* sensu lato genotypes infecting humans—review of current knowledge. Int. J. Parasitol. 44, 9–18.

Alvarez Rojas, C., Ebi, D., Gauci, C.G., Scheerlinck, J.P., Wassermann, M., Jenkins, D.J., Lightowlers, M.W., Romig, T., 2016. Microdiversity of *Echinococcus granulosus* sensu stricto in Australia. Parasitology 143, 1026–1033.

Andreyanov, O.N., 2011. Alveolar echinococcosis at trade animals in the Ryazan region. Russ. Parasitol. J. 3, 7–11.

Armua-Fernandez, M.T., Joekel, D., Schweiger, A., Eichenberger, R.M., Matsumoto, J., Deplazes, P., 2016. Successful intestinal *Echinococcus multilocularis* oncosphere invasion and subsequent hepatic metacestode establishment in resistant RccHan:WIST rats after pharmacological immunosuppression. Parasitology 1–9.

Avcioglu, H., Guven, E., Balkaya, I., Kirman, R., Bia, M.M., Gulbeyen, H., 2016. First molecular characterization of *Echinococcus multilocularis* in Turkey. Vector Borne Zoonotic Dis. 16, 627–629.

Bagrade, G., Kirjusina, M., Vismanis, K., Ozolins, J., 2009. Helminth parasites of the wolf *Canis lupus* from Latvia. J. Helminthol. 83, 63–68.

Balbinotti, H., Santos, G.B., Badaraco, J., Arend, A.C., Graichen, D.A., Haag, K.L., Zaha, A., 2012. *Echinococcus ortleppi* (G5) and *Echinococcus granulosus* sensu stricto (G1) loads in cattle from southern Brazil. Vet. Parasitol. 188, 255–260.

Banks, D.D., 1984. Epidemiology of *Echinococcus granulosus* in Tropical Queensland (Ph.D. thesis). James Cook University of North Queensland.

Barlow, A.M., Gottstein, B., Mueller, N., 2011. *Echinococcus multilocularis* in an imported captive European beaver (*Castor fiber*) in Great Britain. The Vet. Rec. Rec. 169, 339.

Bardonnet, K., Piarroux, R., Dia, L., Schneegans, F., Beurdeley, A., Godot, V., Vuitton, D.A., 2002. Combined eco-epidemiological and molecular biology approaches to assess *Echinococcus granulosus* transmission to humans in Mauritania: occurrence of the 'camel' strain and human cystic echinococcosis. Trans. R. Soc. Trop. Med. Hyg. 96, 383–386.

Barnes, T.S., Hinds, L.A., Jenkins, D.J., Coleman, G.T., 2007a. Precocious development of hydatid cysts in a macropodid host. Int. J. Parasitol. 37, 1379–1389.

Barnes, T.S., Morton, J.M., Coleman, G.T., 2007b. Clustering of hydatid infection in macropodids. Int. J. Parasitol. 37, 943–952.

Barnes, T.S., Goldizen, A.W., Morton, J.M., Coleman, G.T., 2008. Cystic echinococcosis in a wild population of the brush-tailed rock-wallaby (*Petrogale penicillata*), a threatened macropodid. Parasitology 135, 715–723.

Barnes, T.S., Hinds, L.A., Jenkins, D.J., Bielefeldt-Ohmann, H., Lightowlers, M.W., Coleman, G.T., 2011. Comparative pathology of pulmonary hydatid cysts in macropods and sheep. J. Comp. Pathol. 144, 113–122.

Bart, J.M., Abdukader, M., Zhang, Y.L., Lin, R.Y., Wang, Y.H., Nakao, M., Ito, A., Craig, P.S., Piarroux, R., Vuitton, D.A., Wen, H., 2006. Genotyping of human cystic echinococcosis in Xinjiang, PR China. Parasitology 133, 571–579.

Bartel, M.H., Seesee, F.M., Worley, D.E., 1992. Comparison of Montana and Alaska isolates of *Echinococcus multilocularis* in gerbils with observations on the cyst growth, hook characteristics, and host response. J. Parasitol. 78, 529–532.

Beiromvand, M., Akhlaghi, L., Fattahi Massom, S.H., Mobedi, I., Meamar, A.R., Oormazdi, H., Motevalian, A., Razmjou, E., 2011. Detection of *Echinococcus multilocularis* in carnivores in Razavi Khorasan province, Iran using mitochondrial DNA. PLoS Negl. Trop. Dis 5, e1379.

Beiromvand, M., Akhlaghi, L., Fattahi Massom, S.H., Meamar, A.R., Darvish, J., Razmjou, E., 2013. Molecular identification of *Echinococcus multilocularis* infection in small mammals from Northeast, Iran. PLoS Negl. Trop. Dis. 7, e2313.

Berger, K.M., Conner, M.M., 2008. Recolonizing wolves and mesopredator suppression of coyotes: impacts on pronghorn population dynamics. Ecol. Appl. 18, 599–612.

Bessonov, A.S., 1998. *Echinococcus multilocularis* infection in Russia and neighbouring countries. Helminthologia 35, 73–79.

Bessonov, A.S., 2002. Echinococcoses of animals and humans in the Russian Federation. In: Craig, P.S., Pawlowski, Z. (Eds.), Cestode Zoonoses: Echinoccocosis and Cysticercosis - an Emergent and Global Problem. IOS Press, Amsterdam, pp. 91–98.

Beveridge, I., Spratt, D.M., Barker, S.C., Close, R.L., Barker, S.C., Sharman, G.B., 1989. Helminth parasites of rock-wallabies, *Petrogale* spp. (Marsupialia), from Queensland. Aust. Wildl. Res. 16, 273–287.

Beyhan, Y.E., Umur, S., 2011. Molecular characterization and prevalence of cystic echinococcosis in slaughtered water buffaloes in Turkey. Vet. Parasitol. 181, 174–179.

Bino, G., Dolev, A., Yosha, D., Guter, A., King, R., Saltz, D., Kark, S., 2010. Abrupt spatial and numerical responses of overabundant foxes to a reduction in anthropogenic resources. J. Appl. Ecol. 47, 1262–1271.

Boufana, B., Stidworthy, M.F., Bell, S., Chantrey, J., Masters, N., Unwin, S., Wood, R., Lawrence, R.P., Potter, A., McGarry, J., Redrobe, S., Killick, R., Foster, A.P., Mitchell, S., Greenwood, A.G., Sako, Y., Nakao, M., Ito, A., Wyatt, K., Lord, B., Craig, P.S., 2012. *Echinococcus* and *Taenia* spp. from captive mammals in the United Kingdom. Vet. Parasitol. 190, 95–103.

Boufana, B., Qiu, J., Chen, X., Budke, C.M., Campos-Ponce, M., Craig, P.S., 2013a. First report of *Echinococcus shiquicus* in dogs from eastern Qinghai-Tibet plateau region, China. Acta Trop. 127, 21–24.

Boufana, B., Umhang, G., Qiu, J., Chen, X., Lahmar, S., Boue, F., Jenkins, D., Craig, P., 2013b. Development of three PCR assays for the differentiation between *Echinococcus shiquicus, E. granulosus* (G1 genotype), and *E. multilocularis* DNA in the co-endemic region of Qinghai-Tibet plateau, China. Am. J. Trop. Med. Hyg. 88, 795–802.

Boufana, B., Lahmar, S., Rebai, W., Ben Safta, Z., Jebabli, L., Ammar, A., Kachti, M., Aouadi, S., Craig, P.S., 2014. Genetic variability and haplotypes of *Echinococcus* isolates from Tunisia. Trans. R. Soc. Trop. Med. Hyg. 108, 706–714.

Boufana, B., Said, Y., Dhibi, M., Craig, P.S., Lahmar, S., 2015. *Echinococcus granulosus* sensu stricto (s.s.) from the critically endangered antelope *Addax nasomaculatus* in Tunisia. Acta Trop. 152, 112–115.

Bowles, J., Blair, D., McManus, D.P., 1992. Genetic variants within the genus *Echinococcus* identified by mitochondrial DNA sequencing. Mol. Biochem. Parasitol. 54, 165–173.

Bowles, J., McManus, D.P., 1993. NADH dehydrogenase 1 gene sequences compared for species and strains of the genus *Echinococcus*. Int. J. Parasitol. 23, 969–972.

Breyer, I., Georgieva, D., Kurdova, R., Gottstein, B., 2004. *Echinococcus granulosus* strain typing in Bulgaria: the G1 genotype is predominant in intermediate and definitive wild hosts. Parasitol. Res. 93, 127–130.

Brown, B., Copeman, D.B., 2003. Zoonotic importance of parasites in wild dogs caught in the vicinity of Townsville. Aust. Vet. J. 81, 700–702.

Bruzinskaite, R., Sarkunas, M., Torgerson, P.R., Mathis, A., Deplazes, P., 2009. Echinococcosis in pigs and intestinal infection with *Echinococcus* spp. in dogs in southwestern Lithuania. Vet. Parasitol. 160, 237–241.

Bruzinskaite-Schmidhalter, R., Sarkunas, M., Malakauskas, A., Mathis, A., Torgerson, P.R., Deplazes, P., 2012. Helminths of red foxes (*Vulpes vulpes*) and raccoon dogs (*Nyctereutes procyonoides*) in Lithuania. Parasitology 139, 120–127.

Budke, C.M., Campos-Ponce, M., Qian, W., Torgerson, P.R., 2005a. A canine purgation study and risk factor analysis for echinococcosis in a high endemic region of the Tibetan plateau. Vet. Parasitol. 127, 43–49.

Budke, C.M., Jiamin, Q., Craig, P.S., Torgerson, P.R., 2005b. Modeling the transmission of *Echinococcus granulosus* and *Echinococcus multilocularis* in dogs for a high endemic region of the Tibetan plateau. Int. J. Parasitol. 35, 163–170.

Buishi, I., Walters, T., Guildea, Z., Craig, P., Palmer, S., 2005. Reemergence of canine *Echinococcus granulosus* infection, Wales. Emerg. Infect. Dis. 11, 568–571.

Burlet, P., Deplazes, P., Hegglin, D., 2011. Age, season and spatio-temporal factors affecting the prevalence of *Echinococcus multilocularis* and *Taenia taeniaeformis* in *Arvicola terrestris*. Parasit. Vectors 4, 6.

Callait, M.P., 2003. *Marmota marmota*, intermediate host of *Echinococcus multilocularis*: identification of three cases in French Alps. In: Ramousse, R., Allaine, D., Le Berre, M. (Eds.), Marmot and Helath/Adaptive Strategies and Diversity in Marmots. International Network on Marmots, pp. 9–10.

Campbell-Palmer, R., Del Pozo, J., Gottstein, B., Girling, S., Cracknell, J., Schwab, G., Rosell, F., Pizzi, R., 2015. *Echinococcus multilocularis* detection in live Eurasian beavers (*Castor fiber*) using a combination of laparoscopy and abdominal ultrasound under field conditions. PLoS One 10, e0130842.

Capuano, F., Rinaldi, L., Maurelli, M.P., Perugini, A.G., Veneziano, V., Garippa, G., Genchi, C., Musella, V., Cringoli, G., 2006. Cystic echinococcosis in water buffaloes: epidemiological survey and molecular evidence of ovine (G1) and buffalo (G3) strains. Vet. Parasitol. 137, 262–268.

Cardona, G.A., Carmena, D., 2013. A review of the global prevalence, molecular epidemiology and economics of cystic echinococcosis in production animals. Vet. Parasitol. 192, 10–32.

Carmena, D., Cardona, G.A., 2013. Canine echinococcosis: global epidemiology and genotypic diversity. Acta Trop. 128, 441–460.

Castrodale, L.J., Beller, M., Wilson, J.F., Schantz, P.M., McManus, D.P., Zhang, L.H., Fallico, F.G., Sacco, F.D., 2002. Two atypical cases of cystic echinococcosis (*Echinococcus granulosus*) in Alaska, 1999. Am. J. Trop. Med. Hyg. 66, 325–327.

Casulli, A., Interisano, M., Sreter, T., Chitimia, L., Kirkova, Z., La Rosa, G., Pozio, E., 2012. Genetic variability of *Echinococcus granulosus* sensu stricto in Europe inferred by mitochondrial DNA sequences. Infect. Genet. Evol. 12, 377–383.

Catalano, S., Lejeune, M., Liccioli, S., Verocai, G.G., Gesy, K.M., Jenkins, E.J., Kutz, S.J., Fuentealba, C., Duignan, P.J., Massolo, A., 2012. *Echinococcus multilocularis* in urban coyotes, Alberta, Canada. Emerg. Infect. Dis. 18, 1625–1628.

Chaignat, V., Boujon, P., Frey, C.F., Hentrich, B., Muller, N., Gottstein, B., 2015. The brown hare (*Lepus europaeus*) as a novel intermediate host for *Echinococcus multilocularis* in Europe. Parasitol. Res. 114, 3167–3169.

Cirovic, D., Pavlovic, I., Kulisic, Z., Ivetic, V., Penezic, A., Cosic, N., 2012. *Echinococcus multilocularis* in the European beaver (Castor fibre L.) from Serbia: first report. Vet. Rec. 171, 100.

Comte, S., Umhang, G., Raton, V., Boucher, J.-M., Caillot, C., Favier, S., Hormaz, V., Boué, F., Combes, B., 2014. Fox culling against *Echinococcus multilocularis*, reverse consequences. ESCCAP 2014 Echinococcus Abstr. Bookl. 26.

Contesse, P., Hegglin, D., Gloor, S., Bontadina, F., Deplazes, P., 2004. The diet of urban foxes (*Vulpes vulpes*) and the availability of anthropogenic food in the city of Zurich, Switzerland. Mamm. Biol. 69, 81–95.

Corsini, M., Geissbuhler, U., Howard, J., Gottstein, B., Spreng, D., Frey, C.F., 2015. Clinical presentation, diagnosis, therapy and outcome of alveolar echinococcosis in dogs. Vet. Rec. 177, 569.

Craig, P.S., Deshan, L., Macpherson, C.N., Dazhong, S., Reynolds, D., Barnish, G., Gottstein, B., Zhirong, W., 1992. A large focus of alveolar echinococcosis in central China. Lancet 340, 826–831.

Craig, P.S., Giraudoux, P., Shi, D., Bartholomot, B., Barnish, G., Delattre, P., Quere, J.P., Harraga, S., Bao, G., Wang, Y., Lu, F., Ito, A., Vuitton, D.A., 2000. An epidemiological and ecological study of human alveolar echinococcosis transmission in south Gansu, China. Acta Trop. 77, 167–177.

Craig, P.S., 2004. Epidemiology of echinococcosis in western China. In: Torgerson, P., Shaikenov, B. (Eds.), Echinococcosis in Central Asia: Problems and Solutions. Dauir, Zurich/Almaty, pp. 43–58.

Craig, P., Mastin, A., van Kesteren, F., Boufana, B., 2015. *Echinococcus granulosus*: epidemiology and state-of-the-art of diagnostics in animals. Vet. Parasitol. 213, 132–148.

Craig, P.S., Hegglin, D., Lightowlers, M.W., Torgerson, P., Wang, Q., 2016. Echinococcosis: control and prevention. In: Thompson, A. (Ed.), Echinococcus and Echinococcosis, Vol. 96, pp. 55–158.

Cringoli, G., Rinaldi, L., Musella, V., Veneziano, V., Maurelli, M.P., Di Pietro, F., Frisiello, M., Di Pietro, S., 2007. Geo-referencing livestock farms as tool for studying cystic echinococcosis epidemiology in cattle and water buffaloes from southern Italy. Geospat. Health 2, 105–111.

D'Alessandro, A., Rausch, R.L., Morales, G.A., Collet, S., Angel, D., 1981. *Echinococcus* infections in Colombian animals. Am. J. Trop. Med. Hyg. 30, 1263–1276.

D'Alessandro, A., Rausch, R.L., 2008. New aspects of neotropical polycystic (*Echinococcus vogeli*) and unicystic (*Echinococcus oligarthrus*) echinococcosis. Clin. Microbiol. Rev. 21, 380–401.

Dalimi, A., Motamedi, G., Hosseini, M., Mohammadian, B., Malaki, H., Ghamari, Z., Ghaffari Far, F., 2002. Echinococcosis/hydatidosis in western Iran. Vet. Parasitol. 105, 161–171.

Dalimi, A., Sattari, A., Motamedi, G., 2006. A study on intestinal helminthes of dogs, foxes and jackals in the western part of Iran. Vet. Parasitol. 142, 129–133.

Daniel Mwambete, K., Ponce-Gordo, F., Cuesta-Bandera, C., 2004. Genetic identification and host range of the Spanish strains of *Echinococcus granulosus*. Acta Trop. 91, 87–93.

Davidson, R., Simard, M., Kutz, S.J., Kapel, C.M., Hamnes, I.S., Robertson, L.J., 2011. Arctic parasitology: why should we care? Trends Parasitol. 27, 239–245.

Davidson, R.K., Romig, T., Jenkins, E., Tryland, M., Robertson, L.J., 2012. The impact of globalisation on the distribution of *Echinococcus multilocularis*. Trends Parasitol. 28, 239–247.

Davidson, R.K., Lavikainen, A., Konyaev, S., Schurer, J., Miller, A.L., Oksanen, A., Skirnisson, K., Jenkins, E., 2016. *Echinococcus* across the north: current knowledge, future challenges. Food Waterborne Parasitol (in press).

de la Rue, M.L., Dinkel, A., Mackenstedt, U., Romig, T., 2006. New data on *Echinococcus* spp. in southern Brazil. Rev. Inst. Med. Trop. Sao. Paulo 48, 103–104.

Delattre, P., Giraudoux, P., Quéré, J.P., 1990. Conséquence épidémiologique de la réceptivité d'un nouvel hôte intermédiaire du ténia multiloculaire (*Echinococcus multilocularis*) et de la localisation spatiotemporelle des rongeurs infestés. C. R. Acad. Sci. 310, 339–344.

Delattre, P., Giraudoux, P., Baudry, J., Quéré, J.P., Fichet, E., 1996. Effect of landscape structure on Common Vole (*Microtus arvalis*) distribution and abundance at several space scale. Landsc. Ecol. 11, 279–288.

Delattre, P., De Sousa, B., Fichet, E., Quéré, J.P., Giraudoux, P., 1999. Vole outbreaks in a landscape context: evidence from a six year study of *Microtus arvalis*. Landsc. Ecol. 14, 401–412.

Deplazes, P., Eckert, J., 2001. Veterinary aspects of alveolar echinococcosis—a zoonosis of public health significance. Vet. Parasitol. 98, 65–87.

Deplazes, P., Hegglin, D., Gloor, S., Romig, T., 2004. Wilderness in the city: the urbanization of *Echinococcus multilocularis*. Trends Parasitol. 20, 77–84.

Deplazes, P., Grimm, F., Sydler, T., Tanner, I., Kapel, C.M., 2005. Experimental alveolar echinococcosis in pigs, lesion development and serological follow up. Vet. Parasitol. 130, 213–222.

Deplazes, P., van Knapen, F., Schweiger, A., Overgaauw, P.A., 2011. Role of pet dogs and cats in the transmission of helminthic zoonoses in Europe, with a focus on echinococcosis and toxocarosis. Vet. Parasitol. 182, 41–53.

Deplazes, P., 2015. Echinococcoses. In: Beugnet, F., Halos, L. (Eds.), Parasitoses & Vector Borne Diseases of Cats. Merial, Lyon, pp. 3–7.

Deplazes, P., Rinaldi, L., Alvarez Rojas, C.A., Torgerson, P., Harandi, M.F., Romig, T., Antolova, D., Schurer, J., Lahmar, S., Cringoli, G., Magambo, J., Thompson, A., Jenkins, E., 2016. Global distribution of alveolar and cystic echinococcosis. In: Thompson, A. (Ed.), Echinococcus and Echinococcosis, Vol. 95, pp. 315–494.

Dinkel, A., Tackmann, K., Schulze, C., Drewes, W., Romig, T., Mackenstedt, U., 2006. First record of the pig strain (G7) of *Echinococcus granulosus* in Germany. Tagungsband, 22. In: Tagung der Deutschen Gesellschaft für Parasitologie. Institut für Parasitologie und Zoologie, Wien, Austria, ISBN 3-200-00559-9, p. 139.

Do Carmo Perreira Soares, M., Rodrigues, A.L., Moreira Silva, C.A., Brito, E.M., Gomes-Gouvea, M.S., Correa, I.R., Pinho, J.R., Malheiros, A.P., Nunes, H.M., Povoa, M.M., 2013. Anatomo-clinical and molecular description of liver neotropical echinococcosis caused by *Echinococcus oligarthrus* in human host. Acta Trop. 125, 110–114.

Do Carmo Perreira Soares, C., Souza de Souza, A.J., Pinheiro Malheiros, A., Nunes, H.M., Almeida Carneiro, L., Alves, M.M., Farias da Conceicao, B., Gomes-Gouvea, M.S., Marins Povoa, M., 2014. Neotropical echinococcosis: second report of *Echinococcus vogeli* natural infection in its main definitive host, the bush dog (*Speothos venaticus*). Parasitol. Int. 63, 485–487.

Durie, P.H., Riek, R.F., 1952. The role of the dingo and wallaby in the infestation of cattle with hydatids (*Echinococcus granulosus* (Batsch 1786) Rudolphi, 1805) in Queensland. Aust. Vet. J. 48, 249–254.

Eastman, K.L., Worley, D.E., 1979. The muskrat as an intermediate host of *Echinococcus multilocularis* in Montana. J. Parasitol. 65, 34.

Eckert, J., Thompson, R.C.A., Michael, S.S., Kumaratilake, L.M., El-Sawah, H.M., 1989. *Echinococcus granulosus* of camel origin: development in dogs and parasite morphology. Parasitol. Res. 75, 536–544.

Eckert, J., Thompson, R.C.A., Lymbery, A.J., Pawlowski, Z.S., Gottstein, B., Morgan, U.M., 1993. Further evidence for the occurrence of a distinct strain of *Echinococcus granulosus* in European pigs. Parasitol. Res. 79, 42–48.

EFSA, 2015. Scientific opinion: *Echinococcus multilocularis* infection in animals. EFSA J. 13 (12), 4373.

Elchuev, M., 1989. Landscape and ecological characteristics of cestodes in wild Canidae in the Sheki-Zaqatal area of the Azerbaijan SSR. Izvestiia. Acad. Sci. Azerbaijan SSR. Ser. Biol. Sci. 2.

Elmahdi, I.E., Ali, Q.M., Magzoub, M.M., Ibrahim, A.M., Saad, M.B., Romig, T., 2004. Cystic echinococcosis of livestock and humans in central Sudan. Ann. Trop. Med. Parasitol. 98, 473–479.

Eslami, A., Meshgi, B., Jalousian, F., Rahmani, S., Salari, M.A., 2016. Genotype and Phenotype of *Echinococcus granulosus* derived from wild sheep (*Ovis orientalis*) in Iran. Korean J. Parasitol. 54, 55–60.

Fan, Y.L., Lou, Z.Z., Li, L., Yan, H.B., Liu, Q.Y., Zhan, F., Li, J.Q., Liu, C.N., Cai, J.Z., Lei, M.T., Shi, W.G., Yang, Y.R., McManus, D.P., Jia, W.Z., 2016. Genetic diversity in *Echinococcus shiquicus* from the plateau pika (*Ochotona curzoniae*) in Darlag county, Qinghai, China. Infect. Genet. Evol. (accepted).

Fay, F.H., Stephenson, R.O., 1989. Annual, seasonal and habitat-related variation in feeding habits of the arctic fox (*Alopex lagopus*) on St. Lawrence Island, Bering Sea. Can. J. Zool. 67, 1986–1994.

Fedriani, J.M., Fuller, T.K., Sauvajot, R.M., York, E.C., 2000. Competition and intraguild predation among three sympatric carnivores. Oecologia 125, 258–270.

Feldhamer, G.A., Thompson, B.C., Chapman, J.A., 2003. Wild Mammals of North America: Biology, Management and Conservation. JHU Press, pp. 511–540.

Feigley, H.P., Worley, D.E., 1988. Prevalence of larval *Echinococcus multilocularis* in native Montana small mammals and susceptibility of laboratory rodents to a Montana isolate. Proc. Helminthol. Soc. Wash. 55, 105–108.

Fuglei, E., Stien, A., Yoccoz, N.G., Ims, R.A., Eide, N.E., Prestrud, P., Deplazes, P., Oksanen, A., 2008. Spatial distribution of *Echinococcus multilocularis*, Svalbard, Norway. Emerg. Infect. Dis. 14, 73–75.

Gemmell, M.A., 1959. Hydatid disease in Australia VI. Observations on the carnivore of New South Wales as definitive hosts of *Echinococcus granulosus* (Batsch 1786), (Rudolphi1801) and their role in the spread of hydatidiasis in domestic animals. Aust. Vet. J. 35, 450–455.

Gemmell, M.A., Lawson, J.R., Roberts, M.G., 1986. Population dynamics in echinococcosis and cysticercosis: biological parameters of *Echinococcus granulosus* in dogs and sheep. Parasitology 92 (Pt 3), 599–620.

Gemmell, M.A., 1990. Australasian contributions to an understanding of the epidemiology and control of hydatid disease caused by *Echinococcus granulosus*—past, present and future. Int. J. Parasitol. 20, 431—456.

Genov, T., Svilenov, D., Polyakova-Krusteva, O., 1980. The natural occurrence of *Alveococcus multilocularis* in the *Microtus nivalis* in Bulgaria, 33, 981—984.

Gesy, K., Hill, J.E., Schwantje, H., Liccioli, S., Jenkins, E.J., 2013. Establishment of a European-type strain of *Echinococcus multilocularis* in Canadian wildlife. Parasitology 140, 1133—1137.

Gesy, K.M., Schurer, J.M., Massolo, A., Liccioli, S., Elkin, B.T., Alisauskas, R., Jenkins, E.J., 2014. Unexpected diversity of the cestode *Echinococcus multilocularis* in wildlife in Canada. Int. J. Parasitol. Parasites Wildl. 3, 81—87.

Giraudoux, P., Delattre, P., Habert, M., Quéré, J.P., Deblay, S., Defaut, R., Dehamel, R., Moissinet, M.F., Salvi, D., Truchetet, D., 1997. Population dynamics of fossorial water vole (*Arvicola terrestris sherman*): a land usage and landscape perspective. Agric. Ecosyst. Environ. 66, 47—60.

Giraudoux, P., Delattre, P., Takahashi, K., Raoul, F., Quere, J.P., Craig, P., Vuitton, D., 2002. Transmission ecology of *Echinococcus multilocularis* in wildlife: what can be learned from comparative studies and multiscale approaches? In: Craig, P., Pawlowski, Z. (Eds.), Cestode Zoonoses: Echinococcosis and Cysticercosis: An Emergent and Global Problem. IOS Press, Amsterdam, pp. 251—266.

Giraudoux, P., Craig, P.S., Delattre, P., Bao, G., Bartholomot, B., Harraga, S., Quere, J.P., Raoul, F., Wang, Y., Shi, D., Vuitton, D.A., 2003. Interactions between landscape changes and host communities can regulate *Echinococcus multilocularis* transmission. Parasitology 127 (Suppl.), S121—S131.

Giraudoux, P., Raoul, F., Afonso, E., Ziadinov, I., Yang, Y., Li, L., Li, T., Quere, J.P., Feng, X., Wang, Q., Wen, H., Ito, A., Craig, P.S., 2013. Transmission ecosystems of *Echinococcus multilocularis* in China and Central Asia. Parasitology 140, 1655—1666.

Gori, F., Armua-Fernandez, M.T., Milanesi, P., Serafini, M., Magi, M., Deplazes, P., Macchioni, F., 2015. The occurrence of taeniids of wolves in Liguria (northern Italy). Int. J. Parasitol. Parasites Wildl. 4, 252—255.

Gottstein, B., Saucy, F., Wyss, C., Siegenthaler, M., Jacquier, P., Schmitt, M., Brossard, M., Demierre, G., 1996. Investigations on a Swiss area highly endemic for *Echinococcus multilocularis*. Appl. Parasitol. 37, 129—136.

Gottstein, B., Saucy, F., Deplazes, P., Reichen, J., Demierre, G., Busato, A., Zuercher, C., Pugin, P., 2001. Is high prevalence of *Echinococcus multilocularis* in wild and domestic animals associated with disease incidence in humans? Emerg. Infect. Dis. 7, 408—412.

Graber, M., Thal, J., 1980. L'échinococcose des artiodactyles sauvages de la République Centralafricaine: existence probable d'un cycle lion-phacochère. Rev. Elev. Med. Vet. Pays. Trop. 33, 51—59.

Grainger, H.J., Jenkins, D.J., 1996. Transmission of hydatid disease to sheep from wild dogs in Victoria, Australia. Int. J. Parasitol. 26, 1263—1270.

Grenouillet, F., Umhang, G., Arbez-Gindre, F., Mantion, G., Delabrousse, E., Millon, L., Boué, F., 2014. *Echinococcus ortleppi* infections in humans and cattle, France. Emerg. Infect. Dis. 20, 2100—2102.

Guerra, D., Armua-Fernandez, M.T., Silva, M., Bravo, I., Santos, N., Deplazes, P., Madeira de Carvalho, L.M., 2013. Taeniid species of the Iberian wolf (*Canis lupus signatus*) in Portugal with special focus on *Echinococcus* spp. Int. J. Parasitol. Parasites Wildl. 2, 50—53.

Guerra, D., Hegglin, D., Bacciarini, L., Schnyder, M., Deplazes, P., 2014. Stability of the southern European border of *Echinococcus multilocularis* in the Alps: evidence that *Microtus arvalis* is a limiting factor. Parasitology 141, 1593—1602.

Guislain, M.H., Raoul, F., Giraudoux, P., Terrier, M.E., Froment, G., Ferte, H., Poulle, M.L., 2008. Ecological and biological factors involved in the transmission of *Echinococcus multilocularis* in the French Ardennes. J. Helminthol. 82, 143–151.

Hailemariam, Z., Nakao, M., Menkir, S., Lavikainen, A., Yanagida, T., Okamoto, M., Ito, A., 2012. Molecular identification of unilocular hydatid cysts from domestic ungulates in Ethiopia: implications for human infections. Parasitol. Int. 61, 375–377.

Hamel, S., Killengreen, S.T., Henden, J.A., Yoccoz, N.G., Ims, R.A., 2013. Disentangling the importance of interspecific competition, food availability, and habitat in species occupancy: recolonization of the endangered Fennoscandian arctic fox. Biol. Cons. 160, 114–120.

Hanosset, R., Saegerman, C., Adant, S., Massart, L., Losson, B., 2008. *Echinococcus multilocularis* in Belgium: prevalence in red foxes (*Vulpes vulpes*) and in different species of potential intermediate hosts. Vet. Parasitol. 151, 212–217.

Harris, R.E., Revfeim, K.J., Heath, D.D., 1980. Simulating strategies for control of *Echinococcus granulosus, Taenia hydatigena* and *T. ovis*. J. Hyg. 84, 389–404.

Heath, D.D., Zhang, L.H., McManus, D.P., 2005. Short report: inadequacy of yaks as hosts for the sheep dog strain of *Echinococcus granulosus* or for *E. multilocularis*. Am. J. Trop. Med. Hyg. 72, 289–290.

Hegglin, D., Bontadina, F., Contesse, P., Gloor, S., Deplazes, P., 2007. Plasticity of predation behaviour as a putative driving force for parasite life-cycle dynamics: the case of urban foxes and *Echinococcus multilocularis* tapeworm, 21, 552–560.

Hegglin, D., Deplazes, P., 2013. Control of *Echinococcus multilocularis*: strategies, feasibility and cost-benefit analyses. Int. J. Parasitol. 43, 327–337.

Hegglin, D., Bontadina, F., Deplazes, P., 2015. Human-wildlife interactions and zoonotic transmission of *Echinococcus multilocularis*. Trends Parasitol. 31, 167–173.

Henttonen, H., Fuglei, E., Gower, C.N., Haukisalmi, V., Ims, R.A., Niemimaa, J., Yoccoz, N.G., 2001. *Echinococcus multilocularis* on Svalbard: introduction of an intermediate host has enabled the local life-cycle. Parasitology 123, 547–552.

Heptner, V.G., Naumov, N.P., Yurgenson, P.B., Sludskii, A.A., Chirkova, A.F., Bannikov, A.G., 1998. Mammals of the Soviet Union, vol. 2, Part 1A. Smithsonian Institution Libraries and The National Science foundation, Washington DC, USA.

Hildreth, M.B., Sriram, S., Gottstein, B., Wilson, M., Schantz, P.M., 2000. Failure to identify alveolar echinococcosis in trappers from South Dakota in spite of high prevalence of *Echinococcus multilocularis* in wild canids. J. Parasitol. 86, 75–77.

Himsworth, C.G., Skinner, S., Chaban, B., Jenkins, E., Wagner, B.A., Harms, N.J., Leighton, F.A., Thompson, R.C., Hill, J.E., 2010. Multiple zoonotic pathogens identified in canine feces collected from a remote Canadian indigenous community. Am. J. Trop. Med. Hyg. 83, 338–341.

Hirvelä-Koski, V., Haukisalmi, V., Kilpela, S.S., Nylund, M., Koski, P., 2003. *Echinococcus granulosus* in Finland. Vet. Parasitol. 111, 175–192.

Hofer, S., Gloor, S., Muller, U., Mathis, A., Hegglin, D., Deplazes, P., 2000. High prevalence of *Echinococcus multilocularis* in urban red foxes (*Vulpes vulpes*) and voles (*Arvicola terrestris*) in the city of Zurich, Switzerland. Parasitology 120 (Pt 2), 135–142.

Höjgård, S., Sundstrom, K., Christensson, D., Hallgren, G., Hjertqvist, M., Wallensten, A., Vagsholm, I., Wahlstrom, H., 2012. Willingness to pay for compulsory deworming of pets entering Sweden to prevent introduction of *Echinoccoccus multilocularis*. Prev. Vet. Med. 106, 9–23.

Holt, D.W., Hanns, C., O'Hara, T., Burek, K., Frantz, R., 2005. New distribution records of *Echinococcus multilocularis* in the brown lemming from Barrow, Alaska, USA. J. Wildl. Dis. 41, 257–259.

Howard, E.B., Gendron, A.P., 1980. *Echinococcus vogeli* infection in higher primates at the Los Angeles zoo. In: Montali, R.J., Migaki, G. (Eds.), The Comparative Pathology of Zoo Animals. Smithsonian Institution Press, Washington DC, USA, pp. 379–382.
Huang, Y., Heath, D.D., Yang, W., Qiu, J.M., Chen, X.W., Yang, Y., Wang, Q., Li, T.Y., Xiao, Y.F., Qiu, D.C., Xiao, N., Chen, F.X., Ge, S., Se, D., 2008. Epidemiology and risk factor analysis for canine echinococcosis in a Tibetan pastoral area of Sichuan. Chin. J. Parasitol. Parasit. Dis. 26, 245–252.
Hüttner, M., Nakao, M., Wassermann, T., Siefert, L., Boomker, J.D., Dinkel, A., Sako, Y., Mackenstedt, U., Romig, T., Ito, A., 2008. Genetic characterization and phylogenetic position of *Echinococcus felidis* (Cestoda: Taeniidae) from the African lion. Int. J. Parasitol. 38, 861–868.
Hüttner, M., Romig, T., 2009. *Echinococcus* species in African wildlife. Parasitology 136, 1089–1095.
Hüttner, M., Siefert, L., Mackenstedt, U., Romig, T., 2009. A survey of *Echinococcus* species in wild carnivores and livestock in East Africa. Int. J. Parasitol. 39, 1269–1276.
Ibrahim, K., Romig, T., Kern, P., Omer, R.A., 2011. A molecular survey on cystic echinococcosis in Sinnar area, Blue Nile state (Sudan). Chin. Med. J. 124, 2829–2833.
Ishige, M., 1984. Incidence of swine multilocular echinococcosis in Hokkaido. Rep. Hokkaido Inst. Pub. Hlth. 34, 70–71.
Ito, A., Romig, T., Takahashi, K., 2003. Perspective on control options for *Echinococcus multilocularis* with particular reference to Japan. Parasitology 127 (Suppl.), S159–S172.
Ito, A., Chuluunbaatar, G., Yanagida, T., Davaasuren, A., Sumiya, B., Asakawa, M., Ki, T., Nakaya, K., Davaajav, A., Dorjsuren, T., Nakao, M., Sako, Y., 2013. *Echinococcus* species from red foxes, corsac foxes, and wolves in Mongolia. Parasitology 140, 1648–1654.
IUCN, 2016. The IUCN Red List of Threatened Species. Version 2016–1.
Iwaki, T., Hatakeyama, S., Nonaka, N., Miyaji, S., Yokohata, Y., Okamoto, M., Ooi, H., Oku, Y., Kamiya, M., 1993. Survey on larval *Echinococcus multilocularis* and other hepatic helminths in rodents and insectivores in Hokkaido, Japan, from 1985 to 1992. Jpn. J. Parasitol. 42, 502–506.
Jareno, D., Vinuela, J., Luque-Larena, J.J., Arroyo, L., Arroyo, B., Mougeot, F., 2015. Factors associated with the colonization of agricultural areas by common voles *Microtus arvalis* in NW Spain. Biol. Invasions 17, 2315–2327.
Jenkins, D.J., Morris, B., 1991. Unusually heavy infections of *Echinococcus granulosus* in wild dogs in south-eastern Australia. Aust. Vet. J. 68, 36–37.
Jenkins, D.J., Thompson, R.C., 1995. Hydatid cyst development in an experimentally infected wild rabbit. Vet. Rec. 137, 148–149.
Jenkins, D.J., Romig, T., 2000. Efficacy of Droncit® spot-on (praziquantel) 4% w/v against immature and mature *Echinococcus multilocularis* in cats. Int. J. Parasitol. 30, 959–962.
Jenkins, D.J., Macpherson, C.N., 2003. Transmission ecology of *Echinococcus* in wild-life in Australia and Africa. Parasitology 127 (Suppl.), S63–S72.
Jenkins, D.J., Romig, T., 2003. Milbemycin oxime in a new formulation, combined with praziquantel, does not reduce the efficacy of praziquantel against *Echinococcus multilocularis* in cats. J. Helminthol. 77, 367–370.
Jenkins, D.J., Morris, B., 2003. *Echinococcus granulosus* in wildlife in and around the Kosciuszko National Park, south-eastern Australia. Aust. Vet. J. 81, 81–85.
Jenkins, D.J., 2005. Hydatid control in Australia: where it began, what we have achieved and where to from here. Int. J. Parasitol. 35, 733–740.
Jenkins, D.J., 2006. *Echinococcus granulosus* in Australia, widespread and doing well! Parasitol. Int. 55 (Suppl.), S203–S206.
Jenkins, D.J., McKinlay, A., Duolong, H.E., Bradshaw, H., Craig, P.S., 2006. Detection of *Echinococcus granulosus* coproantigens in faeces from naturally infected rural domestic dogs in south eastern Australia. Aust. Vet. J. 84, 12–16.

Jenkins, D.J., Allen, L., Goullet, M., 2008. Encroachment of *Echinococcus granulosus* into urban areas in eastern Queensland, Australia. Aust. Vet. J. 86, 294–300.

Jenkins, D.J., Lievaart, J.J., Boufana, B., Lett, W.S., Bradshaw, H., Armua-Fernandez, M.T., 2014. *Echinococcus granulosus* and other intestinal helminths: current status of prevalence and management in rural dogs of eastern Australia. Aust. Vet. J. 92, 292–298.

Jenkins, E.J., Castrodale, L.J., de Rosemond, S.J., Dixon, B.R., Elmore, S.A., Gesy, K.M., Hoberg, E.P., Polley, L., Schurer, J.M., Simard, M., Thompson, R.C., 2013. Tradition and transition: parasitic zoonoses of people and animals in Alaska, northern Canada, and Greenland. Adv. Parasitol. 82, 33–204.

Jiang, W., Liu, N., Zhang, G., Renqing, P., Xie, F., Li, T., Wang, Z., Wang, X., 2012. Specific detection of *Echinococcus* spp. from the Tibetan fox (*Vulpes ferrilata*) and the red fox (*V. vulpes*) using copro-DNA PCR analysis. Parasitol. Res. 111, 1531–1539.

Jimenez, S., Perez, A., Gil, H., Schantz, P., Ramalle, E., Juste, R., 2002. Progress in control of cystic echinococcosis in La Rioja, Spain: decline in infection prevalences in human and animal hosts and economic costs and benefits. Acta Trop. 83, 213–221.

Johnson, P.M., Spear, R., Beveridge, I., 1998. Mortality in wild and captive rock-wallabies and nailtail wallabies due to hydatid disease caused by *Echinococcus granulosus*. Aust. Mammal. 20, 419–423.

Joly, D.O., Messier, F., 2004. The distribution of *Echinococcus granulosus* in moose: evidence for parasite-induced vulnerability to predation by wolves? Oecologia 140, 586–590.

Kagendo, D., Magambo, J.K., Agola, E.L., Njenga, S.M., Zeyhle, E., Kakundi, E.M., Njeru, P.G., Mbae, C., Muchiri, E.M., Wassermann, M., Kern, P., Romig, T., 2014. A survey for *Echinococcus* spp. of carnivores in six wildlife conservation areas in Kenya. Parasitol. Int. 63, 604–611.

Kamenetzky, L., Gutierrez, A.M., Canova, S.G., Haag, K.L., Guarnera, E.A., Parra, A., Garcia, G.E., Rosenzvit, M.C., 2002. Several strains of *Echinococcus granulosus* infect livestock and humans in Argentina. Infect. Genet. Evol. 2, 129–136.

Kamler, J.F., Ballard, W.G., 2002. A review of native and non-native red foxes in North America. Wildl. Soc. Bull. 30, 370–379.

Kaneko, Y., Nakata, K., Saitoh, T., Stenseth, N.C., Bjornstad, O.N., 1998. The biology of the vole *Clethrionomys rufocanus*: a review. Res. Popul. Ecol. 40, 21–37.

Kapel, C.M., Torgerson, P.R., Thompson, R.C., Deplazes, P., 2006. Reproductive potential of *Echinococcus multilocularis* in experimentally infected foxes, dogs, raccoon dogs and cats. Int. J. Parasitol. 36, 79–86.

Kauhala, K., Holmala, K., Schregel, J., 2007. Seasonal activity patterns and movements of the raccoon dog, a vector of diseases and parasites, in southern Finland. Mamm. Biol. 72, 342–353.

Kauhala, K., Salonen, L., 2012. Does a non-invasive method – latrine surveys – reveal habitat preferences of raccoon dogs and badgers? Mamm. Biol. 77, 264–270.

Kedra, A.H., Tkach, V.V., Swiderski, Z., Pawlowski, Z., Emets, A., Pawlowski, J., 2000. Molecular characterisation of *Echinococcus granulosus* from a wild boar. Acta Parasitol. 45, 121–122.

Kern, P., Menezes da Silva, A., Akhan, O., Müllhaupt, B., Vizcaychipi, K.A., Budke, C., Vuitton, D.A., 2016. The echinococcoses: diagnosis, clinical management and burden of disease. In: Thompson, A. (Ed.), Echinococcus and Echinococcosis, Vol. 96, pp. 259–368.

King, H.W.H., 1959. Transhumant grazing in the snow belt of New South Wales. Aust. Geogr. 7, 129–140.

Kirillova, I., Kirillov, A., 2008. Ecological analysis of cestodes (Cestoda) of murine rodents in the Samarskaya Luka. Povolzhskiy J. Ecol. 1, 20–28.

Knapp, J., Staebler, S., Bart, J.M., Stien, A., Yoccoz, N.G., Drogemuller, C., Gottstein, B., Deplazes, P., 2012. *Echinococcus multilocularis* in Svalbard, Norway: microsatellite genotyping to investigate the origin of a highly focal contamination. Infection, genetics and evolution. J. Mol. Epidemiol. Evol. Genet. Infect. Dis. 12, 1270–1274.

Knapp, J., Gottstein, B., Saarma, U., Millon, L., 2015. Taxonomy, phylogeny and molecular epidemiology of *Echinococcus multilocularis*: from fundamental knowledge to health ecology. Vet. Parasitol. 213, 85–91.

Kondo, N.K., Takahashi, K., Yagi, K., 1986. Winter food of the red fox, *Vulpes vulpes schrencki* Kishida, in the endemic area of multilocular echinococcosis. Bull. Prep. Off. Nemuro Munic. Mus. 1, 23–31.

Konyaev, S.V., Yanagida, T., Ingovatova, G.M., Shoikhet, Y.N., Nakao, M., Sako, Y., Bondarev, A.Y., Ito, A., 2012. Molecular identification of human echinococcosis in the Altai region of Russia. Parasitol. Int. 61, 711–714.

Konyaev, S.V., Yanagida, T., Nakao, M., Ingovatova, G.M., Shoykhet, Y.N., Bondarev, A.Y., Odnokurtsev, V.A., Loskutova, K.S., Lukmanova, G.I., Dokuchaev, N.E., Spiridonov, S., Alshinecky, M.V., Sivkova, T.N., Andreyanov, O.N., Abramov, S.A., Krivopalov, A.V., Karpenko, S.V., Lopatina, N.V., Dupal, T.A., Sako, Y., Ito, A., 2013. Genetic diversity of *Echinococcus* spp. in Russia. Parasitology 140, 1637–1647.

Kritsky, D.C., Leiby, P.D., Miller, G.E., 1977. The natural occurrence of *Echinococcus multilocularis* in the bushy-tailed woodrat, *Neotoma cinerea rupicola*, in Wyoming. Am. J. Trop. Med. Hyg. 26, 1046–1047.

Kumaratilake, L.M., Thompson, R.C.A., 1982. Hydatidosis/echinococcosis. Helminthological Abstracts - Series a. Animals Hum. Helminthol. 51, 233–252.

Kumaratilake, L.M., Thompson, R.C.A., Dunsmore, J.D., 1983. Comparative strobilar development of *Echinococcus granulosus* of sheep origin from different geographical areas of Australia in *in vivo* and *in vitro*. Int. J. Parasitol. 13, 151–156.

Kumaratilake, L.M., Thompson, R.C., Eckert, J., 1986. *Echinococcus granulosus* of equine origin from different countries possess uniform morphological characteristics. Int. J. Parasitol. 16, 529–540.

Lahmar, S., Boufana, B.S., Lahmar, S., Inoubli, S., Guadraoui, M., Dhibi, M., Bradshaw, H., Craig, P.S., 2009. *Echinococcus* in the wild carnivores and stray dogs of northern Tunisia: the results of a pilot survey. Ann. Trop. Med. Parasitol. 103, 323–331.

Lahmar, S., Boufana, B., Boubaker, S.B., Landolsi, F., 2014. Intestinal helminths of golden jackals and red foxes from Tunisia. Vet. Parasitol. 2014, 297–303.

Lalosevic, D., Lalosevic, V., Simin, V., Miljevic, M., Cabrilo, B., Bjelic Cabrilo, O., 2016. Spreading of multilocular echinococcosis in southern Europe: the first record in foxes and jackals in Serbia, Vojvodina Province. Eur. J. Wildl. Res. 62, 793–796.

Lanszki, J., Heltai, M., Szabó, L., 2006. Feeding habits and trophic niche overlap between sympatric golden jackal (*Canis aureus*) and red fox (*Vulpes vulpes*) in the Pannonian ecoregion (Hungary). Can. J. Zool. 84, 1647–1656.

Latif, A.A., Tanveer, A., Maqbool, A., Siddiqi, N., Kyaw-Tanner, M., Traub, R.J., 2010. Morphological and molecular characterisation of *Echinococcus granulosus* in livestock and humans in Punjab, Pakistan. Vet. Parasitol. 170, 44–49.

Laurimaa, L., Suld, K., Moks, E., Valdmann, H., Umhang, G., Knapp, J., Saarma, U., 2015. First report of the zoonotic tapeworm *Echinococcus multilocularis* in raccoon dogs in Estonia, and comparisons with other countries in Europe. Vet. Parasitol. 212, 200–205.

Laurimaa, L., Moks, E., Soe, E., Valdmann, H., Saarma, U., 2016. *Echinococcus multilocularis* and other zoonotic parasites in red foxes in Estonia. Parasitology. http://dx.doi.org/10.1017/S0031182016001013.

Lavikainen, A., Lehtinen, M.J., Meri, T., Hirvela-Koski, V., Meri, S., 2003. Molecular genetic characterization of the Fennoscandian cervid strain, a new genotypic group (G10) of *Echinococcus granulosus*. Parasitology 127, 207–215.

Lavikainen, A., Lehtinen, M.J., Laaksonen, S., Agren, E., Oksanen, A., Meri, S., 2006. Molecular characterization of *Echinococcus* isolates of cervid origin from Finland and Sweden. Parasitology 133, 565–570.

Leiby, P.D., Carney, W.P., Woods, C.E., 1970. Studies on sylvatic echinococcosis. III. Host occurrence and geographic distribution of *Echinococcus multilocularis* in the north central United States. J. Parasitol. 56, 1141–1150.

Leiby, P.D., Kritsky, D.C., 1972. *Echinococcus multilocularis*: a possible domestic life cycle in central North America and its public health implications. J. Parasitol. 58, 1213–1215.

Li, T., Chen, X., Zhen, R., Qiu, J., Qiu, D., Xiao, N., Ito, A., Wang, H., Giraudoux, P., Sako, Y., Nakao, M., Craig, P.S., 2010. Widespread co-endemicity of human cystic and alveolar echinococcosis on the eastern Tibetan Plateau, northwest Sichuan/southeast Qinghai, China. Acta Trop. 113, 248–256.

Liccioli, S., Duignan, P.J., Lejeune, M., Deunk, J., Majid, S., Massolo, A., 2013. A new intermediate host for *Echinococcus multilocularis*: the southern red-backed vole (*Myodes gapperi*) in urban landscape in Calgary, Canada. Parasitol. Int. 62, 355–357.

Liccioli, S., Kutz, S.J., Ruckstuhl, K.E., Massolo, A., 2014. Spatial heterogeneity and temporal variations in *Echinococcus multilocularis* infections in wild hosts in a North American urban setting. Int. J. Parasitol. 44, 457–465.

Liccioli, S., Giraudoux, P., Deplazes, P., Massolo, A., 2015a. Wilderness in the 'city' revisited: different urbes shape transmission of *Echinococcus multilocularis* by altering predator and prey communities. Trends Parasitol. 31, 297–305.

Liccioli, S., Bialowas, C., Ruckstuhl, K.E., Massolo, A., 2015b. Feeding ecology informs parasite epidemiology: prey selection modulates encounter rate with *Echinococcus multilocularis* in urban coyotes. PLoS One 10, e0121646.

Lichtenwalner, A., Adhikari, N., Kantar, L., Jenkins, E., Schurer, J., 2014. *Echinococcus granulosus* genotype G8 in Maine moose (*Alces alces*). Alces 50, 27–33.

Lidetul, D., Hutchinson, G.W., 2007. The prevalence, organ distribution and fertility of cystic echinococcosis in feral pigs in tropical North Queensland, Australia. Onderstepoort J. Vet. Res. 74, 73–79.

Lindström, E., 1989. Food limitation and social regulation in a red fox population. Holarct. Ecol. 12, 70–79.

Liu, Q., Cao, L., Zhang, Y., Xu, D., Shang, L., Wang, X., Wei, F., Xiao, L., Ma, R., Cai, J., Zhao, Q., 2013. Genotypes of *Echinococcus granulosus* in animals from Yushu, Northeastern China. Vector Borne Zoonotic Dis. 13, 134–137.

Lukasik, V.M., Alexander, S.M., 2012. Spatial and temporal variation of coyote (*Canis latrans*) diet in Calgary, Alberta. Cities Environ. 4 (8).

Lymbery, A.J., 2016. Phylogenetic pattern, evolutionary processes and species delimitation in the genus *Echinococcus*. In: Thompson, A. (Ed.), Echinococcus and Echinococcosis, Vol. 95, pp. 111–146.

Lymbery, A.J., Jenkins, E.J., Schurer, J.M., Thompson, R.C., 2015a. *Echinococcus canadensis, E. borealis, and E. intermedius*. What's in a name? Trends Parasitol. 31, 23–29.

Lymbery, A.J., Jenkins, E.J., Schurer, J.M., Thompson, R.C., 2015b. Response to Nakao et al. - is *Echinococcus intermedius* a valid species? Trends Parasitol. 31, 343–344.

Ma, J., Wang, H., Lin, G., Craig, P.S., Ito, A., Cai, Z., Zhang, T., Han, X., Ma, X., Zhang, J., Liu, Y., Zhao, Y., Wang, Y., 2012. Molecular identification of *Echinococcus* species from eastern and southern Qinghai, China, based on the mitochondrial cox1 gene. Parasitol. Res. 111, 179–184.

Ma, J., Wang, H., Lin, G., Zhao, F., Li, C., Zhang, T., Ma, X., Zhang, Y., Hou, Z., Cai, H., Liu, P., Wang, Y., 2015. Surveillance of *Echinococcus* isolates from Qinghai, China. Vet. Parasitol. 207, 44–48.

Ma, S.M., Maillard, S., Zhao, H.L., Huang, X., Wang, H., Geng, P.L., Bart, J.M., Piarroux, R., 2008. Assessment of *Echinococcus granulosus* polymorphism in Qinghai province, People's Republic of China. Parasitol. Res. 102, 1201–1206.

Macdonald, D.W., Reynolds, J.C., 2008. Vulpes Vulpes. The IUCN Red List of Threatened Species 2008 eT23062A9412884. http://dx.doi.org/10.2305/IUCN.UK.2008.RLTS.T23062A9412884.en.

Macpherson, C.N.L., 1983. An active intermediate host role for man in the life cycle of *Echinococcus granulosus* in Turkana, Kenya. Am. J. Trop. Med. Hyg. 32, 397–404.

Macpherson, C.N.L., Wachira, T.M., 1997. Cystic echinococcosis in Africa South of the Sahara. In: Andersen, F.L., Ouhelli, H., Kachani, M. (Eds.), Compendium of Cystic Echinococcosis in Africa and in Middle Eastern Countries with Special Reference to Morocco. Brigham Young University, Provo, Utah, pp. 245–277.

Martinek, K., Kolarova, L., Hapl, E., Literak, I., Uhrin, M., 2001. *Echinococcus multilocularis* in European wolves (*Canis lupus*). Parasitol. Res. 87, 838–839.

Masur, O.E., Fomina, A.S., 2012. Revelation of *Echinococcus multilocularis* (Leuckart, 1856) in the muskrat (*Ondatra zibethicus*) of the delta of the Selenga river (Russia). Russ. J. Biol. Invasions 3, 251–254.

Massolo, A., Liccioli, S., Budke, C., Klein, C., 2014. *Echinococcus multilocularis* in North America: the great unknown. Parasite 21, 73.

Matsumoto, J., Kouguchi, H., Oku, Y., Yagi, K., 2010. Primary alveolar echinococcosis: course of larval development and antibody responses in intermediate host rodents with different genetic backgrounds after oral infection with eggs of *Echinococcus multilocularis*. Parasitol. Int. 59, 435–444.

Mayor, P., Baquedano, L.E., Sanchez, E., Aramburu, J., Gomez-Puerta, L.A., Mamani, V.J., Gavidia, C.M., 2015. Polycystic echinococcosis in pacas, Amazon region, Peru. Emerg. Infect. Dis 21. http://dx.doi.org/10.3201/eid2103.141197.

Mbaya, H., Magambo, J., Njenga, S., Zeyhle, E., Mbae, C., Mulinge, E., Wassermann, M., Kern, P., Romig, T., 2014. *Echinococcus* spp. in central Kenya: a different story. Parasitol. Res. 113, 3789–3794.

McManus, D.P., Zhang, L., Castrodale, L.J., Le, T.H., Pearson, M., Blair, D., 2002. Short report: molecular genetic characterization of an unusually severe case of hydatid disease in Alaska caused by the cervid strain of *Echinococcus granulosus*. Am. J. Trop. Med. Hyg. 67, 296–298.

McManus, D.P., 2010. Echinococcosis with particular reference to southeast Asia. Adv. Parasitol. 72, 267–303.

Mech, L.D., 1966. The Wolves of Isle Royale. Fauna of the National Parks of the USA. Fauna Series 7. Government Printing Office, Washington DC, US.

Meléndez, R.D., Yépez, M.S., Coronado, A., 1984. *Echinococcus oligarthrus* cysts of rabbits in Venezuela. J. Parasitol. 70, 1004–1005.

Melotti, J.R., Muzzall, P.M., O'brien, D.J., Cooley, T.M., Tsao, J.I., 2015. Low prevalence of *Echinococcus multilocularis* in Michigan, U.S.A.: a survey of coyotes (*Canis latrans*), red foxes (*Vulpes vulpes*), and gray foxes (*Urocyon cinereoargenteus*), 2009-2012. Comp. Parasitol. 82, 285–290.

Meneghelli, U.G., Martinelli, A.L., Velludo, M.A., 1990. *Echinococcus vogeli* cysts in paca liver (*Cuniculus paca*) native from the Acre State, Brazil. Rev. Soc. Bras. Med. Trop. 23, 153–155.

Miller, A.L., Olsson, G.E., Walburg, M.R., Sollenberg, S., Skarin, M., Ley, C., Wahlstrom, H., Hoglund, J., 2016. First identification of *Echinococcus multilocularis* in rodent intermediate hosts in Sweden. Int. J. Parasitol. Parasites Wildl. 5, 56–63.

Miyauchi, T., Sakui, M., Ishige, M., Fukumoto, S., Ueda, A., Ito, M., Ohibayashi, M., 1984. A case of multilocular echinococcosis in a horse. Jpn. J. Vet. Res. 32, 171–173.

Mobedi, I., Sadighian, A., 1971. *Echinococcus multilocularis* leuckart, 1863, in red foxes, *Vulpes vulpes* Linn., in Moghan, Azerbaijan Province, northwest of Iran. J. Parasitol. 57, 493.

Mobedi, I., Bray, R.A., Arfaa, F., Movafag, K., 1973. A study on the cestodes of carnivores in the northwest of Iran. J. Helminthol. 47, 277–281.

Moks, E., Jogisalu, I., Valdmann, H., Saarma, U., 2008. First report of *Echinococcus granulosus* G8 in Eurasia and a reappraisal of the phylogenetic relationships of 'genotypes' G5-G10. Parasitology 135, 647–654.

Moro, P.L., Nakao, M., Ito, A., Schantz, P.M., Cavero, C., Cabrera, L., 2009. Molecular identification of *Echinococcus* isolates from Peru. Parasitol. Int. 58, 184–186.

Murdoch, J.D., Munkhzul, T., Buyandelger, S., Reading, R.P., Sillero-Zubiri, C., 2010. Seasonal food habits of corsac and red foxes in Mongolia and the potential for competition. Mamm. Biol. 75, 36–44.

Nakao, M., Xiao, N., Okamoto, M., Yanagida, T., Sako, Y., Ito, A., 2009. Geographic pattern of genetic variation in the fox tapeworm *Echinococcus multilocularis*. Parasitol. Int. 58, 384–389.

Nakao, M., Li, T., Han, X., Ma, X., Xiao, N., Qiu, J., Wang, H., Yanagida, T., Mamuti, W., Wen, H., Moro, P.L., Giraudoux, P., Craig, P.S., Ito, A., 2010. Genetic polymorphisms of *Echinococcus* tapeworms in China as determined by mitochondrial and nuclear DNA sequences. Int. J. Parasitol. 40, 379–385.

Nakao, M., Lavikainen, A., Yanagida, T., Ito, A., 2013a. Phylogenetic systematics of the genus *Echinococcus* (Cestoda: Taeniidae). Int. J. Parasitol. 43, 1017–1029.

Nakao, M., Yanagida, T., Konyaev, S., Lavikainen, A., Odnokurtsev, V.A., Zaikov, V.A., Ito, A., 2013b. Mitochondrial phylogeny of the genus *Echinococcus* (Cestoda: Taeniidae) with emphasis on relationships among *Echinococcus canadensis* genotypes. Parasitology 140, 1625–1636.

Nakao, M., Lavikainen, A., Hoberg, E., 2015. Is *Echinococcus intermedius* a valid species? Trends Parasitol. 31, 342–343.

Nonaka, N., Hirokawa, H., Inoue, T., Nakao, R., Ganzorig, S., Kobayashi, F., Inagaki, M., Egoshi, K., Kamiya, M., Oku, Y., 2008. The first instance of a cat excreting *Echinococcus multilocularis* eggs in Japan. Parasitol. Int. 57, 519–520.

Nonaka, N., Kamiya, M., Kobayashi, F., Ganzorig, S., Ando, S., Yagi, K., Iwaki, T., Inoue, T., Oku, Y., 2009. *Echinococcus multilocularis* infection in pet dogs in Japan. Vector Borne Zoonotic Dis. 9, 201–206.

Obwaller, A., Schneider, R., Walochnik, J., Gollackner, B., Deutz, A., Janitschke, K., Aspock, H., Auer, H., 2004. *Echinococcus granulosus* strain differentiation based on sequence heterogeneity in mitochondrial genes of cytochrome c oxidase-1 and NADH dehydrogenase-1. Parasitology 128, 569–575.

O'Grady, J.P., Yeager, C.H., Esra, G.N., Thomas, W., 1982. Ultrasonic evaluation of echinococcosis in four lowland gorillas. J. Am. Vet. Med. Assoc. 181, 1348–1350.

Ohbayashi, M., Rausch, R.L., Fay, F.H., 1971. On the ecology and distribution of *Echinococcus* spp. (Cestoda: Taeniidae), and characteristics of their development in the intermediae host. II. Comparative studies on the development of larval *E. multilocularis* leuckart, 1863, in the intermediate host. Jpn. J. Vet. Res. 19 (Suppl. 3), 1–53.

Ohbayashi, M., 1996. Host animals of *Echinococcus multilocularis* in Hokkaido. In: Uchino, J., Sato, N. (Eds.), Alveolar Echinococcosis of the Liver. Fujishoin, Sapporo, pp. 59–64.

Okamoto, M., Fujita, O., Arikawa, J., Kurosawa, T., Oku, Y., Kamiya, M., 1992. Natural *Echinococcus multilocularis* infection in a Norway rat, *Rattus norvegicus*, in southern Hokkaido. Jpn. Int. J. Parasitol. 22, 681–684.

Oksanen, A., Lavikainen, A., 2015. *Echinococcus canadensis* transmission in the North. Vet. Parasitol. 213, 182–186.

Oksanen, A., Siles-Lucas, M., Karamon, J., Possenti, A., Conraths, F.J., Romig, T., Wysocki, P., Mannoci, A., Mipatrini, D., La Torre, G., Boufana, B., Casulli, A.,

2016. The geographical distribution and prevalence of *Echinococcus multilocularis* in animals in the European Union and adjacent countries: a systematic review and meta-analysis. Parasit. Vectors 9, 519.

Omer, R.A., Dinkel, A., Romig, T., Mackenstedt, U., Elnahas, A.A., Aradaib, I.E., Ahmed, M.E., Elmalik, K.H., Adam, A., 2010. A molecular survey of cystic echinococcosis in Sudan. Vet. Parasitol. 169, 340–346.

Onac, D., Gyorke, A., Oltean, M., Gavrea, R., Cozma, V., 2013. First detection of *Echinococcus granulosus* G1 and G7 in wild boars (*Sus scrofa*) and red deer (*Cervus elaphus*) in Romania using PCR and PCR-RFLP techniques. Vet. Parasitol. 193, 289–291.

Otero-Abad, B., Torgerson, P.R., 2013. A systematic review of the epidemiology of echinococcosis in domestic and wild animals. PLoS Negl. Trop. Dis. 7, e2249.

Pednekar, R.P., Gatne, M.L., Thompson, R.C., Traub, R.J., 2009. Molecular and morphological characterisation of *Echinococcus* from food producing animals in India. Vet. Parasitol. 165, 58–65.

Peklo, G.N., 2014. Echinococcosis and Trichinellosis in the Northwestern Federal District (Inventories of Biohelminths in Russian Federation). Federal Service for Supervision of Protection Consumer Rights and Welfare. Tyumen Research Institute of Boundary Infectious Pathology, Rospotrebnadzor. Tyumen, 2012-2014.

Petavy, A.F., Deblock, S., Walbaum, S., 1990. The house mouse: a potential intermediate host for *Echinococcus multilocularis* in France. Trans. R. Soc. Trop. Med. Hyg. 84, 571–572.

Petavy, A.F., Tenora, F., Deblock, S., Sergent, V., 2000. *Echinococcus multilocularis* in domestic cats in France. A potential risk factor for alveolar hydatid disease contamination in humans. Vet. Parasitol. 87, 151–156.

Peters, M., Kilwinski, J., Wohlsein, P., Conraths, F.J., 2010. Alveolar echinococcosis in a captive red-necked wallaby (*Macropus rufogriseus*). Berl. Münch. Tierärztl. Wschr. 123, 63–69.

Plesker, R., Nadler, T., Dinkel, A., Romig, T., 2009. A case of an *Echinococcus ortleppi* infestation in a red-shanked douc langur (*Pygathrix nemaeus*) in northern Vietnam. Vietnam. J. Primatol. 3, 75–81.

Pliyeva, A., Uspenskii, A., 2006. Epizootiology processes of echinococcosis in carnivores and agricultural animals. Vet. Pathol. 2, 120–122.

Qiu, J., Chen, H., Che, X., Liu, G., Yue, Y., Liu, D., Zhu, G., 1989. Natural alveolaris Echinococcus infection in yaks and sheep in Shiqu county, Sichuan province. Endem. Dis. Bull. 4, 26–29.

Qiu, J.M., Chen, X.W., Ren, M., Lu, C.X., 1995. Epidemiological study on alveolar hydatid disease in Qinghai-Xizang Plateau. J. Pract. Parasit. Dis. 3.

Raoul, F., Deplazes, P., Nonaka, N., Piarroux, R., Vuitton, D.A., Giraudoux, P., 2001. Assessment of the epidemiological status of *Echinococcus multilocularis* in foxes in France using ELISA coprotests on fox faeces collected in the field. Int. J. Parasitol. 31, 1579–1588.

Raoul, F., Deplazes, P., Rieffel, D., Lambert, J.C., Giraudoux, P., 2010. Predator dietary response to prey density variation and consequences for cestode transmission. Oecologia 164, 129–139.

Raoul, F., Hegglin, D., Giraudoux, P., 2015. Trophic ecology, behaviour and host population dynamics in *Echinococcus multilocularis* transmission. Vet. Parasitol. 213, 162–171.

Rausch, R.L., Williamson, F.S., 1959. Studies on the helminth fauna of Alaska. XXXIV. The parasites of wolves, *Canis lupus* L. J. Parasitol. 45, 395–403.

Rausch, R.L., Nelson, G.S., 1963. A review of the genus *Echinococcus* Rudolphi, 1801. Ann. Trop. Med. Parasitol. 57, 127–135.

Rausch, R.L., Richards, S.H., 1971. Observations on parasite-host relationships of *Echinococcus multilocularis* Leuckart, 1863, in North Dakota. Can. J. Zool 49, 1317–1330.

Rausch, R.L., Bernstein, J.J., 1972. *Echinococcus vogeli* spp. n. (Cestoda: Taeniidae) from the bush dog, *Speothos venaticus* (Lund.). Z. Tropenmed. Parasit. 23, 25–34.

Rausch, R.L., Fay, F.H., 1988. Postoncospheral development and cycle of *Taenia polyacantha* leuckart, 1856 (Cestoda: Taeniidae). Second part. Ann. Parasitol. Hum. Comp. 63, 334–348.

Rausch, R.L., Fay, F.H., Williamson, F.S., 1990a. The ecology of *Echinococcus multilocularis* (Cestoda: Taeniidae) on St. Lawrence Island, Alaska. II. Helminth populations in the definitive host. Ann. Parasitol. Hum. Comp. 65, 131–140.

Rausch, R.L., Wilson, J.F., Schantz, P.M., 1990b. A programme to reduce the risk of infection by *Echinococcus multilocularis*: the use of praziquantel to control the cestode in a village in the hyperendemic region of Alaska. Ann. Trop. Med. Parasitol. 84, 239–250.

Rausch, R.L., 1995. Life cycle patterns and geographic distribution of *Echinococcus* species. In: Thompson, R.C.A., Lymbery, A.J. (Eds.), *Echinococcus* and hydatid disease. CAB International, Wallingford, UK, pp. 89–134.

Rausch, R.L., D'Alessandro, A., 1999. Histogenesis in the metacestode of *Echinococcus vogeli* and mechanism of pathogenesis in polycystic hydatid disease. J. Parasitol. 85, 410–418.

Rausch, R.L., D'Alessandro, A., 2002. The epidemiology of echinococcosis caused by *Echinococcus oligarthrus* and *E. Vogeli* in the neotropics. In: Craig, P., Pawlowski, Z. (Eds.), Cestode Zoonoses: Echinococcosis and Cysticercosis: An Emergent and Global Problem. IOS Press, Amsterdam, Netherlands, pp. 107–113.

Rausch, R.L., Fay, F.H., 2002. Epidemiology of alveolar echinococcosis, with reference to St. Lawrence Island, Bering Sea. In: Craig, P., Pawlowski, Z. (Eds.), Cestode Zoononoses: Echinococcosis and Cysticercosis, an Emergent and Global Problem. IOS Press, Amsterdam, pp. 309–325.

Rausch, R.L., 2003. Cystic echinococcosis in the arctic and sub-arctic. Parasitology 127 (Suppl.), S73–S85.

Reperant, L.A., Hegglin, D., Tanner, I., Fischer, C., Deplazes, P., 2009. Rodents as shared indicators for zoonotic parasites of carnivores in urban environments. Parasitology 136, 329–337.

Robardet, E., Giraudoux, P., Caillot, C., Boue, F., Cliquet, F., Augot, D., Barrat, J., 2008. Infection of foxes by *Echinococccus multilocularis* in urban and suburban areas of Nancy, France: influence of feeding habits and environment. Parasite 15, 77–85.

Robardet, E., Giraudoux, P., Caillot, C., Augot, D., Boue, F., Barrat, J., 2011. Fox defecation behaviour in relation to spatial distribution of voles in an urbanised area: an increasing risk of transmission of *Echinococcus multilocularis*? Int. J. Parasitol. 41, 145–154.

Rojo-Vazquez, F.A., Pardo-Lledias, J., Francos-Von Hunefeld, M., Cordero-Sanchez, M., Alamo-Sanz, R., Hernandez-Gonzalez, A., Brunetti, E., Siles-Lucas, M., 2011. Cystic echinococcosis in Spain: current situation and relevance for other endemic areas in Europe. PLoS Negl. Trop. Dis. 5, e893.

Rolls, E.C., 1984. They All Ran Wild: The Animals and Plants that Plague Australia. Pub. Harper Collins, (Australia) Pty Ltd.

Romig, T., Bilger, B., Dinkel, A., Merli, M., Mackenstedt, U., 1999. *Echinococcus multilocularis* in animal hosts: new data from Western Europe. Helminthologia 36, 185–191.

Romig, T., Thoma, D., Weible, A.K., 2006. *Echinococcus multilocularis* - a zoonosis of anthropogenic environments? J. Helminthol. 80, 207–212.

Romig, T., Omer, R.A., Zeyhle, E., Hüttner, M., Dinkel, A., Siefert, L., Elmahdi, I.E., Magambo, J., Ocaido, M., Menezes, C.N., Ahmed, M.E., Mbae, C., Grobusch, M.P., Kern, P., 2011. Echinococcosis in sub-Saharan Africa: emerging complexity. Vet. Parasitol. 181, 43–47.

Romig, T., Ebi, D., Wassermann, M., 2015. Taxonomy and molecular epidemiology of *Echinococcus granulosus* sensu lato. Vet. Parasitol. 213, 76–84.

Saitoh, T., Takahashi, K., 1998. The role of vole populations in prevalence of the parasite (*Echinococcus multilocularis*) in foxes. Res. Popul. Ecol. 40, 97–105.
Sakui, M., Ishige, M., Fukumoto, S., Ueda, A., Ohbayashi, M., 1984. Spontaneous *Echinococcus multilocularis* infection in swine in north-Eastern Hokkaido, Japan. Jpn. J. Parasitol. 33, 291–296.
Salinas-Lopez, N., Jimenez-Guzman, F., Cruz-Reyes, A., 1996. Presence of *Echinococcus oligarthrus* (Diesing, 1863) Luhe, 1910 in *Lynx rufus texensis* allen, 1895 from San Fernando, Tamaulipas state, in North-east Mexico. Int. J. Parasitol. 26, 793–796.
Sanchez, E., Caceres, O., Naquira, C., Miranda, E., Samudio, F., Fernandes, O., 2012. *Echinococcus granulosus* genotypes circulating in alpacas (*Lama pacos*) and pigs (*Sus scrofa*) from an endemic region in Peru. Mem. Inst. Oswaldo Cruz. 107, 275–278.
Santos, G.B., do Carmo Perreira Soares, M., de, F.B.E.M., Rodrigues, A.L., Siqueira, N.G., Gomes-Gouvea, M.S., Alves, M.M., Carneiro, L.A., Malheiros, A.P., Povoa, M.M., Zaha, A., Haag, K.L., 2012. Mitochondrial and nuclear sequence polymorphisms reveal geographic structuring in Amazonian populations of *Echinococcus vogeli* (Cestoda: Taeniidae). Int. J. Parasitol. 42, 1115–1118.
Sarkari, B., Mansouri, M., Khabisi, S.A., Mowlavi, G., 2015. Molecular characterization and seroprevalence of *Echinococcus granulosus* in wild boars (*Sus scrofa*) in south-western Iran. Ann. Parasitol. 61, 269–273.
Sato, C., Kawase, S., Yano, S., Nagano, H., Fujimoto, S., Kobayashi, N., Miyahara, K., Yamada, K., Sato, M., Kobayashi, Y., 2005. Outbreak of larval *Echinococcus multilocularis* infection in Japanese monkey (*Macaca fuscata*) in a zoo, Hokkaido: western blotting patterns in the infected monkeys. J. Vet. Med. Sci. 67, 133–135.
Schaller, G.B., 1998. Wildlife of the Tibetan Steppe. The University of Chicago Press, USA.
Schantz, P.M., Lord, R.D., de Zavaleta, O., 1972. *Echinococcus* in the South American red fox (*Dusicyon culpaeus*) and the European hare (*Lepus europaeus*) in the province of Neuquen, Argentina. Ann. Trop. Med. Parasitol. 66, 479–485.
Schantz, P.M., Colli, C., 1973. *Echinococcus oligarthrus* (Diesing, 1863) from Geoffroy's cat (*Felis geoffroyi* D'Orbigny y Gervais) in temperate South America. J. Parasitol. 59, 1138–1140.
Schantz, P.M., Colli, C., Cruz-Reyes, A., Prezioso, U., 1976. Sylvatic echinococcosis in Argentina. 2. Susceptibility of wild carnivores to *Echinococcus granulosus* (Batsch 1786) and host induced morphological variation. Tropenmed. Parasitol. 27, 70–78.
Schurer, J., Shury, T., Leighton, F., Jenkins, E., 2013. Surveillance for *Echinococcus canadensis* genotypes in Canadian ungulates. Int. J. Parasitol. Parasites Wildl. 2, 97–101.
Schurer, J.M., Gesy, K.M., Elkin, B.T., Jenkins, E.J., 2014. *Echinococcus multilocularis* and *Echinococcus canadensis* in wolves from western Canada. Parasitology 141, 159–163.
Schurer, J.M., Pawlik, M., Huber, A., Elkin, B., Cluff, D., Pongracz, J., Gesy, K., Wagner, B., Dixon, B., Merks, H., Bal, M., Jenkins, E., 2016. Intestinal parasites of wolves (*Canis lupus*) in northern and western Canada. J. Zool. 94, 643–650.
Schwarz, S., Sutor, A., Staubach, C., Mattis, R., Tackmann, K., Conraths, F.J., 2011. Estimated prevalence of *Echinococcus multilocularis* in raccoon dogs *Nyctereutes procyonoides* in northern Brandenburg, Germany. Curr. Zool. 57, 655–661.
Schweiger, A., Ammann, R.W., Candinas, D., Clavien, P.A., Eckert, J., Gottstein, B., Halkic, N., Muellhaupt, B., Prinz, B.M., Reichen, J., Tarr, P.E., Torgerson, P.R., Deplazes, P., 2007. Human alveolar echinococcosis after fox population increase, Switzerland. Emerg. Infect. Dis. 13, 878–882.
Scioscia, N.P., Beldomenico, P.M., Petrigh, R.S., Pierangeli, N., Denegri, G.M., 2013. Epidemiological studies on *Echinococcus* in Pampas fox (*Lycalopex gymnocercus*) and European hare (*Lepus europaeus*) in Buenos Aires province, Argentina. Parasitol. Res. 112, 3607–3613.

Seimenis, A., 2003. Overview of the epidemiological situation on echinococcosis in the Mediterranean region. Acta Trop. 85, 191–195.

Shaikenov, B.S., Torgerson, P.R., 2002. Distribution of *Echinococcus multilocularis* in Kazakhstan. In: Craig, P., Pawlowski, Z. (Eds.), Cestode Zoonoses: Echinococcosis and Cysticercosis: an Emergent and Global Problem. IOS Press, Amsterdam, pp. 299–307.

Shaikenov, B.S., Torgerson, P.R., Usenbayev, A.E., Baitursynov, K.K., Rysmukhambetova, A.T., Abdybekova, A.M., Karamendin, K.O., 2003. The changing epidemiology of echinococcosis in Kazakhstan due to transformation of farming practices. Acta Trop. 85, 287–293.

Shaikenov, B.S., 2006. Distribution and ecology of *Echinococcus multilocularis* in Central Asia. Parasitol. Int. 55 (Suppl.), S213–219.

Sikó Barabási, S.S., Marosfoi, L., Barabási, Z.S., Cozma, V., 2011. Natural alveolar echinococcosis with *Echinococcus multilocularis* in wild rodents. Sci. Parasitol. 12, 11–21.

Simsek, S., Eroksuz, Y., 2009. Occurrence and molecular characterization of *Echinococcus granulosus* in Turkish mouflon (*Ovis gmelinii anatolica*). Acta Trop. 109, 167–169.

Simsek, S., Cevik, A., 2014. First detection and molecular characterization of *Echinococcus equinus* in a mule in Turkey. Acta Parasitol. 59, 773–777.

Simsek, S., Roinioti, E., Eroksuz, H., 2015. First report of *Echinococcus equinus* in a donkey in Turkey. Korean J. Parasitol. 53, 731–735.

Smyth, J.D., 1977. Strain differences in *Echinococcus granulosus*, with special reference to the status of equine hydatidosis in the United Kingdom. Trans. R. Soc. Trop. Med. Hyg. 71, 93–100.

Sobrino, R., Gonzalez, L.M., Vicente, J., Fernandez de Luco, D., Garate, T., Gortazar, C., 2006. *Echinococcus granulosus* (Cestoda, Taeniidae) in the Iberian wolf. Parasitol. Res. 99, 753–756.

Soriano, S.V., Pierangeli, N.B., Pianciola, L., Mazzeo, M., Lazzarini, L.E., Saiz, M.S., Kossman, A.V., Bergagna, H.F., Chartier, K., Basualdo, J.A., 2010. Molecular characterization of *Echinococcus* isolates indicates goats as reservoir for *Echinococcus canadensis* G6 genotype in Neuquen, Patagonia Argentina. Parasitol. Int. 59, 626–628.

Sousa, O.E., Thatcher, V.E., 1969. Observations on the life-cycle of *Echinococcus oligarthrus* (Diesing, 1863) in the Republic of Panama. Ann. Trop. Med. Parasitol. 63, 165–175.

Statham, M.J., Sacks, B.N., Aubry, K.B., Perrine, J.D., Wisely, S.M., 2012. The origin of recently established red fox populations in the United States: translocations or natural range expansions? J. Mammal. 93, 52–65.

Staubach, C., Thulke, H.H., Tackmann, K., Hugh-Jones, M., Conraths, F.J., 2001. Geographic information system-aided analysis of factors associated with the spatial distribution of *Echinococcus multilocularis* infections of foxes. Am. J. Trop. Med. Hyg. 65, 943–948.

Stieger, C., Hegglin, D., Schwarzenbach, G., Mathis, A., Deplazes, P., 2002. Spatial and temporal aspects of urban transmission of *Echinococcus multilocularis*. Parasitology 124, 631–640.

Süld, K., Valdmann, H., Laurimaa, L., Soe, E., Davison, J., Saarma, U., 2014. An invasive vector of zoonotic disease sustained by anthropogenic resources: the raccoon dog in northern Europe. PLoS One 9, e96358.

Sutor, A., Kauhala, K., Ansorge, H., 2010. Diet of the raccoon dog *Nyctereutes procyonoides* — a canid with an opportunistic foraging strategy. Acta Theriol. 55, 165–176.

Sweatman, G.K., Williams, R.J., 1962. Wild animals in New Zealand as hosts of *Echinococcus granulosus* and other taeniid tapeworms. Trans. R. Soc. N.Z. Zool. 2, 221–250.

Sweatman, G.K., Williams, R.J., 1963. Comparative studies on the biology and morphology of *Echinococcus granulosus* from domestic livestock, moose and reindeer. Parasitology 53, 339–390.

Sydler, T., Mathis, A., Deplazes, P., 1998. *Echinococcus multilocularis* lesions in the livers of pigs kept outdoors in Switzerland. Eur. J. Vet. Pathol. 4, 43–46.

Szell, Z., Marucci, G., Pozio, E., Sreter, T., 2013. *Echinococcus multilocularis* and *Trichinella spiralis* in golden jackals (*Canis aureus*) of Hungary. Vet. Parasitol. 197, 393–396.

Tackmann, K., Löschner, U., Mix, H., Staubach, C., Thulke, H.H., Conraths, F.J., 1998. Spatial distribution patterns of *Echinococcus multilocularis* (Leuckart 1863) (Cestoda: Cyclophyllidea: Taeniidae) among red foxes in an endemic focus in Brandenburg, Germany. Epidemiol. Infect. 120, 101–109.

Takahashi, K., Yagi, K., Ishige, M., Hattori, K., 1986. The natural occurrence of *Echinococcus multilocuilaris* in the house mouse, *Mus musculus* in Hokkaido. Jpn. J. Parasitol. 35, 84.

Takahashi, K., Yagi, K., Uraguchi, K., Kondo, N., 1989. Infection of larval *Echinoccus multilocularis* in red-backed vole *Clethrionomys rufocanus bedfordiae* captured around fox dens. Rep. Hokkaido Inst. Pub. Hlth. 39, 5–9.

Takahashi, K., Uraguchi, K., 1994. Natural infection of larval *Echinococcus multilocularis* in two species of *Soricine shrews, Sorex unguiculatus* and *S. caecutiens*. Jpn. J. Parasitol. 43, 415.

Takahashi, K., Nakata, K., 1995. Note on the first occurrence of larval *Echinococcus multilocularis* in *Clethrionomys rex* in Hokkaido. Jpn. J. Helminthol. 69, 265–266.

Takahashi, K., Uraguchi, K., 1996. Ecological factors influencing prevalence of larval E. multilocularis in vole. In: Uchino, J., Sato, N. (Eds.), Alveolar Echinococcosis: Strategy for Eradication of Alveolar Echinococcosis of the Liver. Fujishoin, Sapporo, pp. 75–77.

Takahashi, K., Mori, T., 2001. Host animals and prevalence of *Echinococcus multilocularis* in Hokkaido. Pub. Hlth. Hokkaido 27, 577–587.

Takahashi, T., Uraguchi, K., Kudo, S., 2005. The epidemiological status of *Echinococcus multilocularis* in animals in Hokkaido, Japan. Mammal. Study 30, S101–S105.

Tang, C.T., Quian, Y.C., Kang, Y.M., Cui, G.W., Lu, H.C., Shu, L.M., Wang, Y.H., Tang, L., 2004. Study on the ecological distribution of alveolar *Echinococcus* in Hulunbeier pasture of inner Mongolia, China. Parasitology 128, 187–194.

Tang, C.T., Wang, Y.H., Peng, W.F., Tang, L., Chen, D., 2006. Alveolar *Echinococcus* species from *Vulpes corsac* in Hulunbeier, Inner Mongolia, China, and differential development of the metacestodes in experimental rodents. J. Parasitol. 92, 719–724.

Tang, C., Cui, G., Qian, Y., Kang, Y., Wang, Y., Peng, W., Lu, H., Chen, D., 2007. Studies on the alveolar *Echinococcus* species in northward Daxingan mountains, Inner Mongolia, China. III. *Echinococcus russicensis* sp. nov. Chin. J. Zoonoses 23, 957–963.

Thatcher, V.E., 1972. Neotropical echinococcosis in Columbia. Ann. Trop. Med. Parasitol. 66, 99–105.

Thompson, R.C.A., Smyth, J.D., 1975. Equine hydatidosis: a review of the current status in Great Britain and the results of an epidemiological survey. Vet. Parasitol. 1, 107–127.

Thompson, R.C.A., Kumaratilake, L.M., Eckert, J., 1984. Observations on *Echinococcus granulosus* of cattle origin in Switzerland. Int. J. Parasitol. 14, 283–291.

Thompson, R.C., Lymbery, A.J., Hobbs, R.P., Elliot, A.D., 1988. Hydatid disease in urban areas of Western Australia: an unusual cycle involving western grey kangaroos (*Macropus fuliginosus*), feral pigs and domestic dogs. Aust. Vet. J. 65, 188–190.

Thompson, R.C.A., Lymbery, A.J., Constantine, C.C., 1995. Variation in *Echinococcus*: towards a taxonomic review of the genus. Adv. Parasitol. 35, 145–176.

Thompson, R.C., McManus, D., 2001. Aetiology: parasites and life-cycles. In: Eckert, J., Gemmell, M.A., Meslin, F.X., Pawowski, Z.S. (Eds.), WHO/OIE Manual on Echinococcosis in Humans and Animals: A Public Health Problem of Global Concern. World Health Organisation/World Organisation of Animal Health, Paris, pp. 1–16.

Thompson, R.C., McManus, D.P., 2002. Towards a taxonomic revision of the genus *Echinococcus*. Trends Parasitol. 18, 452–457.

Thompson, R.C., Deplazes, P., Eckert, J., 2003. Observations on the development of *Echinococcus multilocularis* in cats. J. Parasitol. 89, 1086–1088.
Thompson, R.C., Boxell, A.C., Ralston, B.J., Constantine, C.C., Hobbs, R.P., Shury, T., Olson, M.E., 2006. Molecular and morphological characterization of *Echinococcus* in cervids from North America. Parasitology 132, 439–447.
Tigre, W., Deresa, B., Haile, A., Gabriel, S., Victor, B., Pelt, J.V., Devleesschauwer, B., Vercruysse, J., Dorny, P., 2016. Molecular characterization of *Echinococcus granulosus* s.l. cysts from cattle, camels, goats and pigs in Ethiopia. Vet. Parasitol. 215, 17–21.
Tolnai, Z., Széll, Z., Sréter, T., 2013. Environmental determinants of the spatial distribution of *Echinococcus multilocularis* in Hungary. Vet. Parasitol. 198, 292–297.
Torgerson, P.R., Burtisurnov, K.K., Shaikenov, B.S., Rysmukhambetova, A.T., Abdybekova, A.M., Ussenbayev, A.E., 2003. Modelling the transmission dynamics of *Echinococcus granulosus* in sheep and cattle in Kazakhstan. Vet. Parasitol. 114, 143–153.
Torgerson, P.R., Oguljahan, B., Muminov, A.E., Karaeva, R.R., Kuttubaev, O.T., Aminjanov, M., Shaikenov, B., 2006. Present situation of cystic echinococcosis in Central Asia. Parasitol. Int. 55 (Suppl.), S207–S212.
Torgerson, P.R., Ziadinov, I., Aknazarov, D., Nurgaziev, P., Deplazes, P., 2009. Modelling the age variation of larval protoscoleces of *Echinococcus granulosus* in sheep. Int. J. Parasitol. 39, 1031–1035.
Trachsel, D., Deplazes, P., Mathis, A., 2007. Identification of taeniid eggs in the faeces from carnivores based on multiplex PCR using targets in mitochondrial DNA. Parasitology 134, 911–920.
Tsukada, H., Morishima, Y., Nonaka, N., Oku, Y., Kamiya, M., 2000. Preliminary study of the role of red foxes in *Echinococcus multilocularis* transmission in the urban area of Sapporo, Japan. Parasitology 120 (Pt 4), 423–428.
Tsukada, H., Li, W., Duo, H., Guo, Z., Fu, Y., Peng, M., Shen, X., Jing, J., Yuan, A., Ni, M., He, S., Huang, F., Feng, K., Ishikawa, K., Kobayashi, I., Shinohara, A., Nonaka, N., 2014. A comparison of the diet and fine-scale distribution of sympatric Tibetan and red foxes in Qinghai, P. R. China. Wildl. Biol. 20, 356–361.
Umhang, G., Richomme, C., Boucher, J.M., Guedon, G., Boué, F., 2013. Nutrias and muskrats as bioindicators for the presence of *Echinococcus multilocularis* in new endemic areas. Vet. Parasitol. 197, 283–287.
Umhang, G., Richomme, C., Hormaz, V., Boucher, J.M., Boue, F., 2014. Pigs and wild boar in Corsica harbor *Echinococcus canadensis* G6/7 at levels of concern for public health and local economy. Acta Trop. 133, 64–68.
Umhang, G., Forin-Wiart, M.A., Hormaz, V., Caillot, C., Boucher, J.M., Poulle, M.L., Franck, B., 2015. *Echinococcus multilocularis* detection in the intestines and feces of free-ranging domestic cats (*Felis s. catus*) and European wildcats (*Felis s. silvestris*) from northeastern France. Vet. Parasitol 214, 75–79.
Umhang, G., Lahoreau, J., Hormaz, V., Boucher, J.M., Guenon, A., Montange, D., Grenouillet, F., Boue, F., 2016. Surveillance and management of *Echinococcus multilocularis* in a wildlife park. Parasitol. Int. 65, 245–250.
Urach Monteiro, D., de Azevedo, M.I., Weiblen, C., Correia Ribeiro, T., Emmanoulidis, J., Tonin, A.A., de Avila Botton, S., de la Rue, M., 2016. *Echinococcus granulosus* sensu strict, *Echinococcus canadensis* (G7), and *Echinococcus ortleppi* in fertile hydatid cysts isolated from cattle in southern Brazil. Acta Trop. 164, 41–44.
Uraguchi, K., Takahashi, K., 1999. Ecology of the red fox in Hokkaido - a fiftieth-Anniversary Publication. Hokkaido Inst. Pub. Hlth. 39–48.
Utuk, A.E., Simsek, S., 2013. Molecular characterization of the horse isolate of *Echinococcus granulosus* in Turkey. J. Helminthol. 87, 305–308.
Vande Vusse, F.J., Little, D.E., Callaway, R.B., Ballard, N.B., 1978. Incidence and distribution of *Echinococcus multilocularis* in fox from southern Minnesota. In: Program

and Abstracts, 53rd Annual Meeting, American Association of Parasitologists, Chicago, USA, p. 93.

Vaniscotte, A., Raoul, F., Poulle, M.L., Romig, T., Dinkel, A., Takahashi, K., Guislain, M.H., Moss, J., Tiaoying, L., Wang, Q., Qiu, J., Craig, P.S., Giraudoux, P., 2011. Role of dog behaviour and environmental fecal contamination in transmission of *Echinococcus multilocularis* in Tibetan communities. Parasitology 138, 1316–1329.

Van Kesteren, F., Mastin, A., Mytynova, B., Ziadinov, I., Boufana, B., Torgerson, P.R., Rogan, M.T., Craig, P.S., 2013. Dog ownership, dog behaviour and transmission of *Echinococcus* spp. in the Alay Valley, southern Kyrgyzstan. Parasitology 140, 1674–1684.

Varcasia, A., Canu, S., Lightowlers, M.W., Scala, A., Garippa, G., 2006. Molecular characterization of *Echinococcus granulosus* strains in Sardinia. Parasitol. Res. 98, 273–277.

Varcasia, A., Canu, S., Kogkos, A., Pipia, A.P., Scala, A., Garippa, G., Seimenis, A., 2007. Molecular characterization of *Echinococcus granulosus* in sheep and goats of Peloponnesus, Greece. Parasitol. Res. 101, 1135–1139.

Varcasia, A., Garippa, G., Pipia, A.P., Scala, A., Brianti, E., Giannetto, S., Battelli, G., Poglayen, G., Micagni, G., 2008. Cystic echinococcosis in equids in Italy. Parasitol. Res. 102, 815–818.

Veit, P., Bilger, B., Schad, V., Schafer, J., Frank, W., Lucius, R., 1995. Influence of environmental factors on the infectivity of *Echinococcus multilocularis* eggs. Parasitology 110 (Pt 1), 79–86.

Verster, A., 1962. Hydatidosis in the Republic of South Africa. South Afr. J. Sci. 58, 71–74.

Villalobos, N., Gonzalez, L.M., Morales, J., de Aluja, A.S., Jimenez, M.I., Blanco, M.A., Harrison, L.J., Parkhouse, R.M., Garate, T., 2007. Molecular identification of *Echinococcus granulosus* genotypes (G1 and G7) isolated from pigs in Mexico. Vet. Parasitol. 147, 185–189.

Vizcaychipi, K.A., Helou, M., DeMatteo, K., Macchiaroli, M., Cucher, M., Rosenzvit, M., D'Alessandro, A., 2013. Primera identificacion de *Echinococcus vogeli* en una paca en la provincia des Misiones, Argentina. Rev. Argent. Microbiol. 45, 169–173.

Vogel, H., 1957. Über den *Echinococcus multilocularis* Süddeutschlands. I. Das Bandwurmstadium von Stämmen menschlicher und tierischer Herkunft. Z. Tropenmed. Parasitology 8, 404–454.

Vuitton, D.A., Zhou, H., Bresson-Hadni, S., Wang, Q., Piarroux, M., Raoul, F., Giraudoux, P., 2003. Epidemiology of alveolar echinococcosis with particular reference to China and Europe. Parasitology 127 (Suppl.), S87–S107.

Wachira, T.M., Sattran, M., Zeyhle, E., Njenga, M.K., 1993. Intestinal helminths of public health importance in dogs in Nairobi. East Afr. Med. J. 70, 617–619.

Wandeler, P., Funk, S.M., Largiader, C.R., Gloor, S., Breitenmoser, U., 2003. The city-fox phenomenon: genetic consequences of a recent colonization of urban habitat. Mol. Ecol. 12, 647–656.

Wang, W., Wu, Y., Ding, Z., 1989. The occurence of *Echinococcus multilocularis* Leukart, 1863 in fox and wolf in Tacheng district, Xinjiang. Endem. Dis. Bull. 4, 8–12.

Wang, Z., Wang, X., Liu, X., 2008. Echinococcosis in China, a review of the epidemiology of *Echinococcus* spp. EcoHealth 5, 115–126.

Wassermann, M., Aschenborn, O., Aschenborn, J., Mackenstedt, U., Romig, T., 2015. A sylvatic lifecycle of *Echinococcus equinus* in the Etosha National Park, Namibia. Int. J. Parasitol. Parasites Wildl. 4, 97–103.

Wassermann, M., Woldeyes, D., Gerbi, B.M., Ebi, D., Zeyhle, E., Mackenstedt, U., Petros, B., Tilahun, G., Kern, P., Romig, T., 2016. A novel zoonotic genotype related to *Echinococcus granulosus* sensu stricto from southern Ethiopia. Int. J. Parasitol. 46, 663–668.

Weber, J.M., Aubry, S., 1993. Predation by foxes, *Vulpes vulpes*, on the fossorial form of the water vole, *Arvicola terrestris scherman*, in western Switzerland. J. Zool. 229, 553–559.

Williams, R.J., Sweatman, G.K., 1963. On the transmission, biology and morphology of *Echinococcus granulosus equinus*, a new subspecies of hydatid tapeworm in horses in Great Britain. Parasitology 53, 391–407.

Wilson, J.F., Diddams, A.C., Rausch, R.L., 1968. Cystic hydatid disease in Alaska. A review of 101 autochthonous cases of *Echinococcus granulosus* infection. Am. Rev. Respir. Dis. 98, 1–15.

Wobeser, G., 1971. The occurrence of *Echinococcus multilocularis* (Leukart, 1863) in cats near Saskatoon, Saskatchewan. Can. Vet. J. 12, 65–68.

Woolsey, I.D., Jensen, P.M., Deplazes, P., Kapel, C.M., 2015a. Establishment and development of *Echinococcus multilocularis* metacestodes in the common vole (*Microtus arvalis*) after oral inoculation with parasite eggs. Parasitol. Int. 64, 571–575.

Woolsey, I.D., Bune, N.E., Jensen, P.M., Deplazes, P., Kapel, C.M., 2015b. *Echinococcus multilocularis* infection in the field vole (*Microtus agrestis*): an ecological model for studies on transmission dynamics. Parasitol. Res. 114, 1703–1709.

Woolsey, I.D., Jensen, P.M., Deplazes, P., Kapel, C.M.O., 2016. Peroral *Echinococcus multilocularis* egg inoculation in *Myodes glareolus, Mesocricetus auratus* and *Mus musculus* (CD-1 IGS and C57BL/6j). Int. J. Parasitol. Parasites Wildl. 5, 158–163.

Xiao, N., Qiu, J., Nakao, M., Nakaya, K., Yamasaki, H., Sako, Y., Mamuti, W., Schantz, P.M., Craig, P.S., Ito, A., 2003. Short report: identification of *Echinococcus* species from a yak in the Qinghai-Tibet plateau region of China. Am. J. Trop. Med. Hyg. 69, 445–446.

Xiao, N., Li, T.Y., Qiu, J.M., Nakao, M., Chen, X.W., Nakaya, K., Yamasaki, H., Schantz, P.M., Craig, P.S., Ito, A., 2004. The Tibetan hare *Lepus oiostolus*: a novel intermediate host for *Echinococcus multilocularis*. Parasitol. Res. 92, 352–353.

Xiao, N., Qiu, J., Nakao, M., Li, T., Yang, W., Chen, X., Schantz, P.M., Craig, P.S., Ito, A., 2005. *Echinococcus shiquicus* n. sp., a taeniid cestode from Tibetan fox and plateau pika in China. Int. J. Parasitol. 35, 693–701.

Xiao, N., Qiu, J., Nakao, M., Li, T., Yang, W., Chen, X., Schantz, P.M., Craig, P.S., Ito, A., 2006a. *Echinococcus shiquicus*, A new species from the Qinghai-Tibet plateau region of China: discovery and epidemiological implications. Parasitol. Int. 55 (Suppl.), S233–S236.

Xiao, N., Nakao, M., Qiu, J., Budke, C.M., Giraudoux, P., Craig, P.S., Ito, A., 2006b. Dual infection of animal hosts with different *Echinococcus* species in the eastern Qinghai-Tibet plateau region of China. Am. J. Trop. Med. Hyg. 75, 292–294.

Yagi, K., Takahashi, K., Hattori, K., Seki, N., 1988. Natural infection of *Echinococcus multilocularis* in a racoon dog, *Nyctereutes procyonoides albus* in Hokkaido, Japan. Jpn. J. Parasitol. 37, 79.

Yagi, K., Takahashi, K., Ishige, M., Hattori, K., Sawabe, Y., 1984. Epidemiological survey of multilocular echinococcosis in small mammals in Hokkaido, 1983. Rep. Hokkaido Inst. Pub. Hlth. 34, 72–74.

Yagi, K., Takahashi, K., Ishige, M., Hattori, K., Sawabe, Y., 1985. Epidemiological survey of multilocular echinococcosis in small mammals in Hokkaido, 1984. Rep. Hokkaido Inst. Pub. Hlth. 35, 87–90.

Yamamoto, N., Morishima, Y., Kon, M., Yamaguchi, M., Tanno, S., Koyama, M., Maeno, N., Azuma, H., Mizusawa, H., Kimura, H., Sugiyama, H., Arakawa, K., Kawanaka, M., 2006. The first reported case of a dog infected with *Echinococcus multilocularis* in Saitama prefecture, Japan. Jpn. J. Infect. Dis. 59, 351–352.

Yamano, K., Kouguchi, H., Uraguchi, K., Mukai, T., Shibata, C., Yamamoto, H., Takaesu, N., Ito, M., Makino, Y., Takiguchi, M., Yagi, K., 2014. First detection of *Echinococcus multilocularis* infection in two species of nonhuman primates raised in a zoo: a fatal case in *Cercopithecus diana* and a strongly suspected case of spontaneous recovery in *Macaca nigra*. Parasitol. Int. 63, 621–626.

Yanagida, T., Mohammadzadeh, T., Kamhawi, S., Nakao, M., Sadjjadi, S.M., Hijjawi, N., Abdel-Hafez, S.K., Sako, Y., Okamoto, M., Ito, A., 2012. Genetic polymorphisms of *Echinococcus granulosus* sensu stricto in the Middle East. Parasitol. Int. 61, 599–603.

Yang, Y.R., Rosenzvit, M.C., Zhang, L.H., Zhang, J.Z., McManus, D.P., 2005. Molecular study of *Echinococcus* in west-central China. Parasitology 131, 547–555.

Yang, Y.R., Sun, T., Li, Z., Zhang, J., Teng, J., Liu, X., Liu, R., Zhao, R., Jones, M.K., Wang, Y., Wen, H., Feng, X., Zhao, Q., Zhao, Y., Shi, D., Bartholomot, B., Vuitton, D.A., Pleydell, D., Giraudoux, P., Ito, A., Danson, M.F., Boufana, B., Craig, P.S., Williams, G.M., McManus, D.P., 2006. Community surveys and risk factor analysis of human alveolar and cystic echinococcosis in Ningxia Hui Autonomous Region, China. Bull. World Health Organ. 84, 714–721.

Yang, Y.R., McManus, D.P., Huang, Y., Heath, D.D., 2009a. *Echinococcus granulosus* infection and options for control of cystic echinococcosis in Tibetan communities of Western Sichuan province, China. PLoS Negl. Trop. Dis. 3, e426.

Yang, Y.R., Liu, T., Bai, X., Boufana, B., Craig, P.S., Nakao, M., Ito, A., Zhang, J.Z., Giraudoux, P., McManus, D.P., 2009b. Natural infection of the ground squirrel (*Spermophilus* spp.) with *Echinococcus granulosus* in China. PLoS Negl. Trop. Dis. 3, e518.

Yang, Y.R., Clements, A.C.A., Gray, D.J., Atkinson, J.M., Williams, G.M., Barnes, T.S., McManus, D.P., 2012. Impact of anthropogenic and natural environmental changes of *Echinococcus* transmission in Ningxia Hui Autonomous region, the People's Republic of China. Parasit. Vectors 5, 146.

Yimam, A.E., Nonaka, N., Oku, Y., Kamiya, M., 2002. Prevalence and intensity of *Echinococcus multilocularis* in red foxes (*Vulpes vulpes schrencki*) and raccoon dogs (*Nyctereutes procyonoides albus*) in Otaru City, Hokkaido, Japan. Jpn. J. Vet. Res. 49, 287–296.

Young, E., 1975. Echinococcosis (hydatodosis) in wild animals of the Kruger National Park. J. S. Afr. Vet. Assoc. 46, 285–286.

Zeyhle, E., 1976. Untersuchungen über die Entwicklung von *Echinococcous multilocularis* Leuckart, 1863, in verschiedenen Nagetieren nach oraler Verabreichung von Eiern; zugleich ein Beitrag zur Epidemiologie der Echinococcose - natürliche Infektion beim Fuchs und experimentelle bei der Hauskatze (Diploma thesis). University of Hohenheim, Germany.

Zhang, L.H., Chai, J.J., Jiao, W., Osman, Y., McManus, D.P., 1998. Mitochondrial genomic markers confirm the presence of the camel strain (G6 genotype) of *Echinococcus granulosus* in north-western China. Parasitology 116 (Pt 1), 29–33.

Zhang, W., Zhang, Z., Wu, W., Shi, B., Li, J., Zhou, X., Wen, H., McManus, D.P., 2015. Epidemiology and control of echinococcosis in central Asia, with particular reference to the People's Republic of China. Acta Trop. 141, 235–243.

Zhao, Y.M., Tong, S.X., Jing, T., Chong, S.G., Cai, X.P., Jing, Z.Z., Han, J., 2009. Investigation on echinococcosis in animals in Gannan Tibetan Autonomous prefecture. Zhongguo Ji Sheng Chong Xue Yu Ji Sheng Chong Bing Za Zhi 27, 27–30.

Zhou, H.X., Wen, H., Wang, Y.H., Delattre, P., Quere, J.P., Vuitton, D.A., Craig, P.S., Giraudoux, P., 1998. Experimental susceptibility of *Spermophilus erythrogenys* to *Echinococcus multilocularis*. Ann. Trop. Med. Parasitol. 92, 335–337.

Ziadinov, I., Mathis, A., Trachsel, D., Rysmukhambetova, A., Abdyjaparov, T.A., Kuttubaev, O.T., Deplazes, P., Torgerson, P.R., 2008. Canine echinococcosis in Kyrgyzstan: using prevalence data adjusted for measurement error to develop transmission dynamics models. Int. J. Parasitol. 38, 1179–1190.

Ziadinov, I., Deplazes, P., Mathis, A., Mutunova, B., Abdykerimov, K., Nurgaziev, R., Torgerson, P.R., 2010. Frequency distribution of *Echinococcus multilocularis* and other helminths of foxes in Kyrgyzstan. Vet. Parasitol. 171, 286–292.

CHAPTER SIX

Global Distribution of Alveolar and Cystic Echinococcosis

P. Deplazes*,[1], L. Rinaldi[§], C.A. Alvarez Rojas[¶], P.R. Torgerson*, M.F. Harandi[||], T. Romig[#], D. Antolova, J.M. Schurer[§§,¶¶], S. Lahmar[||||], G. Cringoli[§], J. Magambo[##], R.C.A. Thompson***, E.J. Jenkins[§§]**

*University of Zürich, Zurich, Switzerland
[§]University of Naples Federico II, Napoli, Italy
[¶]The University of Melbourne, Parkville, VIC, Australia
[||]Research centre of Hydatid Disease in Iran, Kerman University of Medical Sciences, Kerman, Iran
[#]University of Hohenheim, Stuttgart, Germany
**Institute of Parasitology SAS, Kosice, Slovak Republic
[§§]University of Saskatchewan, Saskatoon, SK, Canada
[¶¶]University of Washington, Seattle, WA, United States
[||||]National School of Veterinary Medicine, Sidi Thabet, Tunisia
[##]Meru University of Science and Technology, Meru, Kenya
***Murdoch University, Murdoch, WA, Australia
[1]Corresponding author: E-mail: deplazesp@access.uzh.ch

Contents

Advances in Parasitology, Volume 95
ISSN 0065-308X
http://dx.doi.org/10.1016/bs.apar.2016.11.001

Abstract

Alveolar echinococcosis (AE) and cystic echinococcosis (CE) are severe helminthic zoonoses. *Echinococcus multilocularis* (causative agent of AE) is widely distributed in the northern hemisphere where it is typically maintained in a wild animal cycle including canids as definitive hosts and rodents as intermediate hosts. The species *Echinococcus granulosus, Echinococcus ortleppi, Echinococcus canadensis* and *Echinococcus intermedius* are the causative agents of CE with a worldwide distribution and a highly variable human disease burden in the different endemic areas depending upon human behavioural risk factors, the diversity and ecology of animal host assemblages and the genetic diversity within *Echinococcus* species which differ in their zoonotic potential and pathogenicity. Both AE and CE are regarded as neglected zoonoses, with a higher overall burden of disease for CE due to its global distribution and high regional prevalence, but a higher pathogenicity and case fatality rate for AE, especially in Asia. Over the past two decades, numerous studies have addressed the epidemiology and distribution of these *Echinococcus* species worldwide, resulting in better-defined boundaries of the endemic areas. This chapter presents the global distribution of *Echinococcus* species and human AE and CE in maps and summarizes the global data on host assemblages, transmission, prevalence in animal definitive hosts, incidence in people and molecular epidemiology.

1. GENERAL INTRODUCTION

Alveolar echinococcosis (AE) and cystic echinococcosis (CE) are zoonotic diseases caused by *Echinococcus* spp. transmitted from carnivores. The history of these two distinct diseases has been reviewed in chapter 'Historical Aspects of Echinococcosis' by Eckert and Thompson (2017). In this chapter, the causative agents of human CE (a complex of several species with additional genotypes) are referred to using the well-recognized genotype terminology (G1−G10), although a more formal taxonomic nomenclature has now been proposed by Thompson (2017) in chapter 'Biology and Systematics of *Echinococcus*'. Human AE is caused by geographically distinct strains/genotypes of *Echinococcus multilocularis*.

In this chapter, we will focus on the distribution of CE in humans and animal intermediate hosts, as well as intestinal *Echinococcus granulosus* infections in definitive hosts. For *E. multilocularis*, the chapter focuses mainly on the distribution in canid definitive hosts (predominantly foxes outside North America and parts of Asia) and on AE in humans.

Since the last publications focusing on the global distribution of *Echinococcus* spp. (Schantz et al., 1995; Eckert et al., 2001) a considerable amount of evidence is now available documenting in much more detail the changing distribution of both AE and CE. This chapter attempts to give a general overview on the current distribution and epidemiology of both diseases but refers to some key reviews for more detailed information and citations. In addition, the authors present in this review many sources of information which are not easily available (e.g., the so-called grey literature) or had to be translated from original languages (e.g. Russian).

Finally, for more detailed information on the ecology of *Echinococcus* spp., including the different cycles and host ranges involved [see chapter: Ecology and Life Cycle Patterns of *Echinococcus* Species by Romig et al. (2017) and chapter: Echinococcosis: Control and Prevention by Craig et al. (2017)].

The burden of human echinococcosis can be expressed in terms of disability adjusted life years (DALYs). The global burden of disease of AE is estimated to be 18,200 cases per annum, resulting in approximately 666,000 DALYs (37 DALYs per case) (Torgerson et al., 2010). However, 91% of cases and 95% of the DALYs were estimated to be in China. Thus, there are approximately 1600 cases of AE per annum in Europe, Russia and central Asia resulting in 33,000 DALYs or 21 DALYs per case. Survival analyses of French and Swiss AE patients have shown that modern treatments such as resection of liver lesions followed by prolonged therapy with benzimidazoles can result in survival of AE patients similar to those of healthy populations (Torgerson et al., 2008; Piarroux et al., 2011). Where treatment options are available, the burden in terms of DALYs is modest because of the improved prognosis; for example, in Switzerland, there is a total burden of approximately 78 DALYs per annum due to AE, or 3.7 DALYs per case, 10 times less than the global estimate (Torgerson et al., 2008). This is one important factor for the predominance of the global burden in China, where the majority of cases were estimated to occur in resource poor communities on the Tibetan plateau, thus inflating the DALYs per case. In central Asia this is likely to be similar (Torgerson, 2017).

The latest estimate for the global burden of CE is 188,000 new cases per annum resulting in 184,000 DALYs (0.98 DALYs per case (Torgerson et al., 2015). The much lower human health burden of CE compared to AE is entirely due to the low mortality rate of CE relative to AE. However, it is important to note that as the lifecycle of CE in many countries involves livestock intermediate hosts, there can be economic and animal health repercussions beyond that of AE.

A comprehensive review of peer-reviewed literature was undertaken to record the current distribution at global, continental, regional, national, provincial, departmental (or any jurisdiction), administrative levels regarding *E. multilocularis* in wild canids (primarily foxes) and *Echinococcus* spp. causing CE in humans and intermediate hosts.

Relevant data on the prevalence/incidence of *E. multilocularis* and *E. granulosus* were identified through a combination of (1) structured searches of electronic bibliographic databases, (2) additional searches of the 'grey' literature, including unpublished studies and government and international reports (e.g., WHO, OIE, etc.) and (3) direct contact with researchers and health managers.

The online databases PubMed and SCOPUS were used to identify relevant studies for *E. multilocularis* and *E. granulosus s.l.* with the keywords "*Echinococcus multilocularis, Echinococcus granulosus*, alveolar echinococcosis, cystic echinococcosis and current and former country names." No restrictions were placed on publication dates and language. Researchers were contacted when additional information was required or when data needed to be disaggregated.

Studies were included if they provided (1) the number of people/animals surveyed, (2) the number of positive cases, (3) details about the methodology of diagnosis and (4) details of the geographical site where they were conducted, including the administrative level. Studies reporting only prevalence/incidence data without provision of the denominator were also included as these can be used to delineate the limits of transmission of AE or CE.

A repository of data in excel was created to collect all the information and a geographical information system was developed using Arc-GIS 10.2.2 software (ESRI, Redlands, CA, United States) to produce maps.

The data and maps produced for this chapter offer the potential to improve the spatiotemporal targeting of control measures and to enhance the cost-effectiveness of integrated disease control programmes for *E. multilocularis* and *E. granulosus* in definitive and intermediate hosts. However, we should take into account the limits and bias of these maps due to the fact that the real prevalence, incidence and burden of CE and AE are difficult to estimate. This is due to the patchy distribution of CE and AE within transmission areas, the high proportion of asymptomatic infected individuals and symptomatic patients living in resource-poor areas with logistical and/or economic constraints, who never reach medical attention, and the unknown effect of underreporting of diagnosed cases. This last aspect is primarily due to the widespread lack of mandatory notification of echinococcosis overall (reviewed in Rossi et al., 2016).

2. GLOBAL DISTRIBUTION OF *ECHINOCOCCUS MULTILOCULARIS*

2.1 General information

2.1.1 Global distribution

Although generally considered a parasite of the northern hemisphere, criteria for designation as an endemic region for *E. multilocularis* cannot be consistently defined largely due to marked global differences in surveillance effort and detection capacity, as well as variability in host assemblages and prevalence.

The most widely used systems for detection of genetic variability of *E. multilocularis* are the microsatellite EmsB and mitochondrial gene sequences. Due to its high discriminative power, EmsB has been primarily used for the study of local diversity in Europe (Knapp et al., 2008, 2009). Our present understanding of the global genetic structure of the parasite has been initially established using concatenated sequences of the mitochondrial *cob*, *nad2* and *cox1* genes (Nakao et al., 2009; Gesy et al., 2013), as well as by using *cox1* sequences alone (Konyaev et al., 2012a, 2013). Initially, haplotypes of mt DNA sequences were thought to cluster in correlation with geographical regions, forming 'European', 'Asian', 'Mongolian' and 'North American' clades (Nakao et al., 2009). More recent data, still limited by inadequate geographical and numerical sampling coverage, demonstrate that variants of the 'European' clade are widespread at least in western Canada (Gesy et al., 2013; Massolo et al., 2014; Gesy and Jenkins, 2015), 'Asian' haplotypes are reported in western Russia and St. Lawrence Island, Alaska, and isolates clustering most closely with the North American (Alaskan) N1 genotype were found in arctic parts of Siberia (Konyaev et al., 2013), Svalbard, Norway (Knapp et al., 2012) and St. Lawrence Island, Alaska (Nakao et al., 2009) (Table 1). The diversity of *E. multilocularis* in the circumpolar North is recently reviewed in Davidson et al. (2016).

The ecosystems where *E. multilocularis* occurs are as diverse as arctic tundra, high altitude grassland, but also agricultural and urban landscapes [characterized by Romig et al. (2017) in chapter: Ecology and Life Cycle Patterns of *Echinococcus* Species].

Occurrence in definitive hosts (foxes and other canids) was used to define endemicity for areas of Europe and North America. From Turkey, Central Asia and large parts of Russia and China, human data on AE were considered the most reliable source of information. Infections in rodent intermediate hosts were considered as an indication of endemicity

Table 1 Confirmed occurrence of *Echinococcus multilocularis* haplotype assemblages in animal and human hosts

E. multilocularis haplotypes	Confirmed presence	References
Asian haplotypes	Japan (Hokkaido); China (Sichuan); Mongolia, Kazakhstan; Russia (European part, southern Siberia); United States (Alaska, St. Lawrence Island)	Nakao et al. (2009), Ito et al. (2010), and Konyaev et al. (2013)
Mongolian haplotype	China (Inner Mongolia); Mongolia; Russia (Southern Siberia)	Nakao et al. (2009), Ito et al. (2010), and Konyaev et al. (2013)
North American (N1) haplotype	United States (Alaska, St. Lawrence Island); Norway (Svalbard); Russia (Northern Siberia)	Nakao et al. (2009) and Konyaev et al. (2013) Knapp et al. (2012)
North American (N2) haplotype	United States (Indiana, South Dakota, Minnesota)	Nakao et al. (2009) Yamasaki et al. (2008)
	Canada (Saskatchewan)	Gesy and Jenkins (2015)
European haplotypes	Europe (Austria, Belgium, France, Germany, Western Russia, Slovakia); Canada (British Columbia, Alberta, Saskatchewan)	Nakao et al. (2009), Konyaev et al. (2013), Gesy et al. (2013), Massolo et al. (2014), and Gesy and Jenkins (2015)

in a few cases. Fig. 1 depicts the global endemic area of *E. multilocularis* based on data mainly from wild canid definitive hosts.

2.2 *Echinococcus multilocularis* and alveolar echinococcosis in Europe

There are multiple, related mitochondrial haplotypes of *E. multilocularis* in Europe (Nakao et al., 2009), and microsatellite analyses have shown even more significant biogeographic genetic variability (Knapp et al., 2009).

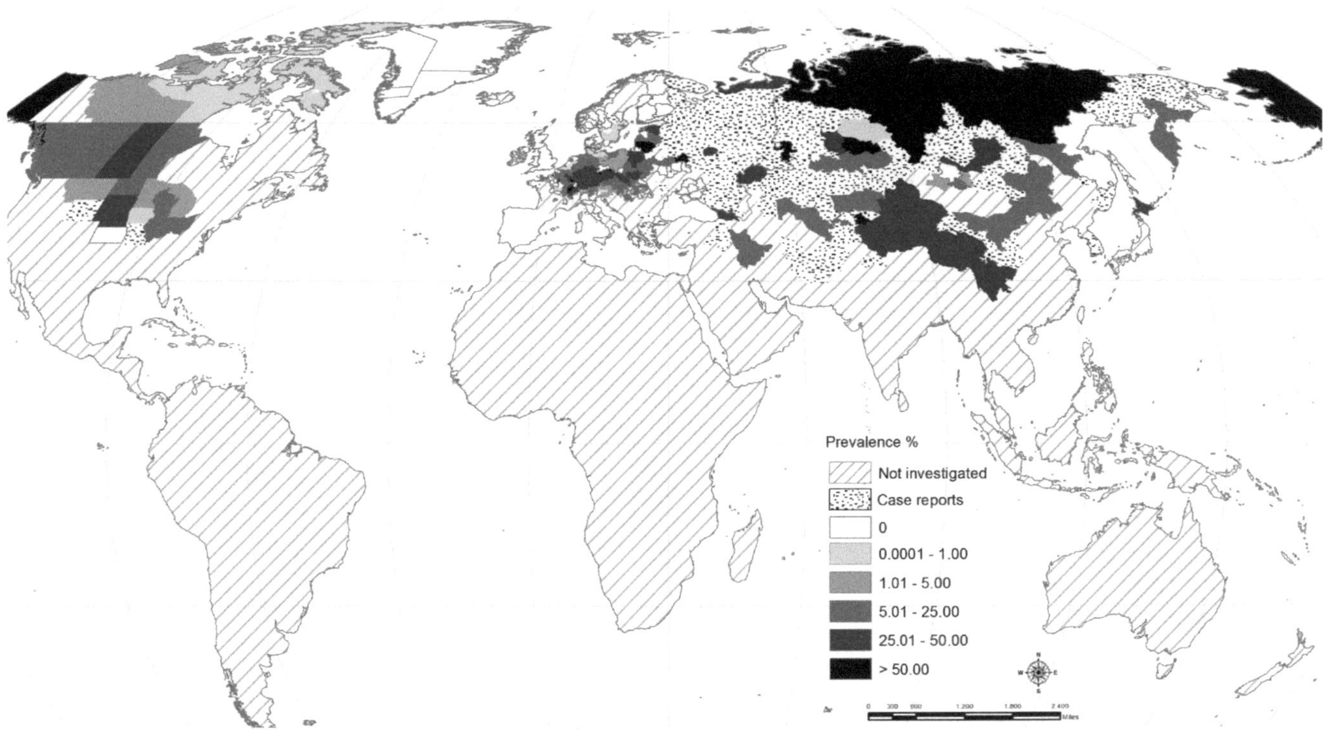

Figure 1 ***Current global distribution and prevalences of Echinococcus multilocularis in the main wild canid definitive hosts.*** In addition, in areas where data of wild canids are missing, case reports of AE in intermediate hosts or in humans as well as infections in other definitive hosts are given as *dotted areas*. For published details of the distribution/prevalence in the different countries/jurisdictions, see detailed maps (Figs. 2–4), Supplementary Material (Tables S1–S3) in Appendix A and text. Note that a positive finding in any host, study or jurisdiction renders the whole jurisdiction endemic.

Convincing evidence for the emergence of AE in Europe has been collected over the last 10 years (Gottstein et al., 2015). An expansion of the Central European endemic area has been observed to the north, west and to the east (see Fig. 2). Data generated in the last 20 years document that the Central European endemic area is connected to old known endemic areas in Eastern Europe and Asia. An emergence of AE has been documented in the Baltic region (Marcinkutė et al., 2015) with high prevalences in foxes and increasing numbers of human AE cases. The changes in parasite transmission have been associated with a drastic increase of fox populations in both rural and urban areas in Europe (Deplazes et al., 2004; Hegglin et al., 2015). Due to the long incubation period, increased incidences of AE can be expected in some known endemic areas and new cases can be expected in areas only recently discovered to be endemic. On the other hand, the epidemiological situation on the southern border of the Central European

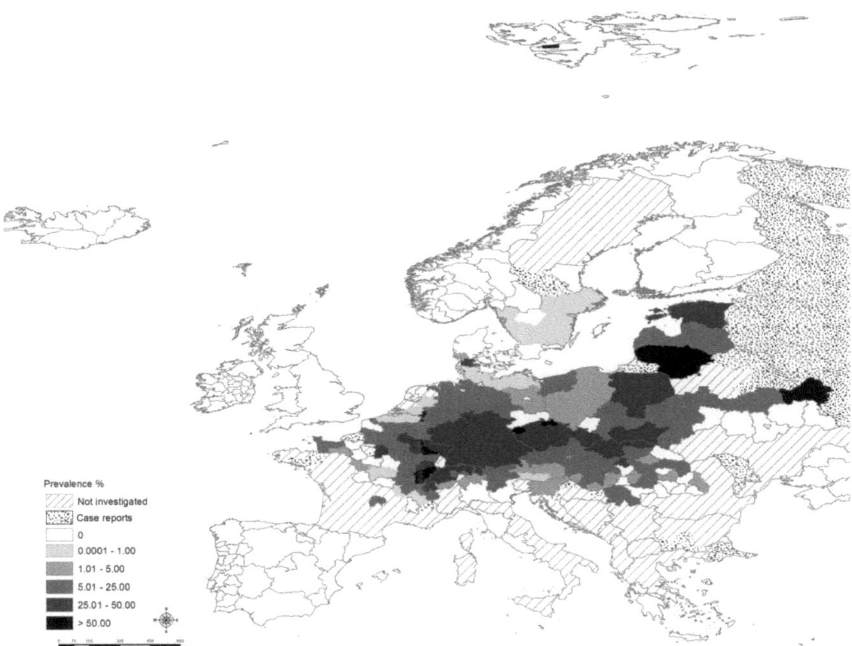

Figure 2 ***Current distribution of Echinococcus multilocularis in foxes (*****Vulpes vulpes, Vulpes lagopus*****) in Europe.*** Areas with case reports of AE in intermediate hosts or humans are approximately given as *dotted areas* if data of definitive hosts were missing. The detailed information (prevalence data in each jurisdiction) is listed in Table S1 of the Supplementary Material.

endemic area (Rausch, 1967) seems to be more stable but the endemicity has not been investigated intensively in all southeastern regions. South Europe is regarded as free of *E. multilocularis* but, in most of the areas, studies focusing on *E. multilocularis* have not been performed. For the situation in the European part of the Russian Federation, see Section 3.3.1.

2.2.1 Central Europe: France, Belgium, The Netherlands, Luxembourg, Germany, Switzerland, Czech Republic, Austria, Northern Italy

2.2.1.1 Transmission and host assemblages

Transmission of *E. multilocularis* in the 'old' endemic area of Central Europe is mainly based on the red fox as definitive host and voles (*Microtus arvalis* and *Arvicola* spp.) as intermediate hosts (Raoul et al., 2015), see chapter: Ecology and Life Cycle Patterns of *Echinococcus* Species by Romig et al. (2017).

There is evidence in many areas of Central Europe not only of a range expansion but also of an increase of the parasite prevalence in foxes (e.g., in Germany and The Netherlands) (Romig, 2002; van der Giessen, 2005). Simultaneous rapid increases of fox population densities with associated prevalence of *E. multilocularis* resulted in an estimated 10-fold increase of the parasite density in southwestern Germany (Romig et al., 2006). Red fox populations have increased throughout Europe in the last decades in all habitats, but especially in urban areas (Deplazes et al., 2004; Hegglin et al., 2015). Transmission of *E. multilocularis* is determined by host ecology (such as predator–prey preferences, defecating behaviour, activity patterns of definitive hosts and host dietary preferences). Key intermediate hosts have been identified for only central study areas in Europe and on Svalbard Island (Raoul et al., 2015); for most of the other areas there are insufficient data concerning rodent population composition and predation by definitive hosts. Therefore, it is difficult to understand the transmission patterns or predict changes in parasite abundance or distribution (such as range expansion). In the last few decades, an expansion of the endemic area was confirmed in the western, northern and eastern parts of Central Europe (Fig. 2); however, it is possible that the parasite was present at low prevalences in the past and is only newly detected in these areas. In some endemic areas, significant increases in prevalence have been demonstrated.

Besides foxes and voles as the key species involved in the cycle, the raccoon dogs (*Nyctereutes procyonoides*) as wild definitive hosts and muskrats (*Ondatra zibethicus*) as intermediate hosts may be of local significance, but their role in maintaining the cycle has not been well studied [see chapter:

Ecology and Life Cycle Patterns of *Echinococcus* Species by Romig et al. (2017)] (Bruzinskaite-Schmidhalter et al., 2012). Domestic dogs with patent infections have been found in several areas and may play an important role for parasite transmission to humans; however, most probably, they are not significant for maintaining the cycle in Central Europe (Deplazes et al., 2011). Cats, with low worm burdens and correspondingly low egg excretion (Kapel et al., 2006), are probably of negligible zoonotic significance in the maintenance of the *E. multilocularis* life cycle (Deplazes, 2015).

The ecology of *E. multilocularis* in rodents has been investigated in depth in a few focal areas in Europe; however, systematic investigations over large areas have not been done. For reviews, see [Raoul et al., 2015; chapter: Ecology and Life Cycle Patterns of *Echinococcus* Species by Romig et al. (2017)].

2.2.1.1.1 Infections in animals In large parts of Central Europe, *E. multilocularis* is present in red fox populations with average prevalences ranging between 0.1% and 50% (Fig. 2), for prevalences see Table S1 in Supplementary Material. A considerable amount of data has been accumulated in the last decade especially for foxes in Germany. Prevalence data for many countries can be retrieved from the EFSA Scientific Opinion on *E. multilocularis* infections in animals (EFSA, 2015b). However, temporal and spatial prevalence variations can be highly significant, and prevalence estimates are furthermore dependant on the population structure, the sampling method, the sample size and the sensitivity of the diagnostic methods. Therefore, differences in prevalences should not be over interpreted. Continuous monitoring of fox populations for *E. multilocularis* is not considered to be justified for the endemic area in Central Europe (Conraths and Deplazes, 2015).

As shown in Fig. 2, lower prevalences are generally found on the border of the endemic area. This might reflect low fox densities (such as in alpine areas) or low population densities of key intermediate hosts, but in most areas, it is not known if the recent findings represent a stable epidemiological situation with ecological factors limiting parasite transmission, and/or range expansion. In one study area in the Southern Alps of Switzerland, a stable epidemiological situation over more than 20 years was described, with the *E. multilocularis* endemic area determined by the southern border of the *M. arvalis* distribution (Guerra et al., 2014). Reports of autochthonous human AE in two patients between 1906 and 1922 from South Tyrol (Hosemann et al., 1928) suggest that recent reports of *E. multilocularis* in foxes in this region is a new detection and not a recent range expansion from the north (Manfredi et al., 2002; Casulli et al., 2005).

In **France**, between 2005 and 2010, a large-scale survey of foxes was conducted in 42 departments (representing almost all of northeastern France). The prevalences varied widely among departments, from 0% to 54% (mean prevalence 17%) (Combes et al., 2012).

Some eastern regions of the **Netherlands** are highly endemic, with prevalence in foxes reaching close to 60% in Limburg (Maas et al., 2014), while no records of infected foxes are known from the western parts of the country (EFSA, 2015b).

High prevalence in foxes is reported from **Luxembourg** and the southern parts of **Belgium**, while prevalence approaches 0% towards the north of **Belgium** (EFSA, 2015b).

In **Switzerland**, a clear gradient is apparent between the well-known high endemic areas in the northern and western cantons where prevalence estimates range from 44% to 53% and low endemic areas in the central (alpine) areas as well as the southern and southwestern cantons (prevalence range 2–12%) (EFSA, 2015b).

For **Austria**, older data show a clear division with high prevalence in the west (Vorarlberg, Tyrol) and low prevalence in the remaining country. Temporal changes are unclear, because surveys in different parts of Austria were done in different time periods (EFSA, 2015b).

In **Italy**, positive records in foxes are limited to the provinces of Trento and Bolzano (South Tyrol), while surveys further south and west (including Lombardy and Aosta) had negative outcomes (EFSA, 2015b).

In **Germany**, investigations of fox infection started in the late 1970s, initially centering on the southern parts of the country which were known to be endemic due to the occurrence of human AE cases (Zeyhle et al., 1990). Up to the present, data on the infection status of close to 100,000 foxes were gathered from all federal states of Germany (except for the cities of Bremen and Hamburg), which show a clear temporal and spatial structuring. Prevalence before 1995 was generally lower than after, and prevalence estimates in the elevated and mountainous regions of southern and central Germany were in both periods higher than in northern and eastern Germany. Studies on fox infection after 1995 resulted in prevalence estimates of 40.4–55.5% (Bavaria), 37.0% (Baden-Württemberg), 2.4% (Brandenburg), 35.9–40.5% (Hesse), 14.4 (Lower Saxony), 16.1–36.8% (North Rhine-Westphalia), 33.8% (Rhineland-Palatinate), 1.4–19.0% (Saxony-Anhalt), 0.0% (Schleswig-Holstein) and 30.9% (Thuringia).

Only older data with negative results are available from Berlin and Saxony [Table S1 and EFSA (2015b)]. *Echinococcus multilocularis* prevalence in urban fox populations is generally lower than in surrounding rural landscapes; examples are Stuttgart [17.3% (Deplazes et al., 2004)] and Munich [20.2% (König and Romig, 2010)]. Direct comparison of prevalence estimates between individual studies is often difficult due to the application of diagnostic methods of different test parameters. Studies, demonstrating a temporal prevalence increase after 1990, are available for the states of Baden-Württemberg (Romig, 2002), Lower Saxony (Berke et al., 2008) and Thuringia (Staubach et al., 2011). Apart from foxes, *E. multilocularis* worms and metacestodes were recorded from raccoon dogs; domestic dogs; domestic and wild cats; several species of voles, muskrats, coypu and various dead-end hosts such as wild boar or zoo animals [lit. in (EFSA, 2015b) and see chapter: Ecology and Life Cycle Patterns of *Echinococcus* Species by Romig et al. (2017)].

In the **Czech Republic** the prevalence of *E. multilocularis* in foxes ranges between 14% and 62%, with an average of 33% (Pijáček, 2011).

Based on recently improved diagnostic strategies, patent *E. multilocularis* infections have been diagnosed in dogs and cats in Switzerland, Czech Republic, Germany (Hegglin and Deplazes, 2013) and France (Umhang et al., 2014b). Prevalence of <1.5% was recorded in privately owned rural and urban pet dogs, but higher prevalences of 3–8% were found in dogs with free access to rodent habitats such as farm dogs and hunting dogs (Hegglin and Deplazes, 2013). Based on the prevalence of 0.3% in a representative dog population, the probability for an individual dog becoming infected at least once during 10 years was estimated at 8.7%. Large investigations of faecal samples sent to a private laboratory revealed *E. multilocularis* eggs in 0.13% of dogs from northern Germany and 0.35% in dogs from southern Germany (Dyachenko et al., 2008). Given that rural dogs with higher infection risk are less likely to receive veterinary care, this figure is probably an underestimate. Considering the total dog population (approximately 5.4×10^6 dogs) in Germany, approximately 13,000 are expected to be infected.

The prevalence of *E. multilocularis* in domestic or wild cats in Europe (determined by necropsy) ranged between 0% and 5.5% (Deplazes, 2015). Cat infections are characterized by low worm burdens and strongly reduced worm development, resulting in lower egg production as compared to foxes or dogs. Prevalences for patent *E. multilocularis* infections of 0.4–5% in cats were determined by PCR in several European countries (Dyachenko et al., 2008; Deplazes et al., 2011; van Asch et al., 2013). Therefore, the role of the cat for maintaining the cycle, or for zoonotic risk, is of low significance (Deplazes, 2015).

Pigs are susceptible to *E. multilocularis* oncosphere invasion but subsequent metacestode development is restricted resulting in dead calcified lesions without protoscoleces (Deplazes et al., 2005). Infections in pigs are therefore of no transmission significance but may represent a marker for the environmental contamination with eggs of *E. multilocularis*, especially pigs fed with grass or animals with outdoor access. In Switzerland, 10% of pigs kept outdoors are presented with liver lesions caused by *E. multilocularis* (Sydler et al., 1998). In Germany, contamination of forage stored in the open or straw litter was estimated to be the most likely source of infection in four farms (Bottcher et al., 2013).

Of veterinary importance is AE in dogs and a variety of captive nonhuman primates. In Central Europe, severe and often lethal liver infections in dogs have been regularly diagnosed since 1988 (Corsini et al., 2015; Deplazes and Eckert, 2001). Such infections can easily be diagnosed by necropsy and histology (as for humans) (Deplazes and Eckert, 2001), and therefore this may indicate an increased exposure of the dog population to this parasite. Based on the relatively large dog population and high prevalences of adult cestode infection in foxes, the susceptibility of dogs for AE is estimated as very low. Similar to dogs, lethal AE has been diagnosed in captive primates (Deplazes and Eckert, 2001), including lemurs (Umhang et al., 2013a), macaques (*Macaca* spp.) (Tappe et al., 2007), western lowland gorillas (Wenker and Hoby, 2011) and chimpanzees (*Pan troglodytes*) (Federer et al., 2016), white-handed gibbons (*Hylobates lar*) and squirrel monkeys (*Saimiri sciureus*). Based on the small numbers of primates exposed and the numerous cases described, the susceptibility of monkeys including apes to AE exacerbated by the infection pressure in captivity is considered to be very high.

2.2.1.1.2 Alveolar echinococcosis in humans Central Europe has been a known core endemic area for AE since the end of the 19th century [for historical review see chapter: Historical Aspects of Echinococcosis by Eckert and Thompson (2017)]. Historically, human AE cases distinct from CE were reported first from the Jura and Alps. Between 1982 and 2000, 599 AE cases (42% in France, 24% in Germany, 21% in Switzerland, 13% in other countries) were registered in the 'EurEchinoReg' project in Central Europe (Kern et al., 2003). For a comprehensive review of the AE situation in Europe, see (Vuitton et al., 2015). In most European countries, human AE and infections in animals are notifiable (not the case for human AE in Denmark, Switzerland and the Netherlands) (Vuitton et al., 2015). For Germany, however, it was estimated that the national surveillance system failed to detect 67% of AE cases over 3 years (Jorgensen et al., 2008).

Most recently there is clear evidence of increasing incidences of AE in **Switzerland**, parts of France, Germany and Austria (Gottstein et al., 2015). Yearly, around 150–200 new AE cases are expected to be diagnosed in Central Europe, with the regional annual incidences (AE cases/10^5 inhabitants) varying between 0 and 8.1 (Vuitton et al., 2015). In Switzerland the overall yearly AE incidence has increased significantly from 0.1 (1993–2000) to 0.26 (2001–05) (Schweiger et al., 2007); thereafter, the incidences persisted at the higher level until 2012. Most of the cases are found in highly populated areas around the largest cities of the country and not in the lower populated mountain areas (Deplazes P., unpublished data). In **Austria**, a pronounced increase in cases was recorded: 2.8 new cases per year between 2000 and 2010 and 13 in 2013. The annual incidence in Vorarlberg with 372,001 residents was 0.08 cases during 1991–2000 which increased to 0.32 cases during 2001–10. In Tyrol (710,048 residents), annual incidences varied from 0.17 (1991–2000) to 0.07 (2001–10) to 0.56 per/10^5 inhabitants in 2011. Only sporadic cases are recorded in the other seven Austrian provinces (Schneider et al., 2011).

In **Germany**, notification of human AE became compulsory in 2001, and the number of AE cases per year (based on data from the Robert-Koch-Institut, 'SurvS tat@RKI 2.0, https://survstat.rki.de') have been increasing steadily from 21 (in 2001) to 47 in 2015. In contrast, the number of CE cases and cases of undetermined echinococcoses has remained rather stable over the same period. The data reported through this system are considered to lead to a substantial underestimate of the true situation (Jorgensen et al., 2008).

In **France**, geographic clustering of AE cases has been well documented based on data from a population-based registry. High-risk areas included the Massif Central and the northeastern regions of the country, where most cases either resided in villages or towns or had farming or gardening activities (Piarroux et al., 2013). A recent study in France based on AE cases between 1982 and 2007 confirmed that the risk of AE was associated with mountain climates/landscapes (Piarroux et al., 2015).

In the **Czech Republic**, the examination of 1892 patients between 1998 and 2014 revealed 20 AE cases, the first two diagnosed in 2007 (Kolářová et al., 2015).

For Switzerland, the annual burden of AE was estimated to be 78 DALYs, or approximately 3.7 DALYs per case, with a median cost of €109,000 per case (Torgerson et al., 2008). Survival analysis on databases of French and Swiss AE patients has shown that advances in diagnosis, staging and state of the art treatments [chapter: *Echinococcus*—Host Interactions at Cellular and Molecular Levels by Brehm and Koziol (2017)] can result in survival of AE patients nearly comparable to the unaffected general population (Torgerson et al., 2008; Piarroux et al., 2011).

2.2.2 Western and Northern Europe: Iceland, Ireland, United Kingdom, Norway, Sweden, Finland and Denmark

2.2.2.1 Transmission and host assemblages

E. multilocularis transmission has been known to be occurring since 1997 in Denmark (DK), in Sweden (SE) since 2011, and on the Svalbard Archipelago of Norway since 1999 (Wahlstrom et al., 2015). Main definitive hosts for *E. multilocularis* in DK and SE are red foxes (possibly raccoon dogs in Jutland, DK). No *E. multilocularis* lesions have been documented in small mammals in the Great Copenhagen area (0/719) (Wahlstrom et al., 2015). Recently, in southern Sweden, for the first time, *E. multilocularis* was detected in voles (*Arvicola amphibius* and *Microtus agrestis*) (Miller et al., 2016).

Transmission in Svalbard is restricted to a limited area of a few square kilometres. The parasite cycles between the only definitive host species, the indigenous arctic fox (*Vulpes lagopus*) and the introduced sibling vole (*Microtus levis*) as intermediate host, in which prevalence can be up to 100% in overwintered males (Henttonen et al., 2001). The *E. multilocularis* haplotype in this area was distinct from the European haplotypes (Table 1). Furthermore, the low polymorphism identified among 27 metacestode isolates supports the hypothesis that *E. multilocularis* was introduced to Svalbard with Arctic foxes and that introduction of the intermediate hosts enabled the local life cycle to become established (Knapp et al., 2012).

2.2.2.2 Infections in animals

The few data published in this area have been comprehensively reviewed (Wahlstrom et al., 2015). In **Denmark**, *E. multilocularis* was detected for the first time with very low worm burdens (<50 worms) in 3/340 foxes (0.9%) from the Greater Copenhagen area during investigations starting in 1997 (Kapel and Saeed, 2000). In 2011, the parasite was confirmed in Southern Jutland in 13/41 foxes (32%, worm burdens up to 1527) and in two raccoon dogs and in 2014 in 4/97 (4.1%) foxes around 100 km north of the first finding in Grindsted (Wahlstrom et al., 2015).

In **Sweden**, a national screening program including more than 3800 foxes was initiated in 2000. Prevalences between 0.1% and 0.9% were detected since 2011 (Fig. 2), with worm burdens up to 1235 (Wahlstrom et al., 2015). Furthermore, south of the first findings, *E. multilocularis* was found in Smaland (Miller et al., 2016). In a recent study of 1566 rodents from four regions in Sweden, *E. multilocularis* metacestodes were detected in the liver of 1 of 187 *M. agrestis* (0.5%) and 8 of 439 (1.8%) *A. amphibius*, but none of 655 *Myodes glareolus* or 285 *Apodemus* spp. were infected (Miller et al., 2016). In both infected species, protoscoleces were present, indicating their competence as intermediate hosts. There was a high local prevalence (6/9 rodents) in an endemic focus (a single field) despite the overall low national prevalence in foxes of <0.1%. In **Norway**, the parasite was not detected in 3405 red foxes investigated between 2000 and 2014. Similarly, 2759 red foxes and 2353 raccoon dogs investigated in **Finland** between 2000 and 2014 were free of *Echinococcus* (Wahlstrom et al., 2015).

So far no data support the endemicity of *E. multilocularis* in **Ireland**, the **United Kingdom** or **Iceland**. In the United Kingdom, the parasite has been found in a captive beaver (Barlow et al., 2011) and a macaque monkey (Boufana et al., 2012), both imported from southern Germany. Presently it is compulsory to treat all dogs and cats that are imported into the UK (https://www.gov.uk/take-pet-abroad/tapeworm-treatment-dogs) or Ireland with praziquantel to prevent the establishment of the parasite (Torgerson and Craig, 2009).

2.2.2.3 Alveolar echinococcosis in humans

As of the end of 2015, no putative autochthonous AE cases have been described in Western and Northern Europe; however, single imported cases in humans were reported from Denmark, United Kingdom and Sweden (Wahlstrom et al., 2015). Interestingly, AE cases have been reported in Russian districts (e.g., Murmansk) and neighbouring eastern Finland (see Section 2.3.1).

2.2.3 Eastern Central Europe: Poland and Baltic countries, Belarus, Ukraine, Moldova, Slovakia, Hungary

2.2.3.1 Transmission and host assemblages

First descriptions of *E. multilocularis* in **Belarus, Ukraine** and **Moldova** go back more than 50 years, with the parasite reported in rodents in Belarus

(Shimalov, 2011), red foxes in Ukraine in 1957 and rodents from Moldova in 1961 (Bessonov, 2002). In **Poland**, *E. multilocularis* was detected for the first time in red foxes in 1994 (Malczewski et al., 1999).

In the countries of the Baltic region (i.e., **Lithuania**), the first report of *E. multilocularis* in animals was described in 2003 in one of the five investigated muskrats (Marcinkutė et al., 2015). Transmission of *E. multilocularis* has so far not been well documented but a *Microtus* sp. was found to be infected in this area (Sarkunas M., personal communication). In definitive hosts, no infections were recorded in 164 foxes and 10 raccoon dogs in 1964 or in 122 foxes and 58 raccoon dogs in 1976 [see (Marcinkutė et al., 2015)]. The parasite was detected for the first time in foxes in 2001 (Bruzinskaite-Schmidhalter et al., 2012). Interestingly, the first case of human AE dates back to 1997 in Lithuania before the first detection in animals. In contrast, in **Latvia** and **Estonia**, *E. multilocularis* was first detected in foxes in 2003 (Marcinkutė et al., 2015).

Detection of *E. multilocularis* in red foxes in the **Czech Republic** and **Slovakia** in the 1990s confirmed that the geographic distribution of the parasite was much wider than that had been previously thought (Dubinsky et al., 1999; Martinek et al., 2001). Later, prevalences up to 27.6% were recorded in red foxes in **Hungary** (Tolnai et al., 2013). Data regarding occurrence of the parasite in intermediate hosts in these three countries are scarce; larval stages were detected in the bank vole (*M. glareolus*) in the Czech Republic and in muskrat (*O. zibethicus*) in Slovakia (Martínek et al., 1998; Miterpáková et al., 2006).

2.2.3.2 Infections in animals

Initial investigations in **Poland** of 2951 red foxes between 1993 and 1998 revealed an overall prevalence of the tapeworm of 2.6%, with the highest regional prevalence (11.8%) recorded in northeastern parts of the country (Malczewski et al., 1999). The results of recent investigations including more than 1500 red foxes from 15 of 16 Polish counties revealed a mean prevalence of 16.5%, ranging from 2% in western and southwestern Poland to 50% in eastern and southern parts of the country (Karamon et al., 2014b). Adult worms were also detected in 2 of the 148 dogs examined (1.4%) (Karamon et al., 2016) and larval stages in 10 (0.8%) of the 1250 pigs with liver lesions (Karamon et al., 2014a).

In **Lithuania**, average countrywide *E. multilocularis* prevalence was 59% in 269 foxes, with worm burdens up to 20,924; the prevalence was 53% in the suburban area of Kaunas (Bruzinskaite-Schmidhalter et al., 2012). In the same study, 8% of 85 raccoon dogs were infected. In two other studies, 2 (0.8%) of 240 and 4 (1.1%) of 360 dogs excreted *E. multilocularis* eggs, further calcified *E. multilocularis* lesions were found in 3 (0.5%) of 685 pig livers (Marcinkutė et al., 2015).

Prevalence of *E. multilocularis* was 31.5% in a recent survey of 108 red foxes in **Estonia**, while only 1.6% of 249 raccoon dogs from the same study area were infected (Laurimaa et al., 2016a,b, 2015). In **Latvia**, *E. multilocularis* was described in foxes, wolves and raccoon dogs (Marcinkutė et al., 2015; Bagrade et al., 2008) and recently in 73 (17%) of 430 foxes and 17 (5.6%) of 305 raccoon dogs, therefore documenting a high endemicity in this country (Bagrade et al., 2014).

In **Belarus**, descriptions of *E. multilocularis* in a variety of rodents go back to the 1960s (Shimalov, 2011). Infected red foxes were found between 2001 and 2003 in the central and the southern part of the country [cited in (Marcinkutė et al., 2015)].

In the **Ukraine**, a well-documented historically endemic area, new data are available from Kharchenko et al. (2008). These authors examined a total of 164 red foxes: 52 originating from the south steppe regions, 93 from the North Polissya regions and 19 from western Ukraine. Only animals from western parts of the Ukraine, namely Volyn and L'viv regions, were infected with *E. multilocularis*.

The distribution of *E. multilocularis* in foxes is well documented in **Slovakia**. According to surveys between 2000 and 2010, the parasite is widespread, and average prevalences in foxes reached 30.3%. The highest prevalences were detected in northern Slovakia and the mountainous areas of central Slovakia, where the prevalence ranged between 30% and 60% (Miterpáková and Dubinský, 2011).

In **Hungary**, surveys conducted in 2008 and 2009 revealed an average prevalence of 11% in 840 red foxes. Positive animals originated from 16/19 Hungarian counties and also from suburban areas of the capital Budapest (Casulli et al., 2010a). In the period of 2012–13, the prevalence decreased to 7.9% in 772 red foxes, with mean intensity of infection of 243 worms/animal (Tolnai et al., 2013).

2.2.3.3 Alveolar echinococcosis in humans

In **Poland**, a total of 121 cases of AE have been confirmed between 1990 and 2011, with a steady increase in annual incidence. While a total of 55 patients were diagnosed in the years 2005–09, 23 new cases were reported in 2010–11 alone. The highest number of patients came from the

Warmińsko-Mazurskie province (Nahorski et al., 2013), where the highest prevalence in red foxes was also documented (Karamon et al., 2014b).

Human AE has been diagnosed in patients from the whole Baltic region. In **Lithuania**, there were 178 patients registered in the period of 1997–2014, and the incidence of the disease (per 10^5 inhabitants) increased from 0.03 in 2004 to 0.57 in 2009 and to 0.74 in 2012 (Marcinkutė et al., 2015). In **Latvia**, 29 cases of AE have been reported between 1996 and 2010 (Tulin et al., 2012). To date, only 13 patients with echinococcosis have been registered in **Estonia**, but the *Echinococcus* species involved was not identified (Marcinkutė et al., 2015); however, according to EFSA (2015a) three AE cases were registered in 2013.

Data on human echinococcosis in **Belarus** are scarce; only two patients with AE were reported, one from Brest and the second from the Mogilev region (Marcinkutė et al., 2015). The real situation with regard to human AE in **Ukraine** and **Moldova** is unknown; no data have been published in the literature so far.

The first two human cases of AE in **Slovakia** were recorded in 2000. In 2004, another two patients were reported. Since then, every year a few new cases of AE have been detected; and to date at least 50 patients with confirmed diagnoses have been documented (Antolova et al., 2014; Antolova D. personal communication).

Although a human case of AE had been reported in **Hungary** in the past (Horvath et al., 2008), the first autochthonous case of the disease was confirmed only recently (Dezsényi et al., 2016).

2.2.4 Southeastern Central Europe: Slovenia, Croatia, Serbia, Romania, Bulgaria, other Balkan countries and European part of Turkey

2.2.4.1 Transmission and host assemblages

Information about the occurrence of *E. multilocularis* in Southeastern Europe is fragmented and needs further investigation. In **Slovenia**, *E. multilocularis* was recently detected in red foxes (Vergles Rataj et al., 2010) and in northern Serbia in red foxes and golden jackals (Lalošević et al., 2016). The absence of *E. multilocularis* from neighbouring regions is likely an artefact for lack of investigations, and systematic studies are particularly required from Croatia, central and southern Serbia, southern Romania and Bulgaria. Indeed, preliminary molecular investigations confirmed positive foxes in northern **Croatia** (Beck R. personal communication). In **Romania**, *E. multilocularis* was for the first time detected in rodents between 1991 and

1995 (Sikó et al., 1995); subsequently, positive red foxes were found in the northwestern part of the country (Siko et al., 2011). In neighbouring **Bulgaria**, the metacestode stages were described in *Chionomys nivalis* and *Myodes glarelous* in the 1980s (Genov et al., 1980; Kolářová, 1999). From the European part of **Turkey** (Thrace) an old report of an infected fox with *E. multilocularis* was convincingly documented (figure shows a mature worm with genital pore anterior to the middle) (Merdivenci, 1963) (for the Asian part of Turkey see Section 2.3.3). Furthermore, a few cases of AE have been proposed from **Greece**, but the diagnostic criteria or the origin of the patient have not been well documented. *Echinococcus multilocularis* has not been described so far in **Bosnia-Herzegovina**, **FYROM** (former Yugoslav Republic of Macedonia), **Kosovo** and **Montenegro**.

Based on the fragmented and often anecdotal nature of the information, the epidemiological situation for *E. multilocularis* in southeastern Central Europe remains largely undetermined.

2.2.4.2 Distribution in animals

In **Slovenia**, several papers suggested that *E. multilocularis* was present in mostly atypical intermediate hosts (cattle, piglets) without convincing evidence [cited in Brglez and Kryštufek (1984)]. For example, histological findings in an unusual rodent intermediate host (*Apodemus flavicollis*) of an agglomeration of 2 mm vesicles filled with reddish (rose) fluid in the liver, mesentery, and other abdominal organs, in the absence of protoscoleces, is not sufficient to document an *E. multilocularis* infection. However, *E. multilocularis* was for the first time found in foxes hunted during 2002–05 in all regions of the country, with an overall prevalence of 2.6% (Vergles Rataj et al., 2010).

In a recent survey in the Vojvodina province of northern **Serbia**, 17.9% of 112 foxes and 14.3% of 28 golden jackals were found infected (Lalošević et al., 2016). An infected beaver that had been found earlier in central Serbia was apparently introduced from southern Germany (Cirovic et al., 2012).

In **Romania**, *E. multilocularis* metacestodes were identified by morphological criteria and the presence of protoscoleces in 0.45% of 442 *C. nivalis*, in 0.44% of 120 *M. arvalis*, in 0.51% of 1172 *Arvicola terrestris* and in 1.67% of 120 *M. glareolus* (Sikó et al., 1995). Between 2007 and 2010, *E. multilocularis* was identified in 27 (4.8%) of 561 red foxes from 15 Transylvanian counties (southeastern Romania), with the highest prevalence (10.5–14.6%) in the counties bordering Hungary and Ukraine (Siko et al., 2011, Fig. 2).

2.2.4.3 Alveolar echinococcosis in humans

In **Slovenia**, a serological study including 1263 patients suspected of having echinococcosis, revealed 9 seropositive cases with a species-specific Western blot; based on these data, a mean annual incidence for seropositive persons of 0.09 cases/10^5 inhabitants was calculated (Logar et al., 2007).

According to Siko et al. (2011), two cases of human AE have been diagnosed in patients from **Romania**, from the northwestern and the central parts of the country. Although Genov et al. (1980) mentioned the finding of human AE in **Bulgaria**, convincing evidence is still lacking (Kolářová, 1999). In **Greece**, sporadic cases have been published earlier and one confirmed case was registered in the European Echinococcosis Register (1996–2000) (Kern et al., 2003), but travel histories are not well documented.

2.3 *Echinococcus multilocularis* and alveolar echinococcosis in Asia

Echinococcus multilocularis is widespread across northern Asia with the endemic region covering most of Russia, central Asia and western China (Torgerson et al., 2010). Most of the burden of AE is in this region with over 90% of the global burden occurring in China alone (Fig. 3, Table S2 in Supplementary Material). In some rural communities, AE prevalences may be over 5% and represent the highest burden of any disease (Budke et al., 2004). As the disease is fatal in the absence of treatment, these represent extraordinarily high incidences of disease which are not reported in official data. Further north, in some districts of Siberia large numbers of human AE cases have been reported since the 1950s. In recent years there is increasing evidence that AE is becoming an increasing public health problem in central Asia, particularly in Kyrgyzstan where hundreds of cases are reported annually (Usubalieva et al., 2013; Raimkylov et al., 2015; Abdybekova et al., 2015).

The parasite is also actively transmitting on the northern Japanese island of Hokkaido. There are only sporadic reports of cases from northern India or Pakistan which appears to be the southern limit of the parasite distribution.

2.3.1 North Asia and the Russian Federation

2.3.1.1 Transmission and host assemblages

Most parts of the **Russian Federation** are considered endemic for *E. multilocularis* (Fig. 3, Table S2 in Supplementary Material). Based on data of human AE (Bessonov, 1998), parts of southern Siberia (e.g., Altai, Omsk and Tomsk), central Yakutia and parts of the Far East are regarded as highly

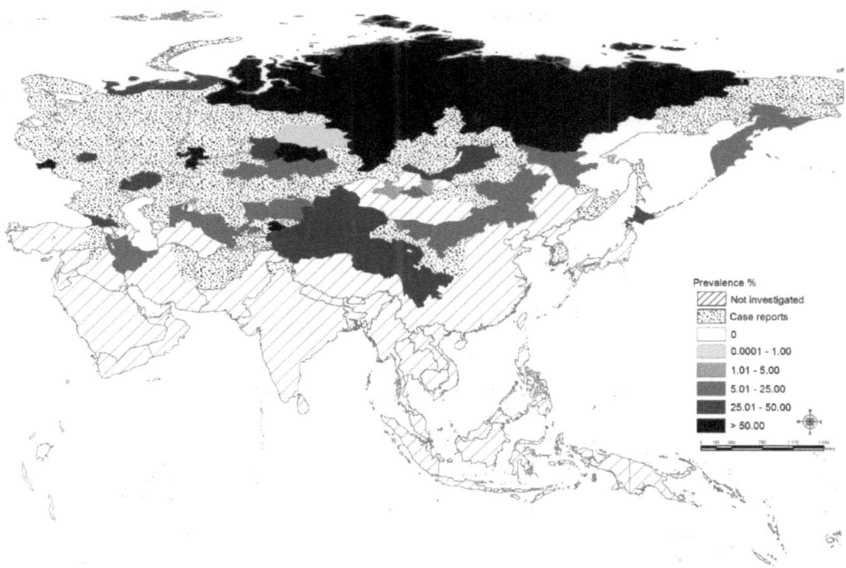

Figure 3 ***Current distribution of Echinococcus multilocularis in foxes (*Vulpes vulpes; Vulpes ferrilata, Vulpes corsac, Vulpes lagopus*) in Asia.*** Areas with case reports of AE in definitive and intermediate hosts or humans are approximately given as *dotted areas* if data of foxes were missing. The detailed information (prevalence data in each jurisdiction) is listed in Table S2 of the Supplementary Material.

endemic areas. In the temperate zones of Russia, red foxes represent the main definitive hosts, although the wolf, raccoon dog and the domestic dog may regionally contribute to the life cycle of *E. multilocularis* (Bessonov, 1998; Konyaev et al., 2012a). In red foxes, high prevalences are reported from Omsk and Novosibirsk areas and from the Tyva steppe bordering Mongolia (Fig. 3). Various voles, hamsters, squirrels (Konyaev et al., 2012a), pikas and bobak marmots have been identified as potential intermediate hosts [see chapter: Ecology and Life Cycle Patterns of *Echinococcus* Species by Romig et al. (2017)] but the ecological significance of these species is not well documented.

In the Russian Federation (see Table 1), haplotypes of four assemblages of *E. multilocularis* have been confirmed: a European haplotype from a captive monkey (Moscow Zoo), an Asian haplotype in western Siberia and European Russia, the Mongolian haplotype on an island of Baikal Lake and in the Altai Republic and the North American (N1) haplotype in Yakutia (Konyaev et al., 2013).

2.3.1.2 Infections in animals

Echinococcus multilocularis has been reported in foxes (*Vulpes vulpes*) in a number of studies in the **Russian Federation** covering a wide geographical area. In Kamchatka a prevalence of 15% was documented (Tranbenkova, 1992), and the prevalences recorded from the Far East were 22.6% (19 of 84 foxes) in the Amur region (Yudin, 2012) and 15–40% in the Buryatia region (Mazur and Fomina, 2012). In the Altay region (southern Siberia), 20 of 91 foxes were infected (Pomamarev and Kostykov, 2012). In Krasnoyarsk, 47 of 54 (Abuladze, 1964) and in Novosibirsk 89 of 162 foxes were infected (Lukashenko and Zorikhina, 1961). In European Russia, prevalences were 33% in foxes from Ryazan (SE of Moscow) (Andreyanov, 2011), 17% from Nenets (Arctic coast of European Russia) (Sorochenko, 1972) and 26% from the Ingushetia region bordering the Caucasus (Pliyeva and Uspenskii, 2006). In Bryansk, one of the westernmost provinces bordering the Ukraine and Belarus, 26 of 51 foxes were reported to be infected by Machulski in 1949, cited in Abuladze (1964).

High prevalences in arctic foxes (*V. lagopus*) have been described in northern Russia, approaching 100% in Yakutia (Kokolova, 2007). A prevalence of 64% in Nenets has been reported (Sorochenko, 1972) and a prevalence of 40% (20 of 50 foxes) was reported in Krasnoyarsk by Mamedov in 1960 (Abuladze, 1964).

Infections in wolves have been reported in Kamchatka (Tranbenkova, 1992) and the Amur region (Yudin, 2012), raccoon dogs in Ryazan (Andreyanov, 2011) and *Lynx lynx* from the Altay region (Pomamarev et al., 2011). Intermediate hosts reported in Russia include muskrat (*O. zibethicus*), suslik (*Spermophilus* spp.), rats (*Rattus norvegicus*), *Apodemus uralensis* and *M. arvalis* from the Caucasus (Pliyeva and Uspenskii, 2006; Kabardiev et al., 2014); lemmings (*Lemmus sibiricus*), house mice (*Mus musculus)* and *Myodes* spp. from Siberia (Kokolova, 2007); and *M. arvalis* and *Apodemus* in south European Russia (Kirillova and Kirillov, 2008).

2.3.1.3 Alveolar echinococcosis in humans

Obtaining accurate data on the number of human cases of AE in the **Russian Federation** is difficult. Data from the sanitary epidemiological service indicate approximately 500–600 cases of echinococcosis per annum but does not report AE and CE separately (Federal Centre of Hygiene and Epidemiology, 2006–10). However, other sources indicate that, between 2007 and 2012, 224 of these echinococcosis cases were AE – 37 per annum (Anonymous, 2013). In contrast, a systematic review of published and unpublished Russian literature and public health reports estimates over 1000 cases per year (Torgerson et al., 2010). This includes a report of 3274 cases of echinococcosis (diagnostic criteria not indicated) in Russia in 2002, representing a threefold increase since 1995 (Bishnevski et al., 2007), which is consistent with the incidence of echinococcosis reported in Russian immigrants residing in Germany (Torgerson, 2017). Several large echinococcosis case studies in Moscow and other cities indicated that the ratio of CE to AE cases was approximately 2:1 [for example Abdullaev et al. (2006) and Lysenko et al. (2006)]. Contemporary and older reports indicate large numbers of cases of AE with a wide geographical distribution, including 147 cases in Novosibirsk (Kabardiev et al., 2014); 52 cases between 2000 and 2010 in Bashkortostan (Pantieleiev et al., 2011); 350 cases between 1961 and 1998 in Kirov in the Volga basin in European Russia (Dhuravliev et al., 2000); 17 cases between 2003 and 2010 in a clinic in Perm, between Moscow and the Urals (Kotelnikova et al., 2011); 197 cases between 1969 and 2013 in one hospital in Tomsk (Zaytsev, 2013); and 38 cases between 2007 and 2012 in Krasnoyarsk district (Rukopisi and Zaitsiev, 2015). In Krasnodar, which borders the Black Sea, both AE and CE are being reported in significant numbers (Anonymous, 2013).

Recent government reports (available at http://rospotrebnadzor.ru) also indicate cases (numbers not reported) in the Belogorod, Bryansk, Chelyabinsk, Mariy-El, Mordovia, Murmansk, Nizhegorod, Omsk, Samara, Saratov, Smolensk, Sverdlovsk, Udmurt, Ulyanovsk and Yaroslav districts. Interestingly, Murmansk district has a border with both Norway and Finland where human AE has not yet been reported.

Furthermore, there is older literature from the time of Soviet administration, including a study in Kamchatka describing 350 cases over an unspecified time period (Nabokov and Vasiliev, 1978), 130 cases from Novosirbisk in the 1960s (Bregadze and Kogan, 1962), and 996 cases between 1950 and 1968 in Sakha (Yakutia) giving an annual incidence of 9.2 cases per 10^5 (Isakov, 1982).

2.3.2 Caucasus and Central Asia: Kazakhstan, Kyrgyzstan, Uzbekistan, Armenia, Azerbaijan and Georgia

2.3.2.1 Transmission and host assemblages

Echinococcus multilocularis has long been known throughout the former Soviet Union. In Central Asia, human AE was thought to be rare and sporadic with few case reports and little surveillance data. However, the first case in

Kyrgyzstan was reported in 1948 (Kuttubaev et al., 2004). Reports of animals date from at least 1957 (Gagarin et al., 1957).

Human AE, along with CE, appears to be emerging in central Asia. After the dissolution of the Soviet Union in 1991, increase in populations of both stray and owned dogs, poverty, and opportunities for scavenging of rodents have been linked to high prevalences of *E. multilocularis* in dogs (Van Kesteren et al., 2013; Ziadinov et al., 2008). This may translate to increased risk of zoonotic transmission due to the close contact between dogs and humans (Torgerson, 2013). Dog owners may have higher odds of being affected by AE than non-dog owners (Torgerson P., personal communication).

2.3.2.2 Infections in animals

There is very limited information on *E. multilocularis* in animals in the three republics in the Caucasus. In 1989, three of 13 wolves examined in the Sheki-Zaqatal region of northern **Azerbaijan** were infected with *E. multilocularis* (Elchuev, 1989). In **Georgia** (Kurashvili, 1966) a prevalence of 30% of *E. multilocularis* was reported in foxes.

In **Kyrgyzstan**, AE was first reported in rodents (Gagarin et al., 1957); subsequently, the parasite was described in *Microtus* spp. and *O. zibethicus* (reviewed by Abdyjaparov and Kuttubaev, 2004). Recent studies have revealed *E. multilocularis* in 20% of dogs in communities with opportunities to scavenge or hunt rodents (Ziadinov et al., 2008). Red foxes, normally considered the natural definitive hosts of this parasite, have been reported to have a prevalence of infection of over 65% in the Naryn region of Kyrgyzstan (Ziadinov et al., 2010). Thus, conditions that allow infection of dogs from the natural cycle favour the onward transmission to humans.

In **Kazakhstan**, *E. multilocularis* infections in foxes and wild rodents have been described for several decades and have long been known to be distributed over most of the country (Abuladze, 1964). In the arid regions of Kazakhstan, characterized by desert or semidesert landscape, foci of AE in rodents are associated with water and are aggregated around oases. Elsewhere, the parasite is more widely distributed in mountainous and steppe habitats (Shaikenov, 2006; Shaikenov et al., 2004). Other reports (see Table S2 in Supplementary Material) estimate 8.3% of foxes infected in the Zhambul region, 23% of foxes infected in the Pavlodar region, 25% of foxes from the Almaty region, and more recently, 28.5% in west Kazakhstan. A prevalence of 5% of *E. multilocularis* in dogs has been described in the mountainous region of south east Kazakhstan (Stefanic et al., 2004; Torgerson, 2013).

In **Uzbekistan**, *E. multilocularis* has been found in 8.6% of foxes from the Amu-Darya delta and lower river valley in Karakalpakstan. The parasite has also been found in small mammals (*Ondatra, Rhombomys* and *Meriones* spp.) from the same area (Kairov, 1976). Older literature documented *E. multilocularis* in 1 of 678 dogs and 19 of 189 foxes (10%) originating from nine districts of Uzbekistan (Sadikov, 1963).

2.3.2.3 Alveolar echinococcosis in humans

There are no available data on human AE from the Caucasus, **Uzbekistan** or **Turkmenistan**.

Human AE has been known in Central Asia for several decades. For example Arslanova (1962) reported 33 cases of AE in 19,837 autopsies performed between 1941 and 1957 in southern **Kazakhstan**. More recently, 46 cases of AE were reported in a single surgical centre in Almaty between 2006 and 2014, and estimates from Kazakhstani expatriats are in the region of 34 cases per annum or 0.2 cases per 10^5 per year (Abdybekova et al., 2015).

In **Kyrgyzstan**, the number of cases of human AE (confirmed by histology on lesions removed at surgery) has increased from 0 to 2 cases per year in the 1990s to 148 cases reported in 2013, or 2.6 cases per 10^5 inhabitants per year. In some districts the incidence is as high as 58 cases per 10^5 per year (Usubalieva et al., 2013; Raimkylov et al., 2015). The estimated human burden of echinococcosis in Kyrgyzstan for 2013 is 11,915 (4.547–29.544) DALYs for AE and a further 1.742 (1.056–2.723) DALY for CE (Counotte et al., 2016). The burden of echinococcosis is roughly one-third of the burden of HIV which was estimated to be 38,870 (21,261–64,297) DALY in 2010 (Ortblad et al., 2013).

In **Tadjikistan**, a surgeon from Dushanbe reported treating 19 AE cases between 2010 and 2013 (Akhmedov et al., 2013). However, these figures are likely prone to underreporting and underdiagnosis (Mathers et al., 2007; Welburn et al., 2015).

2.3.3 *Middle East: e.g., Iran, Turkey*

AE is rare in tropical and subtropical areas of the Middle East (ME). However, in higher latitudes of the region, AE is more frequent and the parasite may be undergoing geographical range expansion to the south

(Geramizadeh et al., 2012). No report of *E. multilocularis* has been documented from **Syria, Jordan, Lebanon, Palestine, Israel** or the **Arabian Peninsula**.

In **Iran**, *E. multilocularis* was first reported in the early 1970s, with 10% of the red foxes infected in Moghan plain of the northwestern parts of the country (Mobedi and Sadighian, 1971). Since then a number of human AE patients have been documented, and prevalence was 22.9% in red foxes and 16% in jackals. However, attempts to identify rodent intermediate hosts have not been successful in the northwest (Zariffard and Massoud, 1998).

New foci of AE have been reported from Khorasan Razavi province (northeastern Iran) in 2007 and 2010 (Fattahi Masoom and Sharifi, 2007; Berenji et al., 2007; Raisolsadat, 2010). A captive spider monkey (*Ateles geoffroyi*) is also believed to have died of AE in the same area (Borji et al., 2012b). All wild carnivores examined, including three foxes, nine jackals, one hyena and one wolf, and five of 77 dogs (6.5%), were positive by PCR (Beiromvand et al., 2011). A number of small mammal intermediate hosts including *Microtus transcaspicus, Ochotona rufescens, M. musculus, Crocidura gmelini* and *Apodemus witherbyi* have been identified in this region (Beiromvand et al., 2013). These findings indicate autochthonous transmission of *E. multilocularis* in the region and document the zoonotic risk.

In **Iraq**, a human case of AE has been reported from the north, in Zakho near the Iraqi–Turkish border (Al-Attar et al., 1983). However, no animal reservoir has yet been identified.

Large parts of **Turkey** are considered as highly endemic for *E. multilocularis*. In animals, there is one report of an infected red fox (Merdivenci, 1963) from the European part of the country (Thrace), and more recently from Central Anatolia (Nevşehir, Kayseri) and Thrace, PCR on fox environmental samples was positive for *E. multilocularis* (Gürler A.T., personal communication). Since 1939, approximately 500 human cases of AE have been documented in more than 60 studies in Turkey. Cases of AE have been frequently reported especially from eastern Anatolia. Between 1980 and 1998 more than 200 new AE patients were identified (Altintas, 1995, 2003, 2008; Miman and Yazar, 2012; Uysal and Paksoy, 1986). During 1980–2000, an incidence of 0.4 cases/10^5 has been recorded in southeastern Anatolia including Diyarbakır, Sanliurfa and Batman provinces (Uzunlar et al., 2003). During the decade 2000–10, 162 human AE cases have been documented, with more than 86% of the patients from eastern and southeastern parts of Turkey (Miman and Yazar, 2012). Based on the relatively rare occurrence of cerebral AE cases (3–4% brain involvement), about

100 liver AE cases are expected to occur annually in Turkey (Torgerson et al., 2010). AE has been recognized as an emerging zoonosis in Turkey and, from a public health point of view, it has been recently made reportable in the country (Altintas, 2008).

2.3.4 North East Asia: China, Mongolia, Korea, Japan

2.3.4.1 China

2.3.4.1.1 Transmission and host assemblages *Echinococcus multilocularis* is present over much of central and western China, but is thought to be absent from eastern China (Fig. 3). Regions known to be endemic for this parasite are inner Mongolia in the north and further north east as far as Heilongjiang. In northwestern China, endemic provinces include Gansu, Qinghai, Ningxia and Xinjiang. In southwestern and central China, the western part of Sichuan province and the eastern part of the Tibet Autonomous Region (TAR) are endemic (Feng et al., 2015).

These endemic areas are mainly in the less densely populated areas of China. Thus, endemic provinces represent nearly two-thirds the area of China, but only 16% of the Chinese population. Within these provinces, *E. multilocularis* transmission risk is variable and seems best predicted by landscape features that support population outbreaks of particular small mammal species rather than species diversity (Giraudoux et al., 2013b). These provinces are also largely coendemic for AE and CE, although at the local level, one may predominate.

The history of AE in China is reviewed by Vuitton et al. (2011). The first human cases of AE were recognized in Xinjiang in the far west of China as early as 1956, with a report of six clinical cases published in 1965. In Qinghai province the first reported case was in 1959. In Gansu there were several reports also published in the 1960s. It was initially believed that the disease was rare and sporadic in China, as elsewhere, but by the 1980s it was clear that there were areas of China with large numbers of human cases. In 1992 a large focus of human AE in central China was published (Craig et al., 1992). Since then there have been numerous reports of AE in some communities in central and western China, where AE is a major cause of disease burden (Budke et al., 2004).

2.3.4.1.2 Infections in animals The first reports of *E. multilocularis* in canids occurred in the 1990s. Prevalence values ranging between 15% and 60% have been reported in red foxes (*V. vulpes*) in Xinjiang, Ningxia (Guyuan) and Qinghai and Inner Mongolia (Jiang, 1998; Wang et al., 2008).

The Tibetan fox (*V. ferrilata*) has been described as an important definitive host in the eastern part of the Tibetan plateau [Qiu et al., 1999; cited in Vuitton et al. (2003)]. In Sichuan province, *E. multilocularis* was detected in 35% of 94 faecal samples of Tibetan foxes collected in the environment (Jiang et al., 2012) and further east, in Qinghai province, 4 of 12 Tibetan foxes were infected (Wang et al., 2008). Evidence is accumulating that the Tibetan fox rather than the red fox is the more important definitive host on the Tibetan plateau (Tsukada et al., 2014). Transmission in the TAR and in other Tibetan regions (Feng et al., 2015) includes a wildlife cycle with Tibetan foxes (*V. ferrilata*) and microtine and/or ochotonid small mammal species (Raoul et al., 2006). Furthermore, the role of domestic dogs in AE transmission in TAR is probably similar to that described in Tibetan communities in northwest Sichuan (Giraudoux et al., 2013a; Vaniscotte et al., 2011).

Isolates of *E. multilocularis* from foxes in Qinghai province are genetically identical to those found further east in Sichuan province (Feng et al., 2013). In the province of Inner Mongolia, it has been reported that 19 of 151 (13%) corsac foxes (*Vulpes corsac*) were infected with *E. multilocularis.*

The parasite has been reported at high prevalence in dogs in a number of studies, particularly from the Tibetan Plateau. In Shiqu County, western Sichuan Province, prevalence was 12% of 372 dogs purged with arecoline, probably closer to 15% when adjusted for the insensitivity of purgation (Budke et al., 2005a, 2005b; Hartnack et al., 2013). Further studies in the same area found 23% of 142 samples of dog faeces positive for *E. multilocularis* (Vaniscotte et al., 2011). In Ganzi County, also in Sichuan province, 8 of 23 dogs were positive for *E. multilocularis* following necropsy (Huang et al., 2008).

Further west in Qinghai region, 16 dogs (of an unreported sample size) were found infected with *E. multilocularis* (Ma et al., 2015a). In Gansu province, 6 of 59 dogs from Zhang County were found to be infected (Shi, 1995). A more recent study from Gansu found four of 74 dogs (5.4%) infected (Zhao et al., 2009). Only one infected dog (of 30 examined) was found in Xinjiang (Zhang et al., 2006).

A number of studies have documented small mammal intermediate hosts infected with *E. multilocularis* over much of the endemic region of China. These include *A. amphibius* in Xinjiang and *Lepus oiostolus* from Sichuan and Qinghai (Wang et al., 2008); *Meriones unguiculatus* in Inner Mongolia; *Microtus ilaeus*, *Lasiopodomys brandti* and *Neodon irene* in Inner Mongolia, Xinjiang and Sichuan (Wang et al., 2008; Tang et al., 2004); *M. musculus* in Xinjiang (Wang et al., 2008); *Eospalax fontanierii* in Ningxi and Gansu and *Ochotona curzoniae* in Sichuan and Qinghai (Wang et al., 2008); 2009); *Ochotona daurica* in Gansu, *Spermophilus dauricus* in Ningxia, and *Spermophilus erythrogenys* in Xinjiang (Wang et al., 2008). Prevalences range from 0.01% for *M. musculus* to approximately 10% for *Ochotona* (Wang et al., 2008).

2.3.4.1.3 Alveolar echinococcosis in humans Over 90% of the global burden of AE occurs in China, with over 16,000 cases annually (Torgerson et al., 2010). Because the main endemic areas of AE are frequently in the economically most disadvantaged and remote areas, precise numbers of human cases are not reported and government surveys grossly underestimate the true number of cases. There are very few studies which directly report numbers or incidence of new cases of AE; one hospital in Xinjiang reported 159 AE cases over a 10-year period (Wang et al., 2015a). The best estimates of incidence are based on ultrasound prevalence studies (Table 2) combined with the size of the populations at risk, the duration of disease, and the case fatality ratio. Where there are inadequate or no treatment options available, which is true for most endemic areas of China, the disease is likely to run a fatal course within 10 years of diagnosis (Torgerson et al., 2008). In total, close to 60,000 individuals have been examined in ultrasound studies with AE prevalences ranging from under 1% to 8%. Regional variation is marked; prevalence was >8% in some individual small communities, while only three cases of AE (of 257,823 people) were detected in Gannan autonomous prefecture in Gansu after the commencement of an echinococcosis control programme (Wang et al., 2015c). Prevalence was 0.5% (114 of 20,730) in children between the ages of 5 and 15 years in Qinghai province, which is remarkably high because of the long latent period of the disease, indicating a high infection pressure from a very early age (Cai et al., 2012). Estimates of the regional annual incidences of AE (Torgerson et al., 2010) are provided in Table 3.

Epidemiological analysis indicates that usually only a small part of the general population is at risk (such as Tibetan pastoralists). In the TAR, which covers 1.23 million square kilometres with a population of 2.81 million (2.3 inhabitants) per square kilometre, reports of AE date back to 1977 and since then 22 AE cases have been documented (Feng et al., 2015). From these data, the incidence of AE was estimated to be between 0.6 and $2.8/10^5$ inhabitants. However, a pilot mass screening in Dingqing in eastern TAR indicated the prevalence of human AE to be 4.7%. Thus, locally, AE might be a serious public health problem

Table 2 Population studies of alveolar echinococcosis (AE) in rural China (all cases were confirmed by diagnostic imaging)

Region	AE cases	Population size studied	References
Gansu	84	2482	Bartholomot et al. (2002)
Gansu (Zhang County)	65	1312	Craig et al. (1992)
Gansu (Minle County)	1[a]	362	Han et al. (2015)
Ningxia (Xija, Guyuan and Haiyuan Counties)	96	4778	Yang et al. (2007)
Ninxia (Xija County)	20	221	Yang et al. (2006)
Qinghai	39	1549	Yu et al. (2008)
Qinghai (Darlag County)	141	1723	Han et al. (2009)
Qinghai (Chindu, Zeko and Garde Counties)	31	3703	Schantz et al. (2003)
Qinghai (Maqing County)	34	1561	Ma et al. (2015b)
Qinghai[b]	114	20,730	Cai et al. (2012)
Qinghai (Zhiduo County)	2	979	Wu et al. (2007)
Sichuan	60	705	Wang et al. (2004)
Sichuan (Ganzu autonomous prefecture)	308	8512	Tiaoying et al. (2005)
Sichuan (Ganzi and Shiqu Counties)	223	7138	Wang et al. (2006)
Tibet Autonomous Region[c]	12	1511	Feng et al. (2015)
Xinjiang (Hobukesar Mongolian Autonomous County)	4	421	Li et al. (2013a)

[a]Additional cases of AE were found by examining records of local hospitals.
[b]Survey only done in children between 5 and 15 years of age.
[c]Three counties of Lhasa prefecture. All the AE cases were in Dingqing County (12 of 232 investigated) with a local prevalence of 5.2%.

in some counties, similar to Tibetan communities in northwest Sichuan and southwest Qinghai provinces. Based on this prevalence of 4.7%, a much higher national prevalence can be estimated as shown in Table 3.

2.3.4.2 Mongolia, Korea, Japan

In **Mongolia**, AE is less common than CE (Ito and Budke, 2015). *Echinococcus multilocularis* has been documented in wild canids (red foxes and wolves) in northern Mongolia (Ito et al., 2013). Apart from one infected vole (*Microtus limnophilus*) from Khovd Province (Gardner et al., 2013), the key intermediate hosts of *E. multilocularis* are unknown and require further study. The first description of AE in humans dates back to 1982

Table 3 Alveolar echinococcosis in China, estimated number of cases per annum in various provinces

Province	Population at risk (million)	Estimated prevalence	Estimated median number of new cases per year	Annual incidence per 10^5 inhabitants
Gansu	3.6	2.9%	7676	2
Heilongjiang[a]			<10	0.02
Inner Mongolia	3.0	0.02%	44	0.2
Qinghai	5.4	1.0%	3766	67
Ningxia	1.2	2.0%	1770	2.7
Sichuan	0.92	3.6%	2390	2.9
Tibet Autonomous Region	2.7	0.1%	172	5.5
Xinjiang	5.8	0.2%	811	3.5

[a]Four cases have been reported in Heilongjiang in 1994 (Wang et al., 2008). Since then there is no additional information from this province.

(Ito and Budke, 2015). So far five cases of AE have been documented in four provinces, mainly in the northwest of Mongolia (Ito and Budke, 2015). Molecular studies from metacestodes of three AE patients revealed both Asian and the Mongolian *E. multilocularis* haplotypes (Ito et al., 2010).

Korea is not considered endemic for *E. multilocularis*. However, a recent report described AE in a 41-year-old woman who had never visited a known endemic area (Kim et al., 2011). Data from North Korea are not available.

In **Japan**, the first human case of AE was diagnosed in 1937 in a woman from Rebun Island, around 9 to 13 years after the introduction of foxes for vole control and fur production (Ito et al., 2003). However, the infection has since been eliminated successfully from the island. A probable independent introduction of Asian haplotypes of *E. multilocularis* happened in eastern Hokkaido, where AE cases in humans have been diagnosed continuously since 1965 (reviewed in Schantz et al., 1995; Ito et al., 2003). Based on longitudinal studies, during the 1980s the parasite dispersed geographically and, during the 1990s, it became increasingly prevalent in the fox population. The whole island of Hokkaido is now regarded as endemic for *E. multilocularis* (Ito et al., 2003). With the urbanization of fox populations, the *E. multilocularis* life cycle has been established in highly populated areas, for example, in Sapporo (Ito et al., 2003). The grey-sided vole (*Myodes rufocanus*) which is present in large populations in forests and scrubland is regarded as the main intermediate host [for details, chapter: Ecology and Life Cycle Patterns of *Echinococcus* Species by Romig et al. (2017)].

Data from recent monitoring programs in Hokkaido revealed intestinal *E. multilocularis* infections in 635/1933 foxes investigated by necropsy during 2009–13 [Japanese Infectious Agents Surveillance (2014) Vol. 35 p.183 (written in Japanese)]. Furthermore, intestinal *E. multilocularis* infections have been detected in raccoon dogs, but the epidemiological role of this species is not known. Between 1966 and 1999, intestinal infections were diagnosed in 99 of 9849 (1%) of domestic dogs investigated, a prevalence similar to European endemic areas. Prevalence of intestinal infections in cats of up to 5.5% are recorded, but are considered of little zoonotic significance due to the absence of mature eggs [cited in Ito et al. (2003)]. Infertile *E. multilocularis* lesions have been systematically detected in a variety of slaughtered animals [e.g., in 0.1% of pigs (n = 18 million) or in 0.15% of horses (n = 15,583) (Takahashi and Mori, 2001, cited in Ito et al. (2003)]. Furthermore, severe and often lethal AE cases have been detected in a variety of monkeys [cited in Ito et al. (2003)]. These animal infections document a high environmental contamination with *E. multilocularis* eggs in the Hokkaido endemic area.

The epidemiology of human AE in Japan has been comprehensively summarized (Ito et al., 2003). From 1937 to 1997, 373 AE cases have been documented, and until 2002, 10–20 new cases were registered per year in a total population of around 6 million in Hokkaido. In a study including 134 human AE patients, cattle and pig farming and the use of well water were identified as risk factors for human infection (Yamamoto et al., 2001). Based on data from the Japanese National Institute of Infectious Diseases (http://www.nih.go.jp/niid/ja/survei/2085-idwr/ydata/5672-report-ja2014-20.html), between 2010 and 2014, on average 16 new cases have been diagnosed in Hokkaido (incidence of 0.3 AE cases per 10^5 inhabitants per year).

2.3.5 South Asia: Afghanistan, Pakistan, India, Nepal and Bhutan

Single AE cases have been documented in the north of South Asia, but the southern distributional limit is still unclear (Fig. 3).

In **India**, an isolated case of AE was reported from a man from the hill regions of Kashmir (Aikat et al., 1978), but no animal reservoir has yet been identified. In a few other case reports from India the patients originated from a known endemic area or no data were given.

Epidemiological investigations are needed to determine the status of AE in **Afghanistan and Pakistan**. A 67-year-old immigrant from Afghanistan was diagnosed with hepatic AE in the UK (Graham et al., 2002). Considering that AE is established in neighbouring northeastern Iran, *E. multilocularis* is probably present in parts of Afghanistan.

For **Nepal**, no evidence was found for the occurrence of AE in a systematic review (Devleesschauwer et al., 2014) and no data are available concerning AE in **Bhutan** (N. K. Thapa, Ministry of Agriculture & Forests, Thimphu, Bhutan).

2.4 Africa

Records of AE in northern Africa are based on macroscopical descriptions, histopathology, radiological imaging and serology of three autochthonous patients who never travelled abroad from Tunisia (Schantz et al., 1995) and Morocco (Maliki et al., 2004), respectively. No positive samples were found in a coprodiagnostic survey of wild canids (Lahmar et al., 2009a), in necropsied jackals and foxes (Lahmar et al., 2014a), or from postmortem examination of livers and lungs from wild boars in the selected areas in Tunisia (Lahmar S., personal communication). Therefore, for the nations south of the Mediterranean Sea, the presence of *E. multilocularis* remains to be confirmed.

2.5 North America

2.5.1 Mexico, United States of America, Canada and Greenland

Within North America, northwestern Canada, northwestern Alaska, and the northcentral United States have long been considered endemic for *E. multilocularis*. Historically, two disjunct regions were recognized: the Northern Tundra Zone (NTZ) and the North Central Region (NCR) (Eckert et al., 2001). The NTZ consists of the northwestern coastal regions of Alaska and of the western Canadian Arctic, essentially corresponding to the range of Arctic fox (*V. lagopus*); however, the parasite has not been reported in the Yukon Territory or in the Canadian Arctic east of Hudson's Bay, possibly due to lack of sampling effort and low prevalence (Jenkins et al., 2013; Rausch, 1995).

The NCR was considered the southern portion of the three Canadian prairie provinces (Alberta, Saskatchewan and Manitoba) and northcentral American States (Montana, North Dakota, South Dakota, Nebraska, Minnesota, Iowa, Wisconsin, Illinois, Michigan, Indiana, Ohio and Missouri) (Fig. 4). Evidence for endemicity of *E. multilocularis* in Wyoming (indicated

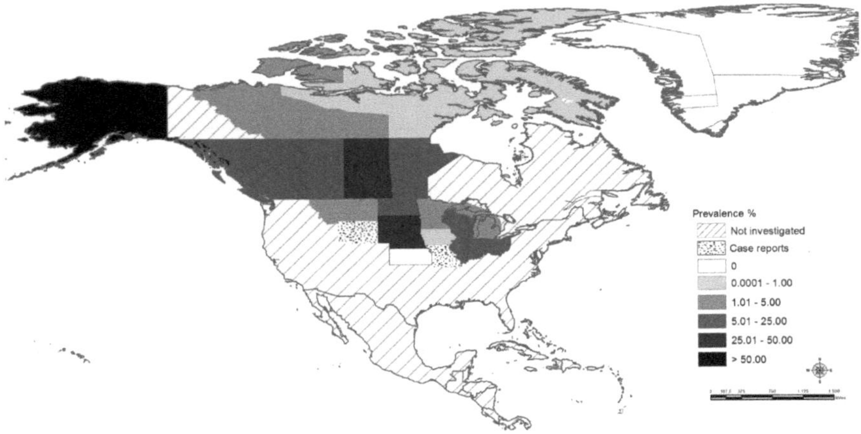

Figure 4 ***Current distribution of Echinococcus multilocularis in foxes (*****Vulpes vulpes, Vulpes lagopus, Urocyon cinereoargenteus*****), coyotes (*****Canis latrans*****) and wolves (*****Canis lupus*****) in North America (Canada, Greenland, United States).*** Areas with case reports of AE in intermediate hosts are approximately given as *dotted areas* if data of definitive hosts were missing. The detailed information (prevalence data in each jurisdiction) is listed in Table S3 of the Supplementary Material.

as case report on map) and Missouri is based on one unusual intermediate host and unpublished data in red fox (*V. vulpes*), respectively (Kritsky et al., 1977; Schantz et al., 1995). Central Indiana remains the southern limit of the parasite based on published data (Melotti et al., 2015).

Definition of the NCR now needs to reflect that the geographic range of the parasite appears to be expanding at western, eastern and northern distributional limits, although it is possible that new detections are, at least in part, a result of increased sampling effort and targeting of definitive hosts (DH) other than foxes. At its western distributional limits in Canada, it was recently detected as AE in a dog in 2009 and, subsequently, adult cestodes were found in coyotes (*Canis latrans*) in central British Columbia (Gesy et al., 2013; Jenkins et al., 2012). At its eastern distributional limits, the parasite has been recently detected as AE in dogs in southern Ontario, Canada, in 2012 and in wild canid definitive hosts in Illinois, Indiana, Ohio, and Michigan in the United States in the 1990s (Melotti et al., 2015; Skelding et al., 2014; Storandt and Kazacos, 1993). North of the northern distributional limit of the NCR in Canada (essentially corresponding to the northern limit of the prairie ecozone), the parasite had been considered absent in the boreal region separating the NTZ and the NCR (Schantz et al., 1995); however, the parasite has recently been detected in wolves (*Canis lupus*) in taiga and boreal regions of the southern Northwest Territories and northern British Columbia (Gesy et al., 2014; Schurer et al., 2014a, 2016).

In **Greenland**, *E. multilocularis* has not been reported. Although arctic foxes are present, the only potential rodent intermediate host (collared lemming, *Dicrostonyx rubricatus*) is not considered particularly suitable (Holt et al., 2005; Jenkins et al., 2013; Rausch, 1995). The parasite has not been reported from the southern United States or **Mexico**. Egg resistance to freezing and susceptibility to warmth and desiccation may limit this parasite to the northern hemisphere, raising the possibility that climate warming in North America may cause this parasite to shift northwards (Mas-Coma et al., 2008; Schiller, 1955; Veit et al., 1995). However, the haplotypes of the parasite present in the NCR are presumably tolerant of warmer temperatures and may continue to expand in distribution where suitable wildlife hosts are present (Jenkins et al., 2011). Red foxes are moving further North and are outcompeting Arctic foxes (Hersteinsson and MacDonald, 1992), perhaps bringing with them more temperate-adapted haplotypes of the parasite.

2.5.1.1 Transmission and host assemblages

Recognition of *E. multilocularis*, initially called *E. sibiricensis*, as a distinct species from *E. granulosus* was accomplished based on laboratory studies in Alaska in the 1950s (Rausch and Schiller, 1951). *Echinococcus multilocularis* was initially thought to be genetically uniform across its circumpolar range, and indeed there is less genetic diversity within *E. multilocularis* than the *E. granulosus* species complex. With the advent of genetic characterization at more discriminatory mitochondrial DNA loci, we are now recognizing greater diversity within *E. multilocularis* around the circumpolar North and within North America (Gesy et al., 2014; Nakao et al., 2009). Although much of this diversity is in the form of single nucleotide polymorphisms, there are distinct genetic groupings and differences, as well as ecological and biological differences between populations of *E. multilocularis* in the NCR and NTZ. Together, this suggests that we may need to revisit the taxonomic status of these populations.

Although morphologically similar, early experimental infections demonstrated distinct developmental differences between NTZ and NCR isolates of *E. multilocularis* (Bartel et al., 1992; Rausch and Richards, 1971). This was supported later by recognition of genetic differences between the NTZ (North American N1 and Asian) and NCR (North American N2) haplotypes of the parasite (Nakao et al., 2009) which belied previous suggestions that the parasite was a recent introduction from the north. Most recently, European haplotypes and the North American N2 and closely related haplotypes have been identified in wildlife in the NCR and other regions of northwestern Canada, suggesting a complex mosaic of endemic and introduced haplotypes of *E. multilocularis* (Gesy and Jenkins, 2015). More work is needed to determine if these haplotypes circulate in different host assemblages and if they differ in zoonotic potential.

Transmission of *E. multilocularis* in North America mainly involves wild canid definitive hosts and rodent intermediate hosts. Rarely, transmission may involve domestic pets and people; for example, on St. Lawrence Island, Alaska, with spillover of the parasite from wildlife to domestic dogs and, through dogs, to people (Rausch and Schiller, 1956). There are distinct ecological groupings in the NTZ and the NCR. In the NTZ, the parasite cycles between Arctic fox as definitive hosts and arvicoline rodents (i.e., lemmings, voles) and shrews as intermediate host. Red fox and coyote are also present as far north as the Arctic coast in western North America (Naughton, 2012), but there has been little surveillance effort in these potential definitive hosts. In the NCR, red fox and coyote serve as definitive hosts and neotomine (deer mice) and arvicoline rodents (primarily voles) as intermediate hosts. Wolves are newly recognized definitive hosts in taiga and boreal regions of North America (Schurer et al., 2016), where the parasite was once thought to be absent due to low rodent densities. Wolves may thus connect the disjunct populations of *E. multilocularis* in the NTZ and NCR, due to their large home range sizes (75–2500 km^2, exceeding 75,000 km^2 in the Arctic) and dispersal distances (50–800 km) (Naughton, 2012).

Translocation of *Echinococcus* spp. into and within North America has likely occurred with introduction of European red fox (Kamler and Ballard, 2002), translocation (illegal) of foxes for hunting enclosures (Davidson et al., 1992), reintroduction of wolves for conservation purposes (Foreyt et al., 2009), and importation of pet dogs from Europe and other regions of the world (Davidson et al., 2012; Jenkins et al., 2012). The latter is particularly concerning (and likely to be ongoing) as there is no mandatory testing or treatment for *Echinococcus* in imported or translocated companion animals in the United States or Canada, in large part due to failure to recognize species and finer scale genetic differences within *Echinococcus* spp. (Lymbery et al., 2015). Regulatory bodies for animal health need to consider genetic diversity within *E. multilocularis* and marked regional differences in prevalence, and consider testing and/or treatment of imported dogs and translocated wild canids to prevent importation of foreign haplotype or dissemination into nonendemic regions (such as the Atlantic provinces and states).

2.5.1.2 Infections in animals

Adult cestodes of *E. multilocularis* have been reported from multiple locations in endemic regions of the continental **United States** and **Canada** from Arctic fox, red fox, coyote and wolves (Table S3 in Supplementary Material). There is only one report of *E. multilocularis* in grey fox (*Urocyon cinereoargenteus*), by Vande Vusse et al., in 1978, cited in Melotti et al. (2015). The most epidemiologically significant wild canid definitive host depends on relative host abundance, home range size and dispersal distance, and infection intensity. By species, study prevalence (from the literature cited in Table S3 in Supplementary Material) was, on average, 25% in red fox (range 0.5–75%), 24% in coyote (range 0.4–44%) and 27% in wolf (8–67%). By region, coyote is likely a more important definitive host than red fox in the NCR, due to higher (and increasing) relative abundance in urban and agricultural regions, larger home ranges (10–190 km^2) and dispersal distances (often more than 100 km) (Naughton, 2012). Increasingly, the parasite is reported at high prevalence and intensity in thriving urban coyote populations in western Canada (Catalano et al., 2012; Gesy, 2012; Gesy and Jenkins, 2015; Hildreth et al., 2000; Leiby et al., 1970; Liccioli et al., 2012; Melotti et al., 2015; Seesee et al., 1983; Storandt and Kazacos, 1993, 2012), paralleling the European situation with red foxes.

Trends in prevalence are difficult to analyse due to highly biased sampling, with intense focus on North Dakota and St. Lawrence Island. Schantz et al. (1995) reported a trend for prevalence to increase over time, but this is greatly influenced by study design and may change dramatically with new molecular techniques for coprodiagnosis. Prevalence varies seasonally, with reports lowest in winter and highest in summer in red fox in North Dakota (Kritsky and Leiby, 1978) and highest in autumn and lowest in spring in Arctic foxes on St. Lawrence Island (Rausch and Fay, 2002). There are also geographic differences. In the United States region of NCR, the highest prevalences (>20%) were reported in South Dakota, Nebraska and Indiana, and lowest (<5%) in Iowa, Michigan and Montana (see Fig. 4 and Table S3 in Supplementary Material). In Canada, prevalence was greater than 20% in canid definitive hosts in Saskatchewan, Alberta, and British Columbia, and less than 5% in two northern territories. Prevalence was much higher in Arctic foxes on islands in Alaska (average 67%, range 32–100%) than on the Alaskan mainland (7.5%, range 2–15%) and in the western Canadian Arctic (1.5%, range 1–2%) (Eaton and Secord, 1979; Gesy et al., 2014; Jenkins et al., 2013; Kirk, 2010).

In a highly endemic island system (St. Lawrence Island), 5–13% of dogs were positive for *E. multilocularis* on examination of intestines at necropsy (Rausch and Fay, 2002), and dogs were considered the primary source of human exposure to the parasite (Stehr-Green et al., 1988). Outside Alaska, neither adult cestodes nor eggs of *E. multilocularis* have been reported from dogs in North America. Molecular characterization of taeniid eggs from faeces found only *Taenia* spp. and *Echinococcus canadensis* in nine taeniid egg positive samples from 1086 shelter dogs from across Canada (Villeneuve et al., 2015). While dogs are theoretically a good definitive host for *E. multilocularis*, so far in North America they are more commonly reported as aberrant intermediate hosts and are thus sentinels of environmental contamination rather than a source. More studies are needed to determine the true prevalence in dogs (as definitive hosts and intermediate hosts) in North America, as well as canine risk factors for exposure (breed, age, urban vs rural, predatory behaviour, immune status, access to areas used by wild canids, etc.).

There have been only two reports of naturally infected cats serving as definitive hosts for *E. multilocularis* in North America, one in Canada (Saskatchewan) and one in the United States (North Dakota) (Leiby and Kritsky, 1972; Wobeser, 1971). Gravid segments were reported in one cat in North Dakota. This reinforces that the parasite is uncommon in cats, and, combined with low infection intensities and fecundity in experimentally exposed cats (Kapel et al., 2006; Rausch and Richards, 1971), suggests that cats are an unlikely source of environmental contamination. There has been some effort to determine if wild felids (bobcats) are infected, but none has been found infected (Storandt et al., 2002).

Data on the incidence of AE in animal intermediate hosts are not routinely collected. The condition is not reportable to animal or human health authorities at the national level in Canada or the United States, although it is annually notifiable to the World Organization for Animal Health – OIE – by laboratories in both countries (http://www.inspection.gc.ca/animals/terrestrial-animals/diseases/annually-notifiable/eng/1305672292490/1305672713247; https://www.aphis.usda.gov/aphis/ourfocus/animalhealth/monitoring-and-surveillance/sa_disease_reporting/ct_disease_list). Most reports are from targeted studies in wild rodents, opportunistic observations from trapped furbearers or case reports in domestic dogs, nonhuman primates in zoos, and people. Low prevalence in intermediate hosts, even in highly endemic areas, is the norm — if detected in rodents, this may be a better indicator of endemicity than detection in wild canid definitive hosts, as the latter have home range sizes of tens to thousands of square kilometres, and dispersal distances of hundreds of kilometres.

In the NTZ, *E. multilocularis* has been reported in ground squirrels (*Urocitellus parryii*), red-back voles (*Myodes rutilus*), tundra voles (*Microtus oeconomus*) and shrews (*Sorex jacksoni*) from St. Lawrence Island (Rausch and Schiller, 1956); reviewed in (Jenkins et al., 2013). Prevalence in voles on St. Lawrence Island ranged from 2 to 63% (reviewed in (Jenkins et al., 2013), and in spring could be as high as 80% before declining due to the 'wash-out' effect from uninfected young of the year (Rausch et al., 1990). The parasite is very rare in rodents in mainland Alaska, although it has been detected in <1% of brown lemmings (*Lemmus trimucronatus*) (Holt et al., 2005). There have been no reports in rodents in the Canadian Arctic; it was not detected in lemmings and voles in a region in the central Canadian Arctic with positive Arctic fox (Gesy et al., 2014).

In the NCR, the parasite has been reported from deer mice (*Peromyscus maniculatus*), house mice (*M. musculus*), meadow voles (*Microtus pennsylvanicus*), southern red backed voles (*Myodes gapperi*), bushy-tailed woodrats (*Neotoma cinerea*), and muskrats (*O. zibethicus*) (Gesy et al., 2013; Gesy and Jenkins, 2015; Holmes et al., 1971; Kritsky et al., 1977; Leiby et al., 1970; Liccioli et al., 2013). In the NCR, prevalence is so low and studies so few that geographic and seasonal trends are not readily apparent. In deer mice, the most successful intermediate hosts in the NCR, prevalence in 11 studies reporting nonzero values ranged from 0.5% to 22%, with an average of 5%. In meadow vole, average prevalence was 3% (range 0.8—6%) in five studies reporting nonzero values (references cited in Table S4 in Supplementary Material).

AE has recently been reported for the first time as a clinical issue in domestic dogs in Canada, with detection of a European haplotype in a dog native to central British Columbia in 2009, followed by detection in a dog native to southern Ontario in 2012, both previously considered nonendemic regions for *E. multilocularis* (Jenkins et al., 2012; Peregrine et al., 2012; Skelding et al., 2014). At present, four dogs with AE have been reported in southern Ontario since 2012 (Peregrine, 2015). Within the Canadian NCR, AE was detected in a dog from Manitoba in 2012, followed by six cases of AE in dogs from Alberta and Saskatchewan from 2014 to 2016 (two cases/year in each province) based on a search of the Prairie Diagnostic Services database in April 2016. This raises the possibility that introduced European haplotypes now established in wild canids may be responsible for the emergence of AE in dogs in both previously and newly endemic regions of Canada. Public health authorities in Canada, and possibly the northern United States, need to be vigilant for new human cases, as dogs could be serving as sentinels of a newly emerging threat to human health.

2.5.1.3 Alveolar echinococcosis in humans

Outside of Alaska, AE has not been considered a mainstream human health issue in North America. Reports of AE in North Americans in the literature are generally limited to case reports, reviews of hospitalization data, and serological/skin test studies (fraught with both false positives and negatives). Unfortunately, AE is not always distinguished from CE, and CE is a far more likely diagnosis. When AE is diagnosed, incomplete travel histories and a very low expected prevalence generally lead to assumptions that most cases are foreign acquired.

In **Canada**, prior to 2013, only one autochthonous case has been reported. This case was an Icelandic immigrant in Manitoba in 1935 (James and Boyd, 1937). In 2013, an immunocompromised patient from rural west-central Alberta presented with AE. The cyst material typed as European, despite no history of travel to Europe (Massolo et al., 2015). A recent review of hospitalization data in Canada found 16 AE cases between 2002 and 2011, most commonly reported in liver and as metastases to multiple sites (Schurer et al., 2015). A similar review found 12 cases of AE between 2001 and 2014, all in regions endemic for *E. multilocularis* (British Columbia, Alberta, Saskatchewan and Ontario) (Massolo et al., 2014). Both studies significantly underestimate the true prevalence of AE, as the type of echinococcosis was not reported in 191 and 251 cases, and because the estimates did not adjust for cases that did not seek medical treatment. Regardless, the geographic distribution of human cases of AE overlaps with the known distribution of *E. multilocularis* in wildlife in Canada, and the possibility of endemic transmission should not be dismissed out of hand.

In the **United States, Alaska** has historically been a hotspot for autochthonous cases of human AE, with a few cases on the northwestern coast of mainland Alaska (Castrodale, 2003). Testing for AE using serology, skin tests, medical imaging and biopsy is reviewed in Jenkins et al. (2013), Rausch and Schiller (1956) and Wilson et al. (1995). The incidence of AE on St. Lawrence Island based on serosurvey has been as high as 98/100,000 (Schantz et al., 1995). From reported data, there were a total of 54 human cases of AE between 1947 and 1986 in Alaska, and no cases of AE from 1987 to 2010 [reviewed in Jenkins et al. (2013)]. Between 2010 and 2014, there were five reported cases, mostly from interior and southeastern regions of Alaska, more likely to be CE than AE (http://epibulletins.dhss.alaska.gov/Bulletin/DisplayClassificationBulletins/42).

Recent characterization of cyst material from voles on St. Lawrence Island (Nakao et al., 2009) and eggs from Arctic foxes on the North Slope of Alaska (Kirk, 2010) as Asian haplotypes of *E. multilocularis* raise the possibility that the hyperendemic focus of AE in northwestern Alaska reflects higher zoonotic potential of Asian haplotype of the parasite, and not simply unique ecological and behavioural risk factors.

Outside Alaska, only one autochthonous case in the United States is reported from a resident of Minnesota in 1977 (Gamble et al., 1979) who was subsequently identified as the N2, or central North American, haplotype (Klein and Massolo, 2015; Yamasaki et al., 2008). This is the first and only report of zoonotic transmission of the N2 haplotype endemic to the NCR; there was no evidence of exposure in a serological survey of high risk people (trappers) in the United States region of the NCR (Hildreth et al., 2000).

Genetic differences between haplotypes of *E. multilocularis* present in the NCR and NTZ (notably St. Lawrence Island) could account for the historical lack of human cases of *E. multilocularis* observed in North America outside Alaska, despite relatively high prevalence in wild canids in the NCR. While it is possible that there is decreased opportunity for human exposure in the NCR, many indigenous and rural inhabitants of this region of North America hunt, trap, consume untreated surface water, keep dogs as pets and working animals, and harvest foods that could be contaminated with faeces of wild canids. Enhanced surveillance, epidemiological investigation and molecular characterization of cestodes from canid definitive hosts, and cysts from animal and human intermediate hosts of AE, are needed to better understand the biological significance of genetic diversity within populations of *E. multilocularis* in North America, and their potential to emerge along with changes in climate, landscape, and the wildlife/human interface.

3. GLOBAL DISTRIBUTION OF *ECHINOCOCCUS* SPP. CAUSING CYSTIC ECHINOCOCCOSIS

The global endemic areas of *Echinococcus* spp. causing CE based on data mainly from domestic intermediate hosts are depicted in Fig. 5.

3.1 North America: Mexico, United States, Canada and Greenland

3.1.1 Introduction and molecular epidemiology

CE in humans and intermediate hosts in North America is caused by wildlife-associated genotypes of *E. canadensis* (primarily in cervid intermediate hosts), and livestock-associated genotypes of *E. granulosus sensu lato* (sheep, swine and cattle intermediate hosts). *Echinococcus canadensis* (G8 and G10 – Table 4) has the greatest geographic distribution and prevalence in Canada, Alaska and the northern latitudes of the contiguous United States, where assemblages involve wolves (*C. lupus*) and cervids (especially moose, *Alces alces*; caribou, *Rangifer tarandus*; and elk, or wapiti, *Cervus canadensis*). *Echinococcus granulosus* (not genotyped) is maintained in a dog–sheep assemblage in western states of the United States. In Mexico, *Echinococcus intermedius* (G7) is maintained in a dog–swine assemblage, and there are isolated reports of *E. granulosus* (G1) and *Echinococcus ortleppi* (G5) (Table 4). While annually notifiable to the World Organisation for Animal Health (OIE) at the laboratory level, there is no formal surveillance in place for CE in animals (especially wildlife) in North America. Domestic livestock are routinely inspected at slaughter in the United States, Canada and Mexico; however, small ruminants (goats and sheep) are often slaughtered and sold at the 'farm gate', and would not be inspected. Wildlife hosts (canids and ungulates) are not routinely inspected, and cysts detected in ungulates at necropsy, the most conclusive method of detection, are often considered incidental findings. Intestines of wild canids are not routinely opened at necropsy due to the risk of zoonotic *Echinococcus* spp. in Canada and the United States.

Historically, human CE in northern North America was associated with the presence of sled dogs, which were a vital source of transport for indigenous groups. The incidence of CE declined as sled dogs were replaced with motorized transport (Rausch, 2003); however, northern and indigenous populations remain overrepresented, possibly due to the presence of large free-roaming dog populations, challenges in preventing dogs from scavenging and hunting and the limited access to commercial dog food or

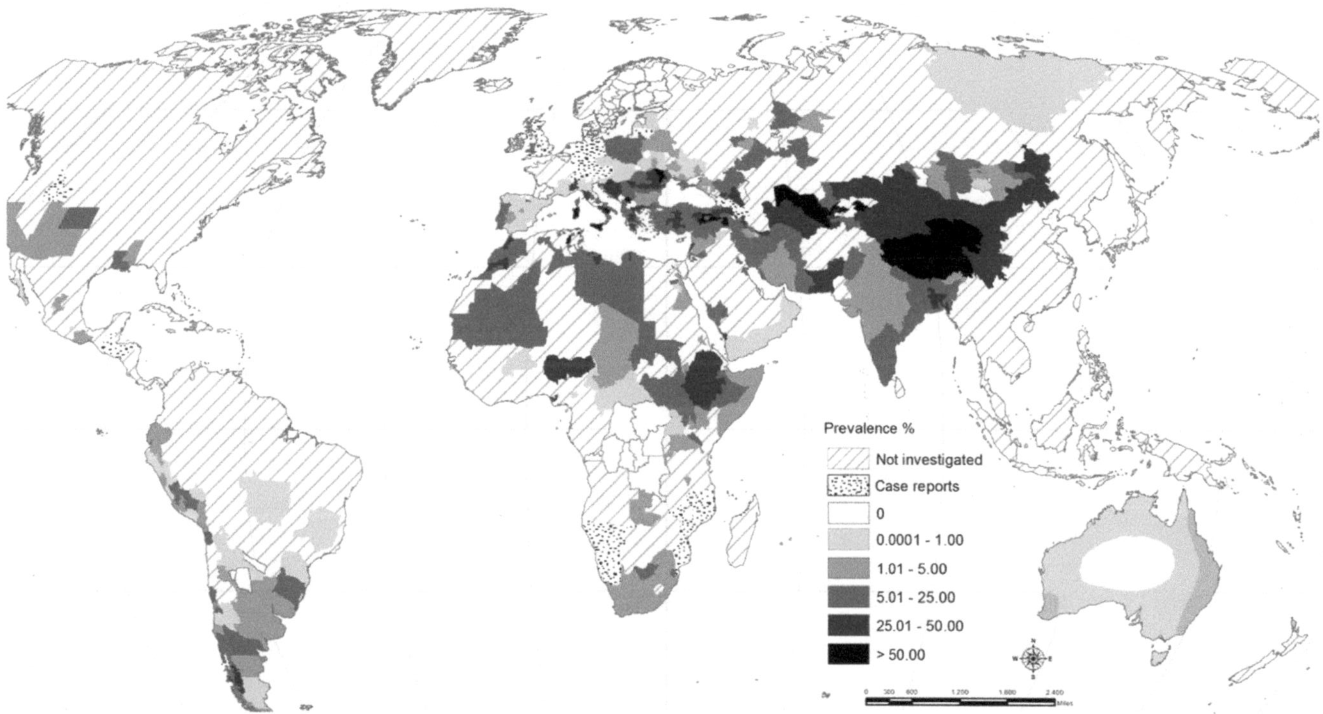

Figure 5 ***Current global distribution of Echinococcus spp. causing cystic echinococcosis in the main domestic intermediate hosts (not included are the cervid genotypes of Echinococcus canadensis and Echinococcus equinus).*** In addition, in areas where data of domestic intermediate hosts are missing, case reports of CE in wild intermediate hosts are given as *dotted areas* (see Fig. 12). For published details of the distribution/prevalence in the different countries/jurisdictions, see detailed maps (Figs. 6—13), Supplementary Material (Tables S5—S14) in Appendix A and text. Note that a positive finding in any host, study or jurisdiction renders the whole jurisdiction endemic.

Table 4 Genotypes of *Echinococcus* spp. causing cystic echinococcosis in North America: *Echinococcus granulosus* (G1), *Echinococcus ortleppi* (G5), *Echinococcus intermedius* (G7) and *Echinococcus canadensis* (G8, G10)

Country	Human	Canids	Cervids	Swine
Canada		G8[1,2] G10[1,2,3,4]	G8[4,5] G10[4,5]	
United States	G8[6]		G8[4,7] G10[4]	Historically present but eradicated
México	G5[8]	G7[9]		G1[10] G7[9,10,11]

[1] (Bryan et al., 2012); [2] (Schurer et al., 2014a); [3] (Himsworth et al., 2010); [4] (Thompson et al., 2006); [5] (Schurer et al., 2013); [6] (McManus et al., 2002); [7] (Lichtenwalner et al., 2014); [8] (Maravilla et al., 2004); [9] (Rodriguez-Prado et al., 2014); [10] (Villalobos et al., 2007); [11] (Cruz-Reyes et al., 2007).

cestocidal treatment. The epidemiological and geographical picture of human CE is somewhat unclear for North American countries because echinococcosis is not nationally reportable in people, and when data are available, species/genotype and source (autochthonous or foreign acquired) of CE in human cases is rarely reported. Efforts to estimate human incidence are also hampered by underdiagnosis, as infected individuals often remain asymptomatic for several years (or indefinitely with the cervid associated strains), before presenting with nonspecific clinical features. Furthermore, only two studies report the molecular characterization of domestically acquired human CE in North America (G8 in Alaska and G5 in Mexico), demonstrating a major gap in our understanding of distribution and risk (Maravilla et al., 2004; McManus et al., 2002).

3.1.2 The United States

Host assemblages and transmission: *Echinococcus canadensis* (G8 and G10) occurs in moose, caribou, white-tailed deer (*Odocoileus virginianus*), mule and black-tailed deer (*Odocoileus hemionus*), and elk, with sporadic or historical reports of *E. granulosus sensu lato* in mountain goats (*Oreamnos americanus*), swine, cattle and horses (Eckert et al., 2001; Hoberg et al., 1994; Jenkins et al., 2013; Ward and Bradshaw, 1956). *Echinococcus canadensis* G10 is present in moose in Washington state, and G8 in moose in Alaska, Minnesota and Maine (Bowles et al., 1994; Lichtenwalner et al., 2014; Thompson et al., 2006). Definitive hosts for *E. granulosus* in the United States include wolves, coyotes (*C. latrans*) and dogs (Eckert et al., 2001). The reintroduction of wolves to Idaho and Wyoming in the 1990s has been associated with an apparent increase of *E. canadensis* in wild cervids and canids, causing considerable conflict between livestock ranchers and conservation biologists (Foreyt et al., 2009). G8 has been identified in a human case of CE in Alaska (McManus et al., 2002). CE is not a major public health concern in the United States, and most recent cases are foreign acquired.

CE in animals: CE is reported in wild ungulates across Alaska and the northern states of the contiguous United States. In the 1950s, the prevalence of CE in Alaskan moose was approximately 20% (N = 124), with infected moose documented in the lower Matanuska River and the Anchorage–Palmer regions (Rausch, 1959). Surveys of caribou in Alaska during the same time period reflected a lower prevalence (0/200, 2/79, 3/63 and 4/67 animals investigated) (Rausch and Williamson, 1959; Rausch, 2003). A recent survey in the state of Maine found hydatid cysts in 39% of 54 moose, which was unexpected because CE was not thought to occur in the northeastern United States due to the absence of wolves (Lichtenwalner et al., 2014). Infected moose have also been observed in Minnesota and the state of Washington (Bowles et al., 1994; Thompson et al., 2006). A 25-year survey of CE in Californian deer reported a prevalence of 1.3% (Brunetti and Rosen, 1970). There are two reports of infected mountain goats in Alaska and in Idaho (Foreyt et al., 2009; Rausch and Williamson, 1959).

Currently in the United States, sheep are the only known intermediate host for livestock genotypes of *E. granulosus*, and the distribution is limited to focal areas of the western United States where incidence is low. In the last 30 years, large scale surveys of Utah slaughterhouses observed CE in 1.6–8.2% of sheep (Crellin et al., 1982; Loveless et al., 1978); this was confirmed by trace-back analysis from another facility reporting infected sheep in Utah (10%; N = 986), Colorado (7%; N = 15), New Mexico (17%; N = 35) and the Navajo reservation (New Mexico and Arizona; 16%; N = 115) (Schantz et al., 1977b). A similar survey of a California slaughterhouse that processed both in-state and out-of-state animals reported an overall prevalence of 4.8% (N = 22,720) (Sawyer et al., 1969). Reports of hepatic CE in swine and cattle in the southeastern United States date to the mid-20th century, when backyard swine production was common, and household dogs had close contact with livestock. Slaughterhouse surveys of swine reported prevalences of 0.9% (N = 33,174) and 6% (N = 8066) in Mississippi, and 5–20% in Louisiana (Hutchison, 1960; Ward and Bradshaw, 1956). Approximately 1% (N = 800) of cattle were infected (Ward and Bradshaw, 1956), but there are no recent records of CE in cattle. Swine CE is no longer thought to occur in the United States, which is likely attributable to improvements in biosafety and the move towards large-scale indoor production. However, the current consumer trend for outdoor, free-range production facilities and the concomitant spread of feral swine across the United States increase the risk for CE to establish in domestic assemblages, indicating the need for monitoring. There are only six American reports of CE in horses, only one of which was confirmed to be autochthonous and originated either in Virginia or Maryland (Hoberg et al., 1994).

The earliest record of an infected wolf was 1933, coinciding with observations of infected moose in Minnesota (Riley, 1933). Two decades later, necropsy of canids in Alaska demonstrated an infection prevalence of 30% (N = 200) for wolves and 4–22% for sled dogs (Jenkins et al., 2013; Rausch and Williamson, 1959; Rausch, 1960), with the latter acting as primary definitive hosts in agriculturally intensive regions (Rausch, 2003). In 2006, the finding of CE in a mountain goat prompted a survey of 123 wolves in Idaho and Montana, revealing an infection prevalence of 63% (Foreyt et al., 2009). Approximately half (53%) of infected wolves harboured more than 20,000 adult cestodes, 13% held between 2000 and 20,000 and 34% had less than 2000. The source of infection for these wolves remains unclear because wolves were historically extirpated from this region. There is some speculation that wolves reintroduced from Canada (Alberta and British Columbia) to the United States (Idaho and Wyoming) in the mid-1990s might have carried the cestode, despite treatment with praziquantel (a cestocidal agent) prior to their move (Foreyt et al., 2009; Johnson, 2001). Alternatively, definitive hosts (such as coyotes) could already have existed in the area, or infected wolves could have migrated naturally from Canada to the United States (Foreyt et al., 2009).

In the contiguous United States, where wolves are largely absent, coyotes likely play a role in transmission of *E. canadensis*. In the 1970s, 4% of 173 coyotes in seven counties in California harboured adult cestodes (Liu et al., 1970). The sympatric distribution of infected black-tailed deer (*Odocoileus hemionus columbianus*) provided evidence for a coyote–deer assemblage in this state (Brunetti and Rosen, 1970; Romano et al., 1974). Negative findings in coyotes from sheep-rearing regions suggest that coyotes may not play an important role in transmitting *E. granulosus* (Butler and Grundmann, 1954). Between 1892 and 1975, *E. granulosus* has been reported in dogs from Washington DC, Georgia, Tennessee, Kentucky, Mississippi, California, Utah and New Mexico (Pappaioanou et al., 1977). Sheep dogs imported from Australia in 1938 likely introduced *E. granulosus* to Utah, after which it spread to surrounding states (Crellin et al., 1982). The maintenance of a dog–sheep assemblage was confirmed by longitudinal surveillance of dogs and sheep in the 1970s, reporting adult cestodes in 11.3% (N = 839) of Utah dogs (Loveless et al., 1978), followed by detection in Navajo dogs (0.7% of 429) in Arizona and New Mexico (Schantz et al., 1977a,b). In the eastern United States where swine infection was historically present, postmortem examination of 9300 dogs identified only four infected animals, and small scale surveys of wild canids revealed no *Echinococcus* positive animals (Hutchison, 1960; Ward, 1965). Recent national surveys of intestinal parasites in dogs did not address *E. granulosus* (Gates and Nolan, 2009; Little et al., 2009).

CE in humans: CE is not an important cause of morbidity or mortality in the United States, and most cases are imported (Donovan et al., 1995). Overall, incidence of CE in people declined significantly in the latter half of the 20th century, which is attributed to the phasing out of sled dogs in Alaska, pathogen reduction in countries of origin for immigrants (e.g., Iceland, New Zealand), the development of cestocidal treatments, and the introduction of control measures in livestock in endemic foci (Katz and Pan, 1958; Loveless et al., 1978; Rausch, 2003). In Alaska, where CE is reportable, 193 cases were reported prior to 1980, but only eight cases were reported after 1990 (Hueffer et al., 2013). A comprehensive review of case reports between 1900 and 1974 identified 123 cases of autochthonous CE in the contiguous United States (Pappaioanou et al., 1977). Maps of the human incidence reflect the rise and decline of *E. granulosus* s.l. circulating in swine in the southeast followed by the introduction and control of CE in sheep in the west (Pappaioanou et al., 1977). Certain cultural groups (Mormons, Navajo and Zuni tribes and American-Basques) experienced higher rates of CE in the western states, largely due to the use of dogs in sheep herding and the practice of feeding sheep carcasses to these dogs (Araujo et al., 1975; Schantz et al., 1977a). Today, autochthonous cases are rare, with sporadic cases in Alaska, California and Utah (Moro and Schantz, 2009). Approximately one to four cases occur among Navajo tribes in New Mexico and Arizona each year (Moro and Schantz, 2009). An examination of death certificates issued for all states between 1990 and 2007 identified echinococcosis (CE and AE) in only 41 deaths and identified a greater frequency in males, Asian/Pacific Islanders, Hispanics, Native Americans and those aged 75 years or older (Bristow et al., 2012).

3.1.3 Canada

Host assemblages and transmission: Only *E. canadensis* (G8, G10) is present in Canada, and G10 appears to be more common in wildlife in western Canada based on limited sampling (Schurer et al., 2013, 2014a). Definitive hosts include wolves, coyotes, and dogs, while intermediate hosts are usually wild or captive-bred cervids. The two genotypes (G8 and G10) of *E. canadensis* occur in sympatric distribution in Canada and are widely distributed across the country, except for the Maritime Provinces, the island of Newfoundland and the High Arctic islands (Schurer et al., 2013; Sweatman, 1952). In intermediate hosts, G8 has been observed in elk (also known as wapiti or *Cervus canadensis*) in Alberta and muskoxen (*Ovibos moschatus*) in Nunavut; while G10 has been observed in moose, elk and caribou in Alberta, Saskatchewan, Manitoba and Quebec (Schurer et al., 2013; Thompson et al., 2006). Wolves infected by *E. canadensis* G8 are found in British Columbia, Saskatchewan, Manitoba and the Northwest Territories, and wolves infected by *E. canadensis* G10 are similarly distributed with the addition of Alberta (Bryan et al., 2012; Schurer et al., 2014a; Thompson et al., 2006). *Echinococcus canadensis* G10 is also present in domestic dogs in Saskatchewan (Himsworth et al., 2010). Coyotes (*C. latrans*) are competent definitive hosts for *E. granulosus* s.l. (presumably *E. canadensis*, but isolates from this host have not yet been characterized genetically). In people, CE is overrepresented in residents in the northern territories, and many cases elsewhere are assumed to be foreign-acquired; overall, it is not currently a major cause of morbidity or mortality in Canada.

CE in animals: Moose and caribou/reindeer (*R. tarandus*) are primary intermediate hosts for CE, and prevalence in these hosts is estimated at 42–47% and 1–21%, respectively, when only studies with sample sizes over 100 animals are considered (Rausch, 2003; Schurer et al., 2013; Sweatman, 1952). The most comprehensive survey of reindeer was conducted in the Northwest Territories in the 1950s, and 9.5% of 1664 animals had hydatid cysts (Choquette et al., 1957). Metacestodes are also commonly observed in elk (11–38%) and white-tailed deer (*O. virginianus*; 0–0.3%); however, cysts in these animals appear to be less fertile than cysts in moose and caribou, suggesting that they play a diminished role in the life cycle (Schurer et al., 2013; Sweatman and Williams, 1963). Other intermediate hosts include black-tailed and mule deer, muskoxen, and American bison (*Bison bison*), although these are less common (Rausch and Williamson, 1959; Schurer et al., 2013; Sweatman and Williams, 1963). Only historic records exist of hydatid cysts in horses and swine, and there are no records of naturally infected cattle or sheep in Canada (Sweatman, 1952; Sweatman and Williams, 1963). Industry losses due to CE are presumably low in

Canada, where captive-raised cervids can be infected, but not domestic livestock (sheep, cattle, goats).

Wolves are the primary definitive host for *E. canadensis* in Canada. Examination of 191 wolves from three western provinces and one territory reported both G8 (6%) and G10 (24%), with a median intensity of 2258 ± 24,397 cestodes (range: 15−149,600) (Schurer et al., 2016). Mixed infections of *E. canadensis* G8−G10 (5%) and G10-*E. multilocularis* (6%) also occurred. These estimates are consistent with older postmortem surveys of wolves (20−24%) in Ontario and the Northwest and Yukon Territories, but are lower than that reported in Alberta (72%) (Choquette et al., 1957; Freeman et al., 1961; Holmes and Podesta, 1968). The most recent helminth surveys of coyotes did not detect *E. granulosus* s.l. in Alberta, Saskatchewan or Newfoundland (Bridger et al., 2009; Catalano et al., 2012; Thompson et al., 2009). Previously, adult cestodes were observed in 0.5−9% of coyotes in Alberta, Manitoba and Ontario (median intensity of 13−675 cestodes per infected animal) (Freeman et al., 1961; Holmes and Podesta, 1968; Samuel et al., 1978). Contemporary surveys of dogs are generally limited to faecal assessments, most of which report only taeniid egg prevalence. Of the four Canadian studies where taeniid eggs were identified to species level, one detected *E. granulosus* s.l. in British Columbia, Alberta and Ontario, one detected *E. canadensis* (G10) in Saskatchewan (6%; N = 155), and two others did not detect *Echinococcus* in Saskatchewan (Himsworth et al., 2010; Schurer et al., 2014a; Villeneuve et al., 2015). A previous postmortem survey of 114 dogs collected from 28 indigenous communities in British Columbia, Alberta and the Northwest Territories detected adult *Echinococcus* spp. cestodes in 28% of animals (Miller, 1953).

CE in humans: Reports of human CE are found as early as 1883, and until the 1950s, most cases occurred in immigrants originating in European countries such as Iceland and Italy, where livestock strains were endemic (Cameron, 1960; Finlayson and Fergus, 1963). After this time, autochthonous cases were routinely detected as incidental findings during chest radiographs for tuberculosis screenings, with higher rates reported for northern and indigenous peoples than any other ethnic group (Lamy et al., 1993; Miller, 1953). Today, as there is no national surveillance, CE prevalence can only be estimated by integrating data from serosurveillance studies, hospitalization data and case reports. Serosurveillance in indigenous communities in Saskatchewan, Nunavut, the Northwest Territories and Quebec show exposure levels ranging from 0% to 48% (Jenkins et al., 2013; Schurer et al., 2014b). According to clinical records, annual CE incidence ranges from 0.72 to 1.4 clinical cases per million people with an overrepresentation of cases in the northern territories, although these rates are acknowledged as underestimates (Gilbert et al., 2010; Schurer et al., 2015). Hospital records suggest that cases of echinococcosis (CE and AE) are distributed across Canada (Massolo et al., 2014; Schurer et al., 2015). Gender (females > males), indigenous ethnicity and residence north of 55°N are considered contemporary risk factors for autochthonous CE (Gilbert et al., 2010; Jenkins et al., 2013; Somily et al., 2005).

Healthcare costs associated with CE have been calculated in Canada and are based on direct costs associated with medical treatment (e.g., medical imaging, surgery, chemotherapy, over-the-counter and prescription medications and hospitalization). On average, the healthcare system pays $8842 CAD (2015) to treat one case of CE. This does not include indirect costs associated with recurrent illness, long-term disability, mortality, income lost due to missed work, the cost to care for sick family members or transportation to obtain medical care, which can be significant for patients in remote or northern regions. Regional programs to prevent CE in people by dosing dog populations with praziquantel at 6-week intervals are not currently feasible from a financial perspective, even when indirect costs such as long-term sequelae are considered (Schurer et al., 2015). However, such prevention programs could become cost-saving in areas of localized outbreaks or in high-risk communities.

3.1.4 Mexico

Host assemblages and transmission: In Mexico, CE is predominantly maintained by the swine–dog assemblage (*E. intermedius,* G7) (Cruz-Reyes et al., 2007; Rodriguez-Prado et al., 2014; Villalobos et al., 2007). *Echinococcus oligarthra* has been found in a bobcat (Salinas-López et al., 1996). Reports of G1, *E. ortleppi* and *E. oligarthra* are uncommon, and there is little evidence to support their widespread distribution in México (Rodriguez-Prado et al., 2014). Only a single autochthonous human case has been reported and involved *E. ortleppi* (cattle strain, G5) (Maravilla et al., 2004). CE is not considered a public health priority in Mexico as the majority of human cases are foreign-acquired, and new cases occur infrequently (Eckert et al., 2001; Steta and Torre, 2009).

CE in animals: Slaughterhouse studies of swine have shown CE prevalence of 0.27% of 40,073 (Vargas Rivera et al., 1995), 6.5% of 2873 (Martínez-Maya et al., 1994) and 5% of 87 (Flisser et al., 2015) in the states of la Paz, Zacatecas and Oaxaca, respectively. The prevalence in cattle is far lower (0.1% of 3079), and CE was not detected in other livestock (sheep and goats) surveyed in the same studies (Martínez Maya et al., 1994; Rodriguez-Prado et al., 2014).

Parasite surveillance in the Federal District and the states of Queretaro, Zacatecas and México demonstrated that dogs were infected with adult *Echinococcus* stages. The infection prevalence was low in all studies (0.49−0.85%), and it is unclear whether wild definitive hosts also contribute to the life cycle (Eguia-Aguilar et al., 2005; Fernández and Cantó, 2002; Rodriguez-Prado et al., 2014).

CE in humans: Between 1990 and 1998, only 33 hospitalized cases with postoperative diagnosis of CE were reported in the yearbook statistics of the Mexican Ministry of Health. A small number of human cases have been reported in the literature in different states of Mexico (Cornejo-Juarez et al., 2013; Cortés Carrasco et al., 2002; Cruz Benítez, 2009; Orea-Martinez et al., 2013; Steta and Torre, 2009; Suarez et al., 1995; Valenzuela Ramos et al., 2010; Villarreal Jiménez et al., 1995). Only 11 cases in the last 60 years were autochthonous (Steta and Torre, 2009). The only molecular data available reported the presence of *E. ortleppi* (G5) in a single human case of CE (Maravilla et al., 2004). Serosurveillance of a suburban population close to a slaughterhouse that detected pigs with CE found that 15% of 200 participants had been exposed (Sánchez-González et al., 1997); however, it is important to consider that cross-reactivity with other cestodes in this kind of test can lead to false positives. Finally, ultrasound screening has shown cases of CE in 0.75% of 401 people in the State of Mexico (Mata-Miranda et al., 2007).

3.2 Central America

CE has been reported sporadically in humans in the past in countries of Central America, such as Guatemala, El Salvador, Honduras, Cuba, Panama (Sanchez et al., 1992; Sousa and Lombardo Ayala, 1965) and Costa Rica (Brenes Madrigal et al., 1977). However, local transmission and molecular data have not been documented for any of these countries. No new information is available regarding human CE in most of these countries, except in Cuba, with recent reports of both autochthonous and imported cases (Escalante et al., 2012; González Núñez et al., 2001).

Previous data described sporadic animal and human cases in Guatemala, El Salvador and Honduras. In recent years, a report of postmortem inspections of livestock in Haiti described CE as affecting mostly pigs (5.2%), but also sheep (2.1%), goats (0.9%) and cattle (0.3%). A high infection prevalence in dogs was detected in the same study (25%) (Blaise and Raccurt, 2007).

3.3 South America: Argentina, Bolivia, Brazil, Chile, Colombia, Ecuador, Paraguay, Peru, Uruguay and Venezuela

3.3.1 Host assemblages, transmission and molecular epidemiology

In South America, CE is known to occur with high prevalence in parts of Argentina (Patagonia, Pampas, Coast), Bolivia (southwest), Brazil (south), Chile (southland central valley regions and extreme south), Peru (central and southern highlands) and Uruguay (see Fig. 6 and Table S5 in the Supplementary Material). The disease has major human health and socioeconomic importance (Fig. 7 and Table S6) and has a high prevalence in livestock.

In South America, CE is maintained by domestic cycles of transmission that involve dogs and herbivores (sheep, swine, cattle, goats, horses and camelids) and multiple species/genotypes including *E. granulosus* (G1—3), *E. ortleppi* (G5) and *E. intermedius* (G6/7) (Table 5). Argentina hosts more strains than any other country, while Peru has the most human CE cases, despite underreporting. Human CE is associated with G1-3 in Argentina, Brazil, Chile, Peru and Bolivia; G5 in Argentina and Brazil; and G6 in Argentina, Chile and Peru. Intermediate hosts for G1—3 include sheep, cattle, goats, alpaca and swine with cattle the intermediate host for G5, goats for G6 and swine for G7.

The reservoir for G6 involves goats (Soriano et al., 2010) and infects humans in the Neuquén region (Guarnera et al., 2004; Kamenetzky et al., 2002), Rio Negro and Buenos Aires (Rosenzvit et al., 1999) and Catamarca. Dogs infected with *E. intermedius* (G6) have been reported in Rio Negro, Neuquén and Catamarca. *Echinococcus ortleppi* has been described in few isolated cases in animals and humans from remote areas of Argentina: Neuquén (South) (Guarnera et al., 2004) and Tucuman (North) (Kamenetzky et al., 2002). Infected animals include two cattle in Santa Fe and two dogs in Catamarca (Kamenetzky et al., 2002). The G7 genotype of *E. intermedius* has been described in pigs from Santa Fe (Rosenzvit et al., 1999) and Buenos Aires (Kamenetzky et al., 2002) and one dog in Neuquén (Kamenetzky et al., 2002); no human cases have been detected. In southern **Brazil**, *E. granulosus* has a sympatric distribution with *E. ortleppi* and with *E. intermedius* (G7) (Rio Grande do Sul State)

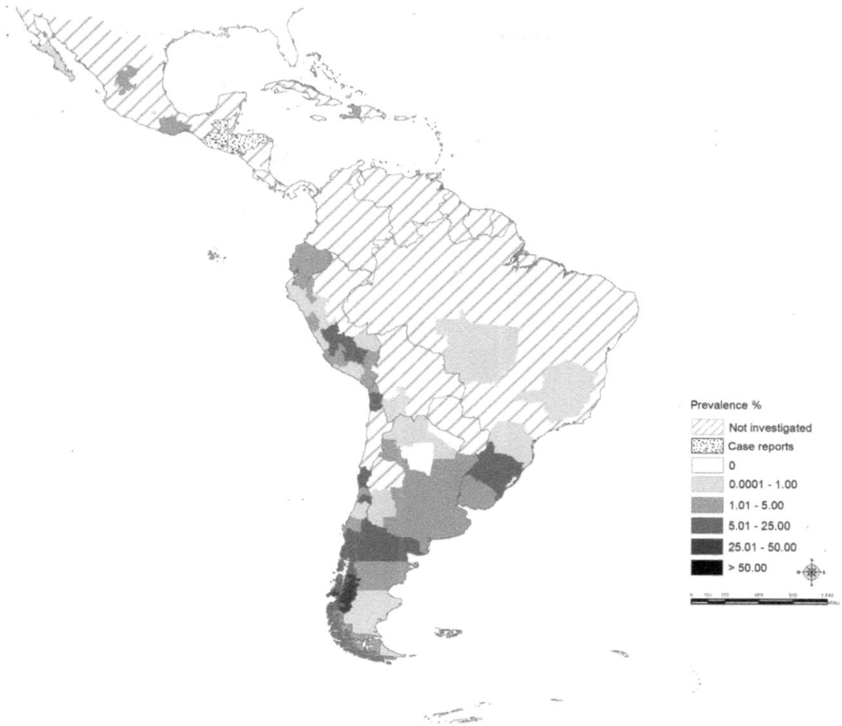

Figure 6 ***Current distribution of Echinococcus spp. causing cystic echinococcosis in domestic intermediate hosts (sheep, cattle and pigs) in Central and South America.*** The detailed information (prevalence data in each jurisdiction) is listed in Table S5 of the Supplementary Material.

(Balbinotti et al., 2012; de la Rue et al., 2006). A single human case of *E. ortleppi* was reported (de la Rue et al., 2011).

In **Chile**, a single case of *E. intermedius* (G6) has been reported in a study in which 19 other CE cases were *E. granulosus* (G1) (Manterola et al., 2008). A second study described the presence of *E. granulosus* in cattle (Espinoza et al., 2014). In **Peru**, *E. granulosus* is responsible for most animal and human infections (Moro et al., 2009; Sánchez et al., 2010; Santivanez et al., 2008); however, a few cases of *E. intermedius* (G6 and G7) have been reported in humans and pigs, respectively (Moro et al., 2009; Sanchez et al., 2012; Santivanez et al., 2008). Molecular data from **Uruguay** is

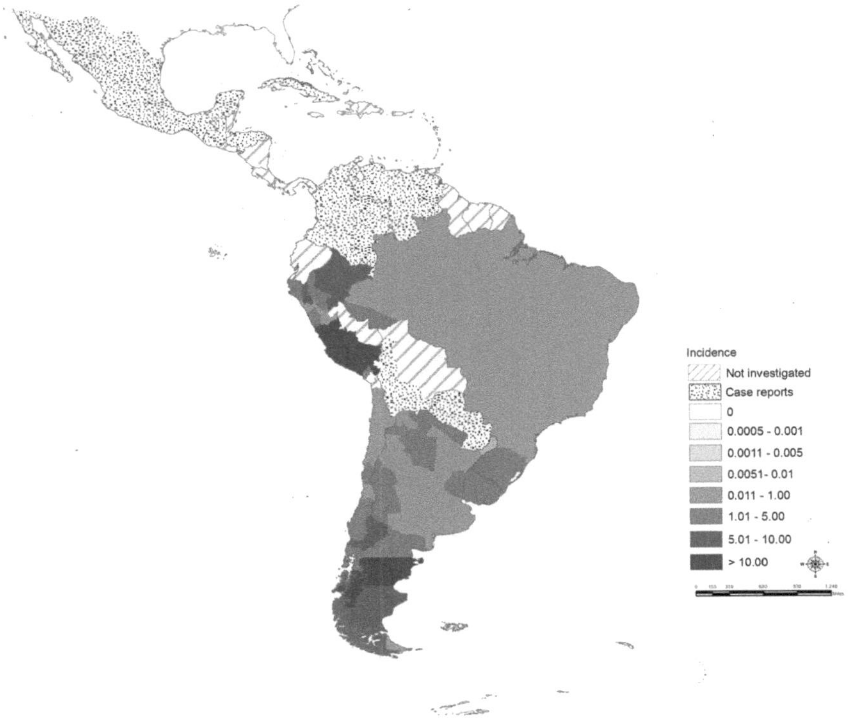

Figure 7 ***Current incidence of human cystic echinococcosis in Central and South America.*** The detailed information (incidence data in each jurisdiction) is listed in Table S6 of the Supplementary Material.

limited to a few samples included in studies reporting the genetic variability of *E. granulosus* in other South American countries. Five samples from cattle have been characterized as *E. granulosus* (G1) (Cucher et al., 2016; Kamenetzky et al., 2002), while two samples from cattle have been identified as *E. ortleppi* (Cucher et al., 2016). A similar situation is true for **Bolivia**, where molecular data is limited to a single report of *E. granulosus* (G1) in a person (Kamenetzky et al., 2002).

3.3.2 Infection in animals

In general, good documentation is available for livestock cases at post-mortem examination in most South American countries (Fig. 6) involved in the initiative for the control of CE led by the Pan American

Table 5 Genotypes of *Echinococcus* spp. causing cystic echinococcosis in South America: *Echinococcus granulosus* (G1−3), *Echinococcus ortleppi* (G5), *Echinococcus intermedius* (G6/7)

Country	Human	Dog	Sheep	Cattle	Goat (Go)/Alpaca (A)	Pig
Argentina	G1[1,2,3,4] G2[2,4] G5[2] G6[2,4]	G1[3,5] G6[5]	G1[4,5] G2[4] G3[5]	G1[1,3]	Go: G1[5] Go: G6[5]	G1[3] G7[4,5]
Brazil	G1[6] G3[6] G5[6]	G1[6] G3[6] G5[6]	G1[8]	G1[3,7,8] G5[3,7,8]		G1[9] G7[9]
Chile	G1[3,10] G6[10]			G1[3,11] G3[11]		
Peru	G1[12,13,14] G6[12,14]		G1[12,13]	G1[12,13]	A: G1[15] Go: G6[12]	G1[15] G7[12,15]
Uruguay				G1[1,3] G5[1]		
Bolivia	G1[3]					

[1] (Cucher et al., 2016); [2] (Guarnera et al., 2004); [3] (Kamenetzky et al., 2002); [4] (Rosenzvit et al., 1999); [5] (Soriano et al., 2010); [6] (de la Rue et al., 2011); [7] (Balbinotti et al., 2012); [8] (de la Rue et al., 2006); [9] (Monteiro et al., 2014); [10] (Manterola et al., 2008); [11] (Espinoza et al., 2014); [12] (Moro et al., 2009); [13] (Sánchez et al., 2010); [14] (Santivanez et al., 2008); [15] (Sanchez et al., 2012).

Health Organization (PAHO). However, the data regarding infection in dogs is scarce and no systematic surveillance is in place. In 2011, official abattoir inspections in **Argentina** (Trezeguet et al., 2011) reported the following prevalence of CE at the national level: 3% of 916,102 sheep, 1.6% of 3,273,864 pigs and 2.9% of 9,010,321 cattle. The most heavily infected regions for sheep were Rio Negro (21.9%), Buenos Aires (9.2%) and Chubut (2.2%), while infected swine were most frequently found in La Pampa (2.2%), Mendoza (1.8%), Buenos Aires (1.7%) and Santa Fe (1.1%). Infected cattle occurred most often in Rio Negro (19.1%), followed by Neuquén and Misiones (15%). Approximately, 2.5% of 1042 dogs sampled from 352 sheep farms from La Pampa, Neuquén, Rio Negro, Chubut, Santa Cruz and Tierra del Fuego were positive on copro ELISA (Cavagion et al., 2005). In Neuquén, the prevalence of infection in rural dogs is 12.4% (Pierangeli et al., 2010).

In **Brazil**, the most affected state is Rio Grande do Sul. There, 12% of cattle, 17% of sheep, and 11% of dogs (coproantigen test) have been found to be infected (de la Rue, 2008). The same report shows low levels of infection in cattle in other states of Brazil: 0.12% in cattle in Parana and 0.002% in Mato Grosso (de la Rue, 2008). Older data also showed a low level of infection in cattle in Minas Gerais (0.18%) between 1990 and 1994 (Sa et al., 1998). The only study of canine echinococcosis reported a prevalence of 27.7% in Santana do Livramento county, an endemic area in the state of Rio Grande do Sul (Farias et al., 2004).

In **Chile**, *E. granulosus* is the second leading cause of livestock viscera condemnation after the liver fluke *Fasciola hepatica*. In 2014, 28.9% of animals presenting an issue at meat inspection were infected with *E. granulosus*. Abattoir inspections report CE in 19% of cattle, 3% of sheep, 0.01% of pigs, 2.3% of horses and 3% of goats. The most affected regions are the Capital (Metropolitan Region), Los Lagos (X region), Araucania (IX region), Los Rios (XIV region), Aysen (XI region) and Magallanes (XII region) (SAG, 2015). In the IV region (Coquimbo), a coproantigen study of dog faeces demonstrated that CE transmission is not exclusive to rural areas [prevalences of 11.7% in city dogs, 5.9% in town dogs and 3.5% in rural dogs (Acosta-Jamett et al., 2010)].

In **Peru**, few studies have reported prevalence in livestock. In 1997, 82% of sheep and 32% of dogs were infected in Junin (Moro et al., 1997). A second study in the same area reported 46% (23/50) of dogs positive on coproantigen testing while 38% (13/34) of sheep were infected at postmortem examination (Moro et al., 1999). A study on abattoir workers and stray dogs from a nonendemic coastal city found 4 of 22 dogs positive while 3 of 32 human had liver CE (Reyes et al., 2012). Between 2009 and 2012, official data reported infection prevalence ranges of 3.61–6.12% for cattle and 6.55–10.44% for sheep (PAHO, 2015).

In **Uruguay**, from 2009 to 2014, the infection prevalence was 2.2–5.9% in sheep and 3.9–7.05% in cattle (PAHO, 2015). Surveillance of rural dogs using a coproantigen test revealed that approximately 4.3% were infected (Irabedra and Salvatella, 2010).

In **Ecuador**, CE has been described in 0.12% of 1658 cattle and 0.5% of 1790 pigs (Torres, 2012). A previous report showed a higher prevalence in pigs (13%) in a different area (Allaico, 2010). A number of other previous reports have been published showing prevalence in pigs (between 1.13% and 5.1%), cattle (0.21–11.8%) and sheep (0.04%) [reviewed in Allaico (2010)]. These data suggest that CE is an important problem in animals in Ecuador.

3.3.3 Infection in humans

Between January 2009 and December 2014, countries involved in the initiative for the control of CE in South America (Argentina, Brazil, Chile, Peru and Uruguay) reported 29,556 cases of CE, with the majority of cases in **Peru** (20,785). CE cases are usually underreported with a higher number in hospital records than reported each year. This is partly influenced by the different systems for notification of the disease in each country (see Fig. 7 and Table S6 in the Supplementary Material). Moreover, notification of human CE is not compulsory in all countries (PAHO, 2015).

In **Argentina,** between 2005 and 2010, the annual incidence was 0.95/100,000, with the highest incidence in Chubut (12.75/10^5 inhabitants), followed by Neuquén (8.14) and Santa Cruz (6.41) (Moral, 2010). Argentina has been a pioneer in South America implementing population ultrasound screenings as a tool to diagnose CE and also to assess the success of control programs. For example, in the Argentinean Patagonia, 87 (0.4%) cases were diagnosed after 22,793 ultrasound scans were performed in children from 6 to 14 years of age from 2000 to 2008 (Del Carpio et al., 2012). In Rio Negro, the overall prevalence of CE based on ultrasound was 7.1% (40/560) in 2009 (Bingham et al., 2014).

In **Brazil**, between 1981 and 1998, 701 CE cases underwent surgery in Rio Grande do Sul (Mardini and Souza, 1999). Between 1999 and 2002, 14, 8, 32 and 2 cases were registered respectively, and none from 2003 until August 2005 [reviewed by de la Rue (2008)]. Between 2009 and 2014, 91 cases were confirmed (PAHO, 2015). In this country the notification of the disease is not compulsory. According to data from the Brazilian government (DATASUS, 2013) there were 110 deaths due to echinococcosis between 1993 and 2013, assuming a case fatality rate of about 2% there would be 5500 cases in this period suggesting an incidence of 0.1/10^5 inhabitants per year for the whole country (DATASUS, 2013). Interestingly, with these data the highest incidence occurs in the state of Acre: 3.3/10^5 inhabitants per year in the west bordering Peru and Amazonas. Since there is no detailed cause of death in this report, it is possible that some of the cases are due to neotropical echinococcosis rather than CE. In Rio Grande do Sul the incidence was calculated at 1.1/10^5 inhabitants per year.

In **Bolivia**, CE is endemic in southwestern parts of the country. However, of the 106 CE cases recently reported (1998–2004), 83% originated from La Paz, suggesting that CE is increasing in eastern regions (Vera et al., 2004). This idea is supported by other reported cases in La Paz (Vera et al., 2006) and Cochabamba (Torrez Salazar et al., 2009). A study of hospitalized cases in children between January 1984 and February 1999 showed an increasing frequency of CE in La Paz city (Tamayo Meneses et al., 2004). Finally, in the locality of Tupiza, an increase in human cases of CE has been observed, with 238/1030 (18.8%) ultrasounds showing signs of CE and 23.9% of dog faeces positive to coproantigens (Villena and Uzqueda, 2011).

In **Chile**, according to the data from the Ministry of Health during 2009–14, the annual incidence was between 1.4 and 1.8/100,000, while the incidence based on the number of discharges in hospitals due to CE in the same period was between 4.68 and 5/100,000 (PAHO, 2015). The disease is present in the whole country (see Fig. 7 and Table S6), with highest prevalence in the regions where livestock production is concentrated including Aysen, Magallanes in the extreme south; Coquimbo in the north–central area and Araucania, Bio Bio, Los Rios and Los Lagos in the south part of the country (Martinez, 2011, 2014).

In **Peru**, there is no compulsory notification of CE; therefore, the number of cases is likely underestimated (PAHO, 2015). Incidence rates can reach between 14 and 43 cases/100,000 in the regions of Pasco, Huancavelica, Junin, Puno and Cusco (Cabrera, 2007). Official reports from the Ministry of Health based on registers at hospitals show an incidence between 7 and 8 cases/100,000 per year (Cabrera, 2007). A study using ultrasound and X-ray to detect CE reported a prevalence of 9.1% amongst 407 people in the central Peruvian Andes (Moro et al., 1997). Another study using ultrasound and serology in the same area reported a prevalence of 9.3% ($N = 214$) (Moro et al., 1999). In Canchayllo, a locality in the Peruvian highlands, ultrasound surveys detected CE in 4.9% of 309 subjects (Moro et al., 2005). Autochthonous transmission of *E. granulosus* occurs in Lima, based on findings of CE in 3/32 workers in unlicensed abattoirs, using a combination of abdominal ultrasound, chest X-rays and serology (Reyes et al., 2012).

In **Uruguay**, the incidence has decreased in recent years based on ultrasound studies in risk areas, from 6.5 per 1000 inhabitants in 2008 to 2.8 in 2013 (Irabedra et al., 2016). The incidence in the endemic areas is estimated between 1.3 and 3.8 cases/10^5 inhabitants (PAHO, 2015).

The remaining South American countries have not received much attention in the reporting of CE and epidemiological studies are limited (see Fig. 7 and Table S6 in the Supplementary Material). It is not known with certainty the incidence, prevalence or burden of CE in Colombia, Venezuela and Paraguay. In **Colombia**, the first case was reported in 1941 (Perez Fontana, 1949). Two other cases were reported in the following decade (Lichtenberger, 1957), and more recently, a single autochthonous case (Gómez et al., 2003). In **Venezuela**, the first case described in 1938 was not thought to be autochthonous (Gómez and Luna, 1938), and only nine autochthonous cases have since been reported (Guanipa et al., 1990). In **Paraguay**, the disease is rare, and there is only one record of autochthonous CE (Rodas et al., 2011).

3.4 Cystic echinococcosis in Europe

3.4.1 Introduction

For south and southeastern Europe, *E. granulosus* (sheep strain, genotypes G1-3) represents the principal causative agent of CE. *Echinococcus intermedius* (pig strain, G7) is the main human CE agent in the Baltic countries (Marcinkutė et al., 2015). Furthermore, two less pathogenic genotypes of *E. canadensis* (G8 and G10) have been documented in northern Europe (Oksanen and Lavikainen, 2015). Across Europe the actual prevalence of CE in animals or humans remains fragmented, partly due to the lack of efficient and dedicated reporting systems. In this regard, a European register was initiated within the FP7 HERACLES project aiming to provide prospective data on the epidemiology and clinical features of human CE (Rossi et al., 2016). Fig. 8 and Table S7 reports the current distribution of *Echinococcus* spp. causing CE in Europe (not including the cervid genotypes of *E. canadensis*). Information (incidence data) of CE in humans in Europe is reported in Table S8 of the Supplementary Material.

3.4.1.1 Western and Northern Europe: Iceland, Ireland, Great Britain, Norway, Sweden, Finland and Denmark

3.4.1.1.1 Host assemblages, transmission and molecular epidemiology After a successful control program, **Iceland** can be regarded as free of CE transmission since decades (Schantz et al., 1995; Sigurdarson, 2010).

The **United Kingdom** is endemic for *E. granulosus* (G1), and there is a long history of CE occurring in humans. For example, a likely case of CE was reported in 1785 at the Edinburgh infirmary (Risse, 2005). More recently the main foci of transmission have been in the Western Isles off the North West coast of Scotland and in Wales. Historically, *Echinococcus equinus* circulated in a typical foxhound-horse cycle (Thompson and Smith, 1975), but transmission is still documented in the UK (see later).

For the north of **Scandinavia**, the current epidemiological situation of the cervid strains (G8 and G10) of *E. canadensis* has recently been reviewed in

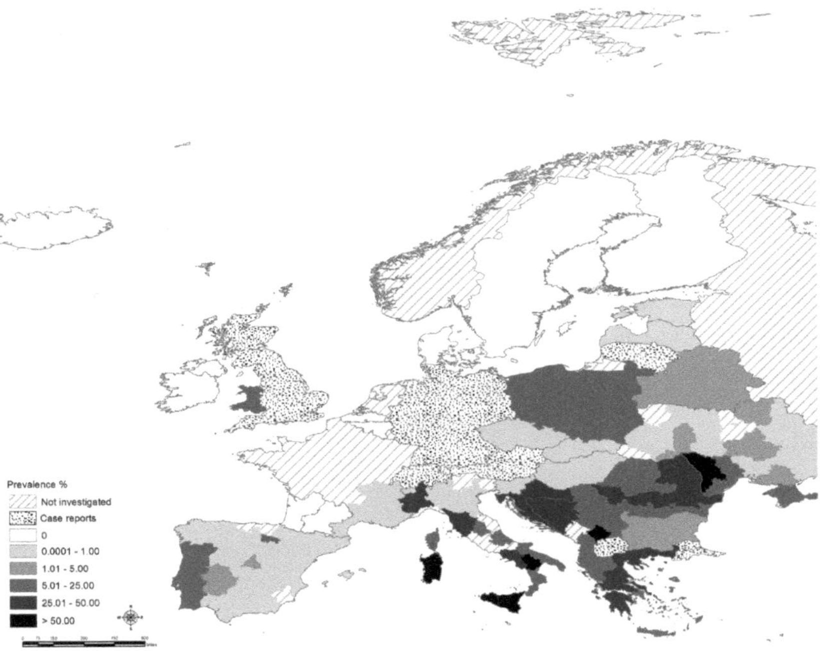

Figure 8 ***Current distribution of Echinococcus spp. causing cystic echinococcosis in domestic intermediate hosts (sheep, cattle, pigs and boar) in Europe.*** The detailed information (prevalence data in each jurisdiction) is listed in Table S7 of the Supplementary Material.

depth (Oksanen and Lavikainen, 2015). The cycle is mainly perpetuated between wolves as definitive and cervids as intermediate hosts (the involvement of working dogs has decreased). *Echinococcus canadensis* (G10) was identified in Finland in wolves, cervids and in one human (Oksanen and Lavikainen, 2015).

3.4.1.1.2 Infections in animals For the **UK** and **Ireland**, there is very limited data on the prevalence of *Echinococcus* spp. in animal hosts (see Fig. 8 and Table S7 in the Supplementary Material). However, in the UK *E. granulosus* (G1–3) has been found in dogs, sheep and cattle whilst *E. equinus* (G4) has been isolated from horses, dogs and from two captive mammals: zebra (*Equus burchellii*) and lemur (*Varecia rubra*) (Boufana et al., 2015b). In **Ireland**, *E. equinus* has been found in horses and dogs (Hatch, 1970; Kumaratilake et al., 1986). Ireland is not believed to be endemic for other species/genotypes of *E. granulosus* s.l. (Torgerson and Budke, 2003). In the north of Scandinavia, *E. canadensis* is found in cervids (1.2% in Finland,

1.6% in Sweden) and wolves (10–46% in Finland) (Oksanen and Lavikainen, 2015).

3.4.1.1.3 Cystic echinococcosis in humans In the **UK**, there were a total of 110 hospital admissions in Scotland with a diagnosis of CE between 1968 and 1989. The highest incidence values were recorded in the Western Isles (annually 2.5 cases per 10^5 inhabitants), Shetland (1.2) and Highland (0.2) (Braddick and Reilly, 1993). Between 1974 and 1983 the average annual incidence in Wales was $0.4/10^5$ with 0.02 cases in England. Within Wales, Powys and particularly Brecknock had the highest incidence of approximately 7 cases per 10^5 inhabitants (Palmer and Biffin, 1987). Because of the high incidence in Powys, a control programme was implemented in 1983 that led to the successful reduction in the cases of human CE in the area (Palmer et al., 1996). Unfortunately, the dog dosing programme was suspended in 1989 and since then there has been evidence of renewed transmission in animal hosts (Mastin et al., 2011). Nevertheless, generally human CE has been decreasing in incidence in the UK. Between 2005 and 2009, 52 cases were reported as confirmed CE ($0.03/10^5$ per year), although many of these cases were diagnosed in immigrants (Halsby et al., 2014). However, these might be underestimates as 227 hospitalizations and 2 deaths were attributed to CE in England between 2005 and 2009. No cases were reported from Scotland during this period. Other data demonstrated 46 cases of CE reported from 2008 to 2012 (European Centre for Disease Prevention and Control, 2013).

Ireland is believed to be nonendemic for *E. granulosus* and no reports of autochthonous cases of CE have been registered, although it is endemic for *E. equinus* (Torgerson and Budke, 2003).

In **Scandinavia,** human CE caused by *E. canadensis* (G8 and G10) involved primarily the lungs causing a relative benign disease; since the 1960s only one CE case caused by *E. canadensis* (G10) has been diagnosed in eastern Finland (Oksanen and Lavikainen, 2015).

3.4.2 Central Europe: Belgium, The Netherlands, Luxembourg, Germany, Switzerland, Austria and Czech Republic

In Austria, Germany and Switzerland, historically *E. ortleppi* (cattle strain, G5) was endemic (Eckert et al., 2001), and there are no recent findings published. Local transmission of *Echinococcus* genotypes causing CE has not been well documented in this area over the last decades. However, several CE cases in humans have been considered to be autochthonous in Austria and

Germany; and therefore parts of these countries have still to be regarded endemic areas with sporadic cases.

3.4.2.1 Infections in animals

No systematic data are available but slaughterhouse reports indicate that CE in livestock is nearly absent. In one case, G7 was identified in a cyst from a pig in **Austria** (H. Auer, personal communication). From **Germany**, *E. granulosus* findings date back to the 1970s, but no current data have been published. Sporadic cases of *E. granulosus* s.s. in sheep have been observed in the vicinity of Stuttgart, but infection sources are not clear (Romig, personal communication).

In **Switzerland**, *E. granulosus* is occasionally diagnosed in imported dogs and in single cases in ruminants. In one case, *E. granulosus* (G1) transmission in a sheep flock was observed (Deplazes P., personal communication).

Occasionally *E. equinus* has been diagnosed in Switzerland, Germany and Belgium but these cases are most probably imported. A case of CE in the lungs of a horse foaled and raised in Germany was documented, but sources of such rare cases are unclear (Blutke et al., 2010). In the **Czech Republic**, in the period of 2005 and 2007, CE was reported in 6 cows, 267 pigs and 33 sheep, considered sporadic relative to the total number of slaughtered animals (Svobodová, 2014).

3.4.2.2 Cystic echinococcosis in humans

Most of the CE cases diagnosed in humans have been classified as imported cases. For example, in the **Netherlands**, nearly 30 cases of CE in patients originating mainly from areas around the Mediterranean Sea are confirmed each year (Herremans et al., 2010). In Switzerland, no autochthonous CE cases are documented, and around 50 imported cases of CE are reported each year in immigrants.

For **Germany** the annual incidence (2001–13) of CE per 10^5 inhabitants was calculated to be 0.05 (of total 552 CE cases, 111 were in the autochthonous populations) (Torgerson, 2017). In a series of 37 CE patients with genotyped *Echinococcus* (35 *E. granulosus* s.s., 2 *E. intermedius* G7) treated in a southern German hospital between 1999 and 2011, only one patient (with G1 infection) originated from central Europe (Wagner, 2014). However, autochthonous infection may occur sporadically: for at least 2 out of 15 CE patients who grew up in Germany, diagnosed 1999–2008, there is no reasonable doubt that they acquired the infection locally (Richter et al., 2009).

In a retrospective investigation of CE in **Austria**, the majority (92%) of 23 autochthonous cases harboured cysts belonging to G7, while immigrants were mainly infected with G1 or G6. All in all, the annual incidence of CE in Austria was estimated to be 0.4 per 10^5 inhabitants (Schneider et al., 2010). Similarly, in the **Czech Republic**, CE is reported occasionally. Since 2005, 10 cases have been treated in two hospitals (Prague, Liberec); of these, only one is likely autochthonous (Stejskal et al., 2014).

3.4.3 Eastern Central Europe: Poland and Baltic countries, Belarus, Ukraine, Moldova, Slovakia, Hungary

3.4.3.1 Host assemblages, transmission and molecular epidemiology

In the Baltic region (**Lithuania, Latvia** and **Estonia**) CE has been known to occur since the beginning of the 20th century. For example, in Estonia the first description dates back to 1904 (Marcinkutė et al., 2015). So far, four genotypes have been identified (Table 6). In **Lithuania**, *E. intermedius* (the pig strain, G7) with a farm dog–pig cycle maintained by home slaughter practices has been predominantly observed in humans, pigs, cattle (sterile cysts) and in dogs (Bruzinskaite et al., 2009). The cervid strains of *E. canadensis* (G8, G10) transmitted by wolves as definitive hosts and cervid species as intermediate hosts have been identified in **Estonia** and **Latvia** (Marcinkutė et al., 2015); (Oksanen and Lavikainen, 2015) and *E. granulosus* (G1) was found in a dog in Estonia (Marcinkutė et al., 2015) but the importance of this cycle is not known. In **Poland**, CE also occurs in animals (mainly pigs) and in humans; in people, *E. intermedius* (G7) and *E. granulosus* (G1) have been confirmed (autochthonous origin could not be excluded in two patients with G1) (Dybicz et al., 2015; Marcinkutė et al., 2015).

In **Moldova**, **Ukraine** and **Belarus**, countries of the former Soviet Union, the presence of *Echinococcus* species causing CE as well as *E. multilocularis* has been known for several decades. The Republic of Moldova is recognized as a CE endemic area, where human disease is considered as a high-priority public health problem. Similarly, in the Ukraine human CE cases are reported regularly, while a few cases have been registered in Belarus.

In **Slovakia and Hungary**, CE had been known to occur regularly in the past, likely attributed to high prevalence in slaughtered pigs on many small breeding farms before the 1990s (Turčeková et al., 2009). Later, economic and social changes caused a decrease in the numbers of family farms,

Table 6 Genotypes of *Echinococcus* spp. causing cystic echinococcosis in eastern Central Europe: *Echinococcus granulosus* (G1–3), *Echinococcus equinus* (G4), *Echinococcus ortleppi* (G5), *Echinococcus intermedius* (G6/7) and *Echinococcus canadensis* (G8, G10) (no data found in the missing countries)

Country	Human	Dog	Wild canids	Cattle	Pig (P), Wild boar (Wb)	Sheep (S), Cervids (C)
Poland	G7 and G1[1]				P: G7[3]	
Estonia		G1[2]	G8[2]			C: G8 and G10[2,4]
			G10[2]			
Latvia			G10[2]			
Lithuania	G7[5]	G7[2,5]		G7[2,5]	P: G7[2,5]	
Ukraine					P, Wb: G7[6]	
Moldova				G1[7]		S: G1[7]
				G3[7]		S: G3[7]
Slovakia	G7[8] and G1–3★				P: G7[8]	

[1] (Dybicz et al., 2015); [2](Marcinkutė et al., 2015); [3] (Karamon et al., 2012); [4] (Moks et al., 2008); [5] (Bruzinskaite et al., 2009); [6] (Kedra et al., 2000); [7] (Umhang et al., 2014a); [8] (Turčeková et al., 2009); ★ Antolová D. unpublished data.

and pig husbandry moved almost exclusively indoors. Moreover, preventive measures and the improvement of veterinary meat inspection contributed to the decrease of human CE cases as well as the decline of CE in slaughtered animals.

3.4.3.2 Infections in animals

In **Lithuania, Estonia and Latvia**, *Echinococcus* species causing CE have been described in the past and were recently reviewed in Marcinkutė et al. (2015).

In **Estonia**, historically CE was reported in pigs, sheep and wild cervids (Marcinkutė et al., 2015). In more recent studies, *E. granulosus* has been found in Estonian wildlife including moose and roe deer (*Capreolus capreolus*) as intermediate hosts, with the grey wolf as definitive host. The Estonian Veterinary and Food Laboratory has only reported a few cases of CE in farm animals, e.g., in 2004, 4 of 444,084 pigs (0.0009%) and 8 of 6202 moose (0.1%) were infected. Cysts were also detected in 2 of 1787 imported reindeer (0.1%) in 2005, and in 1 of 53,903 cattle (0.002%) (Marcinkutė et al., 2015). In 2003, adult *Echinococcus* cestodes were detected in one of 26 grey wolves, and in 2004–05 CE was detected in the lungs of 16 (0.8%) of 2038 hunted moose (Moks et al., 2008). Of these, 11 belonged to genotype G8 and 5 to genotype G10. This was the first record of G8 in Eurasia (Moks et al., 2008).

In **Latvia**, historical data document a pig – dog life cycle which was considered typical for the region. Recent low prevalence (<0.2%) of echinococcosis in farm animals in the period 2004–07 and 2010 and no detected infections in dogs suggest that a cycle in domestic animals is not of relevance in Latvia (Marcinkutė et al., 2015). Intestinal *Echinococcus* stages detected in 1 (2.9%) of 34 wolves were later confirmed to be *E. canadensis* (G10). The populations of wildlife involved in the known *E. canadensis* (G10) life cycle are stable with a tendency to increase in Latvia (Marcinkutė et al., 2015).

In **Lithuania**, historic and recent data document mainly the presence of a pig–dog *Echinococcus* cycle. A recent study (2005–06) performed in the southwestern part of Lithuania detected CE in 13.2% (81/612) of pigs reared in small family farms and in 4.1% of those reared in industrial farms. Molecular analysis of isolated taeniid eggs revealed *Taenia* spp. in 10.8%, *E. intermedius* (G6/7) in 3.8% and *E. multilocularis* in 0.8% of the dogs investigated in the same area. In addition, three samples from the livers of humans, one sterile cyst from a cow, seven samples from pigs, and eggs from faeces of eight dogs were confirmed as *E. intermedius* (G6/7) (Bruzinskaite et al., 2009).

In **Poland**, CE has been documented in the past in pigs (4.5%), beef cattle (0.007%), sheep and goats (18.7%) (Derylo, 1998). According to more recent data, *E. granulosus* s.l. was detected in 4.1% of 267 pigs from southern Poland and the pig strain (G7) seems to be the most common in Poland (Dybicz et al., 2015).

Data from **Belarus**, published in 2000 and 2002, reported the occurrence of *E. granulosus* in 0.15% of cattle, 0.7–1.2% of sheep and 6–7% of pigs as well as in 5–6% of dogs (Malczewski, 2002). Furthermore, *Echinococcus* cestodes were detected in the intestines of 6/52 (11.5%) wolves (Shimalov and Shimalov, 2000).

In Ukraine (northeast), the presence of *Echinococcus* intestinal stages was documented in 6.25% of wolves (Kornyushin et al., 2011) and in 10.9% of dogs (Varodi et al., 2007). CE has been seen in 4/58 (5.2%) of wild boars and 1/400 red deer (0.7%) (Yemets, 2013). The highest prevalence in sheep occurs in the Crimea (21.3%) and Odessa regions (12.2%) (Litvinenko, 2015), whilst in cattle the highest prevalences are seen in Odessa (3.3%) and Kirovohrad regions (2.4%) (Litvinenko, 2012). In pigs the two districts with the highest prevalence are the Khmelnytskyi and Sumy regions (3.1% and 2.7%, respectively) (Litvinenko, 2013). Molecular analyses confirmed the presence of *E. intermedius* (pig strain, G7) in two pigs and a wild boar from Sumy region (Kedra et al., 2000).

In the **Republic of Moldova** a slaughterhouse survey was conducted in 2012 to estimate the prevalence of CE in cattle, sheep and pigs. The infection was highly prevalent in cattle (904/1525, 59.3%) and sheep (3450/5580, 61.9%), while no positive pigs (0/12,700) were found. The prevalence of infection was significantly higher in animals raised in private households than in those from collective farms (Chihai et al., 2016). In both cattle and sheep, only the occurrence of *E. granulosus* G1 and/or G3 was confirmed (Umhang et al., 2014a). In **Slovakia**, the annual prevalence of CE in pigs between 2000 and 2008 ranged from 0.02% to 0.13% (average: 0.08%), with decreasing tendency, especially after 2005 (Turčeková et al., 2009). Few data about the occurrence of intestinal *Echinococcus* spp. in carnivores are available; *E. granulosus*-like cestodes (genotype not determined) have been detected in two golden jackals (*Canis aureus*) in **Hungary** (Takacs et al., 2014).

3.4.3.3 Cystic echinococcosis in humans

For the **Baltic countries**, the historical and actual situations of CE have been reviewed in Marcinkutė et al. (2015). In **Lithuania**, incidences of clinical CE were estimated to be $0.03/10^5$ inhabitants in 1958 and 0.1 in 1960, but later in 2005 the incidence was 0.39 and it remained at a higher level of 1.11–1.15 for the years 2009–13 (Marcinkutė et al., 2015). As no obligatory notification of CE exists in Lithuania, and these data were recorded in two hospitals only, it can be assumed that the situation is underestimated in some districts. Most CE cases were registered from southeastern and northwestern areas of Lithuania with particularly high number of CE cases registered from the Vilnius district (Marcinkutė et al., 2015).

In **Latvia**, during the period 1999–2005, 29 CE cases were registered, but since 2001, an increase in human CE cases has been recorded. In the Infectology Centre of Latvia, 11 new cases were registered ($0.43/10^5$ inhabitants per year) in 2005 and subsequently, the number of diagnosed CE cases rose to 17 in 2008 ($0.77/10^5$), but decreased and remained stable in the period between 2009 and 2012 ($0.27–0.34/10^5$). Data of 93 patients with CE, diagnosed between 2002 and 2012, document the public health significance of CE in Latvia [Laivacuma and Viksna, cited in Marcinkutė et al. (2015)], but since echinococcosis is not a notifiable infectious disease in Latvia, the number of presently reported cases is probably underestimated.

To date, 13 cases of echinococcosis have officially been registered in humans in **Estonia** but AE and CE were not differentiated. However, based on clinical descriptions in two cases, CE can be suggested to be present in the country (Marcinkutė et al., 2015).

In **Poland**, a relatively low incidence of CE has been registered in recent years. In 2009 and 2010, 25 and 34 cases of CE were registered, respectively (incidence = 0.07 and $0.09/10^5$ inhabitants) (WHO, 2015, European Hospital Morbidity Database, World Health Organization Regional Office for Europe: http://data.euro.who.int/hmdb/). Most cases were reported in the Masovian province (33% of all reported cases, incidence $0.23/10^5$ inhabitants), and the smallest number in the Kuyavian-Pomeranian province (1 case, incidence $0.05/10^5$ inhabitants). More cases of echinococcosis (70%) were recorded in the countryside. In both rural and urban areas, women were more affected (86%) than men (Golab and Czarkowski, 2014; Waloch, 2012). Serological investigations (ELISA, Western blot) by the National Institute of Public Health confirmed CE in 162 patients of 5483 persons suspected of echinococcosis in the period of 2003 and 2010 (Wnukowska et al., 2011).

In **Belarus**, the incidence of CE appears quite low. In 2010–11, there were 20 cases reported (Anichkin and Martyniuk, 2012) resulting in an annual incidence of 0.1 cases per 10^5 inhabitants per year. In **Moldova**, CE seems to be of major importance; indeed 1770 human CE patients were reported between 2000 and 2010 (a mean of 177 per year) with an annual incidence of 4.9 CE cases per 10^5 inhabitants (Lungu, 2013).

In **Ukraine**, 2153 human CE cases were recorded between 2000 and 2013 (Litvinenko and Polokokovska, 2015) giving an annual incidence of 0.36 per 10^5 inhabitants. Nearly half the cases (986) were reported from the Odessa region which has an annual incidence of 2.9 cases per 10^5 inhabitants.

In **Slovakia**, human CE occurs sporadically, with a few cases reported to the Slovak Health Authorities every year, although most can be classified only as possible or probable cases. At the Institute of Parasitology of Slovak Academy of Sciences the diagnosis was confirmed in four patients between 2012 and 2015. In a woman from eastern Slovakia, *E. intermedius* (G7) was reported, while *E. granulosus* (G1–3) was recorded in one man. As the patient had travelled several times to Romania in the past, the case is not necessarily autochthonous (Antolovà D., personal communication). In **Hungary**, few case reports of echinococcosis or CE have been published (Casulli et al., 2010a; Csotye et al., 2011) and the origin of these patients is not documented.

3.4.4 Southern Europe: Portugal, Spain, France, Italy, Greece

3.4.4.1 Host assemblages and transmission

Echinococcus granulosus has been known in southern Europe since ancient times but it was only in 1801 that Rudolphi established the genus *Echinococcus*, the name referring to the small, round, 'spiny' protoscoleces found in the cysts (Romig et al., 2015).

Nowadays, CE remains the most important helminth zoonosis in southern Europe, causing serious consequences in terms of public health and the economy due to the considerable morbidity rates both in the public health sector and in livestock industries (Seimenis, 2003). The main way of transmission is through the domestic cycle, involving dogs as definitive hosts and farm livestock (especially small ruminants) as intermediate hosts. Eggs shed by infected dogs (especially free-roaming dogs living in rural areas) remain the most important source of infection for humans and other intermediate hosts. The role of wild animals in transmission has not yet been accurately studied (Seimenis, 2003). However, a wild animal cycle maintained among wild carnivores acting as definitive hosts (Gori et al., 2015; Guberti et al., 2004; Guerra et al., 2013) and wild boars as intermediate hosts have been documented in south-central Spain (Martin-Hernando et al., 2008), central

Italy (Busi et al., 2007) and the island of Corsica in France (Umhang et al., 2014c).

Risk factors for infection of intermediate and definitive animal hosts with *E. granulosus* have been recently reviewed by Otero-Abad and Torgerson (2013). The close proximity of humans to animals, traditional types of husbandry (especially for sheep and goats), clandestine home slaughter with insufficient facilities to destroy infected offal to which dogs have free access, and the high number of stray dogs and sheepdogs are some of the most important factors that allow the spread of CE in southern Europe (Garippa, 2006; Seimenis, 2003). In the Mediterranean area, pastoralism is a major occupation, with transhumance being very common. Usually, each rural family keeps one or more dogs for guarding, herding, hunting and/or companionship. These animals in some areas may be fed offal as food sources and/or have access to the location where animals are slaughtered as well as to livestock rearing areas and carcasses. The ability of dogs to roam freely is one of the most commonly reported risk factors for *E. granulosus* infection (Otero-Abad and Torgerson, 2013). Another important risk factor for *E. granulosus* infection is associated with the dog owner's lack of knowledge about parasite transmission or deficiencies in anthelmintic treatment. Additionally, the cultural and economic background of the owners has been found to be related to infection risk in dogs (Otero-Abad and Torgerson, 2013).

A particular epidemiological situation can be observed in southern Italy (Campania region), including the role of a sheep/dog cycle for the transmission of CE to cattle and water buffalo (Cringoli et al., 2007). In this region, illegal farm-slaughter of large ruminants is nearly absent and cattle and water buffaloes are slaughtered only in modern and efficient abattoirs, where the presence of canids is strictly forbidden. A study using geospatial tools confirmed the role of sheep in CE transmission based on the close proximity of sheep farms to positive bovines and/or water buffaloes, presumably because free-ranging canids with access to infected sheep carcasses contaminated the cattle/buffalo farms with *Echinococcus* eggs (Cringoli et al., 2007).

Control initiatives for CE in southern Europe started in the second half of the 19th century and have since been implemented in whole territories or in selected regions. However, to date CE has been successfully controlled at a national level only in parts of Cyprus (Economides et al., 1998). In other areas, the initiatives led only to a temporary reduction in animal and human CE prevalence (Magnino et al., 2014).

3.4.4.2 Molecular epidemiology

Four *Echinococcus* species including several genotypes have been documented so far to be circulating in southern Europe: *E. granulosus* (G1–3), *E. equinus* (G4), *E. ortleppi* (G5) and *E. intermedius* (G7) (Table 7).

Echinococcus granulosus (G1–3) is the species most frequently identified in domestic animals and humans in southern Europe (Beato et al., 2013; Busi et al., 2007; Gonzalez et al., 2002). Although the typical life cycle patterns involve livestock and domestic dogs, *E. granulosus* is also known to occur in wolves in Italy and Spain (Gonzalez et al., 2002; Gori et al., 2015) and in wild boars in Spain, Italy and Greece (Busi et al., 2007; Mwambete et al., 2004; Varcasia et al., 2007), highlighting that the wild animal cycle must be also taken into account in the epidemiology of *E. granulosus.*

Echinococcus equinus (G4) has been documented in horses from Spain and Italy (Gonzalez et al., 2002; Varcasia et al., 2008a) and *E. ortleppi* (G5) was found in cattle from Italy and France (Casulli et al., 2008; Grenouillet et al., 2014). In a study conducted in France (Casulli et al., 2008; Grenouillet et al., 2014), evidence of liver infections caused by *E. ortleppi* was reported in two humans.

Echinococcus intermedius (G6/7) is prevalent in France, Greece, Italy, Spain and Portugal in goats, pigs and wild boars. Furthermore, a study conducted with hunting dogs in Corsica showed the presence of *E. intermedius* (G6/7) in 1.2% of animals examined (Umhang et al., 2014c).

In southern **France**, the presence of *E. granulosus* (G1–3) in sheep and cattle has been documented (Umhang et al., 2013b) and, in Corsica, Grenouillet et al. (2014) and Umhang et al. (2014c,d) provided evidence of infection with *E. ortleppi* in cattle and humans and *E. intermedius* (G6/7) in pigs, wild boars and hunting dogs (Umhang et al., 2014c,d).

In **Greece**, *E. granulosus* was a dominant species in livestock, especially sheep, goats and buffaloes (Chaligiannis et al., 2015; Varcasia et al., 2007; Roinioti et al., 2016), and also in wild boars (Chaligiannis et al., 2015). Furthermore, *E. intermedius* (G7) was identified in goats (Chaligiannis et al., 2015; Varcasia et al., 2007) and in sheep (Roinioti et al., 2016).

In **Italy**, six genotypes have been identified in different hosts (Table 7), but their distribution was not uniform. *E. granulosus* (G1–3) is the species most frequently identified from intermediate and definitive hosts. In 2008, the presence of G2 in water buffaloes from southern Italy (a Mediterranean area) was reported for the first time (Capuano et al., 2006; Casulli et al., 2008; Rinaldi et al., 2008a), and Calderini et al. (2012) reported the occurrence of the G3 genotype in goats. *Echinococcus granulosus* (G1–3)

Table 7 Genotypes of *Echinococcus* spp. causing cystic echinococcosis in southern Europe: *Echinococcus granulosus* (G1−3), *Echinococcus equinus* (G4), *Echinococcus ortleppi* (G5), *Echinococcus intermedius* (G6/7) and *Echinococcus canadensis* (G8, G10) (no data found in the missing countries)

Country	Human	Dog (D), Wolf (W)	Sheep(S), Goat (G)	Cattle (C), Buffalo (B)	Equine	Pig (P), Wild boar (Wb)
France	G5[1]	G6/7[2]	S: G1−3[3]	C: G1−3[3] C: G5[1]		P and Wb: G6/7[4]
Greece			S: G1−3[5,6,21] S: G7[21] G: G1−3[5] G7[6]	B: G1−3[7]		Wb: G1−3[6]
Italy	G1−3[8]	D: G1−3★ W: G1−3[9]	S: G1−3[8] G: G3[10]	C: G1−3[8,9,10,11] C: G5[8,9,10,11] B: G1−3[9,12,13]	G4[14,15]	P: G1[11] P: G7[11] Wb: G1[8]
Portugal	G1−3[16]	W: G6/7[17]	S: G1−3[16,18] G: G1−3[16]	C: G1−3[16,18] C: G7[16]		
Spain	G1[19]	W: G1[19,20,21]	S: G1[19,20] G: G1[20] G7[20]	C: G1[20]	G4[19,20]	P: G1[19,20] P: G7[19] Wb: G1[20] G7[20]

[1] (Grenouillet et al., 2014); [2] (Umhang et al., 2014c); [3] (Umhang et al., 2013b); [4] (Umhang et al., 2014d); [5] (Chaligiannis et al., 2015); [6] (Varcasia et al., 2007); [7] (Chaabane-Banaoues et al., 2015); [8] (Busi et al., 2007); [9] (Casulli et al., 2008); [10] (Rinaldi et al., 2008b); [11] (Varcasia et al., 2006); [12] (Capuano et al., 2006); [13] (Rinaldi et al., 2008a); [14] (Scala et al., 2006); [15] (Varcasia et al., 2008a); [16] (Beato et al., 2013); [17] (Guerra et al., 2013); [18] (Beato et al., 2010); [19] (Gonzalez et al., 2002); [20] (Mwambete et al., 2004), [21] (Roinioti et al., 2016); ★ Cringoli G. unpublished data.

has been confirmed in dogs (Maurelli et al., 2015) and in humans (Busi et al., 2007) in the Campania region, *E. intermedius* (G7) in pigs in Sardinia (Varcasia et al., 2006), *E. ortleppi* in cattle in the Lombardy region (Casulli et al., 2008), and *E. equinus* in horses of Tuscany, Sardinia and Sicily (Scala et al., 2006; Varcasia et al., 2008a).

In southern **Portugal**, there is a predominance of *E. granulosus* (G1) both in ruminants and in humans (Beato et al., 2010, 2013). In the northern part of the country, *E. intermedius* (G7) in pigs seems to be more prevalent (Guerra et al., 2013). Recent data highlight the finding of the G7 genotype in wolves (Beato et al., 2013; Guerra et al., 2013) and in cattle (fertile cysts) in northern Portugal (Beato et al., 2013; Guerra et al., 2013). In **Spain**, molecular studies showed the existence of three species: *E. granulosus*, *E. equinus* and *E. ortleppi*.

3.4.4.3 Infections in animals

CE was widespread in **Cyprus** before the 1970s. Baseline data show that the parasite was present in 40–100% adult sheep, 20–50% cattle, 27–93% goats and 5–22% pigs (Economides et al., 1998) and very common (up to 40%) among dogs (Polydorou, 1983). A successful campaign of eradication was achieved from 1971 to 1985 including a drastic reduction of the dog population. After the division of Cyprus in 1974, the control program was consolidated in the Greek Cypriot sector and this part of the island is virtually free of CE transmission. In the Turkish Republic of Northern Cyprus, the control program was abandoned and CE remained endemic with a prevalence of about 40% in livestock (Economides and Christofi, 2000).

France is still considered an endemic area for CE (Fig. 8; Table S7 in Supplementary Material), but current prevalence of CE in intermediate hosts remains unknown due to the absence of official data reporting for the last 20 years (Umhang et al., 2013b). In 1989, a nation-wide slaughterhouse survey revealed the following CE prevalences in livestock in southern France: 0.42% in sheep and goats, 0.13% in cattle and 0.009% in pigs (Soule et al., 1995). In a 1994 survey, 0.31% of 43,148 cattle slaughtered in the Midi-Pyrenees were infected (Bichet and Dorchies, 1998). More recently, a cross-sectional study conducted in 2009–10 confirmed the low prevalence of CE in animals in France, with a total of 27 of 725,903 (0.00004%) positive sheep from the Alpes-de-Haute-Provence department and 4 of 138,624 (0.00003%) cattle from the Herault and Haute-Savoie departments (Umhang et al., 2013b). Moreover, 85% of the sheep cysts were fertile while none of the infected cattle exhibited fertile cysts, supporting that cattle do not play a role in transmission of *E. granulosus* (G1) in Europe (Umhang et al., 2013b). The main infected area was the south of France, where sheep breeding and traditional transhumance is concentrated. CE is historically present at high prevalence in Corsica, a French Mediterranean island, with 46% of sheep infected in 1960, and still 15% 20 years later (Umhang et al., 2014d). A recent slaughterhouse survey in Corsica showed a prevalence of 5.9% in pigs, and at the same time a similar prevalence of 4.0% in wild boars was reported (Umhang et al., 2014d). Interestingly, one of the four infected wild boars harboured fertile cysts, which highlights the potential role of this wild species as an intermediate host. More recently, a survey in Corsica revealed *E. intermedius* (G6/7) in 1.2% of hunting dogs (Umhang et al., 2014c).

In **Greece**, CE was widely prevalent long before the 1970s (Sotiraki and Chaligiannis, 2010). The prevalence of infection in farm animals was as high as 82% in cattle, 80% in sheep, 24% in goats and 5% in pigs (Sotiraki et al., 2003). In definitive hosts, prevalences were as follow: sheepdogs 50.4%, sentinel dogs 26.9%, hunting dogs 19.2%, urban stray dogs 9.3% and companion (pet) dogs 0–1% (Sotiraki et al., 2003). A control programme has been in force since 1984, and surveillance in livestock species since 1998 has documented prevalences of 31.3% in sheep, 10.3% in goats, 0.6% in pigs and 0% in cattle (Sotiraki et al., 2003). A survey conducted by Varcasia et al. (2007) in sheep and goats in the Peloponnesus area in southern Greece revealed prevalence values of 30.4% and 14.7%, respectively. Fertile cysts were found in 16.2% and in 7.4% of sheep and goats, respectively. Similar results were obtained by Christodoulopoulos et al. (2008) in sheep flocks in Thessaly (central) Greece from 2002 to 2006, with values of 39.3% (76.7% fertile cysts). Recently, Chaligiannis et al. (2015) conducted an epidemiological survey in different geographical regions of Greece (Thrace, Thessaly, Western and central Macedonia). In the Thrace region, hydatid cysts were found in 33.3% of sheep, while prevalences were 30.3% and 26.1% in sheep of western and central Macedonia, respectively. The highest CE prevalence was found in sheep of Thessaly (53.8%). While not detected in goats from Thrace and western Macedonia, 0.4% of goats and 42% of water buffalo in central Macedonia were infected. Fertile cysts were found in 6.4% of sheep, 3.2% of goats and 7.9% of buffaloes, but not in wild boars.

Regarding **Italy**, *E. granulosus* infection is prevalent in all parts of the country but regional differences (sporadic, endemic and hyperendemic areas) have been identified along a north-south gradient [reviewed in

Garippa (2006)]. CE has a sporadic distribution in northern regions where CE prevalence in farm animals are the lowest registered in Italy (<1%) (Manfredi et al., 2011). In central Italy CE is usually recorded in livestock with lower prevalence levels than in the southern regions. In Abruzzo and Tuscany, CE prevalences in sheep were 22% and 47%, respectively (Garippa, 2006). Regarding southern Italy, in Campania CE prevalence values were 33.3–75.0% in sheep (Cringoli et al., 2007; unpublished data), 10.4% in cattle (Rinaldi et al., 2008b) and 10.5% in buffaloes (Capuano et al., 2006), whereas in Basilicata 67.7% of sheep were affected (Cringoli et al., 2007; unpublished data) and further south in Calabria the prevalence is approximately 15% in sheep (Vincenzo Musella, University Magna Graecia of Catanzaro, Italy, personal communication). The highest CE prevalence was observed in Sardinia and Sicily at 75.0% (Scala et al., 2006) and 57.6% (Giannetto et al., 2004) with a fertility of 10.3% and 9.2%, respectively, whereas CE prevalence in cattle was 41.5% (fertile cysts 2.6%) and 67.1% (fertile cysts 4%) in the same areas (Garippa, 2006). CE has also been reported in different Italian regions in pigs, with a prevalence of 9.4–11.1% (Garippa et al., 2004; Varcasia et al., 2006) and 3.7% in wild boars (Varcasia et al., 2008b). A CE prevalence of less than 1% has been reported in horses (Varcasia et al., 2008a). In dogs, prevalence of *E. granulosus* is generally less than 6% (according to Garippa, 2006; Maurelli et al., 2015; Varcasia et al., 2011). Interestingly, *E. granulosus* was also isolated in 5.9–15% of wolves of northern Italy (Gori et al., 2015; Guberti et al., 2004). The role of the wolf as a definitive host for *E. granulosus* is confirmed in Italy in the Apennines.

In **Portugal**, CE has been recognized as a public health problem and is notifiable since 1987, but epidemiological studies are scarce. In the northern province of Trás-os-Montes, *E. granulosus* prevalences of 8–11%, 30% and 7–12% in swine, small ruminants and dogs were reported, respectively (de Carvalho and Guerra, 2014). Sheep, goats and cattle of the southern regions showed the highest rates of infection (de Carvalho and Guerra, 2014).

In **Spain**, CE is considered endemic, associated mainly with extensive or semiextensive sheep raising in the central part of the country, but *E. equinus* is also present (Carmena et al., 2008). Although specific control programs initiated in the 1980s have led to marked reductions in CE infection rates, the disease remains an important public health problem in the northeastern, central and western parts of the country. Updated data on the current prevalence of *E. granulosus* in definitive and intermediate hosts in Spain was given by Carmena et al. (2008). La Rioja and Aragon were the Autonomous Regions (ARs) with the highest prevalences for ovine/caprine CE, at 22.7%

and 2.0%, respectively. A similar prevalence trend and geographic pattern of disease was observed for bovine CE, with the Bask Country (3.8%) and Extremadura (2.2%) being the ARs with the highest prevalence of disease. In contrast, low CE prevalences (<0.1) in sheep, goats and cows were found in the Mediterranean and Cantabric coastal regions. In pigs, the mean prevalence of CE was 0.03%. Extremadura was the region with the highest infection rates for pigs (0.38%). The overall prevalence of *E. granulosus* infection in sheepdogs in the province of Alava was 8.0% and is considered a public health threat (Benito et al., 2006). As well, 15% of Iberian wolves harboured *E. granulosus* thus confirming the importance of a wild animal cycle (Sobrino et al., 2006).

3.4.4.4 Cystic echinococcosis in humans

CE is still regarded as an important zoonosis with a high burden of disease in southern Europe (see Table S8 in the Supplementary Material). Historically in **Cyprus**, an annual surgical incidence rate of 12.9 per 10^5 inhabitants was recorded. In the 'Greek sector' of the island, parasite transmission was controlled, resulting in very few human CE cases per year, whereas in the 'Turkish sector', the incidence remained high (8 cases per 10^5 inhabitants) (Economides and Christofi, 2000). In **France**, human cases of CE are most often considered to have been imported with immigrants (Umhang et al., 2013b). Corsica is the most important focus of human CE in France with an annual incidence of 1.3 cases per 10^5 inhabitants. In **Portugal**, the incidence of CE is variable with an increase from northern to southern regions, with the highest incidence in the area of Evora (3.2/10^5 inhabitants between 2004 and 2008) (de Morais, 2010). In **Spain**, the highest incidence of the disease occurs in the northeastern, central and western parts of the country, in the range of 1.1−3.4 cases per 10^5 inhabitants. Aragon, La Rioja and Castile-Leòn are the regions with the highest incidence of CE (3.41, 2.46 and 2.02 cases per 10^5 inhabitants). In contrast, coastal areas present relatively low incidence rates of human CE (Carmena et al., 2008). In **Greece**, after the CE control campaign, CE incidence has been dramatically reduced from 12.9 in 1984 (Sotiraki et al., 2003) to 0.25 per 10^5 inhabitants (ECDC, 2010). In **Italy**, human CE continues to be a public health concern with an annual incidence of 1.6/10^5 inhabitants. The highest average incidence rates were observed in Sardinia and Sicily, 6.8/10^5 and 4.0/10^5 respectively, followed by the south regions with an average incidence of 1.9/10^5 but as high as 5.4/10^5 in Basilicata and the central regions with an incidence of 1.07/10^5 inhabitants (1.65 in Latium). The average incidence in the northwest and the northeast of Italy were lower, with a registered incidence of $0.47/10^5$ and $0.36/10^5$ inhabitants, respectively (Brundu et al., 2014).

Each case of CE results in a mean of 0.97 DALYs (Torgerson et al., 2015). With over 1000 cases per year in both Italy (Brundu et al., 2014) and Spain (Herrador et al., 2016), it can be seen that there will be at least 2000 DALYs per annum due to CE in these two countries. These countries dominate the disease burden of CE in southern Europe simply because of their large size and relatively large numbers of cases of CE. The burden in other endemic countries in the region can easily be seen from the relative incidence or numbers of cases reported above.

3.4.5 Southeastern Central Europe: Romania, Bulgaria, Serbia, Croatia, Slovenia, Bosnia and Herzegovina, Kosovo, FYROM and Albania

3.4.5.1 Host assemblages, transmission and molecular epidemiology

CE is a major public health problem in many countries of the Balkan region, particularly in Romania. Predominantly *E. granulosus* (genotypes G1-3) have been confirmed in southeastern Central Europe (Table 8). The first human case of CE was mentioned in **Romania** in medical annals of 1862. Later, the disease was reported regularly throughout the whole country with 3,072 cases registered between 1987 and 1991 (Neghina et al.,

Table 8 Genotypes of *Echinococcus* spp. causing cystic echinococcosis in southeastern Central Europe (no data found in the missing countries): *Echinococcus granulosus* (G1–3), *Echinococcus equinus* (G4), *Echinococcus ortleppi* (G5), *Echinococcus intermedius* (G7)

Country	Human	Dog(D), Wild canids (Wc)	Sheep (S), Red deer (Rd)	Cattle	Swine (S), Wild boar (Wb)
Bulgaria		Wc: G1[1]	S: G1[1]	G1[1]	S: G1[1]
Romania	G1–3, G7[2]		S: G1–3[3] Rd: G7[4]	G1–3[3]	Wb: G7[4]
Serbia	G1, G7[5]		S: G1 cited in[6]	G1, G7 cited in[6]	S: G1, G7 cited in[6]
Kosovo		D: G1–3[7]		G1[9]	
Albania		D: G1[8]			

[1] (Breyer et al., 2004); [2] (Piccoli et al., 2013); [3] (Mitrea et al., 2014); [4] (Onac et al., 2013); [5] (Maillard et al., 2009); [6] (Bobic et al., 2012); [7] (Sherifi et al., 2011); [8] (Xhaxhiu et al., 2011); [9] Sherifi K. and Deplazes P. unpublished data.

2010). The disease is frequently found in domestic and wild intermediate hosts, particularly in cattle, sheep, wild boars and deer.

In **Bulgaria**, CE is still prevalent in all parts of the country. After a control campaign in the 1960s, a considerable decrease in the annual CE incidence (from $6.5/10^5$ inhabitants to $2.0/10^5$ inhabitants) was observed between 1971 and 1982. Later, between 1996 and 2013, the average annual incidence increased to $6.7/10^5$, and Bulgaria is now considered to have the highest morbidity rate of human CE in EU state members (Todorov and Boeva, 1999; Jordanova et al., 2015). G1 has been reported in both definitive and intermediate hosts in the country (Breyer et al., 2004).

The first report of echinococcosis in **Serbia** dates back to 1899. CE was very rarely diagnosed before the World War II, but numbers increased later on, and in 1997 the incidence was estimated to be 0.21 per 10^5 inhabitants (Bobic et al., 2012). So far, *E. granulosus* (G1) has been reported in humans, sheep and swine, and *E. intermedius* (G7) in humans and swine (Colovic, 2009; Maillard et al., 2009) (Table 8). Interestingly, infections in swine were of major significance in the past (Dakkak, 2010); however, recently infections in sheep might be of higher relevance (Bobic et al., 2012). As in other Balkan countries, home slaughtering is still an important risk factor for the transmission of CE in Serbia. According to the Statistical Reports of Serbia, only 30–33% of swine, 40–47% of cattle and 5–7% of sheep were slaughtered in abattoirs after 2005. Investigations of *E. granulosus* in dogs in Serbia have not been conducted/published so far, but they are assumed to serve as definitive hosts.

In **Croatia**, CE is recognized as a public health problem, but data on the spread of the parasite are rather limited. The region of Dalmatia that covers the eastern coast of the Adriatic sea is historically known as an endemic area. The incidence of human CE in Croatia decreased from 28.3 per 10^5 inhabitants between 1940 and 1950 to 4.2 per 10^5 inhabitants in the 1990s, but in some communities, the prevalence in domestic animals increased (Morovic, 1997). Wild boars also play a role in transmission of the parasite in Croatia (Rajkovic-Janje et al., 2002).

Only limited information on CE is available from **Bosnia and Herzegovina**, **Macedonia** and **Kosovo**. In these Balkan countries, the dog populations are not registered, and free roaming and stray dogs have easy access to uncooked offal and carcasses. Furthermore, the lack of knowledge of CE and the absence of strategic deworming of dogs contribute to the persistence of the infection pressure even in hunting and pet dogs. Based on the scarce data available, the predominant intermediate hosts are sheep and cattle (with a high percentage of sterile cysts). In contrast to Serbia, infections in pigs have so far not been documented in the three other Balkan countries. In **Kosovo**, besides professional slaughtering in abattoirs, home slaughtering of sheep and cattle for private meat consumption is traditionally performed not only by farmers but also by a large part of the population in rural and urban areas. Transmission of CE is linked mainly to home slaughtering and the feeding of infected organs to dogs. Many old sheep are slaughtered, particularly during Eid al-Adha (Feast of Sacrifice), and this celebration is followed by a pronounced increase of taeniid infections in dogs, including *E. granulosus* (Alishani et al., 2017).

In **Albania**, *E. granulosus* has been described for more than 60 years in intermediate and definitive hosts. Only recently infected dogs have been identified in the area surrounding Tirana (Xhaxhiu et al., 2011) and interestingly, a high proportion of human CE cases originated from Tirana in the last few years. Reasons for this peri-urban infection risk were suggested to be aggressive urbanization, sanitary and garbage disposal problems, and the presence of sheep and dogs inside landfills and in the city itself (Pilaca et al., 2014).

3.4.5.2 Infections in animals

In **Romania**, CE is frequently found in domestic (Fig. 8 and Table S7 in Supplementary Material) and wild intermediate hosts. Cysts were observed in 29.0–34.4% of cattle and 47.1–54.7% of sheep surveyed. Positive findings were also documented in wild boars (12.4%) and red deer (9.5%) in which *E. granulosus* (G1) and/or *E. intermedius* (G7) were confirmed (Mitrea et al., 2014; Onac et al., 2013).

In **Bulgaria**, CE prevalence values recorded in 2009 were 0.1% in pigs, 5.1% in cattle, 7.0% in sheep and 10.5% in goats (EFSA, 2011). Analysis of 24 isolates from definitive (jackal, wolves) and intermediate hosts (cattle, sheep, pigs) revealed the presence of *E. granulosus* (G1) in Stara Zagora county of Bulgaria (Breyer et al., 2004).

From **Slovenia**, low prevalences of CE in animals were reported (Fig. 8 and Table S7 in Supplementary Material). In 2009, CE was recorded in less than 0.1% of 123,760 cattle and 295,960 pigs examined and none of 9759 sheep, 450 goats and 1426 equines (EFSA, 2011). However, the data from humans in eastern Slovenia (see below) suggest ongoing CE transmission.

In **Serbia**, CE continues to be endemic, mainly in sheep (Fig. 8 and Table S7 in Supplementary Material), but despite predictions, neither official data nor those from clinical studies indicate its re-emergence (Bobic et al.,

2012). In fact, a gradual decrease in the prevalence of CE during the last few decades has been identified in livestock (from 14% to 1% in sheep, from 13% to 2% in cattle and from 9% to 4% in swine (Bobic et al., 2012). However, the authors suggest that underreporting of CE at slaughterhouses, structural and economic reorganization and/or closing of abattoirs and an increase in home slaughtering in the last 20 years have to be considered.

Data about the occurrence of *Echinococcus* spp. in **Croatia** are scarce. CE was commonly found in livestock in Croatia in the 1950s until the 1970s, with prevalence values of 10–40% in cattle, 40–80% in sheep and up to 30% in pigs. In the 1980s, high variation in prevalences in sheep from 5% to 88% was found, and in 1991, 14% of stray dogs were positive for *E. granulosus* (Ecca et al., 2002). In eastern Croatia (Slavonia) CE was detected in 2 of 47 wild boars (*Sus scrofa*) collected in 2000 and 2001 (Rajkovic-Janje et al., 2002), documenting the possible involvement of wild animals in the life cycle.

In **Bosnia and Herzegovina**, a retrospective study based on unpublished data reported a high CE prevalence in cattle (27.2%) and sheep (80.3%). A more recent study confirmed CE in 22% of cattle and 65% of sheep (approximately 2000 animals in total) (Zuko and Obradović, 2014).

Regarding **FYROM (Former Yugoslav Republic of Macedonia)**, CE is thought to be endemic in intermediate hosts and humans, but no published data are available for this country.

In **Kosovo**, the prevalence of *Taenia* spp. and *E. granulosus* (sheep strain, G1–3) was 7.5% and 1.3%, respectively, in naturally infected pet dogs, sheepdogs and hunting and stray dogs (Sherifi et al., 2011). The occurrence of CE in slaughtered cattle from all over the country was high (72%), indicating a high environmental contamination with *E. granulosus* eggs (Hamidi et al., 2010).

In **Albania**, *E. granulosus* intestinal infections in dogs have been documented over more than 60 years (Xhaxhiu et al., 2011). In a recent study (2004–09), necropsy of 111 dogs revealed infections with *Taenia hydatigena* in 16.2% and with *E. granulosus* (G1) in 2.7%, suggesting transmission via free access of dogs to carcasses of intermediate hosts (Xhaxhiu et al., 2011). Studies covering the whole country revealed 1347/6051 (22.2%) of cattle and 168/603 (27.8%) of sheep positive for CE (Zanaj, 1997). According to abattoir surveys, CE has been present in Albania for decades with highly variable prevalences (5–75%), primarily in sheep and cattle rather than goats and swine [citations Xhaxhiu et al. (2011)].

3.4.5.3 Cystic echinococcosis in humans

According to the WHO, the highest annual incidence (per 10^5 inhabitants) of CE between 2001 and 2010 was reported for Bulgaria (range 3.88–9.27), followed by the FYROM (0.3–1.89), Bosnia and Herzegovina (0.32–1.06) and Croatia (0.23–0.81), whereas the lowest incidence was reported for Hungary (0.05–0.13) (Torgerson, 2017).

In **Romania**, 451 cases (including 82 children) were diagnosed during 2004–10 in the counties of Arad, Hunedoara, Caras-Seeverin and Timis. The yearly incidence was $3.4/10^5$ adults and $3.1/10^5$ children (Vlad et al., 2013). Hospital discharge data indicate 1730 cases of CE in 2012 which represents an incidence of 6.6/105 (European Hospital Morbidity Database – http://data.euro.who.int/hmdb/). Although some of these may be recurrent cases, rather than new cases it does indicate the massive scale of human CE in Romania.

In **Bulgaria**, 291–596 human cases of CE per year were recorded in the period of 2003–12, with an annual incidence ranging between $3.95/10^5$ and $8.14/10^5$ inhabitants (Rainova et al., 2014). According to official data from the National Centre of Infectious and Parasitic Diseases, the average incidence of the disease was steady ($4.45/10^5$ inhabitants) in the years 2009–14 (Muhtarov, 2014; Piccoli et al., 2013).

In **Slovenia**, cases of echinococcosis are reported sporadically, and often without differentiation between AE and CE. The number of reported AE/CE cases decreased from nine cases ($0.44/10^5$ inhabitants) in 2009 to eight cases (0.39) in 2010 and 2011, and six cases (0.29) in 2012 and 2013 (EFSA, 2015a). Seroepidemiological surveys performed in 2002–06 on 1323 suspected CE patients revealed 127 (9.6%) seropositive individuals originating mostly from the eastern part of the country, historically known as a CE endemic region (Logar et al., 2008). In 2012, 12 cases ($0.54/10^5$) were reported in the European Hospital Morbidity database.

In **Croatia**, one of the historically high endemic area of the country is Dalmatia although there has been a decrease in the CE incidence from the mid-1950s until 1990 (Morovic, 1997). Since 1965, reporting of CE to the Epidemiology Reference Centre of the Croatian National Institute of Public Health (CNIPH) has been mandatory, and based on these data, the annual number of reported cases of CE in Croatia varied between 4 and 35, with an increasing tendency until 2004 and a decrease thereafter (Tabain et al., 2011). For Croatia, an annual CE incidence (2001–10) of 0.23–0.81 per 10^5 inhabitants was reported (WHO, 2010) and 74 cases are reported in 2013 in the WHO European Hospital Morbidity Database = $1.2/10^5$ http://data.euro.who.int/hmdb/.

A seroepidemiological survey conducted on 540 patients with cystic liver disease revealed seropositive reactions for CE in 3.9% of the patients, thus confirming that CE still circulates in Croatia (Tabain et al., 2011).

For **Serbia**, a recent comprehensive retrospective study (Bobic et al., 2012) reported a total of 409 officially reported human CE cases (1998–2010) in contrast to 820 clinical cases described during this period. The annual incidence of CE in official records of the Institute for Public Health ranged from $0.38/10^5$ to $0.63/10^5$ inhabitants but is likely to be higher due to underreporting. The European Hospital Morbidity Database (http://data.euro.who.int/hmdb/) reported 252 cases of CE in 2013 or $3.2/10^5$ inhabitants. As this is a WHO database (i.e., UN), it may also include data from Kosovo.

No trend in the incidence of infection among adults was observed, but the number of CE cases in children continuously decreased. Females and patients from rural areas were more frequently infected. Differences in the geographic distribution of CE cases with a lower incidence in the central part of Serbia were also documented (Bobic et al., 2012).

In **Kosovo**, 163 CE patients from all over the country (75% living in rural and 25% in urban areas) were treated at the University Clinical Center of Prishtina between 1999 and 2001. Based on these data, a minimal average annual incidence of $2.7/10^5$ inhabitants was calculated (Alishani et al., 2017). However, this incidence probably underestimates the real epidemiological situation as many CE cases have been detected in patients who originated from Kosovo but now live in other European countries such as Germany or Switzerland.

Bosnia and **Herzegovina** is recognized as endemic for CE and after the war (1992–95), human surgical incidence increased, especially in cities that were under siege during that period [cited in Dakkak (2010)]. Furthermore, a recent preliminary study mentioned in Zuko and Obradović (2014) showed high seroprevalences in this country. For **Bosnia** (based on WHO reports), 34 cases per year have been registered (2003–10) resulting in an annual incidence of 0.7 CE cases per 10^5 inhabitants (Torgerson, 2017). Only fragmented information is available for **FYROM**, with an incidence of $0.3-1.89/10^5$ inhabitants recorded in the WHO report (cited in Torgerson (2017)).

In a recent study performed in 2005–11 in **Albania**, 333 CE cases have been diagnosed and treated in the only tertiary (university) medical hospital of the country (UHC, Tirana). Based on these data, a minimal annual incidence of 1.5 clinical CE cases per 10^5 inhabitants was reported (Pilaca et al., 2014).

3.5 Asia (including Eastern Europe)

3.5.1 North Asia: Russian Federation

3.5.1.1 Host assemblages, transmission and molecular epidemiology

Echinococcus species are widespread throughout Russia (see Figs. 9 and 10). They have been described in people, farm animals, dogs and wildlife such as moose, reindeer and wolves (Table 9). As shown in Table 9, *E. granulosus* (genotypes G1–3) and *E. canadensis* (G8, G10) have been confirmed in the European Russia and the Altai region. In Yakutia, G1 has been reported in sheep, and *E. intermedius* (G6) and *E. canadensis* (G8, G10) have been found in wild ungulates (including reindeer and elk) and wolves (Konyaev et al., 2013). An unusual case of CE in a domestic cat from St. Petersburg was caused by *E. granulosus* G1 (Konyaev et al., 2012b). *E. intermedius* (G7) has been reported from Armenia (Snabel et al., 2009).

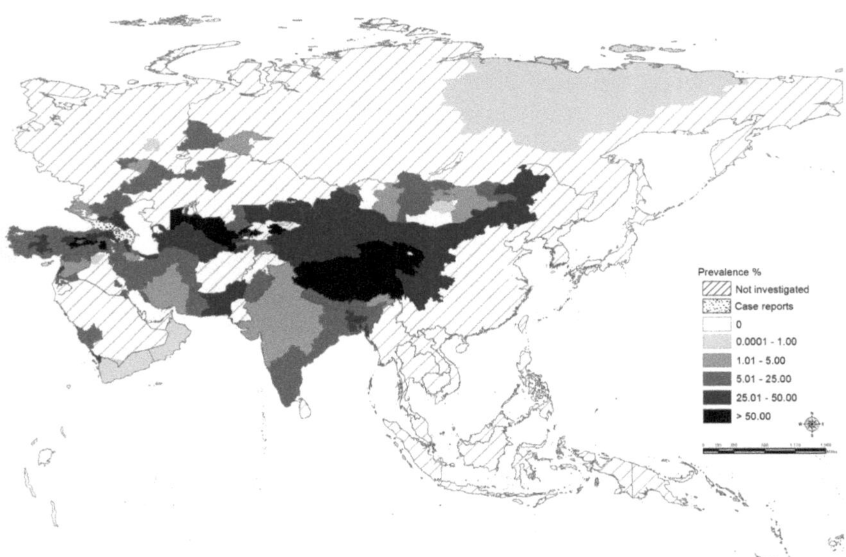

Figure 9 ***Current distribution of Echinococcus spp. causing cystic echinococcosis in domestic intermediate hosts (sheep, cattle, goats, pigs, buffaloes and yaks) in Asia.*** The detailed information (prevalence data in each jurisdiction) is listed in Table S9 of the Supplementary Material.

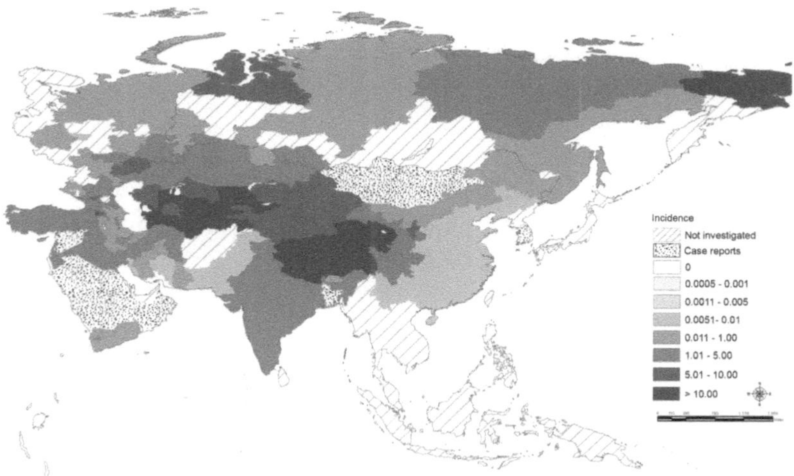

Figure 10 ***Current incidence of human cystic echinococcosis in Asia.*** The detailed information (incidence data in each jurisdiction) is listed in Table S10 of the Supplementary Material.

Table 9 Genotypes of *Echinococcus* spp. causing cystic echinococcosis in North Asia (including the European part of Russia), Central Asia and Caucasus (no data found in the missing countries): *Echinococcus granulosus* (G1–3), *Echinococcus equinus* (G4), *Echinococcus ortleppi* (G5), *Echinococcus intermedius* (G6/7) and *Echinococcus canadensis* (G8, G10)

Country	Human	Wolf (W), Dog (D)	Sheep (S), Goat (G)	Cervids	Pig	Cattle	Other
Russian Federation	G1–3, G6 G10[1,2]	W: G6 and G10[1]	G1–3[1]	G6, G8, G10			Cat: G1[3]
Kazakhstan		D: G1, G6/7[4,5,6]					
Kyrgyzstan		D: G1, G4, G6/7[7]					
Armenia	G1–3[8]		S: G1–3[8]		G7[9]	G1–3[8]	

[1] (Konyaev et al., 2013); [2] (Sharma et al., 2013); [3] (Konyaev et al., 2012b); [4] (Stefanic et al., 2004); [5] (Boufana et al., 2015a); [6] (Trachsel et al., 2007); [7] (Ziadinov et al., 2008); [8] Ebi, Gevorgyan H.S., personal communication; [9] (Snabel et al., 2009).

3.5.1.2 Infections in animals

Infection of animals with CE is widespread in Russia and reflects the wide distribution of human cases observed (see Figs. 9 and 10 and Tables S9 and S10 of the Supplementary Material). In the far south of the Caucasus, prevalences in sheep and cattle are between 20% and 30% (Dyachenko et al., 2014; Fiapsheva et al., 2007; Makhiyeva et al., 2010, 2012; Pliyeva and Uspenskii, 2006; Zhekamukhova et al., 2012). Dogs in this area are also highly infected, with prevalences of up to 70% (Bersanukayeava et al., 2013; Pliyeva and Uspenskii, 2006).

In the district of Kalmykia in Russia, 8.3% of sheep have been reported to be infected (Lazarev, 2010). Infection in cattle in Krasnodar has been reported at 4% (Shevkoplyas and Lopatin, 2009) and 2% in Samara Oblast (Uspenskii et al., 2013). Pigs also appear to be frequently infected in Russia: 3.5% in Krasnodar (Shevkoplyas and Lopatin, 2009) and 11% in Ulyanovsk (Romanov and Mishonkova, 2010).

In the region of the Urals, approximately 20% of sheep are infected in Orenburg (Belimenko and Khristianovsky, 2015) and 7.3% in Astrakhan (Postnova et al., 2010). Sheep cysts from the Ural region were identified as *E. granulosus* s.s. (Konyaev et al., 2013). In cattle from Orenburg and Sverdlovsk, prevalences of 23–47% (Terentyeva, 2007) and 23.7% were registered (Belimenko and Khristianovsky, 2015).

Further east in Siberia, 0.4% of cattle are infected in Yakutia (Kokolova et al., 2012) and 3.4% of pigs in Tyumen (Yamov and Antropov, 2008).

CE is frequently reported in wild ungulates such as reindeer. In Chukot in the far east of Siberia, CE is present in 6–12% of reindeer (*R. tarandus*) (Government reports) while in Yakutia, prevalences of 17% and 5% have been reported for reindeer and moose, respectively (Kokolova et al., 2012). Older reports give a prevalence of 4.4% in reindeer from Nenets (Sorochenko, 1972).

Infected wild carnivores have been reported in several regions of Russia including the wolves and jackals in the Caucasus (Pliyeva and Uspenskii, 2006): 61% of wolves in Yakutia (Kokolova et al., 2012), 42% of wolves in the Altai region (Pomamarev et al., 2011) and 36% in Nenets (Sorochenko, 1972). *Echinococcus* spp. isolates from wolves, domestic reindeer and wild cervids in the Altai region and Yakutia were identified as genotypes G6, G8 and G10 (Konyaev et al., 2013).

3.5.1.3 Cystic echinococcosis in humans

CE is a notifiable disease in Russia and the Federal Center of Hygiene and Epidemiology reports the total number of cases on an annual basis. Between 2006 and 2010, 2863 CE cases were reported, or 573 per annum. However, when the data from the individual Russian districts and the regional centres are combined with local scientific reports, the annual number of cases can be estimated at about 950 (0.66 cases per 10^5 inhabitants per year), which clearly indicates underreporting to the central public health services. In Orenburg, analysis of hospital records (Korneev et al., 2014) detected 20% more cases than reported by the Government. Likewise, case reports in Dagestan by Abdylazizov (2012) and by Shodmonov and Razikov (2015) are double the numbers reported to the Government.

There is considerable variability in the incidence of CE within Russia. Chukotka has an incidence of six cases per 10^5 inhabitants, but since the population is just 50,000 this only represents three cases reported in 2014 (Rospotrebnadzor, 2016). However, the same report states that there are 90 individuals undergoing treatment with confirmed CE or AE and a further 203 suspect cases. Furthermore, 91 cases of CE were detected during ultrasound surveillance (Malishev and Sobolevskaya, 2002). Thus, in this remote and sparsely populated region, CE appears to be a major health issue. In Yamal-Nenets, another remote region of the Russian north, about five cases per 10^5 inhabitants are seen. Elsewhere in Russia, high human incidences are reported in Dagestan and Karachy-Cherkess in the Caucasus (up to five cases per 10^5 inhabitants per year), Orenburg (up to four cases per 10^5 inhabitants per year) and Saratov (about two cases per 10^5 inhabitants per year). However, cases of human CE have been found in virtually every region of Russia. Even in Moscow, about 50 cases a year are reported, although 80% of these are in migrant workers who originate from elsewhere in Russia or from Central Asian countries such as Kyrgyzstan or Uzbekistan (Rospotrebnadzor, 2016).

There are regional variations in the organs affected by CE. For example, in the Astrakhan region, 69% of cases of CE are hepatic (Postnova et al., 2010). In contrast, in the far north of European Russia, most cases are pulmonary echinococcosis (Bikov et al., 2011; Buzinov et al., 2012). There has been little genotyping undertaken on isolates from humans. In the Altai region and Bashkiria. *Echinococcus granulosus* s.s. cysts have been identified in human CE cases, and the G6 genotype was identified in one patients from Altai (Konyaev et al., 2012a, 2013).

3.5.2 *Caucasus and Central Asia: Kazakhstan, Kyrgyzstan, Tadjikistan, Turkmenistan, Uzbekistan, Armenia, Azerbaijan and Georgia*

3.5.2.1 Introduction

CE has long been endemic in the five central Asian republics that were previously part of the Soviet Union. This is a large area stretching from the eastern shore of the Caspian Sea across to the western borders of China. Agricultural land in Central Asia is semi arid and mountain pasture resulting in a predominance of pasture-based livestock production. This provides good conditions for the transmission of livestock reservoir of *Echinococcus* species. Since the collapse of the Soviet Union in 1991, CE has emerged as a major zoonosis with substantial increases in incidence in humans. This is due to the privatization of large collective farms, the abandonment of centralized slaughtering and meat processing facilities, and few resources available for veterinary services. There has also been an increase in the dog population and greater availability of infected offal to dogs through unregulated slaughtering of animals (Shaikenov et al., 2003; Torgerson, 2013).

3.5.2.2 Infections in animals

In **Uzbekistan,** the prevalence of CE in sheep has increased from 45% to 62% between 1990 and 2002 (Aminjanov and Aminjanov, 2004). In **Kazakhstan**, prevalences in sheep have also increased in southern Kazakhstan from 14% in the 1980s to 37% by the year 2000 (Torgerson et al., 2002). Generally, prevalences in sheep have been reported between 24% and 48% in southern Kazakhstan with approximately 7% of cattle infected (Torgerson et al., 2003). In western Kazakhstan, prevalences of 18% in sheep and 19% in cattle have been reported (Vaiyeva et al., 2012). In the Kostanay Oblast in the northern part of Kazakhstan, CE was found in 8% of 826 pigs and in 24% of 3823 cattle. Interestingly, in infected pigs, 68% of cysts were found in the lungs, whilst in cattle 73% of cysts were found in the liver (Suleimanova and Shinkina, 2015). This markedly different distribution of organ affinity between the two species might be consistent with different genotypes circulating in different species assemblages.

In the Naryn region of central **Kyrgyzstan** a prevalence of 64% in sheep was reported in 2006 (Torgerson et al., 2009). In **Tadjikistan** prevalence varies with region. Highly endemic areas are located in the north, with over 50% of sheep infected, whereas central and southern districts generally have around 20% prevalence, and only 7% in areas around the Chinese border. Infection varies directly with age in all regions, with prevalence in young animals often less than 10%, and the prevalence in the oldest animals approaching 80% (Muminov et al., 2004).

A number of studies on dogs in the region have revealed widespread infection. In southern areas of **Kazakhstan** – south Kazakhstan, Dzambul and Almaty Oblasts – prevalence in village dogs was 6% while prevalence in farm dogs was 23% (Torgerson, 2013). In the west Kazakhstan Oblast, 15.9% of 176 dogs were infected (Vaiyeva et al., 2012). *Echinococcus granulosus*-like cestodes have also been found in the intestines of two of 51 (4%) of village dogs from Akmola Oblast (Sultanov et al., 2014). In Naryn Oblast in Kyrgyzstan, 19% of 466 dogs were infected (Ziadinov et al., 2008). In Uzbekistan, 531 dogs were investigated using arecoline purgation. Of these, 279 were farm dogs of which 56 were infected (20.1%). Of the remaining 240 village dogs, 19 were infected (7.9%) (Aminjanov and Aminjanov, 2004). In **Tadjikistan**, one study of 120 dogs reported a prevalence of 15.2% (Muminov et al., 2004). In a second study, 23 of 41 dogs were infected with *E. granulosus* at necropsy (Razikov and Adilova, 2011).

A study of wolves in southern **Kazakhstan** revealed eight of 41 (20%) infected with *E. granulosus* (Abdybekova and Torgerson, 2012).

There are relatively few studies on the *Echinococcus* genotypes in animals in central Asia and Caucasus (Table 9). G1 and G6/7 have been isolated from dogs in southern Kazakhstan (Stefanic et al., 2004; Trachsel et al., 2007). In central Kyrgyzstan (Naryn Oblast), G1 and G6/7 have also been isolated from dogs. In addition, G4 (*E. equinus*) has been reported in a single dog in Kyrgyzstan (Ziadinov et al., 2008).

In **Armenia**, studies in the 1970s and 1980s revealed *E. granulosus* in 49.5% of dogs examined and CE in 16.8–46.9% of cattle and in 22.7–47.0% of pigs (Khachatryan, 2015). In the central province of Kotayk, prevalence in sheep can reach 80% (Gevorgyan H.S., personal communication). In a recent molecular survey, all 204 isolates from sheep (89), cattle (72) and humans (43) belonged to *E. granulosus*, while two cysts from pigs belonged to G7 (Ebi, Gevorgyan H.S., personal communication; Gevorgyan et al., 2006; Snabel et al., 2009).

Georgia and **Azerbaijan** have long been considered endemic for *E. granulosus* with pig, sheep and cattle strains reported (Khachatryan, 2015), but there are no data on the prevalence of infection or the genotypes of parasite in the country.

3.5.2.3 Cystic echinococcosis in humans

At the end of the last decade of the 20th century it became apparent that there was an increased incidence of CE in Central Asia (Shaikenov et al., 2003). Official government statistics in **Kazakhstan** documented an increase in cases of CE from about 200 cases per year until 1994, rising rapidly to approximately 1000 cases per year by the beginning of the 21st century (Torgerson et al., 2002). Recent data suggest that the incidence in Kazakhstan has since stabilized at this higher level (Abdybekova et al., 2015). In Kazakhstan, there is a strong association between human incidence and the main sheep rearing areas, with high incidence seen in the southern districts and in West Kazakhstan Oblast (Torgerson et al., 2006).

Likewise there was similar evidence from neighbouring **Kyrgyzstan** with a continuing rise in cases up until 2013 (Raimkylov et al., 2015). In **Tajikistan**, the number of cases increased from 374 in 1992 to 1875 in 2002 (Muminov et al., 2004). Subsequently, data confirmed a similar situation in **Uzbekistan** and **Turkmenistan** (Torgerson et al., 2006). Furthermore, in Uzbekistan, officially reported cases of CE are a substantial underestimate of the numbers of cases being treated (Hong et al., 2013). A detailed case finding study of all hospitals in Uzbekistan uncovered approximately four times the numbers of cases of CE than reported in government statistics (Nazirov et al., 2002).

For **Armenia**, **Georgia** and **Azerbaijan**, relatively high CE incidences have been estimated. In **Armenia**, there were 1470 CE patients reported from 1997 to 2003, representing incidences in urban residents of 4.6 per 10^5 inhabitants and in rural residents of $8.1/10^5$ inhabitants (Khachatryan, 2015). Based on 234 CE cases registered in 2010, an incidence of $7.8/10^5$ was estimated for Armenia (Torgerson, 2017). In Georgia an average of 68 cases has been registered resulting in an incidence of $1.5/10^5$ inhabitants and in Azerbaijan, an annual average of 85 cases (incidence of $0.93/10^5$ inhabitants) was calculated (Torgerson, 2017). More information on human CE is required from the Caucasus.

3.5.3 Middle East: Iran, Iraq, Israel, Jordan, Kuwait, Lebanon, Oman, Palestine, Qatar, Saudi Arabia, Turkey and Yemen

3.5.3.1 Host assemblages and transmission

About 400 million people are living in the Middle East extending from the eastern Mediterranean coast in the west to the Iranian plateau in the east, and from Turkey in the north to Saudi Arabia and Yemen in the south.

Table 10 Genotypes of *Echinococcus* spp. causing cystic echinococcosis in the Middle East (no data found in the missing countries): *Echinococcus granulosus* (G1–3), *Echinococcus equinus* (G4), *Echinococcus ortleppi* (G5), *Echinococcus intermedius* (G6/7) and *Echinococcus canadensis* (G8, G10)

Country	Human	Dog (D), Wild canids (Wc)	Sheep	Camel	Cattle/Buffalo	Goat	Equine
Iran	G1[1,2,3] G2[2,4] G3[1,2,4,5]	D: G1[6] G2[4,6] G3[2,7] G6[7] Wc: G1[8]	G1[9] G3[2,9,10,11]	G1[4,12] G3[2,10,13] G6[1,4,12]	G1[9] G3[2,10,14] G6[9]	G1[12] G6[15] G7[16]	G1/G6/G4[17,18]
Iraq	G1[19]	na			G1[20]		
Jordan		D: G1[21] G4[21]	G1[22]				
Oman			G1[23]	G1[23] G6[23]	G1[23] G6[23]	G1[23]	
Palestine			G1[24] G2[4,24] G3[2,24]				
Turkey	G1[25,26] G3[2,27] G6[27] G7[25]	D: G1[26]	G1[25,26] G3[2,28] G7[25]	G1[26]	G1[26,28,29] G3[2,28,29]	G1[26,28]	G1[30] G4[31,32]
Yemen	G1[33]						

[1] (Zhang et al., 1998); [2] (Rostami et al., 2015); [3] (Kia et al., 2010); [4] (Harandi et al., 2002); [5] (Pezeshki et al., 2013); [6] (Parsa et al., 2012); [7] (Shariatzadeh et al., 2015); [8] (Beiromvand et al., 2011); [9] (Shahnazi et al., 2011); [10] (Hajialilo et al., 2012); [11] (Pezeshki et al., 2013); [12] (Sharbatkhori et al., 2010); [13] (Sharbatkhori et al., 2011); [14] (Pour et al., 2011); [15] (Rajabloo et al., 2012); [16] (Fadakar et al., 2015); [17] (Eslami et al., 2014); [18] (Sarkari et al., 2015); [19] (Hama et al., 2012); [20] (Hama et al., 2015); [21] (Al-Qaoud et al., 2003); [22] (Yanagida et al., 2012); [23] (Al Kitani et al., 2015); [24] (Adwan et al., 2013); [25] (Snabel et al., 2009); [26] (Utuk et al., 2008); [27] (Simsek et al., 2011); [28] (Vural et al., 2008); [29] (Simsek et al., 2010); [30] (Utuk and Simsek, 2013); [31] (Simsek et al., 2015); [32] (Simsek and Cevik, 2014); [33] (Alam-Eldin et al., 2015).

The region has a diverse geography including arid/semiarid tropical regions and humid temperate areas. Precipitation and temperature vary considerably across the ME and even within countries. The Caspian coast of northern Iran receives precipitation of up to 2000 mm per year while the desert areas of eastern/central Iran receive no rain for years.

The Middle East has been one of the hot spots of CE in the world (Cardona and Carmena, 2013; Dakkak, 2010; Dar and Alkarmi, 1997; Sadjjadi, 2006). Sheep and goats are believed to have been domesticated for the first time in the Fertile Crescent of the Middle East around 8500–11,000 years ago, and this was central to CE establishment in livestock/dog assemblages with spillover into people. Therefore the Middle East can be considered as a historically very old human CE-endemic region (Zeder, 2008).

The parasite circulates primarily among domestic dogs (owned and ownerless stray animals) as definitive hosts and a variety of domestic livestock as intermediate hosts (Dowling et al., 2000; Harandi et al., 2011; Rafiei et al., 2007; Shariatzadeh et al., 2015). However, a wild animal cycle has been documented in Iran and Turkey involving wolf, jackal, hyena and red fox as definitive hosts, and wild sheep (*Ovis orientalis*) and the goitered gazelle (*Gazella subgutturosa*) as intermediate hosts (Dalimi et al., 2002, 2006; Eslami et al., 1981, 2016; Simsek and Eroksuz, 2009).

The stable endemicity of CE in the Middle East is largely due to the diverse number of intermediate host species, with sheep and goats as the principal hosts, followed by camels and cattle. In camels, CE is prevalent with highly fertile and viable cysts (Al Kitani et al., 2015).

As part of the ceremonies related to the annual mass gathering of the Hajj (Muslim pilgrimage to Mecca), millions of livestock (mainly sheep and goats) are imported and sacrificed in Saudi Arabia. This is a defining aspect of CE epidemiology in the Arabian Peninsula.

The following factors are among the major determinants of CE transmission in the region: (1) nomadic/rural lifestyle and traditional livestock husbandry including use of shepherd dogs; (2) food habits including high raw vegetable consumption; (3) unregulated/poor quality abattoirs and widespread practice of home slaughter, particularly in special religious festivals and social events; (4) low awareness of people about transmission and pathogenesis of CE; (5) diverse intermediate host species and high populations of stray dogs.

On the other hand, limiting factors for transmission include mostly arid/semiarid environments with less favourable climate conditions for the survival of *Echinococcus* eggs, avoidance of dogs in Muslim communities and the absence of pig farming due to religious beliefs (Dakkak, 2010; Dar and Alkarmi, 1997; Harandi et al., 2011; Rokni, 2009).

3.5.3.2 Molecular epidemiology

Table 10 summarizes the occurrence of *Echinococcus* species and genotypes causing CE in the Middle East. *Echinococcus granulosus* (G1–3) is the dominant species responsible for most human and animal infections (Harandi et al., 2002; Al-Qaoud et al., 2003; Utuk et al., 2008; Vural et al., 2008; Sharbatkhori et al., 2009), with *E. intermedius* (G6, camel strain) as the second most prevalent species in the region and increasingly recognized in human cases (Rostami et al., 2015; Al Kitani et al., 2015). In addition there have been reports of the occurrence of *E. intermedius* (G7) in Turkey and Iran (Snabel et al., 2009; Eryildiz and Sakru, 2012; Fadakar et al., 2015). In Jordan, most dogs were infected by G1 while a single sample was most similar to *E. equinus* (G4) using RAPD-PCR (Al-Qaoud et al., 2003). More genotype data from the Middle East are required especially from Iraq, Lebanon, Palestine/Israel, Syria and the Persian Gulf countries.

3.5.3.3 Infections in animals

A distinct feature of cystic echinococcosis in the Middle East is the diverse number of intermediate host species involved in transmission; however, sheep and goats are the principal intermediate hosts. *Echinococcus granulosus* (G1–3) has been shown to perpetuate in at least 13 species of mammals in this region [i.e., sheep, goat, cattle, buffalo, one- and two-humped camels, horse, donkey, pig, wild sheep (*Ovis orientalis*), goitered gazelle (*Gazalla subgutturosa* and free-ranging Baboon (*Papio hamadryas*)]. Such a diverse intermediate host system is a fundamental element for the stable endemicity of CE in the Middle East.

No recent data are available on *Echinococcus* infections in dogs from countries in the **Arabian Peninsula**. Older investigations indicated that 15% and 23% of dogs were infected in **Saudi Arabia** and **Kuwait**, respectively (Dar and Alkarmi, 1997; Hassounah and Behbehani, 1976). Several studies from **Saudi Arabia** document CE in sheep, goats, cattle and camel. In the Al-Baha region in western Saudi Arabia, 12.6% of sheep, 6.6% of goats, 8.3% of cattle and 32.9% of camels were infected. Prevalences were significantly higher in the Al-Baha region than in Al-Mekwah and Al-Aqiq, likely due to the more suitable climatic conditions (lower temperature and higher precipitation) in the Al-Baha area (Ibrahim, 2010). In Al-Taif municipal abattoir, 13.5% of sheep and 6.1% of goats were infected, with a higher prevalence in the imported sheep than the local animals (Hayajneh et al., 2014). In

Oman, low prevalence values of CE have been reported from camels (5.3%), cattle (0.6%), sheep (0.07%) and goats (0.03%). Higher fertility was observed in cysts from camels compared to those from sheep and goat (Al Kitani et al., 2015). Prevalence was higher in locally bred livestock than imported animals. In the late 1980s, CE was found in 39.6% of camels in **Kuwait** with higher pulmonary infection (63%) and higher cyst fertility than the hepatic cysts (Abdul-Salam and Farah, 1988). Updated information is required from other countries in the Arabian Peninsula.

In **Palestine** and **Israel**, 18.3% of 93 dogs were positive on copro-PCR in three districts of Al-Khalil (Hebron), Tubas and Jenin (Al-Jawabreh et al., 2015). In three independent studies in northern towns of Yirka and Tamra near the Lebanese border, 7.9%, 14.2% and 10.7% of dogs were found infected based on arecoline purgation (Nahmias et al., 1991; Furth et al., 1994; Hoida et al., 1998). In the latter study, despite 10.7% infection in 56 dogs belonging to Muslim and Druze communities, none of the 150 dogs' excrement from Jewish communities was found infected, underlining the significance of sociocultural parameters and animal husbandry systems in transmission of CE. A recent study from Israel and Palestine reported CE in 9% of sheep slaughtered in abattoirs in Nablus, Jenin and Tubas (El-Ibrahim, 2009). In the northern city of Tamra, 5.9% of 874 sheep and 5.3% of 616 goats were infected (Hoida et al., 1998). In neighbouring Yirka, 75% of sheep and goats were slaughtered at individual households with only 25% at a Yirka abattoir (Nahmias et al., 1991).

In **Lebanon**, 17–32.9% of dogs were historically infected (Araj and Mourad, 2014), but recent infection data are not available. The prevalence of CE has been estimated at 6.6% and >41% in sheep and cattle, respectively (Araj and Mourad, 2014).

Very limited information is available from **Syria** and new investigations are required after the ongoing war. The prevalence of *Echinococcus* infection in dogs was estimated at 9–15%, and prevalences of CE in livestock ranged between 5 and 17%, highest in the northern areas of the country (Dakkak, 2010).

Three major studies have been carried out in **Jordan** since the 1980s indicating *E. granulosus* prevalences in dogs of 14%, 9.4% and 29.5% (Ajlouni et al., 1984; El-Shehabi et al., 1999; Al-Qaoud et al., 2003). Fourteen of twenty-five dogs (56%) were infected around a disposal area of a municipal slaughterhouse, where condemned organs were readily available. In five regions of **Jordan**, CE prevalence was 12.9% in sheep, 12.7% in goats, 0.9% in cattle and 11% in camels (Kamhawi, 1995). CE was found in 44.9% of camels slaughtered in Al-Ramtha abattoir in northern Jordan (Sharrif et al., 1998). In Irbid governate, CE was found in 16.9% of 130 donkeys (Mukbel et al., 2000).

High prevalences have been reported for dogs in **Iraq** in the past. In the 1980s and 1990s, double-digit prevalences have been reported from several provinces in northern, central and southern Iraq, i.e., 20% in Al-Tamim, 38% in Diala, 56% in Theqar and 25% in Baghdad (Molan, 1993; Molan and Baban, 1992; Tarish et al., 1986). The prevalence of *E. granulosus* in dogs decreased from 70.4% in 1991 to 24.3% in 1998 in the northern province of Arbil (Saeed et al., 2000). Different studies in northern **Iraq** reported CE in 2–15% of sheep, 0.5–6.3% of goats, 0.5–10.9% of cattle (Saeed et al., 2000; Jarjees and Al-Bakri, 2012; Meerkhan and Abdullah, 2012) and 20–72% of camels (Molan, 1993).

Several studies on canine echinococcosis in **Turkey** indicated endemicity across the country, ranging from 1% in Aydin (western Turkey) to more than 40% in Kars (northeastern Turkey) (Altintas, 2003; Kuru et al., 2013). In the southern city of Antakya near the Syrian border, 8.9% of owned dogs were positive in a coproantigen test with a highest prevalence of positive dogs in free ranging populations with no deworming history (Guzel et al., 2008). The prevalence of CE in livestock varied from 3.5% to 58.6%, depending on different geographical locations and specimens examined (Altintas, 2008). CE was more prevalent in sheep (46.4%) in Van (Oguz and Deger, 2013) than in cattle (33.9%) in Erzurum (Simsek et al., 2010), goats (22.1%) in Burdur (Umur, 2003) or buffaloes (10.2%) in the central Black Sea Region (Beyhan and Umur, 2011).

In **Iran**, canine echinococcosis has been reported from across the country. The dog population in Iran has been estimated at 3.5–11.5 million animals and more than 70–90% are ownerless stray dogs (Harandi et al., 2011). Using arecoline purgation, 27.2% of 390 farm dogs examined in 13 provinces of Iran were infected with *E. granulosus* (Eslami and Hosseini, 1998). In five western provinces the prevalences ranged between 9% and 31% (Abdi et al., 2013; Dalimi et al., 2002). In the northwestern and northeastern parts of the country, the prevalence of canine echinococcosis is between 17% and 20% (Beiromvand et al., 2011; Shariatzadeh et al., 2015). In Mashhad (largest city in northeast of Iran) 22% of dogs were found infected with *E. granulosus* (Razmi et al., 2006). In central and southern parts of the country, the prevalence ranges from 6.8% in Kerman (in the southeast) to 36.2% in Shiraz (in the south) and 55.7% in Kashan in central Iran (Arbabi and Hooshyar, 2006; Mehrabani et al., 1999; Sharifi and Tasbiti, 1994). While the prevalence of *E. granulosus* in dogs of the northern city of Sari was revealed as 46.7% in 1993, 15 years later, none of 50 dogs examined in the same locality was infected after the establishment of a well-equipped abattoir in the region and improved awareness of the local people about the consequences of home slaughter (Gholami et al., 2011). In **Iran**, CE in livestock constitutes a major economic and public health problem. In a 5-year study of abattoirs in nine districts in the northeastern province of Khorasan, an estimated 52.2% of the livestock viscera condemnations were due to CE (Borji and Parandeh, 2010). The corresponding figure in a similar 5-year study in southwestern province of Khuzestan was 29.2% (Borji et al., 2012a). Several surveys in Iran have indicated CE prevalence in livestock ranging from 1.3 to 74.4% in sheep, 0.4—37.8% in goats, 1.3—40.1% in cattle, 4.3—31.9% in buffaloes, 8.8—35.5% in camels and 2% in donkeys (Ahmadi, 2005; Ahmadi and Meshkehkar, 2011; Eslami et al., 2014; Samavatian et al., 2009). In the northern Caspian coastal provinces, prevalence of CE ranged between 14.6 and 65.2% in sheep, 10.1—37.8% in goats and 12—41% in cattle (Mansoorlakooraj et al., 2011; Ziaei et al., 2011). In northwestern areas of the country, prevalence of CE ranged between 22.2 and 74.4% in sheep, 20—25% in goats and 28.3—38.3% in cattle (Daryani et al., 2007; Mirzaei et al., 2015). In five western provinces of Iran, the prevalence of CE has been reported as 11.1% in sheep, 6.3% in goats and 16.4% in cattle (Dalimi et al., 2002). Lower prevalences have been reported in central and eastern areas of the country where the climate is arid/semiarid with lower precipitation (Ansari-Lari, 2005; Arbabi and Hooshyar, 2006; Fakhar and Sadjjadi, 2007). In a study conducted in the southeastern province of Kerman, the prevalence of CE in sheep and goats was markedly reduced after an extensive dog-culling program (Sharifi et al., 1996).

3.5.3.4 Cystic echinococcosis in humans

Human CE is widespread in the Middle East and is principally considered a disease of rural areas, mainly affecting farmers and livestock herders. However, CE has been increasingly reported in patients living in urban and peri-urban areas (Ok et al., 2007; Dar and Alkarmi, 1997). Most of the data available for human CE in the Middle East are sporadic hospital-based reports as well as seroprevalence studies in different localities. The need for a regional and a national CE registry system as well as regular community-based ultrasound surveys is becoming apparent in the region. Consequently, Turkey, Iran and Palestine have recently joined the European Register of CE (Rossi et al., 2016).

In the Arabian Peninsula, most human infections are reported from **Yemen**, **Saudi Arabia** and **Oman**. In **Oman**, the Salalah region in the southern province of Dhofar is the main endemic focus of CE (Scrimgeour et al., 1999). In central **Saudi Arabia**, 117 CE patients have been described with a male to female ratio of 1.7:1 and a mean age of 40.9 years (Fahim and Al Salamah, 2007). In a retrospective study in a major general hospital in **Qatar**, 32 human CE patients were recorded between 2000 and 2013, but only 3 patients were Qataris. It has been suggested that no local transmission of CE occurs in Qatar, where no stray dogs are present and the abattoirs are under strict veterinary supervision (Al-Ani et al., 2014). In **Yemen**, 796 cases were treated in the five main hospitals in Sana between 2001 and 2008, with a trend for increasing numbers. Females accounted for 61% of cases (Al-Shibani et al., 2012).

In **Palestine** and **Israel**, several studies indicated that CE is endemic in the West Bank, especially towns in the Tamra and Yirka in the north and in Bedouin communities of Negev desert in the south. The incidence of CE has been calculated as 1.2–3.1 per 10^5 inhabitants in the West Bank and $1.0/10^5$ in Gaza (Abu-Hasan et al., 2002; Al-Jawabreh et al., 2015). A significant difference in CE incidence has been documented between Bedouin (2.7%) and Jewish (0.4%) populations in the south, attributed to differences in lifestyle and living conditions as well as socioeconomic status (Ben-Shimol et al., 2016).

CE has been highly endemic in **Lebanon**, with an incidence of $3.8/10^5$ inhabitants and a 2:1 ratio of Christians to Muslims described in the 1950s (Matossian et al., 1977). However, the incidence seems to be decreasing in recent years, although no hard evidence is currently available in the literature (Araj and Mourad, 2014).

In **Jordan**, according to an investigation in 18 major hospitals, the mean annual surgical incidence of CE has been estimated at $2.9/10^5$ inhabitants (Kamhawi, 1995) and it was estimated that CE surgery constitutes 0.1–1.2% of total surgical operations (Nasrieh et al., 2003). Source of domestic water was described as a significant risk factor for human CE in Jordan (Dowling et al., 2000). It has been suggested that in Muslim countries dog ownership is not a major risk factor for human infection because the major risk in these countries arises from high levels of environmental contamination with dog faeces due to high numbers of ownerless stray dogs (Dowling et al., 2000; Harandi et al., 2011). In the case of **Syria**, current prevalence data regarding human infection are not available and detailed epidemiological studies are required from this country. A single report described more than 20 cases of pulmonary CE undergoing surgery per year in a university hospital in Damascus (Darwish, 2006).

In **Iraq**, CE surgery constitutes about 2% of all surgical operations documented in a retrospective study in two main hospitals in the northern city of Arbil. Between 1990 and 1998 the incidence of CE was estimated at $2.0/10^5$ inhabitants (Saeed et al., 2000). In a 20-year period between 1986 and 2006, 763 patients underwent surgery for pulmonary/thoracic CE in a teaching hospital in Baghdad (Shehatha et al., 2009). In Sulaimani province in the north, a total of 98 cases of CE were hospitalized between 2006 and 2011, indicating an incidence of $2/10^5$ inhabitants (Hama et al., 2014).

In **Turkey**, official reports from the Ministry of Health document more than 52,000 patients undergoing CE-related surgery between 1990 and 2005 (approximately based on US studies 3257 patients per year). The mean annual incidence of CE in Turkey has been estimated at 0.8–2.0 per 10^5 population (Altintas, 2008). However, in some regions, higher incidence rates up to 6.4 per 10^5 have been recorded (Gonlugur et al., 2009). Based on hospital records between 2001 and 2005, most of the human CE patients in Turkey have been reported from central Anatolia (38.6%) and Aegean/Mediterranean regions (33.0%) (Dakkak, 2010). In central Anatolia seroprevalences of 2.7% and 0.9% have been recorded in patients in rural areas around Kayseri using ELISA, IFA and Western blotting, respectively (Yazar et al., 2006). A community-based ultrasound survey of Turkish children in the Aegean province of Manisa indicated 0.3% prevalence of CE while 8.9% and 10.1% seroprevalence was observed in the same individuals using ELISA and IHA, respectively (Kilimcioglu et al., 2006). Other ultrasound-based studies in Elazig and Manisa revealed CE in 0.2% and 0.15% of children (Bakal et al., 2012; Ok et al., 2007). In Aydin, a prevalence of 0.47% based on ultrasound was recorded in all age groups (Ertabaklar et al., 2012).

In **Iran**, recent investigations indicate that CE is present in both rural and urban areas across the country. The number of CE surgeries in Iran has been calculated as 1295 cases per year with a mean annual surgical incidence of $1.6/10^5$ inhabitants (Fasihi Harandi et al., 2012). Surgical incidence of CE ranges from $0.61/10^5$ to $2.6/10^5$ in different geographical areas of the country (Rostami Nejad et al., 2007; Tavakoli et al., 2008; Vejdani et al., 2013).

Asymptomatic CE has been investigated in two community-based ultrasonography surveys as well as seroepidemiological studies. In nomadic populations in the south (Saberi-Firouzi et al., 1998), ultrasound results revealed a prevalence of 1.8% while 13.8% of the individuals were seropositive on ELISA. In rural areas of the southeastern province of Kerman, CE was found in 0.2% of 1140 individuals using ultrasonography, while 7.3% of the samples were seropositive by ELISA (Harandi et al., 2011).

Human and animal CE imposes major public health and economic burdens to Middle Eastern societies. The monetary burden of CE has been estimated at 0.01–0.04% of the gross domestic product of several countries. The cost of animal CE has been estimated at US$ 89.2 million in Turkey (Sariozkan and Yalcin, 2009) and US$ 232.3 million for human and animal CE in Iran (Fasihi Harandi et al., 2012).

3.5.4 South Asia: Afghanistan, Pakistan, India, Bhutan, Nepal, Bangladesh, Sri Lanka, Maldives

3.5.4.1 Host assemblages and transmission

Occurrence of *Echinococcus* spp. causing CE has been documented in South Asia in different regions in the past (Eckert et al., 2001; Schantz et al., 1995). For **Afghanistan**, no transmission data are available; however, CE was described in a US Marine after deployment to Afghanistan (Kronmann et al., 2008) and an immigrant from Afghanistan in the US (Carter et al., 2009); this shows that CE is likely transmitted in this area. Moreover, 77/150 dogs from Kabul (73%) were positive for *E. granulosus,* with one dog infected with 152,700 worms (Le Riche et al., 1988). In **Pakistan**, transmission in the sheep-dog assemblage is important and the lack of abattoirs in rural communities and home slaughter of livestock is common, especially on religious occasions. Lack of public awareness concerning CE transmission, poor hygienic conditions and improper disposal of offal support the access of *Echinococcus* cysts to dogs (Latif et al., 2010).

In **India**, transmission is strongly related to cultural, educational, socio-economical and agricultural factors (Traub et al., 2005). Uncontrolled home slaughter, especially for religious events, is common. Free access of dogs to slaughter waste, improper garbage disposal and presence of stray animals (dogs, cattle) have been identified as risk factors for *Echinococcus* transmission (Singh et al., 2014a). Furthermore, at the local abattoir level, the lack of legislation for meat inspection and safe offal disposal contribute to the maintenance of domestic cycles of transmission (Irshadullah et al., 1989). Buffaloes and cattle are generally considered the most significant intermediate hosts for sustaining the life cycle (Pednekar et al., 2009). Moreover, a wide range of susceptible captive and wild animals may be involved in transmission which has so far not been studied in detail. Despite numerous case reports in humans, especially for India, few epidemiological studies have been performed in southern Asia and so far no large regional or national control programs have been initiated (Traub et al., 2005).

In **Nepal**, CE was first described 1973 in buffaloes, goats, sheep and pigs slaughtered in Kathmandu. A later study by the same first author documented 47 human cases of CE in three hospitals between 1985 and 1990 (Joshi et al., 1997). Stray dogs from areas where livestock are slaughtered (Devleesschauwer et al., 2014) play a major role in the transmission, with cattle and buffalos the most important intermediate hosts. In older studies, 13% of owned dogs were fed with butcher's waste or other raw meats/offal, and 20% of 134 households in a community health survey fed their dogs raw meat and offal (Joshi et al., 1997). Therefore, both stray and owned dogs are important in the transmission of CE in Nepal.

Only preliminary data are available from **Bhutan**, where CE has been documented in cattle and yaks. As home slaughtering is not commonly practised in Buddhist culture and the small sheep population is not used for meat production, CE transmission occurs focally and is probably linked to local risk factors such as sheep production in remote areas and scavenging opportunities by stray dogs with access to meat waste from butcher shops and slaughterhouses (Thapa et al., 2017).

No transmission data are available for **Bangladesh**. An old report describes infection in 62.5% of dogs from Bangladesh, with seasonal variation (Islam, 1980). More recently, active transmission of CE appears to be occurring in Bangladesh due to findings from hospitals in Dhaka, between 2002 and 2011 (Karim et al., 2015). CE has also been reported in cattle (Islam, 1982; Islam and Rahman, 1975). In **Sri Lanka** CE has been reported sporadically in the past.

3.5.4.2 Molecular epidemiology

At least five *Echinococcus* genotypes causing CE are present in South Asia in a variety of hosts (Table 11). *E. granulosus* genotypes (G1 and G3) seem to be the predominant genotypes circulating in sheep and buffaloes and are probably of most zoonotic significance; however, G5 and G6 have been detected in humans.

3.5.4.3 Infections in animals

Prevalence studies, mostly in intermediate hosts, from the last 10 years indicate that CE is endemic in most parts of South Asia (Fig. 9 and Table S9 in the Supplementary Material).

In **Pakistan**, CE in animals is well documented in the literature and older published data show prevalence ranging from 5% to 46% (Iqbal et al., 1989; Khan and Haseeb, 1984; Shafiq et al., 2005). A more recent

Table 11 Genotypes of *Echinococcus* spp. causing cystic echinococcosis in South Asia (no data found in the missing countries): *Echinococcus granulosus* (G1−3), *Echinococcus equinus* (G4), *Echinococcus ortleppi* (G5), *Echinococcus intermedius* (G6/7) and *Echinococcus canadensis* (G8, G10)

Country	Human	Canids	Sheep/ Goat	Camel	Pig	Yak	Cattle/ Buffaloes
India	G1, G3, G5, G6[1]	G1[2]	G1[3,4,5] G2[6] G3[3,4]		G3[3,4] G5[4]		G1−3[3,4,5] G5[4,5] G2[6]
Nepal	G1−3[7,8]	G1	G1, G5[8]		G1−3		G1, G5[8]
Bhutan	G1−3[12]	G1−3[12]				G1−3[9]	G1−3, G5[9]
Pakistan	G1[10,11]		G1, G3[10,11]	G1[10,11]			G1, G3[10,11]
Afghanistan	G6[10]						

[1] (Sharma et al., 2013); [2] (Singh et al., 2014b); [3] (Singh, 2011); [4] (Pednekar et al., 2009); [5] (Gudewar et al., 2009); [6] (Bhattacharya et al., 2007); [7] (Devleesschauwer et al., 2014); [8] (Zhang et al., 2000); [9] (Thapa et al., 2017); [10] (Alvarez Rojas et al., 2014); [11] (Latif et al., 2010); [12] Thapa N.K. and Deplazes P., unpublished data.

study in Punjab documents a reduction of the prevalence, with the following values: camels (17.3%; 95% fertile cysts), sheep (7.5%, 86.4% fertile cysts), buffaloes (7.2%; 84.3% fertile cysts), goats (5.5%; 79.1% fertile cysts) and cattle (5.2%; 75.3% fertile cysts) (Latif et al., 2010). Both G1 and G3 were detected in goats, camels and cattle in Punjab (Latif et al., 2010).

In **India**, the prevalence of intestinal *E. granulosus* in ownerless stray dog populations varies from 3.5% in villages to 33% in towns. Prevalence was higher in stray dogs near slaughterhouses compared to stray dogs of other areas. In Kashmir, dogs in seminomadic pastoral communities in hilly areas had a higher prevalence of *E. granulosus* (35%) than that corresponded to stray dogs in urban areas with access to slaughterhouses (11−17%) [for citations see Traub et al. (2005)]. The role of wolves, foxes and jackals as definitive hosts of *E. granulosus* has not been well documented. In India, the prevalence of CE in intermediate hosts varies considerably with age of the animals and region; however, there is a tendency for decreasing prevalence over the last three decades, especially in urban areas (Pednekar et al., 2009; Singh, 2011; Singh et al., 2014a). CE has been diagnosed in livestock (sheep, goats, cattle and buffaloes) throughout the country, and camels in Aligarh (Irshadullah et al., 1989). High prevalence and high rates of fertility of the cysts in cattle and buffaloes are explained by the older age at which the animals are slaughtered and host-adapted strains (G5, G3) (Pednekar et al., 2009). Moreover, a wide range of susceptible captive and free-ranging wildlife such as deer, wild buffaloes and wild boar may serve as intermediate hosts in a wild animal cycle which has not been studied in detail so far. Total annual median economic losses due to CE in India have been estimated to be US$ 212.35 million, of which 89% was related to cattle and buffalo disease (Singh et al., 2014a).

In **Nepal** (Kathmandu), a study in 17 slaughtering sites revealed CE prevalence in 5% of water buffaloes, 3% of goats, 8% of sheep and 7% of pigs, and further studies in buffaloes in the country revealed CE prevalence of 12–26% (Manandhar et al., 2006). Not surprisingly, several studies have identified taeniid eggs in faeces of stray dogs (for example, 17.6%) (Manandhar et al., 2006), and 15% (3/20) of dogs were infected with *E. granulosus* at necropsy (Joshi et al., 1997).

In **Bhutan**, 10/138 (7.2%) of faecal samples collected in the environment from community dogs in highly populated areas around slaughterhouses or butchers contained eggs of *Echinococcus* species causing CE (Thapa et al., 2017). Furthermore, a survey in Bhutan found CE in 9% of 291 cattle, 26% of yaks and 0% of 167 pigs; most of the cysts isolated were sterile (Thapa N.K., personal communication).

In **Bangladesh**, CE in humans has been reported previously throughout the country (Karim et al., 2015). In abattoirs of the Comilla and Brahmon Baria region CE was diagnosed in 30% of 1460 cattle, 9% of 620 buffaloes, 17% of 460 sheep and 36% of 970 goats (Kabir et al., 2010). A survey in 1975–78 on 611 dogs originating from different parts of the country revealed high variations in prevalence, the highest (76%) being recorded in dogs in and around the slaughterhouses (Islam, 1980).

3.5.4.4 Cystic echinococcosis in humans

Cases of CE in **Pakistan** have been reported in the medical literature and retrospective studies of local hospital records and reports. These reports identified over 470 human CE cases in a 10-year period in Karachi and Punjab [see Latif et al. (2010)]. Despite the predominance of G3 in livestock, the two characterized cysts from humans were G1 (Latif et al., 2010).

In **India** numerous reports of unusual or complicated CE cases document the medical importance of CE in the area. A hospital-based study in north India (Khurana et al., 2007; Singh et al., 2014a) estimated the yearly total number of diagnosed cases without surgery to be 17,075 and the total number of diagnosed cases with surgical/interventional procedure to be 5646. Based on a population of around 1.2 billion, an approximate incidence of 1.878 per 10^5 inhabitants can be estimated. Economic annual median losses for CE in humans have been estimated at approximately

US$ 8.75 million (Singh et al., 2014a). Both the health and economic burden of CE are likely underestimates due to underreporting and underdiagnosis of cases. The burden of CE in India has been estimated as approximately 21,000 DALYs (Budke et al., 2006).

In Nepal, several human case reports document the endemicity of CE in the country (Devleesschauwer et al., 2014). In this study, an annual incidence of 145 CE cases was estimated for Nepal with an annual burden of 251 DALYs.

In **Bhutan**, no published data are available. However, hospital records from the Jigme Dorji Wangchuk National referral Hospital in Thimphu confirmed 53 CE cases that underwent surgery between 2006 and 2013. As for other countries, these data likely underestimate the real incidence; in 2013 alone, six cases of CE were confirmed in the same hospital by histopathological records (Pelden, S., personal communication). Based on these data, an approximate minimal incidence of around one CE case per 10^5 inhabitants can be estimated, but further studies are needed.

In **Bangladesh**, CE has been reported from nearly all geographic areas (Karim et al., 2015). A retrospective hospital-based study in Dhaka between 2002 and 2011 revealed 130 patients (70.8% females), all originating from the northern parts of the country and the majority (76.2%) from rural areas (Karim et al., 2015).

3.5.5 East Asia: China, Mongolia, Korea, Japan

3.5.5.1 Molecular epidemiology and parasite transmission

Occurrence of CE has been documented in East Asia in the past (Eckert et al., 2001; Schantz et al., 1995), and a considerable amount of recent information is available from China (see later).

The G1–3, G6, G7 and G10 genotypes have all been found in China and Mongolia (Table 12). *Echinococcus granulosus* (G1–3) is by far the most commonly isolated and widespread genotype and has been found in humans, sheep, yaks, cattle and dogs, and there are single reports from a camel and from a ground squirrel. *Echinococcus intermedius* (G6) has been found in cattle, camels, humans and dogs in Xinjiang; a human case in Sichuan and goats in Qinghai and in wolves in Mongolia. *Echinococcus intermedius* (G7) has been isolated from four human cases in Heilongjiang in the northeast of China. *Echinococcus canadensis* (G10) has also been isolated from human patients in Heilongjiang and in Mongolia. To date there appears to be no reports of *E. equinus* (G4) or *E. ortleppi* (G5) from East Asia.

Table 12 Genotypes of *Echinococcus* spp. causing cystic echinococcosis in East Asia (no data found in the missing countries): *Echinococcus granulosus* (G1–3), *Echinococcus intermedius* (G6/7) and *Echinococcus canadensis* (G8, G10)

Country	Human ([a])	Dog (D) Wolf (W) ([a])	Sheep (S) Goat (G) ([a])	Cattle (C) Yak (Y) Camel (Cm) ([a])
China: Heilongjiang	G1 (6)[1] G7 (4)[1] G10 (1)[2]			
China: Gansu	G1 (1)[3]		S: G1 (12)[3]	
China: Ningxia	G1 (7)[3]		S: G1 (13)[3]	
China: Sichuan	G1 (45)[4,5,6] G3 (2)[4] G6 (1)[6]		S: G1 (4)[3]	Y: G1 (15): G1[3,4,5]
China: Qinghai	G1 (37)[7]	D: G1 (4)[7] D: G1–3 (4)[8]	S: G1 (102)[3,4,5,6,7,9] G: G6 (2)[9]	C: G1 (13)[3,7] Y: G1 (26)[4,9]
China: Xinjiang	G1 (4)[10] G3 (3)[11,12]	D: G6 (3)[12]	S: G1 (17)[10]	C: G1 (7)[10] C: G6 (2)[10] Cm: G1 (1)[10] Cm: G6 (9)[10,11]
Tibet AR			S: G1 (32)[4,5,6]	Y: G1 (1)[4]
Japan	G1[b]			C: G1–3[b,13]
Mongolia	G1 (46), G6/7 (36), G10 (5)[14,15]	D: G1 (5)[6] W: G10 (3), G6/7 (2)[16]		

[1] (Zheng et al., 2014); [2] (Yang et al., 2015); [3] (Yang et al., 2005); [4] (Yan et al., 2013); [5] (Zhong et al., 2014); [6] (Wang et al., 2015b); [7] (Ma et al., 2008); [8] (Boufana et al., 2015a); [9] (Liu et al., 2013); [10] (Zhang et al., 1998); [11] (Ma et al., 2009); [12] (Bart et al., 2006); [13] (Nakamura et al., 2011); [14] (Ito et al., 2014); [15] (Jabbar et al., 2011); [16] (Ito et al., 2013).
[a]Number of cases.
[b]Imported cases.

3.5.5.2 China

3.5.5.2.1 Infections in animals The distribution of CE in intermediate hosts in China is shown in Fig. 9 (see also Table S9 in the Supplementary Material). CE is highly endemic through much of western China, in Inner Mongolia in the north and further northeast as far as Heilongjiang. In northwestern China, endemic provinces include Gansu, Qinghai, Ningxia and Xinjiang. In southwestern and central China, the western part of Sichuan province and the TAR are endemic. These provinces are also coendemic for AE, although at the local level one or other of the diseases may predominate and CE tends to be more widely distributed with a greater number of cases. The Autonomous Province of **Xinjiang** has some of the most comprehensive data for animals. In 2014, 1801 and 834 sheep of 18,374 examined were infected with hepatic and lung cysts, respectively (Nusilaiti and Yan Hao, 2016). Only in the Changjizhou prefecture was the parasite absent in sheep. Other prefectures had prevalences in sheep ranging from 4.4% (liver cysts) and 1.2% (lung cyst) (n = 9544) in Akesu prefecture to 42.4% (liver cysts) and 34% (lung cysts) (n = 887) in Tacheng Prefecture. Hepatic CE was present in 362 of 3380 cattle and lung cysts were present in 253 of 3380 examined. CE in cattle was absent in Urumqi City, Bozhou, Aletai and Changjizhou Prefectures. Other prefectures had prevalences ranging from 3.3% for hepatic cysts (n = 66, with no lung cysts detected) in Hetian Prefecture to 22.8% (liver cysts) and 14.8% (lung cysts) (n = 189) in Kezhou Prefecture. Of 3842 dogs examined, 378 (9.8%) were positive for taeniid eggs and copro-antigen positive for *Echinococcus*. Canine echinococcosis was not found in Tacheng, Bozhou, Aletai and Yili Prefectures and Urumchi city. Elsewhere, prevalences ranged from 1.1% of 880 dogs in Kashi Prefecture to 25% of 757 dogs in Bazhou Prefecture (Nusilaiti and Yan Hao, 2016).

In **Qinghai**, very high prevalences of CE (78% of 136) have been reported in yaks in Jiuzo county (Yu et al., 2008) and a mean of 48% of 5211 yaks from the Haibi, Haixi Mongol and Yushi, Golog Autonomous Tibetan Prefectures (Cai et al., 2012). High prevalences in sheep are also reported from these districts: 83% of 115 in Jiuzi County (Yu et al., 2008) and 46% of 9468 in the various autonomous Tibetan prefectures (Cai et al., 2012). The prevalence in dogs in these districts was approximately 40% of 1078 (Cai et al., 2012; Yu et al., 2008).

In **Gansu,** in the Gannan Tibetan Autonomous Prefecture, 11% (of 113) sheep, 20% (of 634) yaks and 23% (of 74) dogs were infected (Zhao et al., 2009). In Sunan County, 23% (of 502) sheep were infected (Niu and An, 2012).

In western **Sichuan**, 51% (of 429) yaks were infected (He et al., 2000). More recent reports indicate 21% of 315 yaks and 30% of 210 sheep were infected in the Garze Tibetan Autonomous Prefecture (Guo et al., 2012). Studies of dogs include prevalences of 14.5% (n = 302) (Guo et al., 2012) and 11.1% (n = 365) (Hartnack et al., 2013).

In **Ningxia**, CE was found in 0–9% of sheep in three separate studies (Cleary et al., 2014; Ma et al., 2014; Wu, 2015) and Wang et al. (2008) reported in the same area prevalences of 52% in sheep, 3% in goats, 81% in cattle, 24% in pigs and 19% in camels.

In the **TAR**, CE is common in yak, cattle, sheep and goats (reviewed Feng et al., 2015), and G1 was isolated from sheep and yaks (Table 12). In **Inner Mongolia**, prevalences of CE of 15–48% in sheep and 24–35% in camels have been reported (reviewed by Wang et al., 2008).

3.5.5.2.2 Cystic echinococcosis in humans Cases of human CE have been reported from nearly every province in China (Wang et al., 2010; Table S10) (Fig. 10). However, in eastern China, only few cases are reported and these may be the result of internal migration from the more highly endemic regions further west. The disease is notifiable, and official reports from those seeking hospital treatment are useful in showing the geographical extent of the endemic region, but even hospital treated cases suffer from underreporting. Prevalence studies undertaken using imaging techniques (i.e., ultrasound) give a very different impression of the extent and intensity of disease occurrence compared to hospital reports or official data. Many, if not most, community cases go untreated (and unreported) due to remoteness from medical facilities and poverty. Incidence data based on hospital cases report numbers of cases presenting for treatment over a certain time period. In contrast, community studies using imaging techniques give prevalence data (a snapshot of the proportion of the population suffering from the disease at the time of the study). However, prevalence can be related to incidence (when the prevalence is low and the prevalence and population size is stable) by the relationship: prevalence = incidence × duration (Freeman and Hutchison, 1980).

There are a few data on the likely case fatality ratio of CE in the absence of treatment. If it is relatively low, the duration can be assumed to be the residual life expectancy from the time of diagnosis, which in most studies is between 35 and 40 years, giving a residual life expectancy of 42 years using the latest Chinese life table. This gives an opportunity to compare estimated incidence from the community studies to reported incidence data, giving an indication of the degree of underreporting. Using this methodology, incidence in the populations or districts at risk has been estimated whilst reports based on hospital treated cases are reported alongside as incidence (Fig. 10; Table S10). However, even this approach will underestimate the incidence of new cases as there will be some parasite-induced mortality and ultrasound surveillance will not detect subjects who have pulmonary CE.

In both the nationally reported numbers of cases and those reported at regional or prefecture level, there has been a tendency for an increasing numbers of cases over time. For example, the total number of cases of echinococcosis reported in China was 931 in 2004, increasing to over 5800 in 2008 (Wang et al., 2010), and those in Xinjiang increased from less than 400 in 2005 to 1434 in 2013 (Osman et al., 2014). This is likely because China is rapidly improving the medical and public health programmes and widespread routine surveillance has become a feature of the echinococcosis control programmes now being implemented. Surveillance systems are also continuously improving. This has inevitably resulted in increased numbers of cases being both treated and reported. In Fig. 10 and Table S10 (Supplementary Material) incidence data compiled from officially notified data or from hospital records are given alongside prevalence data from mass ultrasound surveillance, which are then converted to an estimate of incidence per 10^5 inhabitants per year. Where possible, the data are only for CE, with AE data removed. For some incidence data, it was specified only as echinococcosis, although these tended to be in districts where CE is much more common than AE, with the exception of western Sichuan. Data are variably reported at the county, prefecture and province levels.

In **Sichuan** province, most cases of CE originate from the Garze Autonomous Prefecture with perhaps 10% coming from the neighbouring Aba Autonomous Prefecture. Both of these districts are on the Tibetan plateau with a total population of just under 2 million. CE is rare in the rest of Sichuan. Garze Autonomous prefecture is also intensely endemic for AE. Compilation of ultrasound surveillance data from a number of large studies indicated that 334 of 10,186 individuals investigated had lesions of CE (3.2%) with 311 having lesions of AE (3.1%). The mean age of participants was 39 years with a residual mean life expectancy of 42 years. Assuming the life expectancy of AE is just 10 years (Torgerson et al., 2008), the estimated incidences of CE and AE are approximately 78 and 305 cases per 10^5 inhabitants, respectively. The National Infectious Diseases Reporting System reported 9127 cases of echinococcosis between 2007 and 2013, or a mean annual incidence of 173 cases per 10^5 (Wei et al., 2014). These data do not differentiate CE and AE, but indicate that the national reporting system may only be reporting 45% of cases, with the remaining either not seeking treatment, or being under reported by the system.

In the **TAR** CE has been reported from four of seven prefectures, although it is likely that CE is endemic across the entire region. The most intensely endemic district seems to be around Lhasa, where hospital records indicated 526 cases of CE between 2007 and 2013, which translates into an annual incidence of 75 per 10^5 inhabitants (Zheng et al., 2014). A community study found nearly 10% of participants with lesions of CE, which is consistent with the large numbers being treated (Feng et al., 2015). Elsewhere in Nagqu Prefecture, the annual incidence of hospital treated cases has been reported at 27 per 10^5 (Gong et al., 2001), whilst in Chandu over 5% of participants in a community study had CE lesions (Feng et al., 2015).

In **Gansu**, ultrasound surveillance, hospital records and reported data indicate that the incidence of CE is between 2 and 10 cases per 10^5 inhabitants per year.

Qinghai appears to be one of the most intensely endemic provinces for CE, with a large number of cases in each prefecture. Officially notified data reported just 363 cases in 2011 (Li et al., 2013b). However, this is not consistent with large ultrasound screening programmes which have consistently found between 1% and 10% of the population infected with CE and would represent an incidence of 5–10 times that officially reported.

In **Xinjiang**, CE cases are reported from every prefecture, although mostly in the north of the province. Prefecture level incidences range from one to two per 10^5 inhabitants per year in Kashgar and Hotim to just under 20 per 10^5 inhabitants in the Bartal Mongol Autonomous Prefecture [data estimated for 2013, based on the overall trend for Xinjiang (Osman et al., 2014)]. For Xinjiang in total, the surgical incidence was reported at 6.5 cases per 10^5 inhabitants for 2013. As part of the echinococcosis control programme, there have been some large scale ultrasound surveillance programmes. In Emin and Habahe Counties (in Techeng and Altay Prefectures), prevalence of CE was under 1% of over 200,000 subjects, representing over half the total population (Yang et al., 2015). This suggests a lower incidence than the prefecture wide incidence officially reported. In contrast, in a number of counties of the Bortala, Bayingolin Mongol, Kazakh, Altay and Techeng (autonomous) Prefectures, ultrasound prevalences were between 1.5% and 6%, which represents an incidence much higher than official data (Feng, 2012).

In **Ningxia**, officially notified data suggest an annual surgical incidence for the province of 5–7 cases per 10^5 inhabitants per year (Li et al., 2013b; Wang et al., 2010). Hospital records suggest that Xiji, Tongxin and Hiayuan Counties have the highest incidence (Yang et al., 2008).

In **Heilongjiang** and **Inner Mongolia**, only province wide data are available. In both provinces, the anual incidence is approximately 0.5–0.7 per 10^5 inhabitants.

Other provinces sporadically report cases of CE but in very small numbers. In 2013, there were just 162 cases reported from provinces outside the core endemic area of western China, representing an anual incidence of 0.01 cases per 10^5 inhabitants (Li et al., 2013b).

3.5.5.3 Mongolia, Korea, Japan

In Mongolia as in many endemic areas in Asia the lack of confirmation of CE and AE cases and underreporting of human cases have to be considered (Tserennadmid and Enkhjargal, 2000). The documentation of echinococcosis cases goes back to the 1950s in Mongolia; for a recent comprehensive review presenting the historical and current situation of echinococcosis in Mongolia, see Ito and Budke (2015). In **Mongolia**, the life cycle of *E. granulosus* (G1) is typically maintained between dogs and livestock due to the traditional nomadic lifestyle, but the epidemiology of the other genotypes, G10 and G6/7, is not well described so far.

Even though **Korea** is not an endemic area for CE, imported cases are probably increasing in light of the expanding economy and the fact that a high number of Koreans travel or work abroad. At present, **Japan** is not considered as an endemic area of CE and no transmission of the parasite has been documented (Guo et al., 2011). Interestingly, for decades, CE has been detected in cattle imported from foreign countries and fattened in Japan.

Molecular data: Analyses of human CE cases in Mongolia revealed the genotypes G1, G6/7 and G10 (Jabbar et al., 2011; Ito et al., 2014). In wolves, G10 and G6/7 have been identified, but no data concerning isolates originating from ruminants are available (Table 12).

3.5.5.3.1 Infections in animals In a small study in **Mongolia**, 5 out of 14 necropsied dogs were infected with *E. granulosus* (G1) in Bulgan Province (Wang et al., 2005). Only wolves have been identified as definitive hosts of *E. canadensis* (G6/7 and G10) (Ito et al., 2013). Few data are available concerning CE in livestock in **Mongolia**. In a recent study covering all provinces, seroprevalence was 3.6% in 590 sheep, 5.9% in 779 cattle and 9.2% in 338 goats using recombinant Antigen B (8/1) (rAgB8/1) (Chinchuluun et al., 2014). Camels are an important livestock species in some regions of Mongolia and have been found to be infected with *E. granulosus* [cited in Ito and Budke (2015)].

In **Japan**, a recent abattoir survey was conducted annually in Miyazaki (southeastern coast of Kyushu, Japan) over a period of 5 years on 9500 imported cattle from Australia. Hydatid cysts (G1—3) were detected in about 1.8% of cattle (Guo et al., 2011). CE has rarely been reported in Japan, and all confirmed CE cases during the past two decades were imported (Nakamura et al., 2011).

3.5.5.3.2 Cystic echinococcosis in humans Some hospital-based reports of human echinococcosis managed mainly by surgeons and a few community-based screening studies document the occurrence of CE in

Mongolia (Ito and Budke, 2015). For example, CE was diagnosed in 18% of surgical cases treated at the First Hospital of Ulaanbaatar in 1993 [Cross, 1995; cited in Ito and Budke (2015)]. To date, the majority of surgical echinococcosis cases have been confirmed to be CE, except for five cases of AE. Several serological community surveys revealed seroprevalences of echinococcosis to be 2.1−11.7% (Ito and Budke, 2015). Despite questionable specificities of the tests used for serology, this relatively high prevalence documents a high endemicity of CE/AE throughout the country, and further studies are needed to estimate the actual burden of CE for Mongolia.

In **Korea**, 33 CE cases have been reported in the literature; these patients included 25 Koreans, 7 Uzbeks, and 1 Mongolian (2 Korean patients with unclear origin of infection, 31 considered as imported) (Choi et al., 2014).

3.5.6 South East Asia: Indonesia, Vietnam, the Philippines, Malaysia, Thailand and the Lao People's Democratic Republic

There is no evidence of transmission cycles that maintain the causative agents of CE in Southeast (SE) Asia, encompassing Indonesia, Vietnam, the Philippines, Malaysia, Thailand and the Lao People's Democratic Republic (Lao PDR) (Craig, 2004; Eckert et al., 2001; McManus, 2010; Schantz et al., 1995). If CE does exist, transmission is infrequent and restricted to wild animal hosts which have yet to be identified. In this respect, the recent report from Vietnam of a locally acquired case of *E. ortleppi* in a captive primate, that must have acquired the infection locally, is of particular interest (Plesker et al., 2009).

3.5.6.1 Infections in animals

Interest in the potential endemicity of *E. granulosus* in SE Asia was raised following a report in 1974 of natural infection in a dog from Sulawesi in **Indonesia** (Carney et al., 1974). However, a subsequent survey of 63 dogs from Sulawesi found no evidence of *Echinococcus* infection. Data on CE in livestock is not available so far. The possibility of indigenous wildlife cycle(s) was strengthened by the report of a locally acquired case of CE in a primate, a red-shanked douc langur (*Pygathrix nemaeus*) in northern Vietnam in 2008 (Plesker et al., 2009). The species was *E. ortleppi* and, although locally acquired, it was unexpected, as the nearest region where *E. ortleppi* has been previously recorded is the Indian subcontinent (Thompson and McManus, 2002). It is not known if a transmission cycle of *E. ortleppi* is autochthonous in northern Vietnam, or whether the parasite was introduced from elsewhere; for example, in introduced cattle. However, there is no evidence that domestic dogs in Vietnam play a role in the transmission of *E. ortleppi*.

3.5.6.2 Cystic echinococcosis in humans

Sporadic cases of human CE, either imported or seemingly locally acquired, have been reported from various countries in SE Asia but there is no information on the epidemiology of these infections (Eckert et al., 2001; Schantz et al., 1995; Wiwanitki, 2004). A serological survey undertaken on 903 inhabitants of Sulawesi did not provide evidence of CE (Palmieri et al., 1984).

3.6 Australia and New Zealand

3.6.1 Host assemblages, transmission and molecular epidemiology

The endemicity of *Echinococcus* in Australia and New Zealand (see Fig. 11) is a relatively recent occurrence that demonstrates the importance of

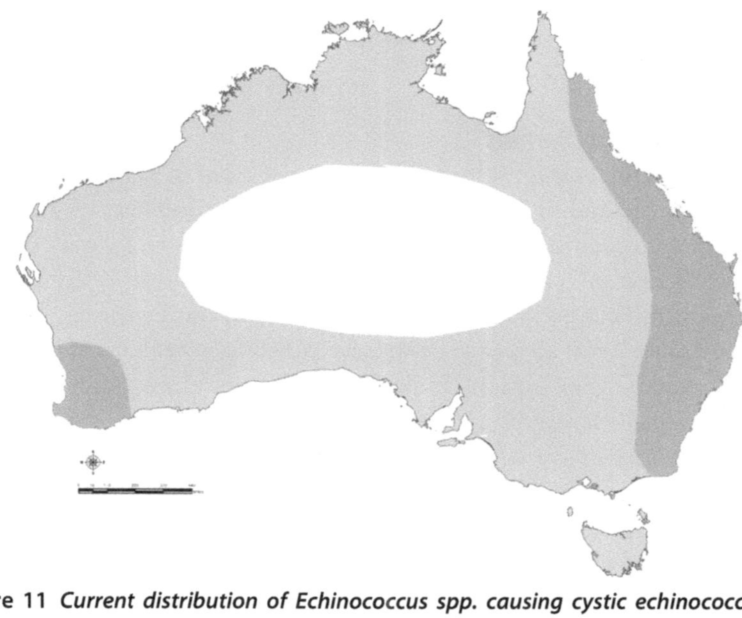

Figure 11 ***Current distribution of Echinococcus spp. causing cystic echinococcosis in domestic intermediate hosts (sheep and cattle) in Australia and in the island state of Tasmania.*** Modified version of the map by (Thompson and Jenkins, 2014) depicting areas of high, low and no transmission of *Echinococcus granulosus* in Australia (dark grey, high transmission; grey, low transmission; white no transmission). For details of the transmission rate in domestic definitive hosts (e.g., dogs) and intermediate hosts (sheep and cattle) in each jurisdiction, see the text.

anthropogenic factors in perpetuating the transmission of the parasite (Thompson, 2013). *Echinococcus* was not present in either country until European settlement when *E. granulosus* was introduced with livestock, principally sheep, by early settlers (Gemmell, 1990). Subsequently, anthropogenic factors were responsible for the transmission of the parasite in a domestic cycle involving sheep and dogs leading to high levels of infection in humans, sheep, and to lesser extent, cattle and dogs, as well as the establishment of a wild animal cycle involving dingoes and macropod marsupials on the mainland of Australia. The control programmes that were instigated were highly successful in both New Zealand and Australia and have been used as models in many other countries where *E. granulosus* is perpetuated in domestic cycles involving dogs and livestock. The historical aspects of establishment, perpetuation and control have been reported in many publications [e.g., Gemmell (1990); Schantz et al. (1995); Beard et al. (2001)] and will not be reiterated here. The current situation in both countries will be summarized.

The Ministry of Agriculture and Forestry declared **New Zealand** provisionally free of CE in 2002 (Anonymous, 2012). It is not clear how long an absence of cases of CE is required before provisional status will change to officially free.

3.6.2 Infections in animals

Fig. 11 reports the current distribution of *E. granulosus* in domestic intermediate hosts (sheep and cattle) in **Australia** and in the island state of Tasmania. Data on the prevalence/incidence of CE in livestock are no longer collected on a routine basis on mainland Australia. *Echinococcus* cysts still occur in sheep on the mainland but prevalence has declined steadily during the last 30 years, and sheep most at risk appear to be those exposed to potential spillover from the wild animal cycle (Jenkins et al., 2014; Thompson and Jenkins, 2014). Recent national surveys in domestic dogs in both urban and rural areas of mainland Australia have demonstrated that the parasite is very uncommon and restricted to rural areas (Palmer et al., 2008; Jenkins et al., 2014). The only routine surveillance of CE undertaken in Australia occurs in Tasmania where provisional elimination of CE was declared in 1996. However, although there are no locally acquired cases of CE in sheep reported, recent reports of *E. granulosus* infections in cattle and dogs suggest transmission is still occurring, albeit at low levels (Jenkins et al., 2014).

A wild animal cycle of transmission is perpetuated on mainland Australia involving dingoes and macropod marsupials. This is confined to eastern and the southwest of Australia (Jenkins and Macpherson, 2003). Because of potential spillover to domestic dogs and sheep, the presence of this cycle will prevent elimination of CE on mainland Australia. The fox, although susceptible to infection, is epidemiologically insignificant in transmission [for details see chapter: Ecology and Life Cycle Patterns of *Echinococcus* Species by Romig et al. (2017)].

3.6.3 Cystic echinococcosis in humans

Notifications of human CE on the mainland of Australia are sporadic and unreliable. New cases continue to be identified but as with the last hospital-based survey (Jenkins and Power, 1996), a significant proportion is in recently arrived immigrants who contracted infection before migrating to Australia (Thompson and Jenkins, 2014). No new human cases have been reported in Tasmania (O'Hern and Cooley, 2013).

3.7 Africa

3.7.1 North Africa: Morocco, Algeria, Tunisia, Libya and Egypt

3.7.1.1 Host assemblages, transmission and molecular epidemiology

For the last two decades, CE has been one of the most important zoonotic diseases affecting North African countries bordering the Mediterranean Sea (Schantz et al., 1995; Ecca et al., 2002; Sadjjadi, 2006; Dakkak, 2010; Hotez et al., 2012; Carmena and Cardona, 2014). The causative agent, *E. granulosus* s.l. has major human health and socioeconomic importance, as well as a significant impact on livestock production (Budke et al., 2006; Torgerson et al., 2014). Annual incidence rates in people usually vary from 5 to 10 cases per 10^5 inhabitants, with minimum and maximum rates of 1 and 25 reported in highly endemic foci, respectively (Torgerson and Macpherson, 2011).

CE is predominantly maintained in these countries by a rural domestic transmission cycle involving dogs as definitive hosts and sheep, cattle, goats, camels, dromedaries and donkeys as intermediate hosts. Clandestine and home-slaughtering (e.g., during weddings and religious feasts) contribute to a secondary suburban and/or urban domestic cycle involving owned and stray dogs that have access to offal of intermediate hosts (Besbes et al., 2003). Wild animals, i.e., golden wolves (*Canis anthus*) as definitive host and wild boars (*S. scrofa*) and antelope (*Addax nasomaculatus*) as intermediate hosts were also infected in the northwest and central regions of Tunisia (Lahmar et al., 2009a). Red fox (*Vulpes v. atlantica*) is a less frequent wild definitive host for *E. granulosus* G1 in Tunisia, with only two cases reported (Lahmar et al., 2009a). The wild animal cycle re-enforces the rural domestic cycle through canid predation of domestic herbivores or environmental contamination of *E. granulosus* eggs in areas inhabited by livestock.

The persistence and spread of CE in **Tunisia** is closely associated with the following epidemiological factors: (1) contact between humans/ livestock and infected dogs; (2) large dog to human population ratios (4: 1, 6:1 and 12:1 in rural, suburban and urban areas, respectively); (3) transhumant livestock migration to northern areas in spring/summer with guard dogs predominantly fed livestock offal; (4) routine home slaughtering during social or religious observances; (5) inadequate hygiene in the rural areas and (6) insufficient abattoir equipment to destroy infected offal. Similar factors are present in **Morocco**, where the high prevalence of CE in humans and animals is influenced by (1) an abundance of dogs (1.8 dogs per household); (2) poor knowledge of CE (50% of people are aware of the disease) and (3) inadequate abattoir infrastructure and hygiene (El Berbri et al., 2015a).

The main drivers for CE transmission in **Morocco** (Kachani et al., 2003; Azlaf et al., 2007; Dakkak, 2010) have identified the high number of stray dogs and close proximity of un-dewormed owned dogs living with humans and livestock; behaviours and practices of the rural population; poor human hygiene practice; the infected environment (soil, food, drinking water) with *Echinococcus* eggs; animal slaughtering in fields or at home for family and religious occasions; abattoirs lacking appropriate facilities and hygiene to which free-roaming dogs have easily access. In **Algeria**, CE is primarily maintained by a synanthropic cycle involving domestic dogs and livestock (sheep and cattle), where dogs are fed offal discarded from animals slaughtered for human consumption (Kouidri et al., 2012).

The same host assemblage is observed in **Libya**, where livestock are slaughtered in a clandestine and uncontrolled manner, providing dogs with a ready source of infection (Buishi et al., 2005). In this country, *Echinococcus* spp. are transmitted primarily through a rural pastoralist life cycle between dogs and ruminants (sheep, camels, cattle and goats) (Elmajdoub and Rahman, 2015); no reports are found on the wildlife cycle. In **Egypt**, CE is a disease of pastoral communities due to the abundance of stray dogs and common home slaughter/street practices. Various livestock are major reservoirs for the human infection (Omar et al., 2013).

So far, despite attempts to control slaughtering practices and to promulgate health education, no regional or national control programs have been established in North Africa.

3.7.1.2 Molecular epidemiology

Table 13 summarizes *Echinococcus* genotypes identified in North Africa. In **Tunisia**, three species and four genotypes – *E. granulosus* (G1 and G3), *E. equinus* (G4) and *E. intermedius* (G6) – have been identified. The most prevalent genotype (G1) occurs in a wide range of domestic livestock species as well as in humans; whereas the buffalo strain (G3) occurs only in cattle and humans. Mixed infections, involving more than one genotype in a single host, have been observed in people (G1, G6) (Oudni-M'rad et al., 2016) and donkeys (G1, G4). Other hosts include dromedaries in southern Tunisia, infected by G1 and G6, and wild canids (golden wolf and red fox), infected by G1. In **Morocco**, sheep and cattle are infected by G1, G2 and G3; G1 also infects humans, goats, camels and equids. In **Libya**, the most common genotype, G1 occurs in humans, sheep, cattle, camels and dogs, whereas G6 occurs in cattle and camels. Three genotypes occur in **Algeria**: G1 (prominent in humans, sheep, camels and cattle/buffalo), G2 (humans, sheep and camels), and G6 (camels). In **Egypt**, five genotypes are reported: G1 (camels, sheep), G6 (buffalo, camels, sheep), G4 (equids), G5 (camels) and G6/G7 (pigs). People in Egypt have been infected by three genotypes (G1, G6 and G7). The molecular epidemiology of *Echinococcus* spp. and genotypes in wildlife is poorly documented.

3.7.1.3 Infections in animals

The current distribution of *E. granulosus* in domestic intermediate hosts (sheep, cattle, goats and camels) in Africa is shown in Fig. 12.

In **Tunisia**, *Echinococcus* infections are widespread in dogs (prevalence above 20%) and domestic ruminants, with prevalence levels of 16.4; 8.56; 2.8; 5.9; 8.4% reported in sheep, cattle, goats, camels and equines, respectively [Fig. 12 and Table S11 and S12 of the Supplementary Material] (Lahmar et al., 2013, 2014b). The rural domestic cycle is primarily maintained by sheep and cattle intermediate hosts in north and central regions and by camels in the south. Ultrasound screening of 1039 sheep in Zaghouan (northeastern region) reported a CE prevalence of 40.4% (Lahmar et al., 2007); other surveys have demonstrated a direct correlation of CE infection to age with prevalence ranging from 9.5% in sheep less than one year of age to 96.3% in sheep greater than 5 years of age (Lahmar et al., 1999). In Benguerden (southwestern region), 10.1% of camels were observed with CE (Lahmar et al., 2004). Among wild hosts, 9.7% of golden wolves (Lahmar et al., 2014a) and 18.8% of wild boars (Lahmar S, unpublished) were found infected.

Table 13 Genotypes of *Echinococcus* spp. causing cystic echinococcosis in North Africa (no data found in the missing countries): *Echinococcus granulosus* (G1–3), *Echinococcus equinus* (G4), *Echinococcus ortleppi* (G5), *Echinococcus intermedius* (G6/7)

Country	Human	Dog (D), Wild canids (Wc)	Sheep	Camel	Cattle/Buffalo	Equine	Goat (G), Antelope (A)	Swine (S), Wild boar (Wb)
Morocco	G1[1]		G1[3] G2[3] G3[3]	G1[2]	G1[2,3] G2[3] G3[3]	G1[2]	G: G1[2]	
Algeria	G1[1,4,5] G2[5]		G1[1,4,5,6] G2[5]	G1[5] G2[5] G6[1,4,5]	G1[1,4,5,6]			
Tunisia	G1[6,7,8,9,10,11,12] G3[9] G6[12]	D: G1[8,14,18,28] Wc: G1[8,14,18,28]	G1[6,7,9,13,14]	G1[14,15] G6[7,14]	G1[6,7,9,13,14]	G1[14,16] G4[14,16]	G: G1[14,16] A: G1[17]	Wb: G1[14]
Libya	G1[19,20]	D: G1[18]	G1[18,19]	G1[19] G6[20]	G1[19,20] G6[20]			
Egypt	G1[21,22] G6[22,25,26,27] G7[27]		G1[4,21,23] G6[23]	G1[21,23] G5[23] G6[22,23,25,27]	G6[23]	G4[24]		S: G6[22,27] S: G7[27]

[1] (Bardonnet et al., 2003); [2] (Azlaf, 2007); [3] (El Berbri et al., 2015b); [4] (Bart et al., 2004); [5] (Maillard et al., 2007); [6] (Boubaker et al., 2013); [7] (M'Rad et al., 2005); [8] (Lahmar et al., 2009b); [9] (M'Rad et al., 2010); [10] (M'Rad et al., 2012); [11] (Schneider et al., 2010); [12] (Oudni-M'rad et al., 2016); [13] (Farjallah et al., 2007); [14] (Boufana et al., 2014); [15] (Lahmar et al., 2004); [16] (Lahmar et al., 2014b); [17] (Boufana et al., 2015c); [18] (Boufana et al., 2015a); [19] (Tashani et al., 2002); [20] (Abushhewa et al., 2010); [21] (Abd El Baki et al., 2009); [22] (Abdel Aaty et al., 2012); [23] (Amer et al., 2015); [24] (Aboelhadid et al., 2013); [25] (Khalifa et al., 2014); [26] (Alvarez Rojas et al., 2014); [27] (Alam-Eldin et al., 2015); [28] (Lahmar et al., 2009a).

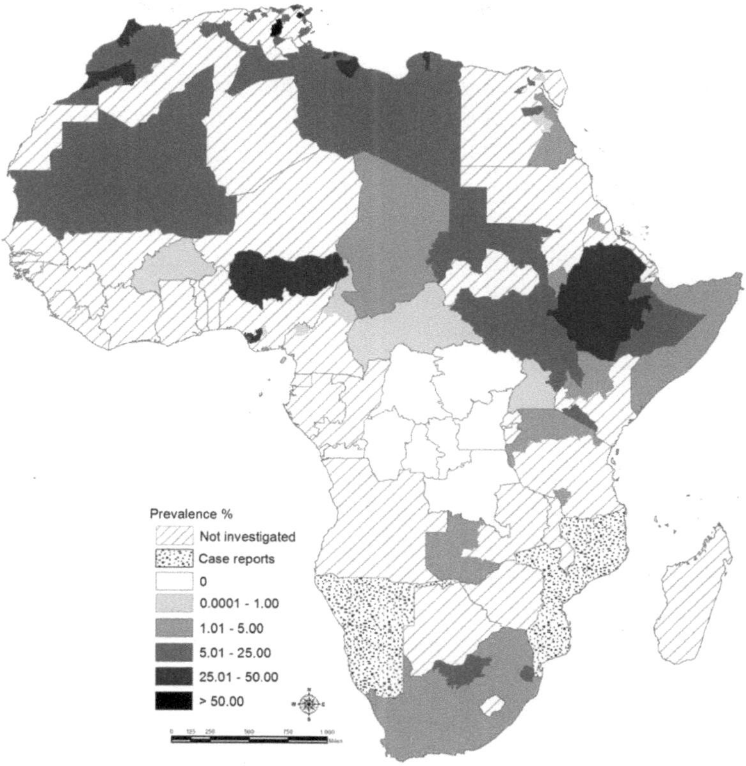

Figure 12 ***Current distribution of Echinococcus spp. causing cystic echinococcosis in domestic intermediate hosts (sheep, cattle, goats and camels) in Africa.*** Areas with case reports of CE in wild intermediate hosts are approximately given if data of domestic intermediate hosts was missing. The detailed information (prevalence data in each jurisdiction) is listed in Tables S11 and S12 of the Supplementary Material.

In **Morocco**, *E. granulosus* is highly endemic in dogs, with prevalence levels as high as 48.4%, 55.4%, and 58.8% in south-central, southern and northwestern areas, respectively. Cattle also experience high levels of CE: 37.6% and 42.9% in northwestern regions (in Loukkos and Sidi Kacem, respectively) (Azlaf and Dakkak, 2006; El Berbri et al., 2015b) and 48.7% in south-central regions (Middle Atlas) (Azlaf and Dakkak, 2006). Sheep are at highest risk of infection in the northwestern areas (Sidi Kacem and Loukkos) (10.9% and 31.65%, respectively) (El Berbri et al., 2015b; Azlaf and Dakkak, 2006). Although cysts from cattle and sheep have similar fertility levels (54.9% and 50.3%, respectively), sheep represent the main source of infection to dogs because they are most commonly slaughtered in the abattoirs of all provinces in Morocco due to their greater population; while cattle are generally slaughtered when they are older than sheep and goats. Control measures targeting CE in both sheep and cattle are appropriate in this context (El Berbri et al., 2015b). A lower disease burden is reported for goats (1.88%), equids (17.8%) in central regions and camels (12%) in the south (Azlaf and Dakkak, 2006).

In **Algeria**, the prevalence of *E. granulosus* in dogs ranges from 15.5% in Batna to 42% in Constantine (Bentounsi et al., 2009). In cattle, CE is considered highly endemic in Tebessa (89.7%), Djelfa (70%) and Tiaret (25.6%), with high levels of cyst fertility reported (Hamrat et al., 2011; Kouidri et al., 2012; Ouchène et al., 2014). Seventy-eight percent of sheep are infected in Tebessa; however, this appears to be an anomaly as sheep experience lower levels of infection than cattle in other regions. This might be explained by differences in age of the slaughtered sheep (young animals) compared to cattle (adult animals). Sheep play an important role in perpetuating and disseminating of CE due to their high infection rates and fertility, while goats do not seem to be very important in transmission dynamics as the majority of cysts were sterile (Kouidri et al., 2013). Camels are also important intermediate hosts, as indicated by the high prevalence (~25%) and fertility (100%) of CE caused by *E. intermedius* (G6) in Ouargla (Bardonnet et al., 2003) and Adrar. Equines (Bardonnet et al., 2003) and wild boars (Ouchène et al., 2014) contribute to parasite transmission in the domestic and wild animal cycles. Domestic dogs are the only known definitive host for *Echinococcus* spp. in Algeria. Golden wolves are a possible wild host, as they have been observed with *E. granulosus* cestodes elsewhere, but the only postmortem study of wild canids did not report the presence of *Echinococcus* species (Jore d'Arces, 1953).

In northwest **Libya** (Tripoli), the infection rate of *Echinococcus* spp. in necropsied stray dogs has been reported as high as 60% (Awan et al., 1990); however, a more recent postmortem study reported prevalences between 26% and 35% (Buishi et al., 2005). Information on canine echinococcosis is not available for other parts of the country. Prevalence of CE in livestock is reported for sheep (55.9%), cattle (28.57%) and goats (40%) (Ekhnefer, 2014), as well as camels (>35%) in eastern (Al-Jabal Al-Akhdar) and north-eastern (Benghazi) regions (Ibrahem and Craig, 1998). Livestock CE is lower in north–central regions (Sirte) (Kassem et al., 2013). Hydatid cysts collected from sheep and camels demonstrate high fertility rates (>80%), suggesting that these intermediate hosts are important to the maintenance of *Echinococcus* (Elmajdoub and Rahman, 2015). No information on CE in wild animal hosts in Libya is available.

In **Egypt**, *E. granulosus* is common in stray dogs with prevalence levels of 5% in Dakahlia, 16% in Cairo and 1.8% in Giza (El Shazly et al., 2007; Mazyad et al., 2007; Haridy et al., 2008). Recent studies report a CE prevalence of 18.9% in camels, 12.7% in cattle, 13% in goats, 14.1% in equines and 7.8% in sheep (El-Madawy et al., 2011; Omar et al., 2013; Mahdy et al., 2014). Pigs and buffaloes are less likely to be infected than other intermediate hosts; however, prevalences of 4.46% and 1.4%, respectively, have been reported (Hamdy et al., 1980; Yassien et al., 2013). Fertility of G1 cysts is highest in sheep (72.7%) and camels (79.2%) (El-Madawy et al., 2011; Mahdy et al., 2014).

3.7.1.4 Cystic echinococcosis in humans

Incidence rates of CE are given in Fig. 13 (see also Table S13 of the Supplementary Material). Despite a decrease in mean annual incidence (surgical rate) from 15 to 12.6 cases per 10^5 inhabitants between 1985 and 2005, **Tunisia** remains the highest endemic North African country (Chahed et al., 2015). Annual losses associated with CE in people and animals are estimated at US$ 19 million for this nation (Majorowski et al., 2005). Northwestern and west-central governorates experience the highest burden of CE, which exacerbates issues related to poverty.

CE is a significant public health problem in **Morocco** where a total of 1700 surgical cases of human CE are treated (5.2 cases per 10^5 inhabitants) in the whole country (Budke et al., 2006; El Berbri et al., 2015b). Surgeries are repeated in 3% of cases and a mortality of 3% is reported. Expenses due to CE surgery are estimated to a total cost of US$ 2,550,000.00 (Herrador et al., 2016).

The incidence of human CE in **Algeria** did not change from 1997 to 2008 with approximately 2.1 surgical cases per 10^5 inhabitants, corresponding to 700 cases recorded (Institut National de la Santé Publique Algérien. Relevé épidémiologique annuelle vol. XIII, 2008). The minimum cost for surgical intervention is of US$ 18,200 per case corresponding to a total annual cost of US$ 127,400 (Hamrat et al., 2011). Hydatid disease continues to be a public health concern in Algeria.

Human CE is considered one of the most important parasitic infections in **Libya**. Prevalence ranges from 1.7% in the north (Shambesh et al., 1999) to 9% in the northwest (Nalut) (Mohamed et al., 2014). The highest levels occur in the northeast (Benghazi), where rural residents are disproportionally

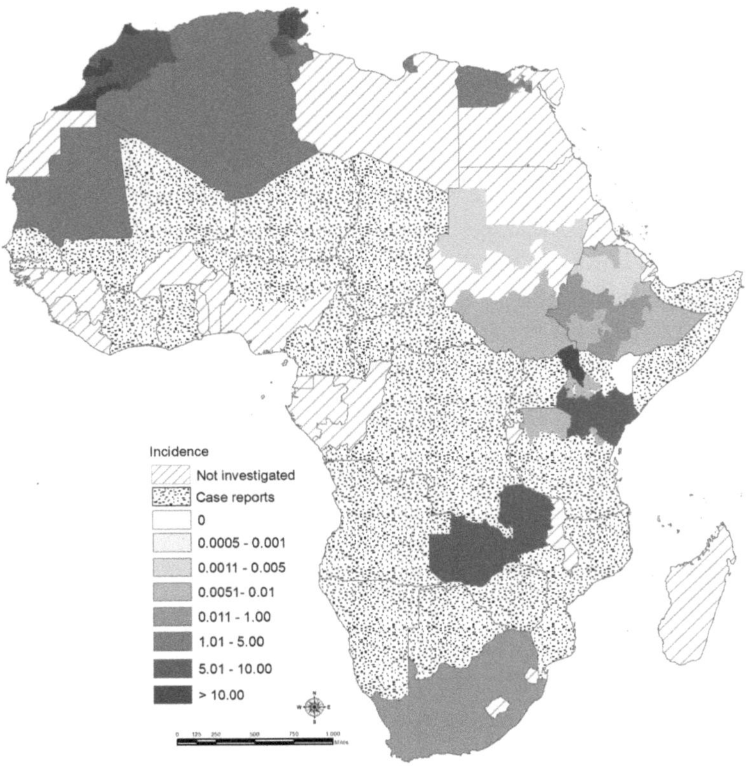

Figure 13 ***Current incidence of human cystic echinococcosis in Africa.*** The detailed information (incidence data in each jurisdiction) is listed in Table S13 of the Supplementary Material. Further prevalence data based on ultrasound surveys in different ethnic groups in eastern Africa are reported in Table 15.

affected compared to their urban counterparts (53.5% versus 46.5%, respectively) (El-Gidaafi and Kassem, 2013). Economic losses due the CE in human and animal populations are not available.

In **Egypt**, human CE is endemic, but the prevalence is low. The annual incidence rate of surgical intervention ranges from 0.05 to 2.6 cases per 10^5 persons in Cairo and Matrouh, respectively. Surveillance for CE using ultrasound and serology recorded 44 new human CE cases per year during 1997–99 with cases occurring in Alexandria, Cairo, Giza, Nile Delta and North Sinaï, Matrouh (Kandeel et al., 2004). Egypt is surrounded by countries where CE is endemic and highly prevalent, suggesting CE could increase in Egypt if current control measures were relaxed. The socioeconomic impact of CE in Egypt has not been estimated.

3.7.2 Sub-Saharan Africa

3.7.2.1 Introduction and molecular epidemiology

Compared to other continents, sub-Saharan Africa is noted for the largest diversity of *Echinococcus* spp. and genotypes causing CE (Table 14). *Echinococcus granulosus* s.s. is widespread except for the arid northern regions (Sahara and Sahel zones) both in domestic and wild animals [for details on the ecology of species occurring in this endemic area are presented in chapter 'Ecology and Life Cycle Patterns of *Echinococcus* Species' by Romig et al. (2017)]. Areas of high prevalence in animals (Fig. 12) (Tables S11 and S12) usually correlate with high incidence of human CE (Fig. 13) (Table S13 in the Supplementary Material). *Echinococcus felidis* is widespread in eastern and southern Africa, but may be restricted to a wildlife cycle between lions, hyenas and warthogs; to date, there is no indication of any involvement of domestic animals or humans. *Echinococcus equinus* is known from a recently recognized wildlife cycle involving lions, wild canids and zebras, which is probably widespread in southern Africa, but may be absent elsewhere; no infection of domestic animals have been reported so far, and no human cases are known worldwide. *Echinococcus ortleppi* is widespread (from Sudan to South Africa), only seems to be frequent in some countries of southern Africa, where traditional cattle pastoralism is practised; only one human case (from South Africa) has ever been found. *Echinococcus intermedius* (G6/7, pig and camel strain) is dominant in the arid North (Mauritania, Sudan, and probably the countries between including the northern regions of West African countries like Ghana or Nigeria), where transmission appears to be based on camels. However, the camel strain G6 is also found outside the camel raising regions (from southern Kenya to South Africa), where goats may be important for the life cycle, and where the involvement of wildlife is also known.

3.7.2.2 Northern arid zone: Mauritania to Sudan

3.7.2.2.1 Host assemblages and transmission Except for Sudan [where research on CE dates back to the 1950s (Omer et al., 2011)] and Mauritania, data from this vast region are restricted to few reports on human cases in Mali, Niger and Chad and — mostly older — dog and livestock infection data from Mali and Chad. Data from wild carnivores are limited; only one study documents *Echinococcus* infections in African golden wolves (*C. anthus*) (Troncy and Graber, 1969). CE seems to be frequent in camels, but only moderately frequent to rare in other livestock species, while human cases are widespread at moderate to low numbers [see Supplementary

Table 14 Genotypes of *Echinococcus* spp. causing cystic echinococcosis in sub-Saharan Africa: *Echinococcus granulosus* (G1–3), *Echinococcus equinus* (G4), *Echinococcus ortleppi* (G5), *Echinococcus intermedius* (G6/7) and *Echinococcus felidis* (Ef)

Country	Human	Dog (D), Wild carnivores (Wc)	Sheep	Camel	Cattle	Pig (P), Warthog (Wa)	Goat	Wild ruminants (Wr), Wild equines (We)
Mauritania	G6[1,2]			G6[1,2]	G6[1]			
Mali		D: G6[26]						
Ghana	G6[23]							
Sudan	G6[4]		G6[3,4,5]	G5[6] G6[3,4,5]	G5[4,25] G6[3,4,5]		G6[4]	
South Sudan	G1–3[14] G6[4]		G6[4]		G5[3,4] G6/7[4]		G6/7[4]	
Ethiopia			G1–3[1,9]	G1–3[8,9] G6/7[8,9]	G1–3[1,7,8,9] G5[7,8] G6/7[7,9]	P: G1[8] P: G5[8]	G1[8] G6/7[8]	
Somalia				G6[10]				
Kenya	G1–3[3,11,12,13] G6[3,12,13]	D: G1[11] Wc: G1–3[16] Wc: Ef[16]	G1–3[3,11,12,15,17] G5[17]	G1–3[3,11,12,17] G6/7[3,11,12,17]	G1–3[3,11,12,15,17] G5[15,17] G6/7[3,11,12,17]	P: G1[3] P: G5[3] P: G6[3]	G1–3[3,11,12,15,17] G5[17] G6/7[3,11,12,15,17]	Wr: G1–3[16]
Uganda		Wc: Ef[12]				Wa: G1[12] Wa: Ef[12]		
Zambia					G5[18]			
Namibia		Wc: G1–3[21] Wc: G4[19,21] D: G6/7[21] D: Ef[20]			G5[20]	Wa: Ef[21]		Wr: G6/7[21] We: G4[19] We: G5[20]
South Africa	G1–3[24] G5[24] G6/7[24]	Wc: Ef[21,22]						

[1] (Maillard et al., 2007); [2] (Maillard et al., 2009); [3] (Dinkel et al., 2004); [4] (Omer et al., 2010); [5] (Ibrahim et al., 2011); [6] (Ahmed et al., 2013); [7] (Romig et al., 2011); [8] (Tigre et al., 2016); [9] (Hailemariam et al., 2012); [10] (Bowles et al., 1992); [11] (Wachira et al., 1993); [12] (Huttner et al., 2009); [13] (Casulli et al., 2010b); [14] (Romig T., personal communication); [15] (Addy et al., 2012); [16] (Kagendo et al., 2014); [17] (Mbaya et al., 2014); [18] (Banda, personal communication); [19] (Wassermann et al., 2015); [20] (Obwaller et al., 2004); [21] (Aschenborn J., personal communication); [22] (Hüttner et al., 2008); [23] (Schneider et al., 2010); [24] (Mogoye et al., 2013); [25] (Omer R.A., personal communication); [26] (Mauti et al., 2016).

Material Table S11 (cattle, sheep and goats), Table S12 (camels)]. The camel strain (G6) appears to be the dominating CE agent in the entire region, although comprehensive molecular surveys have only been done in Sudan so far (Kamal et al., 2011; Omer et al., 2010). The limited number of human cases in this region (compared to eastern Africa, for example) has been hypothetically linked to a lower infectivity or pathogenicity of that genotype for humans, compared to the genotypes of *E. granulosus* which may be absent or rare in this region.

3.7.2.2.2 Infections in animals In **Mauritania** (Nouakchott), *Echinococcus* was diagnosed in 14% of 120 dogs (Salem et al., 2011). In a recent study in **Mali** (Bamako) 1 of 118 dogs was infected with the camel strain G6 (Mauti et al., 2016). In the central region of **Chad** (Troncy and Graber, 1969), 3.7% of 117 dogs were found positive for *Echinococcus* infections. Two studies documented prevalences of 3% (33 dogs) and 51% (49 dogs) in central **Sudan** (El-Badawi et al., 1979; Saad and Magzoub, 1986).

Prevalence data (of uncharacterized cysts) in Mauritania are high in camels (30.1%), but lower in cattle (5.5%) and sheep (6.5%), based on an unspecified but apparently large sample of slaughtered animals (Salem et al., 2011). This is closely similar to data from Chad, Sudan and the Turkana region of Kenya (Tables S11 and S12). A limited number of isolates from Mauritania, Mali and the overwhelming majority of a large sample from Sudan (98.7% of 532 cysts) were characterized as camel strain G6 (Table 14). The high prevalence in camels might indicate a higher susceptibility of this animal for the prevailing parasite taxon, but data have to be interpreted with care since camels are usually slaughtered at a far older age than other livestock. In cattle-raising areas of eastern Sudan, the White Nile region and South Darfur (Sudan), *E. ortleppi* seems to be of sporadic occurrence in cattle (Omer et al., 2010; Omer R.A., personal communication). In an older survey of wild herbivores in Chad (Graber et al., 1969), in 1 of 14 investigated warthogs (*Phacochoerus africanus*) and 1 of 9 scimitar-horned oryx antelopes (*Oryx dammah*) (now extinct in the wild) CE was diagnosed, but in the same study, 8 elephants (*Loxodonta africana*), 18 lelwel hartebeests (*Alcelaphus lelwel*), 7 korrigum (*Damaliscus korrigum*), 2 common duiker (*Sylvicapra grimmia*), 3 oribi (*Ourebia ourebi*), 4 bohor reedbucks (*Redunca redunca*), 7 kobs (*Kobus kob*), 14 waterbucks (*Kobus ellipsiprymnus defassa*), 36 dorcas gazelles (*Gazella dorcas*), 17 red-fronted gazelles (*Eudorcas rufifrons*), 9 dama gazelles (*Nanger dama*), 9 roan antelopes (*Hippotragus *equinus*), 1 addax (*A. nasomaculatus*), 2 greater kudu (*Strepsiceros cottoni*), and 4 African buffaloes (*Syncerus caffer*) were negative (Table S14C).

3.7.2.2.3 Cystic echinococcosis in humans For **Mauritania**, the surgical incidence was estimated at $1.6/10^5$ inhabitants with indications for a possible increase in recent years (Salem et al., 2011). All genotyped cysts from patients belonged to G6 (Bardonnet et al., 2002). The same is true for three patients in **Sudan**, who originated from the Nuba mountains, Darfur and Khartoum, respectively (Omer et al., 2010). Three ultrasound surveys for human CE have been done in the region, all in Sudan. In Tamboul area of central Sudan, 1 of 300 rural residents (0.33%) was infected (Elmahdi et al., 2004), while in the same area a larger study covering 1055 people resulted in a prevalence of 1.04%. The third survey was done in the Nuba mountains, resulting in 0.32% prevalence among 2182 people (Ahmed et al., 2010). The data from Tamboul area might not be representative for a wider region, since the study area serves as a camel slaughter place, where dogs are known to have extremely high rates of infection (Elmahdi, personal communication). From other countries in the arid zone south of the Sahara, 28 CE patients have been identified from rural areas of **Niger** (Develoux et al., 1991), a total of 12 cases are reported from **Mali** during a 40-year period (Carayon and Robert, 1962; Yena et al., 2002), and there are three published cases from Chad (Sirol and Lefevre, 1971).

3.7.2.3 East Africa: South Sudan, Ethiopia, Eritrea, Somalia, Uganda, Kenya

Host assemblages, transmission and molecular epidemiology: Data on CE in East Africa date back to the 1960s, but most research activities started in the early 1980s. The first surveys centred on the Turkana and Maasai regions of Kenya and Tanzania, which were noted for extremely high incidences of human CE. Today we recognize a hyperendemic region with locally >5% prevalence of human CE that covers northwestern Kenya (Turkana), northeastern Uganda, the southeastern part of South Sudan and adjacent areas of southwestern Ethiopia. A second focus exists in Maasailand at the border of Kenya and Tanzania. All highly endemic areas are inhabited by traditional pastoralists of various ethnic groups that live in contact with large numbers of dogs. However, these hyperendemic foci, although receiving most research attention, are not representative for all of eastern Africa, as CE in humans and/or livestock is apparently far less prevalent, e.g., in central and northeastern Kenya or Somalia. Despite intense research in some foci (see Table S13 in the Supplementary Material), large parts of eastern Africa are practically devoid of CE data, e.g., Rwanda, Burundi and Tanzania south of the Maasai area.

Echinococcus granulosus (G1–3) is the most frequent cause of CE in this region (Table 14), mainly transmitted in a sheep–dog cycle. The camel strain G6 is also widespread, particularly in the more arid parts of northern Kenya and eastern Ethiopia where it affects camels at high prevalences. *Echinococcus ortleppi* is widespread across eastern Africa, but occurs only sporadically in cattle, pigs and goats. *Echinococcus felidis*, as a wildlife parasite, is widespread at least in conservation areas of Kenya and Uganda, apparently without impact on coexisting livestock and humans.

3.7.2.3.1 Infections in animals In sheep, CE is highly prevalent throughout **Ethiopia**, but only moderately frequent in **Kenya** (see Table S11 in the Supplementary Material). *Echinococcus granulosus* is responsible for the majority of infection in sheep in Kenya, and all fertile cysts of sheep in Maasailand (prevalence 16.5%) belonged to this parasite (Addy et al., 2012; Romig et al., 2011). This is probably also true for Ethiopia, although few Ethiopian sheep samples were ever characterized (Hailemariam et al., 2012; Maillard et al., 2007). The spatial heterogeneity of CE prevalence in sheep is therefore difficult to explain and might be linked to environmental factors, as prevalence in small stock of the semiarid Turkana area of Kenya is lower than in the climatically much more favourable Maasailand (Table S11). Sheep sampled in the northern and central parts of **South Sudan** as well as in **Somalia** were rarely infected, and only infertile cysts of *E. intermedius* G6/7 were found; possibly, *E. granulosus* does not occur in the region (Macchioni et al., 1984; Omer et al., 2010). In goats, CE prevalence is generally lower, and the majority of fertile cysts in **Kenya** and **Ethiopia** belonged to *E. intermedius* G6/7 or to *E. ortleppi*. Goats may therefore be a key species to maintain the 'camel strain' in Africa, south of the camel husbandry region, and may be an important additional host for the 'cattle strain'. Prevalence in cattle is highly variable across eastern Africa, peaking at values around 50% in Maasailand and some regions of Ethiopia (Table S11). Cattle are most frequently infected with *E. granulosus* (G1–3), but the majority of cysts are nonfertile. However, fertility rate of *E. granulosus* in cattle seems to differ between regions (Addy et al., 2012; Tigre et al., 2016), which may be related to variants of the parasite, breeds of cattle or age of animals at slaughter [chapter: Ecology and Life Cycle Patterns of *Echinococcus* Species by Romig et al. (2017)]. In contrast, *E. ortleppi* and the camel strain G6 occur far less frequently in cattle, but where they are present, cysts show high fertility (Mbaya et al., 2014; Omer et al., 2010). The sporadic presence of the cattle-adapted *E. ortleppi* in most areas has been tentatively explained by the fact that cattle are often sold alive and slaughtered at abattoirs far from their origin, so that the local dog populations have restricted access to cattle offal (Addy et al., 2012). CE prevalence in camels is extremely high in northwestern Kenya, much less so in northeastern Kenya, eastern Ethiopia and Somalia (Table S12), for reasons unknown. In Kenya, the majority of camel cysts belonged to G6/7, but fertile cysts of *E. granulosus* (G1−3) were also present at lower numbers (Dinkel et al., 2004; Mbaya et al., 2014); the reverse situation was found in eastern Ethiopia (Tigre et al., 2016). There are no prevalence data for pigs in eastern Africa, except for two older studies with negative results from Ethiopia (Table S11). Cysts obtained by opportunistic sampling in western Kenya and Ethiopia belonged to *E. granulosus* (G1-3), *E. ortleppi* and *E. intermedius* G6/7 (Dinkel et al., 2004; Tigre et al., 2016).

A large number of **wild animal** species have been recorded as definitive or intermediate hosts for *Echinococcus* spp. in eastern Africa (Tables S14B and C). Without molecular identification of the parasites it is difficult to appreciate which of these records represent spill overs from domestic lifecycles in areas where livestock and wildlife coexist (e.g., southern Kenya), and which represent transmission systems based on wildlife species. To date, it is clear that *E. felidis* (the former 'lion strain') occurs in most larger national parks and game reserves of **Uganda** and **Kenya** (Queen Elizabeth, Maasai Mara, Nairobi, Samburu, Meru, Tsavo). Lions are frequently infected, although prevalence estimates were only attempted for Queen Elizabeth National Park in Uganda (72%) (Table S14B wild animals). Due to uncertainties in the methodology only a fraction of taeniid eggs from environmental faecal samples could be characterized to species level. In these areas spotted hyenas also seem to be secondary definitive hosts for *E. felidis* (Kagendo et al., 2014). Cysts of this parasite have only been identified from a warthog in Uganda (Huttner et al., 2009). The host range of the parasite may be restricted to wildlife, as no *E. felidis* was identified among 394 cysts of human, sheep, goat, cattle and camel origin from Kenya (Huttner et al., 2009). In contrast, *E. granulosus* (G1−3) is a frequent parasite of livestock and humans in eastern Africa and was also found to be widespread in the lion and spotted hyena populations of four conservation areas of Kenya (Kagendo et al., 2014). Whether this is a spill over from domestic animals is yet unclear. To date, this parasite has only been confirmed in two species of wild herbivores as potential intermediate hosts: cysts were found in 4 of 353 surveyed wildebeest from the Maasai Mara conservation area in southwestern Kenya and in one warthog in Queen Elizabeth National Park in southwestern Uganda (Huttner et al., 2009; Kagendo et al., 2014).

3.7.2.3.2 Cystic echinococcosis in humans The largest number of CE cases in eastern Africa originates from a coherent region covering southeastern **South Sudan**, southwestern **Ethiopia** and northwestern **Kenya** (Romig et al., 2011). As a common feature, the region is inhabited by ethnic groups practising traditional nomadic or seminomadic pastoralism. Older estimates of surgical CE incidences for the Turkana region in Kenya were 40–98/10^5 inhabitants (Clement et al., 2000) with a peak value of 220 in the northwestern part of the region (French and Nelson, 1982). The public health impact of CE on the Turkana pastoralists is confirmed by older and recent ultrasound surveys giving prevalence estimates of 2.5–3.0% (Table 15). This hyperendemic focus extends to neighbouring ethnic groups in South Sudan (Table 15) and southwestern **Ethiopia**, where clinical prevalences of 4.8% (Fuller and Fuller, 1981) and ultrasound prevalences of 2.9% (Table 15) were reported from the Dassanetch and Nyangatom peoples. Low surgical incidence of 0.18/10^5 in Central Ethiopia (2008–12) (Assefa et al., 2015) and ultrasound prevalences of 0.0–0.1% (Table 15) in western and northeastern Kenya show that hyperendemicity of CE in East Africa is limited to the region described earlier. The only exception appears to be an additional focus among the Maasai pastoralists in southern Kenya and northern Tanzania, where ultrasound prevalences of 0.5–1.1% and a clinical incidence of 10/10^5 have been estimated (Table S13; Ernest et al., 2010; Zeyhle, personal communication). In addition to the countries mentioned above, human CE cases have been reported from **Uganda**, **Rwanda** and Somalia (Owor and Bitakaramire, 1975; Babady et al., 2009).

3.7.2.4 West and Central Africa: Nigeria, Burkina Faso, Cameroon, the Central African Republic, Democratic Republic of Congo

3.7.2.4.1 Host assemblages, transmission and molecular epidemiology West and Central Africa are the least known regions of sub-Saharan Africa concerning CE in humans or animals. Our current knowledge is based on few, mostly older, animal surveys and scattered reports of human cases or case series. The abundance of CE appears to be highly unequal, being frequent in livestock of the savanna and semidesert zones of northern West and Central Africa, and being absent or of sporadic occurrence in high-rainfall areas of southern West Africa and the Congo

Table 15 Ultrasound surveys for human cystic echinococcosis in eastern Africa

Country (region)	Ethnic group	Prevalence (%)	Number of humans examined	Reference
South Sudan (southeastern)	Toposa	3.1	378	Macpherson et al. (1989)
South Sudan (southeastern)	Toposa	3.5	1010	Magambo et al. (1998)
South Sudan (southeastern)	Bouya	1.3	996	Magambo et al. (1996)
South Sudan (southeastern)	Dinka	1.2	165	Magambo et al. (1996)
South Sudan (southeastern)	Latuka	0.1	954	Magambo et al. (1996)
South Sudan (southeastern)	Didinga	0.0	1031	Magambo et al. (1996)
South Sudan (southeastern)	Nuer	0.0	210	Magambo et al. (1996)
South Sudan (southern)	Mundari	0.7	610	Barclay et al. (2013)
Ethiopia (southern)	Nyangatom	2.9	1334	Macpherson et al. (1989)
Ethiopia (southern)	Borana	1.8	119	Macpherson et al. (1989)
Ethiopia (southern)	Hamar	0.7	369	Macpherson et al. (1989)
Ethiopia (southern)	Hamar	0.7	990	Klungsoyr et al. (1993)
Ethiopia (southern)	Dassanetch	0	267	Macpherson et al. (1989)
Kenya (northwestern)	Turkana	2.6 (0–5.6)	10,491	Macpherson et al. (1989)
Kenya (northwestern)	Turkana	3.0	10,458	Zeyhle, personal communication (2015)
Kenya (western)	Pokot	0.1	2389	Macpherson et al. (1989)
Kenya (northeastern)	Gabbra	0	38	Macpherson et al. (1989)
Kenya (northeastern)	Somali	0	1252	Macpherson et al. (1989)
Kenya (northeastern)	Samburu	0	368	Macpherson et al. (1989)
Kenya (northeastern)	Rendille	0	710	Macpherson et al. (1989)
Kenya (southern)	Maasai	1.0	2577	Zeyhle, personal communication (2015)
Tanzania (northern)	Maasai	1.1	959	Macpherson et al. (1989)

basin. Nothing is known about the *Echinococcus* spp. in this region, only one human case from Ghana infected with G6 (Schneider et al., 2010) indicates that the predominance of this taxon in the northern arid zone (see Section 4.7.2.2) may extend southwards into western and central Africa.

3.7.2.4.2 Infections in animals In northern **Nigeria**, Dada (1980) found 2.4% of 549 dogs infected with *Echinococcus* and in eastern Nigeria 4.4% of 182 dogs were infected (Okolo, 1986). CE prevalence in northern Nigeria is extremely high in camels (>50%), but moderate to low in sheep, goats and cattle. According to one study (Arene, 1985), there is an additional focus in the **Niger** delta with extremely high prevalence of 31.6%, 24.4%, 42.2%, and 55.9% in groups of 320 cattle, sheep, goats and pigs, respectively. **Burkina Faso**, northern **Cameroon**, the **Central African Republic** and the northeast of the **Democratic Republic of Congo** seem to be little affected by CE, but data are old and limited (Table S11). No information on livestock CE exists from other countries of the region. One extensive wildlife study in Central Africa from the 1960s indicates some involvement of wild mammals (including lions) in CE transmission (Graber and Thal, 1979). No CE was diagnosed in small numbers of potential intermediate hosts examined (Graber et al., 1969), but potential wild host species are now largely eradicated in this region and are unlikely to play any significant role.

3.7.2.4.3 Cystic echinococcosis in humans There are no published surveys on human CE from western and central Africa, but a number of case reports and case series exist from **Senegal** (Hane et al., 1989), **Ivory Coast** (Schmidt et al., 1978), **Ghana** (DeMarais et al., 1992; Schneider et al., 2010), northern **Nigeria** (Afonja et al., 1972; Dada et al., 1980), **Cameroon** (Ankouane et al., 2013), the **Central African Republic** (Develoux et al., 2011) and from what is today the **Democratic Republic of Congo** (De Meulemeester and Dardenne, 1958). These records indicate that CE is widespread. Like CE in livestock, the impact on public health is likely to differ among the different countries, but available data do not allow further conclusions.

3.7.2.5 Southern Africa: Angola, Zambia, Mozambique, Zimbabwe, Namibia and South Africa

3.7.2.5.1 Host assemblages, transmission and molecular epidemiology Large-scale surveys on CE in livestock have been done in South Africa in the 1960s (centering on the studies of Anna Verster), indicating that the parasites are widespread at moderate to low levels across the country. In addition to that, surprisingly few prevalence data are available from southern Africa. Even in the Republic of South Africa, the impact on the human population was only recently estimated, and only isolated case reports indicate the presence of the parasite in other countries of the region. Concerning the *Echinococcus* species involved, extensive taxonomic work based on morphology of adult worms derived from domestic animals and wildlife has been done in South Africa in the 1960s, indicating a high diversity (Verster and Collins, 1966). Recent molecular studies confirmed the presence of the genotypes G1–3, G4, G5, G6 and *E. felidis* in southern Africa (Table 14), making it the most diverse region in sub-Saharan Africa. However, most of these data derive from opportunistic sampling, and the relative impact of the various parasites on livestock, wildlife and humans and their spatial distribution are still far from clear.

3.7.2.5.2 Cystic echinococcosis in animals Comprehensive livestock data from **South Africa** published by Verster and Collins (1966) showed uniformly low to moderate prevalences at that time in sheep (1.0–2.0%), goats (0.8–2.6%), cattle (1.8–8.4%) and pigs (0.4–3.5%) with little geographical variation (Table S11). Older cattle data from **Swaziland** (Mitchell, 1977) and a recent retrospective and prospective study from western **Zambia** (Banda et al., 2013) indicate similar prevalence levels there. There is hardly any information on the impact of different *Echinococcus* spp. on southern African livestock. Opportunistic samples from Zambian and northern **Namibian** cattle were all *E. ortleppi* (Banda F., Aschenborn, personal communication), and earlier morphological studies from South Africa suggest that this agent may be also widespread in the region (Verster and Collins, 1966). One report of *E. ortleppi* from a Namibian zebra suggests a spill over into wild hosts species (Obwaller et al., 2004). A large number of wildlife species had earlier been identified as carriers of *Echinococcus* sp. (Table S14 in the Supplementary Material). *Echinococcus felidis*, described morphologically from South African lions (Ortlepp, 1937), was also characterized molecularly based on specimens from that country (Huttner et al., 2009). A wildlife cycle of *E. equinus* was recently identified involving lions, black-backed jackals and plains zebras in Etosha National Park, Namibia (Wassermann et al., 2015). Older reports on frequent zebra infection in Kruger National Park (South Africa) (Young, 1975) indicate a wide distribution in the latter species. Furthermore, *E. intermedius* G6/7 seems to be widespread in various wild mammal hosts in Namibia (Aschenborn, personal communication).

3.7.2.5.3 Cystic echinococcosis in humans A number of case studies confirms the presence of human CE across southern Africa, including **South Africa**, **Namibia**, **Botswana**, **Angola**, **Zimbabwe** and **Mozambique** (Wahlers et al., 2012; Rossouw et al., 1992; Hajek et al., 2004; WHO, 2010; Bordon et al., 1989; Chopdat et al., 2007). Human CE is generally considered to be rare in the region, although some of the published case series are extensive (e.g., 80 patients over a 3-year period in a treatment centre of Botswana) (Hajek et al., 2004). There is a lack of precise data on the impact of CE on the human population. Clinical data from western Zambia gave an annual incidence of $9/10^5$ inhabitants in the years 2006–10 (Banda, 2013), while a recent retrospective study in South Africa reported 137 new cases per year as a most conservative estimate, which corresponds approximately to a countrywide annual incidence of $0.3/10^5$ (Wahlers et al., 2011). Out of a countrywide sample of 32 human cyst isolates that were genotyped, *E. granulosus* G1–3 was the most frequent cause of human CE in South Africa (81%), followed by *E. intermedius* G6/7 (16%) and *E. ortleppi* (3%). The record of the latter species from a single patient represents the first instance of human CE caused by *E. ortleppi* in Africa (Mogoye et al., 2013).

ACKNOWLEDGEMENTS

Paola Pepe and **Vincenzo Musella** (Department of Veterinary Medicine and Animal Productions, University of Naples Federico II, Italy) for the work on the geographic maps; **Rajnish Sharma** (Department of Veterinary Microbiology, University of Saskatchewan) and **Balbir Dhaliwal** (Guru Angad Dev Veterinary and Animal Sciences University, Ludhiana, India) for epidemiological information concerning India, **Nirmal K. Thapa** (Ministry of Agriculture and Forests, Thimphu, Bhutan) for insider information of Bhutan and surrounding countries and **Valentina Racca** (Institute of Parasitology, University of Zurich, Switzerland) for help with the manuscript preparation.

APPENDIX A. SUPPLEMENTARY DATA

Electronic supplementary material associated with this article can be found, in the online version, at http://dx.doi.org/10.1016/bs.apar.2016.11.001. For authorized users.

REFERENCES

Abd El Baki, M.H., El Missiry, A.M., Abd El Aaty, H.E., Mohamad, A.A., Aminou, H.A., 2009. Detection of G1 genotype of human cystic echinococcosis in Egypt. J. Egypt Soc. Parasitol. 39, 711–721.

Abdel Aaty, H.E., Abdel-Hameed, D.M., Alam-Eldin, Y.H., El-Shennawy, S.F., Aminou, H.A., Makled, S.S., Darweesh, S.K., 2012. Molecular genotyping of *Echinococcus granulosus* in animal and human isolates from Egypt. Acta Trop. 121, 125–128.

Abdi, J., Taherikalani, M., Asadolahi, K., Emaneini, M., 2013. Echinococcosis/Hydatidosis in Ilam province, Western Iran. Iran. J. Parasitol. 8, 417–422.

Abdullaev, A.G., Shishlo, L.A., Adrianov, S.O., Rodionova, T.V., 2006. Clinical and laboratory symptoms of space-occupying hepatic lesions and their prognostic value. Khirurgiia 19–23.

Abdul-Salam, J.M., Farah, M.A., 1988. Hydatidosis in camels in Kuwait. Parasitol. Res. 74, 267–270.

Abdybekova, A.M., Torgerson, P.R., 2012. Frequency distributions of helminths of wolves in Kazakhstan. Vet. Parasitol. 184, 348–351.

Abdybekova, A., Sultanov, A., Karatayev, B., Zhumabayeva, A., Shapiyeva, Z., Yeshmuratov, T., Toksanbayev, D., Shalkeev, R., Torgerson, P.R., 2015. Epidemiology of echinococcosis in Kazakhstan: an update. J. Helminthol. 89, 647–650.

Abdyjaparov, T., Kuttubaev, O., 2004. Alveolar echinococcosis in rodents of mountainous pastures of Kyrgystan. In: Echinococcosis in Central Asia: Problems and Solutions. Dauir Publishing House, Almaty, Kazakhstan.

Abdylazizov, A., 2012. The current status of echinococcosis morbidity in Dagestan. Mod. Technol. 8, 51–53.

Aboelhadid, S.M., El-Dakhly, K.M., Yanai, T., Fukushi, H., Hassanin, K.M., 2013. Molecular characterization of *Echinococcus granulosus* in Egyptian donkeys. Vet. Parasitol. 193, 292–296.

Abu-Hasan, N., Daragmeh, M., Adwan, K., Al-Qaoud, K., Abdel-Hafez, S.K., 2002. Human cystic echinococcosis in the West Bank of Palestine: surgical incidence and seroepidemiological study. Parasitol. Res. 88, 107–112.

Abuladze, K.I., 1964. *Taenia* tapeworms in animals and man and the diseases they cause. In: Skrjabin, K.I. (Ed.), Essentials of Cestodology, Academy of Sciences of the USSR, USSR. (English Translation, Israel Programme for Scientific Translations, Jerusalem, 1970.

Abushhewa, M.H., Abushhiwa, M.H., Nolan, M.J., Jex, A.R., Campbell, B.E., Jabbar, A., Gasser, R.B., 2010. Genetic classification of *Echinococcus granulosus* cysts from humans, cattle and camels in Libya using mutation scanning-based analysis of mitochondrial loci. Mol. Cell Probes 24, 346–351.

Acosta-Jamett, G., Cleaveland, S., Bronsvoort, B.M., Cunningham, A.A., Bradshaw, H., Craig, P.S., 2010. *Echinococcus granulosus* infection in domestic dogs in urban and rural areas of the Coquimbo region, north-central Chile. Vet. Parasitol. 169, 117–122.

Addy, F., Alakonya, A., Wamae, N., Magambo, J., Mbae, C., Mulinge, E., Zeyhle, E., Wassermann, M., Kern, P., Romig, T., 2012. Prevalence and diversity of cystic echinococcosis in livestock in Maasailand, Kenya. Parasitol. Res. 111, 2289–2294.

Adwan, G., Adwan, K., Bdir, S., Abuseir, S., 2013. Molecular characterization of *Echinococcus granulosus* isolated from sheep in Palestine. Exp. Parasitol. 134, 195–199.

Afonja, A.O., Sofowora, E.O., Kolawole, T.M., Junaid, T.A., Fatunmbi, B.O., 1972. Hydatid disease of the lung in Nigeria: case report. J. Trop. Med. Hyg. 75, 224–226.

Ahmadi, N.A., Meshkehkar, M., 2011. An abattoir-based study on the prevalence and economic losses due to cystic echinococcosis in slaughtered herbivores in Ahwaz, south-western Iran. J. Helminthol. 85, 33–39.

Ahmadi, N.A., 2005. Hydatidosis in camels (*Camelus dromedarius*) and their potential role in the epidemiology of *Echinococcus granulosus* in Iran. J. Helminthol. 79, 119–125.
Ahmed, M., Ali, I., Abdelrahim, M., Ahmed, Y., Alhakeim, S., 2010. Screening for human cystic echinococcosis in Sudan, a field survey using portable ultrasound. In: 1st International Congress on Pathogens at the Human-Animal Interface (ICOPHAI). Ethiopia, Addis Ababa.
Ahmed, M.E., Eltom, K.H., Musa, N.O., Ali, I.A., Elamin, F.M., Grobusch, M.P., Aradaib, I.E., 2013. First Report on Circulation of *Echinococcus ortleppi* in the one humped camel (*Camelus dromedaries*). Sudan. BMC Vet. Res. 9, 1–6.
Aikat, B.K., Bhusnurmath, S.R., Cadersa, M., Chhuttani, P.N., Mitra, S.K., 1978. *Echinococcus multilocoularis* infection in India: first case report proved at autopsy. Trans. R. Soc. Trop. Med. Hyg. 72, 619–621.
Ajlouni, A.Q., Saliba, E.K., Disi, A.M., 1984. Intestinal cestodes of stray dogs in Jordan. Z. Parasitenkd 70, 203–210.
Akhmedov, S., Rasulov, N., Raddzabov, A., Safarov, B., Tabarov, Z., 2013. Surgical treatment of alveolar echinococcosis of the liver, complicated by jaundice. In: Actual Problems of Surgical Hepatology. In 20th Annual International Congress of the Association of Hepatic Surgeons of the Countries of the NIS (Donetsk), pp. 80–81.
Al Kitani, F.A., Al Riyami, S., Al Yahyai, S., Al Awahi, A.H., Al Aawali, M., Hussain, M.H., 2015. Abattoir based surveillance of cystic echinococcosis (CE) in the Sultanate of Oman during 2010–2013. Vet. Parasitol. 211, 208–215.
Alam-Eldin, Y.H., Abdel Aaty, H.E., Ahmed, M.A., 2015. Molecular characterization of cystic echinococcosis: first record of G7 in Egypt and G1 in Yemen. Acta Parasitol./ Witold Stefanski Inst. Parasitol. Warszawa, Pol. 60, 662–665.
Al-Ani, A.M., Khan, F.Y., Elzouki, A.N., Hajri, M.A., Ibrahim, W., 2014. Epidemiology of hydatid disease in Qatar: a hospital based study from 2000 to 2013. Asian Pac. J. Trop. Med. 7S1, S85–S87.
Al-Attar, H.K., Al-Irhayim, B., Al-Habbal, M.J., 1983. Alveolar hydatid disease of the liver: first case report from man in Iraq. Ann. Trop. Med. Parasitol. 77, 595–597.
Alishani, M., Sherifi, K., Rexhepi, A., Hamidi, A., Armua-Fernandez, M., Grimm, F., Hegglin, D., Deplazes, P., 2017. The Impact of Socio-cultural Factors on Transmission of *Taenia* spp. *and Echinococcus granulosus* in Kosovo (in press).
Al-Jawabreh, A., Dumaidi, K., Ereqat, S., Nasereddin, A., Al-Jawabreh, H., Azmi, K., Al-Laham, N., Abdeen, Z., 2015. Incidence of *Echinococcus granulosus* in domestic dogs in Palestine as revealed by copro-PCR. PLoS Negl. Trop. Dis. 9, e0003934.
Allaico, S., 2010. Determinacion de hidatidosis en cerdos faenados en el camal de Azogues. Universidad de Cuenca.
Al-Qaoud, K.M., Abdel-Hafez, S.K., Craig, P.S., 2003. Canine echinococcosis in northern Jordan: increased prevalence and dominance of sheep/dog strain. Parasitol. Res. 90, 187–191.
Al-Shibani, L.A., Al-Eryani, S.M., Azazy, A.A., Al-Mekhlafi, A.M., 2012. Cases of hydatidosis in patients referred to governmental hospitals for cyst removal in Sana'a City, Republic of Yemen. Trop. Biomed. 29, 18–23.
Altintas, N., 1995. The pathology of the echinococcosis (50 cases) and the current *Echinococcus* problem in Turkey. Türkiye Ektopatoloji Dergisi 1, 55–58.
Altintas, N., 2003. Past to present: echinococcosis in Turkey. Acta Trop. 85, 105–112.
Altintas, N., 2008. Parasitic zoonotic diseases in Turkey. Vet. Ital. 44, 633–646.
Alvarez Rojas, C.A., Romig, T., Lightowlers, M.W., 2014. *Echinococcus granulosus* sensu lato genotypes infecting humans—review of current knowledge. Int. J. Parasitol. 44, 9–18.
Amer, S., Helal, I.B., Kamau, E., Feng, Y., Xiao, L., 2015. Molecular characterization of *Echinococcus granulosus sensu lato* from farm animals in Egypt. PLoS One 10, e0118509.

Aminjanov, M., Aminjanov, S., 2004. Echinococcosis and research in Uzbekistan. In: Echinococcosis in Central Asia: Problems and Solutions. Dauir Publishing House, Almaty, Kazakhstan, pp. 13–19.
Andreyanov, O., 2011. Alveolar echinococcosis in fur animals from Ryazan district. Russian Parasitol. J. 3, 7–11.
Anichkin, V., Martinok, V., 2012. Parasitic diseases in the Republic of Belarus and in the Republic of Yemen (literature review). Probl. Health Ecol. 4, 10–17.
Ankouane, A., Ngo, B., Monabang, C., Djapa, R., Tchoumi, E., Njoya, O., Ndjitoyap, E., McManus, D.P., Malonga, E., 2013. Intracystic bleeding of a solitary hydatid cyst: a rare complication of a rare disease in central Africa. A case report. Case Rep. Clin. Med. 2, 163–166.
Anonymous, 2012. In: Health, M.O. (Ed.), Hydatid Disease. Communicable Disease Control Manual – Hydatid Disease, pp. 1–3 (New Zealand).
Anonymous, 2013. About the Incidence of Cystic and Alveolar Echinococcosis in the Russian Federation.
Ansari-Lari, M., 2005. A retrospective survey of hydatidosis in livestock in Shiraz, Iran, based on abattoir data during 1999–2004. Vet. Parasitol. 133, 119–123.
Antolova, D., Miterpakova, M., Radonak, J., Hudackova, D., Szilagyiova, M., Zacek, M., 2014. Alveolar echinococcosis in a highly endemic area of Northern Slovakia between 2000 and 2013. Euro Surveill. 19.
Araj, G.F., Mourad, Y., 2014. Hydatid disease: the Lebanese contribution. J. Med. Liban. 62, 217–226.
Araujo, F.P., Schwabe, C.W., Sawyer, J.C., Davis, W.G., 1975. Hydatid disease transmission in California. A study of the Basque connection. Am. J. Epidemiol. 102, 291–302.
Arbabi, M., Hooshyar, H., 2006. Survey of echinococcosis and hydatidosis in Kashan region, central Iran. Iran. J. Publ. Health 35, 75–81.
Arene, F.O., 1985. Prevalence of hydatid cysts in domestic livestock in the Niger Delta. Trop. Anim. Health Prod. 17, 3–5.
Arslanova, A., 1962. Human echinococcosis and alveococcosis in the southern zone of Kazakhstan. In: Proceedings of a Scientific Conference in Jambul and Alma-Ata Jambul and Alma-Ata.
Assefa, H., Mulate, B., Nazir, S., Alemayehu, A., 2015. Cystic echinococcosis amongst small ruminants and humans in central Ethiopia. Onderstepoort J. Vet. Res. 82, E1–E7.
Awan, M.A., Gusbi, A.M., Beesley, W.N., 1990. Echinococcosis in Libya. III. Further studies on the prevalence of *Echinococcus granulosus* in dogs. Ann. Trop. Med. Parasitol. 84, 473–475.
Azlaf, R., Dakkak, A., 2006. Epidemiological study of the cystic echinococcosis in Morocco. Vet. Parasitol. 137, 83–93.
Azlaf, R., Dakkak, A., Chentoufi, A., El Berrahmani, M., 2007. Modelling the transmission of *Echinococcus granulosus* in dogs in the northwest and in the southwest of Morocco. Vet. Parasitol. 145, 297–303.
Azlaf, R., 2007. Echinococcosis/Hydatidosis in Morocco: Epidemiology, Modeling, *E. granulosus* Genotyping and Molecular Analysis. Morocco, Rabat.
Babady, N.E., Pritt, B.S., Walker, R.C., Rosenblatt, E.E., Binnicker, M.I., 2009. A 48-year old Somali woman with hip pain. Clin. Infect. Dis. 49, 764–765, 803–805.
Bagrade, G., Šnábel, V., Romig, T., Ozoliņ, J., Hüttner, M., Miterpáková, M., Švecová, D., Dubinský, P., 2008. *Echinococcus multilocularis* is a frequent parasite of red foxes (*Vulpes vulpes*) in Latvia. Helminthologia 45, 157–161.
Bagrade, G., Esīte, Z., Deksne, G., 2014. Data on *Echinococcus multilocularis* in fox and racoon dog population in Latvia during 2010–2013 season of rabies eradication and control programme. In: In 9th Baltic Theriological Conference, 16–18 October 2014 (Daugavpils), p. 34.

Bakal, U., Kazez, A., Akyol, M., Kocakoc, E., Simsek, S., 2012. A portable ultrasound based screening study on the prevalence and risk factors of cystic echinococcosis in primary school children in East Turkey. Acta Trop. 123, 91–95.

Balbinotti, H., Santos, G.B., Badaraco, J., Arend, A.C., Graichen, D.A., Haag, K.L., Zaha, A., 2012. *Echinococcus ortleppi* (G5) and *Echinococcus granulosus sensu stricto* (G1) loads in cattle from Southern Brazil. Vet. Parasitol. 188, 255–260.

Banda, F., Nalubamba, K.S., Muma, J.B., Munyeme, M., Munang, M., Mweemba Munang'andu, H., 2013. A cross-sectional study investigating cystic hydatidosis in slaughtered cattle of western province in Zambia. ISRN Parasitol. 2013 http://dx.doi.org/10.5402/2013/468163. ID: 468163.

Banda, F., 2013. Prevalence and Risk Factors of Cystic Echinococcosis in Cattle and Humans in Western Province of Zambia. University of Zambia, Lusaka.

Barclay, S., Joseph, J., Timothy, F., Mounir, L., Robert, N., Brooker, S., Sidhu, P., Kolaczinski, J., 2013. Cystic echinococcosis in Mundari tribe-members of South Sudan. Path Glob. Health 107, 293–298.

Bardonnet, K., Piarroux, R., Dia, L., Schneegans, F., Beurdeley, A., Godot, V., Vuitton, D.A., 2002. Combined eco-epidemiological and molecular biology approaches to assess *Echinococcus granulosus* transmission to humans in Mauritania: occurrence of the 'camel' strain and human cystic echinococcosis. Trans. R. Soc. Trop. Med. Hyg. 96, 383–386.

Bardonnet, K., Benchikh-Elfegoun, M.C., Bart, J.M., Harraga, S., Hannache, N., Haddad, S., Dumon, H., Vuitton, D.A., Piarroux, R., 2003. Cystic echinococcosis in Algeria: cattle act as reservoirs of a sheep strain and may contribute to human contamination. Vet. Parasitol. 116, 35–44.

Barlow, A.M., Gottstein, B., Mueller, N., 2011. *Echinococcus multilocularis* in an imported captive European beaver (*Castor fiber*) in Great Britain. Vet. Rec. 169, 339.

Bart, J.M., Bardonnet, K., Elfegoun, M.C., Dumon, H., Dia, L., Vuitton, D.A., Piarroux, R., 2004. *Echinococcus granulosus* strain typing in North Africa: comparison of eight nuclear and mitochondrial DNA fragments. Parasitology 128, 229–234.

Bart, J.M., Knapp, J., Gottstein, B., El-Garch, F., Giraudoux, P., Glowatzki, M.L., Berthoud, H., Maillard, S., Piarroux, R., 2006. EmsB, a tandem repeated multi-loci microsatellite, new tool to investigate the genetic diversity of *Echinococcus multilocularis*. Infect. Genet. Evol. 6, 390–400.

Bartel, M.H., Seesee, F.M., Worley, D.E., 1992. Comparison of Montana and Alaska isolates of *Echinococcus multilocularis* in gerbils with observations on the cyst growth, hook characteristics, and host response. J. Parasitol. 78, 529–532.

Bartholomot, G., Vuitton, D.A., Harraga, S., Shi, D.Z., Giraudoux, P., Barnish, G., Wang, Y.H., MacPherson, C.N., Craig, P.S., 2002. Combined ultrasound and serologic screening for hepatic alveolar echinococcosis in central China. Am. J. Trop. Med. Hyg. 66, 23–29.

Beard, T., Bramble, A., Middleton, M., 2001. Eradication in Our Time: A Log Book of the Hydatid Control Program 1962–1996. Department of Primary Industry, Water and Environment, Hobart.

Beato, S., Parreira, R., Calado, M., Gracio, M.A., 2010. Apparent dominance of the G1-G3 genetic cluster of *Echinococcus granulosus* strains in the central inland region of Portugal. Parasitol. Int. 59, 638–642.

Beato, S., Parreira, R., Roque, C., Goncalves, M., Silva, L., Maurelli, M.P., Cringoli, G., Gracio, M.A., 2013. *Echinococcus granulosus* in Portugal: the first report of the G7 genotype in cattle. Vet. Parasitol. 198, 235–239.

Beiromvand, M., Akhlaghi, L., Fattahi Massom, S.H., Mobedi, I., Meamar, A.R., Oormazdi, H., Motevalian, A., Razmjou, E., 2011. Detection of *Echinococcus multilocularis* in carnivores in Razavi Khorasan province, Iran using mitochondrial DNA. PLoS Negl. Trop. Dis. 5, e1379.

Beiromvand, M., Akhlaghi, L., Fattahi Massom, S.H., Meamar, A.R., Darvish, J., Razmjou, E., 2013. Molecular identification of *Echinococcus multilocularis* infection in small mammals from northeast, Iran. PLoS Negl. Trop. Dis. 7, e2313.

Belimenko, V., Khristianovsky, P., 2015. Monitoring echinococcosis farm animals in the southern Urals. inary. J. Farm Animals 1, 26–27.

Benito, A., Carmena, D., Joseph, L., Martinez, J., Guisantes, J.A., 2006. Dog echinococcosis in northern Spain: comparison of coproantigen and serum antibody assays with coprological exam. Vet. Parasitol. 142, 102–111.

Ben-Shimol, S., Sagi, O., Houri, O., Bazarsky, E., Berkowitz, A., Bulkowstein, S., Barrett, C., Greenberg, D., 2016. Cystic echinococcosis in Southern Israel. Acta Parasitol. 61, 178–186.

Bentounsi, B., Meradi, S., Ayachi, A., Cabaret, J., 2009. Cestodes of untreated large stray dog populations in Algeria: a reservoir for herbivore and human parasitic diseases. Open Vet. Sci. J. 3, 64–67.

Berenji, F., Mirsadraii, S., Asadi, L., Marofi, A., Fata, A.M., Shahi, M., 2007. In: The First Reported Case of Alveolar Hydatid Cyst in Khorasan Province. National Congress on Hydatid Cyst, Yasuj, Iran.

Berke, O., Romig, T., von Keyserlingk, M., 2008. Emergence of *Echinococcus multilocularis* among red foxes in northern Germany, 1991–2005. Vet. Parasitol. 155, 319–322.

Bersanukayeava, R., Shakhbiev, I., Tokhayeva, A., Tkhakakhova, A., Shakhbiev, K., Beliyev, S., 2013. Relationship in the epizootology of echinococcosis sheep and dogs with regard to different altitude zones in the northern Caucasus. Theory Pract. Parasit. Dis. Animals 14.

Besbes, M., Sellami, H., Cheikhrouhou, F., Makni, F., Ayadi, A., 2003. Clandestine slaughtering in Tunisia: investigation on the knowledge and practices of butchers concerning hydatidosis. Bull. Soc. Pathol. Exot. 96, 320–322.

Bessonov, A.S., 1998. *Echinococcus multilocularis* infection in Russia and neighbouring countries. Helminthologia 35, 73–78.

Bessonov, A., 2002. Echinococcoses of animals and humans in the Russian Federation. In: Craig, P., Pawlowski, Z. (Eds.), Cestode Zoonoses: Echinoccocosis and Cysticercosis – An Emergent and Global Problem. IOS Press, Ohmsha, pp. 91–98.

Beyhan, Y.E., Umur, S., 2011. Molecular characterization and prevalence of cystic echinococcosis in slaughtered water buffaloes in Turkey. Vet. Parasitol. 181, 174–179.

Bhattacharya, D., Bera, A.K., Bera, B.C., Maity, A., Das, S.K., 2007. Genotypic characterisation of Indian cattle, buffalo and sheep isolates of *Echinococcus granulosus*. Vet. Parasitol. 143, 371–374.

Bichet, H., Dorchies, P., 1998. Estimation of the prevalence of bovine hydatid cyst in the south Pyrenees. Parasite 5, 61–68.

Bikov, V.P., Baranov, S.N., Fedoseev, V.F., 2011. Case Histories of Surgical Patients. Arkhangel, 162 pp.

Bingham, G.M., Budke, C.M., Larrieu, E., Del Carpio, M., Mujica, G., Slater, M.R., Moguillansky, S., 2014. A community-based study to examine the epidemiology of human cystic echinococcosis in Rio Negro Province, Argentina. Acta Trop. 136, 81–88.

Bishnevski, A., Koroleva, N., Bishevskaya, G., 2007. The protocol of the 244th session of the pulmonary section of the surgical society of Moscow and Moscow Oblast – 11.04.06. Probl. Tuberc. Lung Dis. 2.

Blaise, J., Raccurt, C.P., 2007. Hepatobiliary fascioliasis and echinococcosis/hydatidosis in domestic animals in Haiti. Revue Sci. Tech. Int. Office Epizootics 26, 741–746.

Blutke, A., Hamel, D., Huttner, M., Gehlen, H., Romig, T., Pfister, K., Hermanns, W., 2010. Cystic echinococcosis due to *Echinococcus equinus* in a horse from southern Germany. J. Vet. Diagn. Invest. 22, 458–462.

Bobic, B., Nikolic, A., Radivojevic, S.K., Klun, I., Djurkovic-Djakovic, O., 2012. Echinococcosis in Serbia: an issue for the 21st century? Foodborne Pathog. Dis. 9, 967–973.

Bordon, L.M., Tadzimirwa, E., Sawyer, K., 1989. Hydatid disease in Zimbabwe: a case report. Cent. Afr. J. Med. 35, 531–534.

Borji, H., Parandeh, S., 2010. The abattoir condemnation of meat because of parasitic infection, and its economic importance: results of a retrospective study in north-eastern Iran. Ann. Trop. Med. Parasitol. 104, 641–647.

Borji, H., Azizzadeh, M., Kamelli, M., 2012a. A retrospective study of abattoir condemnation due to parasitic infections: economic importance in Ahwaz, southwestern Iran. J. Parasitol. 98, 954–957.

Borji, H., Emami, M., Maleki, M., Razmi, G., Mehrjerdi, H.K., Moghaddas, E., 2012b. Alveolar echinococcosis infection in a monkey (*Ateles geoffroyi*) in Mashhad, Iran. Iran. J. Publ. Health 41, 111–116.

Bottcher, D., Bangoura, B., Schmaschke, R., Muller, K., Fischer, S., Vobis, V., Meiler, H., Wolf, G., Koller, A., Kramer, S., Overhoff, M., Gawlowska, S., Schoon, H.A., 2013. Diagnostics and epidemiology of alveolar echinococcosis in slaughtered pigs from large-scale husbandries in Germany. Parasitol. Res. 112, 629–636.

Boubaker, G., Macchiaroli, N., Prada, L., Cucher, M.A., Rosenzvit, M.C., Ziadinov, I., Deplazes, P., Saarma, U., Babba, H., Gottstein, B., Spiliotis, M., 2013. A multiplex PCR for the simultaneous detection and genotyping of the *Echinococcus granulosus* complex. PLoS Negl. Trop. Dis. 7, e2017.

Boufana, B., Stidworthy, M.F., Bell, S., Chantrey, J., Masters, N., Unwin, S., Wood, R., Lawrence, R.P., Potter, A., McGarry, J., Redrobe, S., Killick, R., Foster, A.P., Mitchell, S., Greenwood, A.G., Sako, Y., Nakao, M., Ito, A., Wyatt, K., Lord, B., Craig, P.S., 2012. *Echinococcus* and *Taenia* spp. from captive mammals in the United Kingdom. Vet. Parasitol. 190, 95–103.

Boufana, B., Lahmar, S., Rebai, W., Ben Safta, Z., Jebabli, L., Ammar, A., Kachti, M., Aouadi, S., Craig, P.S., 2014. Genetic variability and haplotypes of *Echinococcus* isolates from Tunisia. Trans. R. Soc. Trop. Med. Hyg. 108, 706–714.

Boufana, B., Lett, W., Lahmar, S., Griffiths, A., Jenkins, D.J., Buishi, I., Engliez, S.A., Alrefadi, M.A., Eljaki, A.A., Elmestiri, F.M., Reyes, M.M., Pointing, S., Al-Hindi, A., Torgerson, P.R., Okamoto, M., Craig, P.S., 2015a. Canine echinococcosis: genetic diversity of *Echinococcus granulosus* sensu stricto (s.s.) from definitive hosts. J. Helminthol. 89, 689–698.

Boufana, B., Lett, W.S., Lahmar, S., Buishi, I., Bodell, A.J., Varcasia, A., Casulli, A., Beeching, N.J., Campbell, F., Terlizzo, M., McManus, D.P., Craig, P.S., 2015b. *Echinococcus equinus* and *Echinococcus granulosus* sensu stricto from the United Kingdom: genetic diversity and haplotypic variation. Int. J. Parasitol. 45, 161–166.

Boufana, B., Said, Y., Dhibi, M., Craig, P.S., Lahmar, S., 2015c. *Echinococcus granulosus* sensu stricto (s.s.) from the critically endangered antelope *Addax nasomaculatus* in Tunisia. Acta Trop. 152, 112–115.

Bowles, J., Blair, D., McManus, D.P., 1992. Genetic variants within the genus *Echinococcus* identified by mitochondrial DNA sequencing. Mol. Biochem. Parasitol. 54, 165–173.

Bowles, J., Blair, D., McManus, D.P., 1994. Molecular genetic characterization of the cervid strain ('northern form') of *Echinococcus granulosus*. Parasitology 109, 215–221.

Braddick, M.R., Reilly, W.J., 1993. Human hydatid disease leading to hospital admission in Scotland 1968–1989. Health Bull. 51, 80–85.

Bregadze, T.L., Kogan, A.S., 1962. Clinico-anatomical classification of alveolar echinococcosis. Med. Parasitol. Parasit. Bolezni 31, 538–542.

Brehm, K., Koziol, U., 2017. *Echinococcus*–host interactions at cellular and molecular levels. In: Thompson, A. (Ed.), *Echinococcus* and Echinococcosis, Part A, vol. 95, pp. 147–212.

Brenes Madrigal, R.R., Sousa, O.E., Bonilla, M.A., Achit, S.M., 1977. 1st case of human hepatic hydatidosis in Costa Rica. Rev. Cuba. Med. Trop. 29, 5–8.
Breyer, I., Georgieva, D., Kurdova, R., Gottstein, B., 2004. *Echinococcus granulosus* strain typing in Bulgaria: the G1 genotype is predominant in intermediate and definitive wild hosts. Parasitol. Res. 93, 127–130.
Brglez, J., Kryštufek, B., 1984. Metacestod *Echinococcus multilocularis* (Leuckart, 1863) pri *Apodemus flavlcoli* (Melchior) v Sloveniji. Zb. Biotehn. Fak. Vet. 21, 173–176.
Bridger, K.E., Baggs, E.M., Finney-Crawley, J., 2009. Endoparasites of the coyote (*Canis latrans*), a recent migrant to insular newfoundland. J. Wildl. Dis. 45, 1221–1226.
Bristow, B.N., Lee, S., Shafir, S., Sorvillo, F., 2012. Human echinococcosis mortality in the United States, 1990–2007. PLoS Negl. Trop. Dis. 6, e1524.
Brundu, D., Piseddu, T., Stegel, G., Masu, G., Ledda, S., Masala, G., 2014. Retrospective study of human cystic echinococcosis in Italy based on the analysis of hospital discharge records between 2001 and 2012. Acta Trop. 140, 91–96.
Brunetti, O.A., Rosen, M.N., 1970. Prevalence of *Echinococcus granulosus* hydatid in California deer. J. Parasitol. 56, 1138–1140.
Bruzinskaite, R., Sarkunas, M., Torgerson, P.R., Mathis, A., Deplazes, P., 2009. Echinococcosis in pigs and intestinal infection with *Echinococcus* spp. in dogs in southwestern Lithuania. Vet. Parasitol. 160, 237–241.
Bruzinskaite-Schmidhalter, R., Sarkunas, M., Malakauskas, A., Mathis, A., Torgerson, P.R., Deplazes, P., 2012. Helminths of red foxes (*Vulpes vulpes*) and raccoon dogs (*Nyctereutes procyonoides*) in Lithuania. Parasitology 139, 120–127.
Bryan, H.M., Darimont, C.T., Hill, J.E., Paquet, P.C., Thompson, R.C.A., Wagner, B., Smits, J.E., 2012. Seasonal and biogeographical patterns of gastrointestinal parasites in large carnivores: wolves in a coastal archipelago. Parasitology 139, 781–790.
Budke, C.M., Jiamin, Q., Zinsstag, J., Qian, W., Torgerson, P.R., 2004. Use of disability adjusted life years in the estimation of the disease burden of echinococcosis for a high endemic region of the Tibetan plateau. Am. J. Trop. Med. Hyg. 71, 56–64.
Budke, C.M., Campos-Ponce, M., Qian, W., Torgerson, P.R., 2005a. A canine purgation study and risk factor analysis for echinococcosis in a high endemic region of the Tibetan plateau. Vet. Parasitol. 127, 43–49.
Budke, C.M., Jiamin, Q., Craig, P.S., Torgerson, P.R., 2005b. Modeling the transmission of *Echinococcus granulosus* and *Echinococcus multilocularis* in dogs for a high endemic region of the Tibetan plateau. Int. J. Parasitol. 35, 163–170.
Budke, C.M., Deplazes, P., Torgerson, P.R., 2006. Global socioeconomic impact of cystic echinococcosis. Emerg. Infect. Dis. 12, 296–303.
Buishi, I.E., Njoroge, E.M., Bouamra, O., Craig, P.S., 2005. Canine echinococcosis in northwest Libya: assessment of coproantigen ELISA, and a survey of infection with analysis of risk-factors. Vet. Parasitol. 130, 223–232.
Busi, M., Snabel, V., Varcasia, A., Garippa, G., Perrone, V., De Liberato, C., D'Amelio, S., 2007. Genetic variation within and between G1 and G3 genotypes of *Echinococcus granulosus* in Italy revealed by multilocus DNA sequencing. Vet. Parasitol. 150, 75–83.
Butler, J.M., Grundmann, A.W., 1954. The intestinal helminths of the coyote *Canis latrans* Say, in Utah. J. Parasitol. 40, 440–443.
Buzinov, R., Parfenova, E., Gudkov, A., Unguryanu, T., 2012. Assessment of soil epidemic hazard in Arkhangelsk region. Hum. Ecol. 4, 3–10.
Cabrera, R., 2007. In: Ministerio de salud, G.D.P. (Ed.), Equinococosis quística en el Perú: necesidades de prevención y control de una enfermedad postergada. Lima.
Cai, H.X., Guan, Y.Y., Wang, H., Wu, W.P., Han, X.M., Ma, X., Wang, L.Y., 2012. Geographical distribution of echinococcosis among children in Qinghai Province. Chin. J. Parasitol. Parasit. Dis. 30, 127–130.

Calderini, P., Gabrielli, S., Cancrini, G., 2012. Is the goat a new host for the G3 Indian buffalo strain of *Echinococcus granulosus?* Sci. World J. 2012.
Cameron, T., 1960. The incidence and diagnosis of hydatid cyst in Canada, *Echinococcus granulosas* var. *canadensis*. Parassitologia 2, 381–390.
Capuano, F., Rinaldi, L., Maurelli, M.P., Perugini, A.G., Veneziano, V., Garippa, G., Genchi, C., Musella, V., Cringoli, G., 2006. Cystic echinococcosis in water buffaloes: epidemiological survey and molecular evidence of ovine (G1) and buffalo (G3) strains. Vet. Parasitol. 137, 262–268.
Carayon, A., Robert, H., 1962. Surgical aspects of helminthiases in West Africa. (Ascariasis, dracunculosis, filariasis, bilharziasis). VIII. Surgical aspects of rare helminthiasis in West Africa. (Distomiasis, cysticercosis, hydatidosis, sparganosis, porocephalosis). Med. Trop. 22, 361–375 concl.
Cardona, G.A., Carmena, D., 2013. A review of the global prevalence, molecular epidemiology and economics of cystic echinococcosis in production animals. Vet. Parasitol. 192, 10–32.
Carmena, D., Cardona, G.A., 2014. Echinococcosis in wild carnivorous species: epidemiology, genotypic diversity, and implications for veterinary public health. Vet. Parasitol. 202, 69–94.
Carmena, D., Sanchez-Serrano, L.P., Barbero-Martinez, I., 2008. *Echinococcus granulosus* infection in Spain. Zoonoses Public Health 55, 156–165.
Carney, W.P., Cross, J.H., Wheeling, C.H., Purnomo, Sudomo, M., Simandjuntak, G., 1974. Natural infection of *Echinococcus granulosus* in dogs from Sulawesi, Indonesia. Southeast Asian J. Trop. Med. Public Health 5, 385–389.
Carter, C., Bonatti, H., Hranjec, T., Barroso, L.F., Donowitz, G., Sawyer, R.G., Schirmer, B., 2009. Epigastric cystic echinococcus involving stomach, liver, diaphragm, and spleen in an immigrant from Afghanistan. Surg. Infect. 10, 453–456.
Castrodale, L., 2003. Echinococcosis in Alaska. State of Alaska Epidemiology Bulletin, Anchorage, AK.
Casulli, A., Manfredi, M.T., La Rosa, G., Di Cerbo, A.R., Dinkel, A., Romig, T., Deplazes, P., Genchi, C., Pozio, E., 2005. *Echinococcus multilocularis* in red foxes (*Vulpes vulpes*) of the Italian Alpine region: is there a focus of autochthonous transmission? Int. J. Parasitol. 35, 1079–1083.
Casulli, A., Manfredi, M.T., La Rosa, G., Cerbo, A.R., Genchi, C., Pozio, E., 2008. *Echinococcus ortleppi* and *E. granulosus* G1, G2 and G3 genotypes in Italian bovines. Vet. Parasitol. 155, 168–172.
Casulli, A., Szell, Z., Pozio, E., Sreter, T., 2010a. Spatial distribution and genetic diversity of *Echinococcus multilocularis* in Hungary. Vet. Parasitol. 174, 241–246.
Casulli, A., Zeyhle, E., Brunetti, E., Pozio, E., Meroni, V., Genco, F., Filice, C., 2010b. Molecular evidence of the camel strain (G6 genotype) of *Echinococcus granulosus* in humans from Turkana, Kenya. Trans. R. Soc. Trop. Med. Hyg. 104, 29–32.
Catalano, S., Lejeune, M., Liccioli, S., Verocai, G.G., Gesy, K.M., Jenkins, E.J., Kutz, S.J., Fuentealba, C., Duignan, P.J., Massolo, A., 2012. *Echinococcus multilocularis* in urban coyotes, Alberta, Canada. Emerg. Infect. Dis. 18, 1625–1628.
Cavagion, L., Perez, A., Santillan, G., Zanini, F., Jensen, O., Saldia, L., Diaz, M., Cantoni, G., Herrero, E., Costa, M.T., Volpe, M., Araya, D., Rubianes, N.A., Aguado, C., Meglia, G., Guarnera, E., Larrieu, E., 2005. Diagnosis of cystic echinococcosis on sheep farms in the south of Argentina: areas with a control program. Vet. Parasitol. 128, 73–81.
Chaabane-Banaoues, R., Oudni-M'rad, M., Cabaret, J., M'Rad, S., Mezhoud, H., Babba, H., 2015. Infection of dogs with *Echinococcus granulosus*: causes and consequences in an hyperendemic area. Parasit. Vectors 8, 231.

Chahed, M.K., Bellali, H., Ben Alaya, N., Aoun, K., Zouari, B., 2015. High risk areas for echinococcosis-hydatidosis in Tunisia. Tunis. Med. 93, 33–37.
Chaligiannis, I., Maillard, S., Boubaker, G., Spiliotis, M., Saratsis, A., Gottstein, B., Sotiraki, S., 2015. *Echinococcus granulosus* infection dynamics in livestock of Greece. Acta Trop. 150, 64–70.
Chihai, O., Umhang, G., Erhan, D., Boue, F., Talambuta, N., Rusu, S., Zamornea, M., 2016. Slaughterhouse survey of cystic echinococcosis in cattle and sheep from the Republic of Moldova. J. Helminthol. 90, 279–283.
Chinchuluun, B., Sako, Y., Khatanbaatar, I., Bayarmaa, B., Lkhagvatseren, S., Battsetseg, G., Yanagida, T., Itoh, S., Temuulen, D., Budke, C.M., Ito, A., Batsukh, Z., 2014. A survey of seropositivity to antigen B, an immunodiagnostic antigen for human cystic echinococcosis, in domestic animals in Mongolia. Parasitol. Int. 63, 324–326.
Choi, H., Park, J.Y., Kim, J.H., Moon du, G., Lee, J.G., Bae, J.H., 2014. Primary renal hydatid cyst: mis-interpretation as a renal malignancy. Korean J. Parasitol. 52, 295–298.
Chopdat, N., Menezes, C.N., John, M.-A., Mahomed, N., Grobusch, M.P., 2007. A gardener who coughed up blood. Lancet 370, 1520.
Choquette, L.P.E., Whitten, L.K., Rankin, G., Seal, C.M., 1957. Notes on parasites found in reindeer (*Rangifer tarandus*) in Canada. Can. J. Comp. Med. Vet. Sci. 21, 199–203.
Christodoulopoulos, G., Theodoropoulos, G., Petrakos, G., 2008. Epidemiological survey of cestode-larva disease in Greek sheep flocks. Vet. Parasitol. 153, 368–373.
Cirovic, D., Pavlovic, I., Kulisic, Z., Ivetic, V., Penezic, A., Cosic, N., 2012. *Echinococcus multilocularis* in the European beaver (*Castor fibre* L.) from Serbia: first report. Vet. Rec. 171, 100.
Cleary, E., Barnes, T.S., Xu, Y., Zhao, H., Clements, A.C., Gray, D.J., McManus, D.P., Atkinson, J.A., Williams, G.M., Yang, Y., 2014. Impact of "Grain to Green" programme on echinococcosis infection in Ningxia Hui autonomous region of China. Vet. Parasitol. 205, 523–531.
Clement, M., Posada, D., Crandall, K.A., 2000. TCS: a computer program to estimate gene genealogies. Mol. Ecol. 9, 1657–1659.
Colovic, I., 2009. Morphological and Genetic Characteristics of *Echinococcus granulosus* Strains from Patients with Cystic Liver Hydatidosis. University of Belgrade, Belgrade.
Combes, B., Comte, S., Raton, V., Raoul, F., Boué, F., Umhang, G., Favier, S., Dunoyer, C., Woronoff, N., Giraudoux, P., 2012. Westward spread of *Echinococcus multilocularis* in foxes, France, 2005–2010. Emerg. Infect. Dis. 18, 2059.
Conraths, F.J., Deplazes, P., 2015. *Echinococcus multilocularis*: epidemiology, surveillance and state-of-the-art diagnostics from a veterinary public health perspective. Vet. Parasitol. 213, 149–161.
Cornejo-Juarez, P., Espinosa-Altamirano, A., Ibarra-del-Rio, M., Pacheco-Bravo, I., Volkow-Fernandez, P., 2013. Cystic echinococcosis: a disease mimicking cancer in a non-endemic country report of two cases. West Indian Med. J. 62, 266–269.
Corsini, M., Geissbuhler, U., Howard, J., Gottstein, B., Spreng, D., Frey, C.F., 2015. Clinical presentation, diagnosis, therapy and outcome of alveolar echinococcosis in dogs. Vet. Rec. 177, 569.
Cortés Carrasco, A., Martínez Hernández, N., Parraguirre Martínez, S., 2002. Hidatidosis por estudio citológico e histológico. Presentación de un caso. Rev. Hosp. Gral 5, 42–45.
Counotte, M.J., Minbaeva, G., Usubalieva, J., Abdykerimov, K., Torgerson, P.R., 2016. The burden of zoonoses in Kyrgyzstan. PLoS Negl. Trop. Dis. 10, e0004831.
Craig, P.S., Deshan, L., Macpherson, C.N., Dazhong, S., Reynolds, D., Barnish, G., Gottstein, B., Zhirong, W., 1992. A large focus of alveolar echinococcosis in central China. Lancet 340, 826–831.

Craig, P., 2004. Epidemiology of echinococcosis in western China. In: Torgerson, P., Shaikenov, B. (Eds.), Echinococcosis in Central Asia: Problems and Solutions. Dauir, Zurich.

Craig, P.S., Hegglin, D., Lightowlers, M.W., Torgerson, P.R., Wang, Q., 2017. *Echinococcosis*: control and prevention. In: Thompson, R.C.A., Deplazes, P., Lymbery, A.J. (Eds.), *Echinococcus* and Echinococcosis, Part B, vol. 96, pp. 55–158.

Crellin, J.R., Andersen, F.L., Schantz, P.M., Condie, S.J., 1982. Possible factors influencing distribution and prevalence of *Echinococcus granulosus* in Utah. Am. J. Epidemiol. 116, 463–474.

Cringoli, G., Rinaldi, L., Musella, V., Veneziano, V., Maurelli, M.P., Di Pietro, F., Frisiello, M., Di Pietro, S., 2007. Geo-referencing livestock farms as tool for studying cystic echinococcosis epidemiology in cattle and water buffaloes from southern Italy. Geospatial Health 2, 105–111.

Cruz Benítez, L., 2009. Tratamiento de un quiste hidatídico en un paciente con abdomen agudo. Comunicación de un caso y revisión bibliográfica. Rev. Espec. Médico-Quirúrgicas 14, 93–99.

Cruz-Reyes, A., Constantine, C.C., Boxell, A.C., Hobbs, R.P., Thompson, R.C.A., 2007. *Echinococcus granulosus* from Mexican pigs is the same strain as that in Polish pigs. J. Helminthol. 81, 287–292.

Csotye, J., Sisak, K., Bardocz, L., Toth, K., 2011. Pathological femoral neck fracture caused by an *Echinococcus* cyst of the vastus lateralis—case report. BMC Infect. Dis. 11, 103.

Cucher, M.A., Macchiaroli, N., Baldi, G., Camicia, F., Prada, L., Maldonado, L., Avila, H.G., Fox, A., Gutierrez, A., Negro, P., Lopez, R., Jensen, O., Rosenzvit, M., Kamenetzky, L., 2016. Cystic echinococcosis in South America: systematic review of species and genotypes of *Echinococcus granulosus sensu lato* in humans and natural domestic hosts. Trop. Med. Int. Health 21, 166–175.

Dada, B.J., Adegboye, D.S., Mohammed, A.N., 1980. The epidemiology of *Echinococcus* infection in Kano State, Nigeria. Ann. Trop. Med. Parasitol. 74, 515–517.

Dada, B.J., 1980. Taeniasis, cysticercosis and echinococcosis/hydatidosis in Nigeria: IV—prevalence of *Echinococcus granulosus* infection in stray dogs. J. Helminthol. 54, 299–301.

Dakkak, A., 2010. Echinococcosis/hydatidosis: a severe threat in Mediterranean countries. Vet. Parasitol. 174, 2–11.

Dalimi, A., Motamedi, G., Hosseini, M., Mohammadian, B., Malaki, H., Ghamari, Z., Ghaffari Far, F., 2002. *Echinococcosis*/hydatidosis in western Iran. Vet. Parasitol. 105, 161–171.

Dalimi, A., Sattari, A., Motamedi, G., 2006. A study on intestinal helminthes of dogs, foxes and jackals in the western part of Iran. Vet. Parasitol. 142, 129–133.

Dar, F.K., Alkarmi, T., 1997. Public health aspects of cystic echinococcosis in the Arab countries. Acta Trop. 67, 125–132.

Darwish, B., 2006. Clinical and radiological manifestations of 206 patients with pulmonary hydatidosis over a ten-year period. Prim. Care Respir. J. 15, 246–251.

Daryani, A., Alaei, R., Arab, R., Sharif, M., Dehghan, M.H., Ziaei, H., 2007. The prevalence, intensity and viability of hydatid cysts in slaughtered animals in the Ardabil province of Northwest Iran. J. Helminthol. 81, 13–17.

DATASUS, 2013. Sistema de Informações sobre Mortalidade — SIM. Consolidação da base de dados de 2011. In: Government, B. (Ed.).

Davidson, W.R., Appel, M.J., Doster, G.L., Baker, O.E., Brown, J.F., 1992. Diseases and parasites of red foxes, gray foxes, and coyotes from commercial sources selling to fox-chasing enclosures. J. Wildl. Dis. 28, 581–589.

Davidson, R.K., Romig, T., Jenkins, E., Tryland, M., Robertson, L.J., 2012. The impact of globalisation on the distribution of *Echinococcus multilocularis*. Trends Parasitol. 28, 239–247.

Davidson, R.K., Lavikainen, A., Konyaev, S., Schurer, J., Miller, A.L., Oksanen, A., Skirnisson, K., Jenkins, E., 2016. *Echinococcus* across the North: current knowledge, future challenges. Food Waterborne Parasitol. 4, 39–53.

de Carvalho, L., Guerra, D., 2014. Animal and human echinococcosis/hydatidosis: updates on the epidemiological status of the disease in Portugal. In: ESCCAP Conference Echinococcus 2014. 8–9th October 2014, Lithuania, p. 50.

de la Rue, M.L., Dinkel, A., Mackenstedt, U., Romig, T., 2006. New data on *Echinococcus* spp. in Southern Brazil. Rev. Inst. Med. Trop. Sao Paulo 48, 103–104.

de la Rue, M.L., Takano, K., Brochado, J.F., Costa, C.V., Soares, A.G., Yamano, K., Yagi, K., Katoh, Y., Takahashi, K., 2011. Infection of humans and animals with *Echinococcus granulosus* (G1 and G3 strains) and *E. ortleppi* in Southern Brazil. Vet. Parasitol. 177, 97–103.

de la Rue, M.L., 2008. Cystic echinococcosis in southern Brazil. Rev. Inst. Med. Trop. Sao Paulo 50, 53–56.

De Meulemeester, J., Dardenne, G., 1958. Hydatid cyst of the liver in a European living in the Belgian Congo. Acta Chir Belg 57, 283–295.

de Morais, J.D., 2010. The rise and decline of human hydatid disease in Portugal: historical and epidemiological analysis. Med. Int. 17, 246–256.

Del Carpio, M., Mercapide, C.H., Salvitti, J.C., Uchiumi, L., Sustercic, J., Panomarenko, H., Moguilensky, J., Herrero, E., Talmon, G., Volpe, M., Araya, D., Mujica, G., Calabro, A., Mancini, S., Chiosso, C., Labanchi, J.L., Saad, R., Goblirsch, S., Brunetti, E., Larrieu, E., 2012. Early diagnosis, treatment and follow-up of cystic echinococcosis in remote rural areas in Patagonia: impact of ultrasound training of non-specialists. PLoS Negl. Trop. Dis. 6, e1444.

DeMarais, P., Flaherty, J.P., Arnow, P.M., 1992. Echinococcal infection in a Ghanaian patient. Lancet 339, 1357–1358.

Deplazes, P., Eckert, J., 2001. Veterinary aspects of alveolar echinococcosis – a zoonosis of public health significance. Vet. Parasitol. 98, 65–87.

Deplazes, P., Hegglin, D., Gloor, S., Romig, T., 2004. Wilderness in the city: the urbanization of *Echinococcus multilocularis*. Trends Parasitol. 20, 77–84.

Deplazes, P., Grimm, F., Sydler, T., Tanner, I., Kapel, C.M., 2005. Experimental alveolar echinococcosis in pigs, lesion development and serological follow up. Vet. Parasitol. 130, 213–222.

Deplazes, P., van Knapen, F., Schweiger, A., Overgaauw, P.A., 2011. Role of pet dogs and cats in the transmission of helminthic zoonoses in Europe, with a focus on echinococcosis and toxocarosis. Vet. Parasitol. 182, 41–53.

Deplazes, P., 2015. Echinococcoses. In: Beugnet, F., Halos, L. (Eds.), Parasitoses & Vector Borne Diseases of Cats. Merial, Lyon, pp. 3–7.

Derylo, A.S.P., 1998. Prevalence of echinococcosis in swine, beef-cattle, sheep and goats in Poland during 1995–1997. Wiadom Parazytik 44, 413.

Develoux, M., Audoin, J., Lamothe, F., Gali, A., Warter, A., 1991. Human hydatidosis in Niger. J. Trop. Med. Hyg. 94, 423–424.

Develoux, M., Enache-Angoulvant, A., Gounant, V., Brian, E., Khalil, A., Bazelly, B., Hennequin, C., 2011. Hepatic and pulmonary cystic echinococcosis in a patient from the Central African Republic. Travel Med. Infect. Dis. 9, 88–90.

Devleesschauwer, B., Ale, A., Torgerson, P.R., Praet, N., Maertens de Noordhout, C., Pandey, B.D., Pun, S.B., Lake, R., Vercruysse, J., Joshi, D.D., Havelaar, A.H., Duchateau, L., Dorny, P., Speybroeck, N., 2014. The burden of parasitic zoonoses in Nepal: a systematic review. PLoS Negl. Trop. Dis. 8, e2634.

Dezsényi, B., Strausz, T., Makrai, Z., Csomor, J., Danka, J., Kern, P., Rezza, G., Barth, T.F.E., Casulli, A., June 28, 2016. Autochthonous human alveolar echinococcosis in a Hungarian patient. Infection 1–4.

Dhuravliev, V., Yanchenko, V., Porokhnenko, A., 2000. Metastasis of the lungs from alveolar echinococcosis of the liver. Ann. Surg. Hepatol. 5, 10–13.

Dinkel, A., Njoroge, E.M., Zimmermann, A., Walz, M., Zeyhle, E., Elmahdi, I.E., Mackenstedt, U., Romig, T., 2004. A PCR system for detection of species and genotypes of the *Echinococcus granulosus*-complex, with reference to the epidemiological situation in eastern Africa. Int. J. Parasitol. 34, 645–653.

Donovan, S.M., Mickiewicz, N., Meyer, R.D., Panosian, C.B., 1995. Imported echinococcosis in southern California. Am. J. Trop. Med. Hyg. 53, 668–671.

Dowling, P.M., Abo-Shehada, M.N., Torgerson, P.R., 2000. Risk factors associated with human cystic echinococcosis in Jordan: results of a case-control study. Ann. Trop. Med. Parasitol. 94, 69–75.

Dubinsky, P., Svobodova, V., Turcekova, L., Literak, I., Martinek, K., Reiterova, K., Kolarova, L., Klimes, J., Mrlik, V., 1999. *Echinococcus multilocularis* in Slovak Republic: The first record in red foxes (*Vulpes vulpes*). Helminthologia 36, 105–110.

Dyachenko, V., Pantchev, N., Gawlowska, S., Vrhovec, M.G., Bauer, C., 2008. *Echinococcus multilocularis* infections in domestic dogs and cats from Germany and other European countries. Vet. Parasitol. 157, 244–253.

Dyachenko, Y.V., Maslennikova, A.C., Mikhailenko, V.V., Lutsuk, S.N., Tolokonnikov, V.P., 2014. Infestation slaugther ruminants *Echinococcus granulosus* in Republic North Osetia-Alania and some sanitary-microbiological and patomorthological characteristics of internal organs. Mod. Prob. Sci. Educ. 6.

Dybicz, M., Borkowski, P.K., Dabrowska, J., Chomicz, L., 2015. Cases of *Echinococcus granulosus sensu stricto* isolated from polish patients: imported or indigenous? Biomed. Res. Int. 2015, 728321.

Eaton, R.D., Secord, D.C., 1979. Some intestinal parasites of Arctic fox, Banks Island, N. W. T. Can. J. Comp. Med. 43, 229–230.

Ecca, A.R., Gabriele, F., Bortoletti, G., Palmas, C., 2002. Cystic echinococcosis in the Mediterranean Basin. In: Cestode Zoonoses: Echinococcosis and Cysticercosis. IOS Press Amsterdam, pp. 41–56.

ECDC, 2010. European Centre for Disease Prevention and Control.

Eckert, J., Schantz, P., Gasser, R., Torgerson, P., Bessonov, A., Movsessian, S., Thakur, A., Grimm, F., Nikogossian, M.A., 2001. Geographic distribution and prevalence. In: Eckert, J., Gemmell, M., Meslin, F.-X., Pawlowski, Z. (Eds.), WHOI/OIE Manual on Echinococcosis in Humans and Animals: A Public Health Problem of Global Concern. World Organisation for Animal Health, Paris, pp. 100–142.

Eckert, J., Thompson, R.C.A., 2017. Historical aspects of echinococcosis. In: Thompson, R.C.A., Deplazes, P., Lymbery, A.J. (Eds.), *Echinococcus* and Echinococcosis, Part A, vol. 95, pp. 1–64.

Economides, P., Christofi, G., 2000. Evaluation of control programmes for echinococcosis/ hydatidosis in Cyprus. Revue Sci. Tech. Int. Office Epizootics 19, 784–792.

Economides, P., Christofi, G., Gemmell, M.A., 1998. Control of *Echinococcus granulosus* in Cyprus and comparison with other island models. Vet. Parasitol. 79, 151–163.

EFSA, 2011. The European Union summary report on trends and sources of zoonoses, zoonotic agents and food-borne outbreaks in 2009. EFSA J. 9, 2090.

EFSA, 2015a. The European Union summary report. Trends and sources of zoonoses, zoonotic agents and food-borne outbreaks in 2013. EFSA J. 13, 165.

EFSA, 2015b. Opinion on *Echinococcus multilocularis* infections in animals. EFSA J. 13, 4373.

Eguia-Aguilar, P., Cruz-Reyes, A., Martinez-Maya, J.J., 2005. Ecological analysis and description of the intestinal helminths present in dogs in Mexico City. Vet. Parasitol. 127, 139–146.

Ekhnefer, A., 2014. Prevalence of cystic echinococcosis infestation sheep, goats and cattle in the Bayda city – Libya. Am. J. Res. Comm. 2, 11–22.

El Berbri, I., Ducrotoy, M.J., Petavy, A.F., Fassifihri, O., Shaw, A.P., Bouslikhane, M., Boue, F., Welburn, S.C., Dakkak, A., 2015a. Knowledge, attitudes and practices with regard to the presence, transmission, impact, and control of cystic echinococcosis in Sidi Kacem Province, Morocco. Infect. Dis. Poverty 4, 48.
El Berbri, I., Pétavy, A., Umhang, G., Bouslikhane, M., Fassi Fihri, O., Boué, F., Dakkak, A., 2015b. Epidemiological investigations on cystic echinococcosis in North-West (Sidi Kacem Province) Morocco: infection in ruminants. Adv. Epidemiol. 1−9.
El Shazly, A.M., Awad, S.E., Nagaty, I.M., Morsy, T.A., 2007. Echinococcosis in dogs in urban and rural areas in Dakahlia Governorate. Egypt. J. Egypt Soc. Parasitol. 37, 483−492.
El-Badawi, E.K., Eisa, A.M., Slepenev, N.K., Saad, M.B., 1979. Hydatidosis of domestic animals in the central region of the Sudan. In: Bulletin of Animal Health and Production in Africa. Bulletin des sante et Production Animales en Afrique, vol. 27, pp. 249−251.
Elchuev, M., 1989. Landscape and ecological characteristics of cestodes in wild canidae in the Sheki-Zaqatal area of the Azerbaijan SSR. Izvestiia. Acad. Sci. Azerbaijan SSR. Ser. Biol. Sci. 2.
El-Gidaafi, A., Kassem, H., 2013. Human hydatidosis in Benghazi, Libya: an epidemiological study of surgical cases between 2000 and 2007. Egypt. J. Exp. Biol. (Zoo) 9, 91−94.
El-Ibrahim, J.H., 2009. Prevalence of Sheep hydatidosis in North West Bank-Palestine. An-Najah National University.
El-Madawy, R., Khalifa, N., Afify, J., 2011. Epidemiological and molecular studies of hydatid cyst in slaughtered cattle and sheep in Toukh. Egypt. Benha Vet. Med. J. 1, 95−101.
Elmahdi, I.E., Ali, Q.M., Magzoub, M.M., Ibrahim, A.M., Saad, M.B., Romig, T., 2004. Cystic echinococcosis of livestock and humans in central Sudan. Ann. Trop. Med. Parasitol. 98, 473−479.
Elmajdoub, L., Rahman, W., 2015. Prevalence of hydatid cysts in slaughtered animals from different areas of Libya. Open J. Vet. Med. 5, 1−10.
El-Shehabi, F.S., Abdel-Hafez, S.K., Kamhawi, S.A., 1999. Prevalence of intestinal helminths of dogs and foxes from Jordan. Parasitol. Res. 85, 928−934.
Ernest, E., Nonga, H.E., Kynsieri, N., Cleaveland, S., 2010. A retrospective survey of human hydatidosis based on hospital records during the period 1990−2003 in Ngorongoro, Tanzania. Zoonoses Public Health 57, e124−e129. http://dx.doi.org/10.1111/j.1863-2378.2009.01297.
Ertabaklar, H., Dayanir, Y., Ertug, S., 2012. Research to investigate human cystic echinococcosis with ultrasound and serologic methods and educational studies in different provinces of Aydin/Turkey. Turk. Parazitol. Derg. 36, 142−146.
Eryildiz, C., Sakru, N., 2012. Molecular characterization of human and animal isolates of *Echinococcus granulosus* in the Thrace region, Turkey. Balk. Med. J. 29, 261−267.
Escalante, G.E., Espinosa Gómez, I., GonzÃ¡lez Ricardo, Y., 2012. Hidatidosis pulmonar en niño. Presentación de un caso. MediSur 10, 310−313.
Eslami, A., Hosseini, S.H., 1998. *Echinococcus granulosus* infection of farm dogs of Iran. Parasitol. Res. 84, 205−207.
Eslami, A., Rahbari, S., Meydani, M., 1981. Cestodes and trematodes of wild sheep, *Ovis ammon orientalis*, and Goitered Gazelle, *Gazella subgutturosa*, in Iran. Vet. Parasitol. 8, 99−101.
Eslami, A., Shayan, P., Bokaei, S., 2014. Morphological and genetic characteristics of the liver hydatid cyst of a donkey with Iran origin. Iran. J. Parasitol. 9, 302−310.
Eslami, A., Meshgi, B., Jalousian, F., Rahmani, S., Salari, M.A., 2016. Genotype and phenotype of *Echinococcus granulosus* derived from wild sheep (*Ovis orientalis*) in Iran. Korean J. Parasitol. 54, 55−60.

Espinoza, S., Salas, A.M., Vargas, A., Freire, V., Diaz, E., Sanchez, G., Venegas, J., 2014. Detection of the G3 genotype of *Echinococcus granulosus* from hydatid cysts of Chilean cattle using COX1 and ND1 mitochondrial markers. Parasitol. Res. 113, 139–147.

European Centre for Disease Prevention and Control, 2013. Annual Epidemiological Report Reporting on 2011 Surveillance Data and 2012 Epidemic Intelligence Data (Stockholm).

Fadakar, B., Tabatabaei, N., Borji, H., Naghibi, A., 2015. Genotyping of *Echinococcus granulosus* from goats and sheep indicating G7 genotype in goats in the Northeast of Iran. Vet. Parasitol. 214, 204–207.

Fahim, F., Al Salamah, S.M., 2007. Cystic echinococcosis in Central Saudi Arabia: a 5-year experience. Turk J. Gastroenterol. 18, 22–27.

Fakhar, M., Sadjjadi, S.M., 2007. Prevalence of hydatidosis in slaughtered herbivores in Qom province, central part of Iran. Vet. Res. Commun. 31, 993–997.

Farias, L.N., Malgor, R., Cassaravilla, C., Bragança, C., de la Rue, M.L., 2004. Echinococcosis in southern Brazil: efforts toward implementation of a control program in Santana do Livramento, Rio Grande do Sul. Rev. Med. Trop. 46, 153–156.

Farjallah, S., Busi, M., Mahjoub, M.O., Slimane, B.B., Said, K., D'Amelio, S., 2007. Molecular characterization of *Echinococcus granulosus* in Tunisia and Mauritania by mitochondrial rrnS gene sequencing. Parassitologia 49, 239–246.

Fasihi Harandi, M., Budke, C.M., Rostami, S., 2012. The monetary burden of cystic echinococcosis in Iran. PLoS Negl. Trop. Dis. 6, e1915.

Fattahi Masoom, S., Sharifi, N., 2007. Alveolar Hydatid disease: a case report. Med. J. Islam Repub. Iran. 20, 205–207.

Federer, K., Armua-Fernandez, M.T., Gori, F., Hoby, S., Wenker, C., Deplazes, P., 2016. Food contaminated with taeniid (*Taenia, Echinococcus*) eggs as a risk for metacestode infections in captive primates. Int. J. Parasitol. Parasit. Wildl. 5, 249–253.

Feng, K., Huang, F.Q., Duo, H., Fu, Y., Shen, X.Y., Peng, M., Li, W., 2013. Species identification of *Echinococcus* isolates collected from canines and Tibetan foxes in Chengduo County, Qinghai Province. Chin. J. Parasitol. Parasit. Dis. 31, 185–187.

Feng, X., Qi, X., Yang, L., Duan, X., Fang, B., Gongsang, Q., Bartholomot, B., Vuitton, D.A., Wen, H., Craig, P.S., 2015. Human cystic and alveolar echinococcosis in the Tibet Autonomous Region (TAR), China. J. Helminthol. 89, 671–679.

Feng, X., 2012. Immunodiagnosis of Human and Canine Echinococcosis and Community Studies in Northwestern China. University of Salford, UK, Salford.

Fernández, C.F., Cantó, A.G.J., 2002. Intestinal helminth frequency in stray dogs sacrificed in the City of Queretaro, in the State of Queretaro, in Mexico. Veterinaria Mex. 33, 247–253.

Fiapsheva, A., Kanokova, A., Bittirov, A., Chilaev, S., 2007. Distribution of echinococcosis in cattle in Kabardino-Balkarian Republic. Russ. Parasitol. J. 2, 1–5.

Finlayson, D.M., Fergus, A., 1963. Hydatid disease in Toronto. Can. Med. Assoc. J. 88, 1004–1009.

Flisser, A., Maravilla, P., Mata-Miranda, P., Martinez-Hernandez, F., 2015. Echinococcosis in Mexico — a story worth sharing. In: Rodriguez-Morales, A. (Ed.), Current Topics in Echinococcosis.

Foreyt, W.J., Drew, M.L., Atkinson, M., McCauley, D., 2009. *Echinococcus granulosus* in gray wolves and ungulates in Idaho and Montana, USA. J. Wildl. Dis. 45, 1208–1212.

Freeman, J., Hutchison, G.B., 1980. Prevalence, incidence and duration. Am. J. Epidemiol. 112, 707–723.

Freeman, R.S., Adorjan, A., Pimlott, D.H., 1961. Cestodes of wolves, coyotes, and coyote-dog hybrids in Ontario. Can. J. Zool. 39, 527.

French, C.M., Nelson, G.S., 1982. Hydatid disease in the Turkana District of Kenya II. A study in medical geography. Ann. Trop. Med. Parasitol. 76, 439–457.

Fuller, G.K., Fuller, D.C., 1981. Hydatid disease in Ethiopia: clinical survey with some immunodiagnostic test results. Am. J. Trop. Med. Hyg. 30, 645–652.

Furth, M., Hoida, G., Nahmias, J., Greenberg, Z., Barzilay, A., Goldsmith, R.S., el-On, J., 1994. The development of new foci of echinococcosis in northern Israel: prevalence in domestic animals. J. Helminthol. 68, 45–47.
Gagarin, V., Steshenko, V., Tokobaev, M., 1957. Role of rodents in the distribution of helminth zoonoses. Works Inst. Zool. Parasitol. Kyrg. Acad. Sci. 6, 159–161.
Gamble, W.G., Segal, M., Schantz, P.M., Rausch, R.L., 1979. Alveolar hydatid disease in Minnesota. First human case acquired in the contiguous United States. JAMA 241, 904–907.
Gardner, S., Dursahinhan, A., Racz, G., Batsaikhan, N., Ganzorig, S., Tinnin, D., Damdinbazar, D., Wood, C., Peterson, A., Alandia, E., Molericona, J., Salazar-Bravo, J., 2013. Sylvatic Species of *Echinococcus* from Rodent Intermediate Hosts in Asia and South America, 318. Museum of Texas Tech University, pp. 1–13.
Garippa, G., Battelli, G., Cringoli, G., Giangaspero, A., Giannetto, S., Manfredi, M.T., 2004. Animal echinococcosis in Italy: epidemiological update. Parassitologia 46, 33–38.
Garippa, G., 2006. Updates on cystic echinococcosis (CE) in Italy. Parassitologia 48, 57–59.
Gates, M.C., Nolan, T.J., 2009. Endoparasite prevalence and recurrence across different age groups of dogs and cats. Vet. Parasitol. 166, 153–158.
Gemmell, M.A., 1990. Australasian contributions to an understanding of the epidemiology and control of hydatid disease caused by *Echinococcus granulosus*—past, present and future. Int. J. Parasitol. 20, 431–456.
Genov, T., Svilenov, D., Polyakova-Krusteva, O., 1980. The natural occurrence of *Alveococcus multilocularis* in the *Microtus nivalis* in Bulgaria. R. Acad. Bulg. Sci. 33.
Geramizadeh, B., Nikeghbalian, S., Malekhosseini, S.A., 2012. Alveolar echinococcosis of the liver: report of three cases from different geographic areas of Iran. Hepat. Mon. 12, e6143.
Gesy, K.M., Jenkins, E.J., 2015. Introduced and native haplotypes of *Echinococcus multilocularis* in wildlife in Saskatchewan, Canada. J. Wildl. Dis. 51, 743–748.
Gesy, K., Hill, J.E., Schwantje, H., Liccioli, S., Jenkins, E.J., 2013. Establishment of a European-type strain of *Echinococcus multilocularis* in Canadian wildlife. Parasitology 140, 1133–1137.
Gesy, K.M., Schurer, J.M., Massolo, A., Liccioli, S., Elkin, B.T., Alisauskas, R., Jenkins, E.J., 2014. Unexpected diversity of the cestode *Echinococcus multilocularis* in wildlife in Canada. Int. J. Parasitol. Parasit. Wildl. 3, 81–87.
Gesy, K., 2012. The Geographic Distribution and Genetic Variation of *Echinococcus multilocularis* in Canada. University of Saskatchewan.
Gevorgyan, H.S., Dinkel, A., Mackenstedt, U., Romig, T., 2006. Cystic Echinococcosis in Armenia: Host Species and Echinococcus Genotypes. Deutsche Gesellschaft für Parasitologie, Wien, p. 47.
Gholami, S., Daryani, A., Sharif, M., Mobedi, I., Amouei, A., 2011. A Comparison of the Prevalence of *Echinococcus granulosus* in stray dogs in Sari city, in 1991 and 2007. J. Mazandaran Univ. Med. Sci. 20, 70–75.
Giannetto, S., Poglayen, G., Brianti, E., Sorgi, C., Gaglio, G., Canu, S., Virga, A., 2004. An epidemiological updating on cystic echinococcosis in cattle and sheep in Sicily, Italy. Parassitologia 46, 423–424.
Gilbert, N.L., Dare, O.K., Libman, M.D., Muchaal, P.K., Ogden, N.H., 2010. Hospitalization for trichinellosis and echinococcosis in Canada, 2001–2005: the tip of the iceberg. Can. J. Public Health 101.
Giraudoux, P., Raoul, F., Afonso, E., Ziadinov, I., Yang, Y., Li, L., Li, T., Quere, J.P., Feng, X., Wang, Q., Wen, H., Ito, A., Craig, P.S., 2013a. Transmission ecosystems of *Echinococcus multilocularis* in China and Central Asia. Parasitology 140, 1655–1666.

Giraudoux, P., Raoul, F., Pleydell, D., Li, T., Han, X., Qiu, J., Xie, Y., Wang, H., Ito, A., Craig, P.S., 2013b. Drivers of *Echinococcus multilocularis* transmission in China: small mammal diversity, landscape or climate? PLoS Negl. Trop. Dis. 7, e2045.

Golab, E., Czarkowski, M.P., 2014. Echinococcosis and cysticercosis in Poland in 2012. Przegl. Epidemiol. 68, 279–282, 379–281.

Gómez, L., Luna, G., 1938. Primeros casos de quiste hidatídico del pulmón encontrados en Venezuela. In: Memorias del I Congreso Venezolano de la Tuberculosis, Caracas-Venezuela, pp. 573–576.

Gómez, G., Córdoba, E., Córdoba, A., 2003. Quiste hidatídico hepático. Rev. Colomb. Gastroenterol. 18, 183–186.

Gong, X.H., Tswang, R.G., Ze, Y., 2001. Clinical analysis of 709 cases of cystic echinococcosis in Xizang autonomous region. Chin. J. Parasitol. Parasit. Dis. 19, 128.

Gonlugur, U., Ozcelik, S., Gonlugur, T.E., Arici, S., Celiksoz, A., Elagoz, S., Cevit, R., 2009. The retrospective annual surgical incidence of cystic echinococcosis in Sivas, Turkey. Zoonoses Public Health 56, 209–214.

Gonzalez, L.M., Daniel-Mwambete, K., Montero, E., Rosenzvit, M.C., McManus, D.P., Garate, T., Cuesta-Bandera, C., 2002. Further molecular discrimination of Spanish strains of *Echinococcus granulosus*. Exp. Parasitol. 102, 46–56.

González Núñez, I., Díaz Jidy, M., Núñez, F.Á., González Díaz, O.M., 2001. Infección por *Echinococcus granulosus* (quiste hidatídico): Reporte de un caso. Rev. Cuba. Med. Trop. 53, 217–221.

Gori, F., Armua-Fernandez, M.T., Milanesi, P., Serafini, M., Magi, M., Deplazes, P., Macchioni, F., 2015. The occurrence of taeniids of wolves in Liguria (northern Italy). Int. J. Parasitol. Parasit. Wildl. 4, 252–255.

Gottstein, B., Stojkovic, M., Vuitton, D.A., Millon, L., Marcinkutė, A., Deplazes, P., 2015. Threat of alveolar echinococcosis to public health—a challenge for Europe. Trends Parasitol. 31, 407–412.

Graber, M., Thal, J., 1979. Anoplocephalides parasites de l'intestin et des hervibores sauvages d'Afrique Centrale. Revue d'elevage et de medecine veterinaire des pays tropicaux 32, 371–378.

Graber, M., Troncy, P., Tabo, R., Service, J., Oumatie, O., 1969. Echinococcosis-hydatidosis in Central Africa. I. Echinococcosis of domestic and wild animals. Revue d'elevage et de medecine veterinaire des pays tropicaux 22, 55–67.

Graham, J.C., Gunn, M., Hudson, M., Orr, K.E., Craig, P.S., 2002. A mass in the liver. J. Infect. 45, 121–122.

Grenouillet, F., Umhang, G., Arbez-Gindre, F., Mantion, G., Delabrousse, E., Millon, L., Boue, F., 2014. *Echinococcus ortleppi* infections in humans and cattle, France. Emerg. Infect. Dis. 20, 2100–2102.

Guanipa, N., Ehacín-Bonilla, L., Finol, F., Cano, G., 1990. Quiste hidatidico pulmonar. presentacion oe un caso. Invest. Clin. 31, 105–113.

Guarnera, E.A., Parra, A., Kamenetzky, L., Garcia, G., Gutierre, A., 2004. Cystic echinococcosis in Argentina: evolution of metacestode and clinical expression in various *Echinococcus granulosus* strains. Acta Trop. 92, 153–159.

Guberti, V., Bolognini, M., Lanfranchi, P., Battelli, G., 2004. *Echinococcus granulosus* in the wolf in Italy. Parassitologia 46, 425–427.

Gudewar, J., Pan, D., Bera, A.K., Das, S.K., Konar, A., Rao, J.R., Tiwari, A.K., Bhattacharya, D., 2009. Molecular characterization of *Echinococcus granulosus* of Indian animal isolates on the basis of nuclear and mitochondrial genotype. Mol. Biol. Rep. 36, 1381–1385.

Guerra, D., Armua-Fernandez, M.T., Silva, M., Bravo, I., Santos, N., Deplazes, P., Carvalho, L.M., 2013. Taeniid species of the Iberian wolf (*Canis lupus signatus*) in Portugal with special focus on *Echinococcus* spp. Int. J. Parasitol. Parasit. Wildl. 2, 50–53.

Guerra, D., Hegglin, D., Bacciarini, L., Schnyder, M., Deplazes, P., 2014. Stability of the southern European border of *Echinococcus multilocularis* in the Alps: evidence that *Microtus arvalis* is a limiting factor. Parasitology 141, 1–10.

Guo, Z.H., Kubo, M., Kudo, M., Nibe, K., Horii, Y., Nonaka, N., 2011. Growth and genotypes of *Echinococcus granulosus* found in cattle imported from Australia and fattened in Japan. Parasitol. Int. 60, 498–502.

Guo, L., Yang, A., Zhang, Z., Hou, W., Chen, D., Wen, H., Mao, G., Zhou, J., Wu, X., 2012. An epidemiological survey of livestock in Sichuan Province. Chin. J. Vet. Med. 48, 25–27.

Guzel, M., Yaman, M., Koltas, I., Demirkazik, M., Aktas, H., 2008. Detection of *Echinococcus granulosus* coproantigens in dogs from Antakya Province, Turkey. Helminthologia 45, 150–153.

Hailemariam, Z., Nakao, M., Menkir, S., Lavikainen, A., Yanagida, T., Okamoto, M., Ito, A., 2012. Molecular identification of unilocular hydatid cysts from domestic ungulates in Ethiopia: implications for human infections. Parasitol. Int. 61, 375–377.

Hajek, M., Novak, K., Chudacek, Z., 2004. *Echinococcus* cysts (hydatids). Rozhl. Chir. 83, 227–235.

Hajialilo, E., Harandi, M.F., Sharbatkhori, M., Mirhendi, H., Rostami, S., 2012. Genetic characterization of *Echinococcus granulosus* in camels, cattle and sheep from the south-east of Iran indicates the presence of the G3 genotype. J. Helminthol. 86, 263–270.

Halsby, K.D., Walsh, A.L., Smith, R., Said, B., Kirkbride, H., Smyth, B., Browning, L., Larkin, L., Morgan, D., 2014. The health burden of orphan zoonotic disease in the United Kingdom, 2005–2009. Zoonoses Public Health 61, 39–47.

Hama, A.A., Mero, W.M., Jubrael, J.M., 2012. Molecular characterization of *E. granulosus*, first report of sheep strain in Kurdistan-Iraq. In: 2nd International Conference on Ecological, Environmental and Biological Sciences (EEBS™ 2012), pp. 13–14.

Hama, A., Mero, W., Jubrael, J., 2014. Sero-prevalence and incidence rate of hydatid disease in Sulaimani Province/Kurdistan-Iraq. In: Cihan University, First International Scientific Conference (Cihan University), pp. 1–5.

Hama, A.A., Hassan, Z.I., Salih Mero, W.M., Interisano, M., Boufana, B., Casulli, A., 2015. A morphologically unusual *Echinococcus granulosus* (G1 genotype) cyst in a cow from Kurdistan-Iraq. Epidemiology (Sunnyvale) 2015 (5), 2.

Hamdy, I.I., Mikhail, E.G., Soliman, A.A., Hamed, H.H., 1980. A study on hydatidosis in some animals in Egypt. J. Egypt Soc. Parasitol. 10, 43–51.

Hamidi, A., Sherifi, K., Kylejmani, D., Latifi, F., Leps, J., Fries, R., 2010. Fasciolosis, dicroceliosis, cystic echinococcosis and cysticerkosis by slaughtered cattle in Kosovo [In German]. Rundsch. Fleischhyg Leb. 4, 115–117.

Hamrat, K., Achour, Y., Yacin, G., Cozma, V., 2011. Epidemiologic study of hydatidosis in the steppe regions of Djelfa. Algeria. Sci. Parasitol. 12, 177–183.

Han, X.M., Wang, H., Cai, H.X., Ma, X., Liu, Y.F., Wei, B.H., Ito, A., Craig, P.S., 2009. Epidemiological survey on echinococcosis in Darlag County of Qinghai Province. Chin. J. Parasitol. Parasit. Dis. 27, 22–26.

Han, J., Bao, G., Zhang, D., Gao, P., Wu, T., Craig, P., Giraudoux, P., Chen, X., Xin, Q., He, L., Chen, G., Jing, T., 2015. A newly discovered epidemic area of *Echinococcus multilocularis* in west Gansu province in China. PLoS One 10, e0132731.

Hane, A.A., Ndir, M., Badiane, M., Ayad, M., Kane, P.A., 1989. Pleuro-pulmonary parasitosis cases observed in Dakar. Bull. Int. Union against Tuberc. Lung Dis. 64, 31–32.

Harandi, M.F., Hobbs, R.P., Adams, P.J., Mobedi, I., Morgan-Ryan, U.M., Thompson, R.C.A., 2002. Molecular and morphological characterization of *Echinococcus granulosus* of human and animal origin in Iran. Parasitology 125, 367–373.

Harandi, M.F., Moazezi, S.S., Saba, M., Grimm, F., Kamyabi, H., Sheikhzadeh, F., Sharifi, I., Deplazes, P., 2011. Sonographical and serological survey of human cystic echinococcosis and analysis of risk factors associated with seroconversion in rural communities of Kerman, Iran. Zoonoses Public Health 58, 582–588.

Haridy, F.M., Holw, S.A., Hassan, A.A., Morsy, T.A., 2008. Cystic hydatidosis: a zoonotic silent health problem. J. Egypt Soc. Parasitol. 38, 635–644.

Hartnack, S., Budke, C.M., Craig, P.S., Jiamin, Q., Boufana, B., Campos-Ponce, M., Torgerson, P.R., 2013. Latent-class methods to evaluate diagnostics tests for *Echinococcus* infections in dogs. PLoS Negl. Trop. Dis. 7, e2068.

Hassounah, O., Behbehani, K., 1976. The epidemiology of *Echinococcus* infection in Kuwait. J. Helminthol. 50, 65–73.

Hatch, C., 1970. *Echinococcus granulosus equinus* in Irish dogs. Vet. Rec. 86, 632–633.

Hayajneh, F., Althomali, A., Nasr, A., 2014. Prevalence and characterization of hydatidosis in animals slaughtered at Al Taif abattoir, Kingdom of Saudi Arabia. Open J. Anim. Sci. 4, 38–41.

He, J.G., Qiu, J.M., Liu, F.J., Chen, X.W., Liu, D.L., Chen, W.D., 2000. Epidemiological survey on hydatidosis in Tibetan region of western Sichuan. II: infection situation among domestic and wild animals. Chin. J. Zoonoses 16, 62–65.

Hegglin, D., Deplazes, P., 2013. Control of *Echinococcus multilocularis*: strategies, feasibility and cost-benefit analyses. Int. J. Parasitol. 43, 327–337.

Hegglin, D., Bontadina, F., Deplazes, P., 2015. Human-wildlife interactions and zoonotic transmission of *Echinococcus multilocularis*. Trends Parasitol. 31, 167–173.

Henttonen, H., Fuglei, E., Gower, C.N., Haukisalmi, V., Ims, R.A., Niemimaa, J., Yoccoz, N.G., 2001. *Echinococcus multilocularis* on Svalbard: introduction of an intermediate host has enabled the local life-cycle. Parasitology 123, 547–552.

Herrador, Z., Siles-Lucas, M., Aparicio, P., Lopez-Velez, R., Gherasim, A., Garate, T., Benito, A., 2016. Cystic echinococcosis epidemiology 1 in Spain based on 2 hospitalization records 1997–2012. PLoS Negl. Trop. Dis 22 (10), e0004942. http://dx.doi.org/10.1371/journal.pntd.0004942 eCollection 2016.

Herremans, T., Verweij, J.J., Schipper, H.G., Casparie, M., van Lieshout, L., Pinelli, E., Kortbeek, T., 2010. Decline of echinococcosis in the Netherlands; 1997–2008. Ned. Tijdschr. Geneeskd. 154, A2297.

Hersteinsson, P., MacDonald, D.W., 1992. Interspecific competition and the geographical distribution of red and arctic foxes *Vulpes vulpes* and *Alopex lagopus*. Oikos 64, 505–515.

Hildreth, M.B., Sriram, S., Gottstein, B., Wilson, M., Schantz, P.M., 2000. Failure to identify alveolar echinococcosis in trappers from South Dakota in spite of high prevalence of *Echinococcus multilocularis* in wild canids. J. Parasitol. 86, 75–77.

Himsworth, C.G., Jenkins, E., Hill, J.E., Nsungu, M., Ndao, M., Andrew Thompson, R.C.A., Covacin, C., Ash, A., Wagner, B.A., McConnell, A., Leighton, F.A., Skinner, S., 2010. Emergence of sylvatic *Echinococcus granulosus* as a parasitic zoonosis of public health concern in an indigenous community in Canada. Am. J. Trop. Med. Hyg. 82, 643–645.

Hoberg, E.P., Miller, S., Brown, M.A., 1994. *Echinococcus granulosus* (Taeniidae) and autochthonous echinococcosis in a North American horse. J. Parasitol. 80, 141–144.

Hoida, G., Greenberg, Z., Furth, M., Malsha, Y., Craig, P.S., Schantz, P.M., Sneir, R., el-On, J., 1998. An epidemiological survey of *Echinococcus granulosus* and other helminths in animal populations in northern Israel. J. Helminthol. 72, 127–131.

Holmes, J.C., Podesta, R., 1968. The helminths of wolves and coyotes from the forested regions of Alberta. Can. J. Zool. 46, 1193–1204.

Holmes, J.C., Mahrt, J.L., Samuel, W.M., 1971. The occurrence of *Echinococcus multilocularis* Leuckart, 1863 in Alberta. Can. J. Zool. 49, 575–576.

Holt, D.W., Hanns, C., O'Hara, T., Burek, K., Frantz, R., 2005. New distribution records of *Echinococcus multilocularis* in the brown lemming from Barrow, Alaska, USA. J. Wildl. Dis. 41, 257–259.

Hong, S.T., Jin, Y., Anvarov, K., Khadjibaev, A., Hong, S., Ahmedov, Y., Otaboev, U., 2013. Infection status of hydatid cysts in humans and sheep in Uzbekistan. Korean J. Parasitol. 51, 383–385.

Horvath, A., Patonay, A., Banhegyi, D., Szlavik, J., Balazs, G., Gorog, D., Werling, K., 2008. The first case of human alveolar echinococcosis in Hungary. Orv. Hetil. 149, 795–799.
Hosemann, G., Schwarz, E., Lehmann, J.C., Posselt, A., 1928. Die Echinokokken Krankheit. In: Küttner, H. (Ed.), Neue Deutsche Chirurgie, 40. Ferdinand Enke Verlag, Stuttgart, Germany, pp. 103–107.
Hotez, P.J., Savioli, L., Fenwick, A., 2012. Neglected tropical diseases of the Middle East and North Africa: review of their prevalence, distribution, and opportunities for control. PLoS Negl. Trop. Dis. 6, e1475.
Huang, Y., Heath, D.D., Yang, W., Qiu, J.M., Chen, X.W., Yang, Y., Wang, Q., Li, T.Y., Xiao, Y.F., Qiu, D.C., Xiao, N., Chen, F.X., Ge, S., Se, D., 2008. Epidemiology and risk factor analysis for canine echinococcosis in a Tibetan pastoral area of Sichuan. Chin. J. Parasitol. Parasit. Dis. 26, 245–252.
Hueffer, K., Parkinson, A.J., Gerlach, R., Berner, J., 2013. Zoonotic infections in Alaska: disease prevalence, potential impact of climate change and recommended actions for earlier disease detection, research, prevention and control. Int. J. Circumpolar Health 72.
Hutchison, W.F., 1960. Studies on the hydatid worm, *Echinococcus granulosus*. II. Prevalence in Mississippi. Am. J. Trop. Med. Hyg. 9, 612–615.
Hüttner, M., Nakao, M., Wassermann, T., Siefert, L., Boomker, J.D.F., Dinkel, A., Sako, Y., Mackenstedt, U., Romig, T., Ito, A., 2008. Genetic characterization and phylogenetic position of *Echinococcus felidis* Ortlepp, 1937 (cestoda: taeniidae) from the African lion. Int. J. Parasitol. 38, 861–868.
Huttner, M., Siefert, L., Mackenstedt, U., Romig, T., 2009. A survey of Echinococcus species in wild carnivores and livestock in East Africa. Int. J. Parasitol. 39, 1269–1276.
Ibrahem, M.M., Craig, P.S., 1998. Prevalence of cystic echinococcosis in camels (*Camelus dromedarius*) in Libya. J. Helminthol. 72, 27–31.
Ibrahim, K., Thomas, R., Peter, K., Omer, R.A., 2011. A molecular survey on cystic echinococcosis in Sinnar area, Blue Nile state (Sudan). Chin. Med. J. 124, 2829–2833.
Ibrahim, M.M., 2010. Study of cystic echinococcosis in slaughtered animals in Al Baha region, Saudi Arabia: interaction between some biotic and abiotic factors. Acta Trop. 113, 26–33.
Iqbal, Z., Hayat, C., Hayat, B., Khan, M.N., 1989. Prevalence, organ distribution and economics of hydatidosis in meat animals at Faisalabad [Pakistan] abattoir. Pak. Vet. J.
Irabedra, P., Salvatella, R., 2010. El proyecto subregional cono sur de control y vigilancia de la hidatidos. Rev. Peru. Med. Exp. Salud Pública 27, 598–603.
Irabedra, P., Ferreira, C., Sayes, J., Elola, S., Rodriguez, M., Morel, N., Segura, S., Santos, E.D., Guisantes, J.A., 2016. Control programme for cystic echinococcosis in Uruguay. Mem. Inst. Oswaldo Cruz 111, 372–377.
Irshadullah, M., Nizami, W.A., Macpherson, C.N., 1989. Observations on the suitability and importance of the domestic intermediate hosts of *Echinococcus granulosus* in Uttar Pradesh, India. J. Helminthol. 63, 39–45.
Isakov, S., 1982. The focus of alveococcosis in Yakutia. Parasitology 330–332.
Islam, N., Rahman, M., 1975. Echinococcosis in Bangladesh. Trop. Geogr. Med. 27, 305–306.
Islam, A.W., 1980. *Echinococcus granulosus* in dogs in Bangladesh. Am. J. Vet. Res. 41, 415–416.
Islam, A.W., 1982. The prevalence of hydatid cysts in slaughtered cattle in Bangladesh. J. Helminthol. 56, 247–250.
Ito, A., Budke, C.M., 2015. The present situation of echinococcoses in Mongolia. J. Helminthol. 89, 680–688.

Ito, A., Romig, T., Takahashi, K., 2003. Perspective on control options for *Echinococcus multilocularis* with particular reference to Japan. Parasitology 127, S159–S172.
Ito, A., Agvaandaram, G., Bat-Ochir, O.E., Chuluunbaatar, B., Gonchigsenghe, N., Yanagida, T., Sako, Y., Myadagsuren, N., Dorjsuren, T., Nakaya, K., Nakao, M., Ishikawa, Y., Davaajav, A., Dulmaa, N., 2010. Histopathological, serological, and molecular confirmation of indigenous alveolar echinococcosis cases in Mongolia. Am. J. Trop. Med. Hyg. 82, 266–269.
Ito, A., Chuluunbaatar, G., Yanagida, T., Davaasuren, A., Sumiya, B., Asakawa, M., Ki, T., Nakaya, K., Davaajav, A., Dorjsuren, T., Nakao, M., Sako, Y., 2013. *Echinococcus* species from red foxes, corsac foxes, and wolves in Mongolia. Parasitology 140, 1648–1654.
Ito, A., Dorjsuren, T., Davaasuren, A., Yanagida, T., Sako, Y., Nakaya, K., Nakao, M., Bat-Ochir, O.E., Ayushkhuu, T., Bazarragchaa, N., Gonchigsengee, N., Li, T., Agvaandaram, G., Davaajav, A., Boldbaatar, C., Chuluunbaatar, G., 2014. Cystic echinococcoses in Mongolia: molecular identification, serology and risk factors. PLoS Negl. Trop. Dis. 8, e2937.
Jabbar, A., Narankhajid, M., Nolan, M.J., Jex, A.R., Campbell, B.E., Gasser, R.B., 2011. A first insight into the genotypes of *Echinococcus granulosus* from humans in Mongolia. Mol. Cell Probes 25, 49–54.
James, E., Boyd, W., 1937. *Echinococcus alveolaris* (With a report case). Can. Med. Assoc. J. 108.
Jarjees, M., Al-Bakri, H., 2012. Incidence of hydatidosis in slaughtered livestock at Mosul, Iraq. Iraqi J. Vet. Sci. 26, 21–25.
Jenkins, D.J., Macpherson, C.N., 2003. Transmission ecology of *Echinococcus* in wild-life in Australia and Africa. Parasitology 127 (Suppl.), S63–S72.
Jenkins, D.J., Power, K., 1996. Human hydatidosis in New South Wales and the Australian Capital Territory, 1987–1992. Med. J. Aust. 164, 18–21.
Jenkins, E.J., Schurer, J.M., Gesy, K.M., 2011. Old problems on a new playing field: helminth zoonoses transmitted among dogs, wildlife, and people in a changing northern climate. Vet. Parasitol. 182.
Jenkins, E.J., Peregrine, A.S., Hill, J.E., Somers, C., Gesy, K., Barnes, B., Gottstein, B., Polley, L., 2012. Detection of European strain of *Echinococcus multilocularis* in North America. Emerg. Infect. Dis. 18, 1010–1012.
Jenkins, E.J., Castrodale, L.J., de Rosemond, S.J.C., Dixon, B.R., Elmore, S.A., Gesy, K.M., Hoberg, E.P., Polley, L., Schurer, J.M., Simard, M., Thompson, R.C.A., 2013. Tradition and transition: parasitic zoonoses of people and animals in Alaska, Northern Canada, and Greenland. Adv. Parasitol. 82, 33–204.
Jenkins, D.J., Lievaart, J.J., Boufana, B., Lett, W.S., Bradshaw, H., Armua-Fernandez, M.T., 2014. *Echinococcus granulosus* and other intestinal helminths: current status of prevalence and management in rural dogs of eastern Australia. Aust. Vet. J. 92, 292–298.
Jiang, W., Liu, N., Zhang, G., Renqing, P., Xie, F., Li, T., Wang, Z., Wang, X., 2012. Specific detection of *Echinococcus* spp. from the Tibetan fox (*Vulpes ferrilata*) and the red fox (*V. vulpes*) using copro-DNA PCR analysis. Parasitol. Res. 111, 1531–1539.
Jiang, C., 1998. Alveolar echinococcosis in China. Chin. Med. J. 111, 470–475.
Johnson, M.R., 2001. Health aspects of gray wolf restoration. In: Maehr, D.S., Noss, R.F., Larkin, F.L. (Eds.), Ecological and sociological challenges in the 21st century. Island Press, Washington, DC, pp. 163–166.
Jordanova, D.P., Harizanov, R.N., Kaftandjiev, I.T., Rainova, I.G., Kantardjiev, T.V., 2015. Cystic echinococcosis in Bulgaria 1996–2013, with emphasis on childhood infections. Eur. J. Clin. Microbiol. Infect. Dis. 34, 1423–1428.
Jore d'Arces, P., 1953. L'échinococcose en Algérie. Bull. Off. Int. Epiz 40, 45–51.

Jorgensen, P., an der Heiden, M., Kern, P., Schoneberg, I., Krause, G., Alpers, K., 2008. Underreporting of human alveolar echinococcosis. Ger. Emerg. Infect. Dis. 14, 935–937.

Joshi, D.D., Joshi, A.B., Joshi, H., 1997. Epidemiology of echinococcosis in Nepal. Southeast Asian J. Trop. Med. Public Health 28 (Suppl. 1), 26–31.

Kabardiev, S.S., Bittirov, A.M., Gazimagomedov, M.G., Musayev, Z.G., Bittirova, A.A., 2014. Character of the distribution of alveococcosis in the foothill zone of the north Caucases in different species of murine. Successes Contemp. Sci. 12, 541–542.

Kabir, M., Eliyas, M., Hashem, M., Mohiuddin, M., Miazi, O., 2010. Prevalence of zoonotic parasitic diseases of domestic animals in different abattoir of Comilla and Brahman Baria region in Bangladesh. Univ. J. Zool. Rajshahi Univ. 28, 21–25. Available: http://www.cabdirect.org/abstracts/20123221150.html.

Kachani, M., Macpherson, C.N., Lyagoubi, M., Berrada, M., Bouslikhane, M., Kachani, F., El Hasnaoui, M., 2003. Public health education/importance and experience from the field. Educational impact of community-based ultrasound screening surveys. Acta Trop. 85, 263–269.

Kagendo, D., Magambo, J., Agola, E.L., Njenga, S.M., Zeyhle, E., Mulinge, E., Gitonga, P., Mbae, C., Muchiri, E., Wassermann, M., Kern, P., Romig, T., 2014. A survey for *Echinococcus* spp. of carnivores in six wildlife conservation areas in Kenya. Parasitol. Int. 63, 604–611.

Kairov, I., 1976. *Echinococcus multilocularis* in Karakalpakstan. Vest. Karakal Fil. Akad. Nauk. Uzb. SSR 4, 20–25.

Kamal, I., Romig, T., Kern, P., Omer, R.A., 2011. A molecular survey on cystic echinococcosis in Sinnar area, Blue Nile state (Sudan). Chin. Med. J. 124, 2829–2833.

Kamenetzky, L., Gutierrez, A.M., Canova, S.G., Haag, K.L., Guarnera, E.A., Parra, A., Garcia, G.E., Rosenzvit, M.C., 2002. Several strains of *Echinococcus granulosus* infect livestock and humans in Argentina. Infect. Genet. Evol. 2, 129–136.

Kamhawi, S., 1995. A retrospective study of human cystic echinococcosis in Jordan. Ann. Trop. Med. Parasitol. 89, 409–414.

Kamler, J.F., Ballard, W.B., 2002. A review of native and nonnative red foxes in North America. Wildl. Soc. Bull. (1973–2006) 30, 370–379.

Kandeel, A., Ahmed, E.S., Helmy, H., El Setouhy, M., Craig, P.S., Ramzy, R.M., 2004. A retrospective hospital study of human cystic echinococcosis in Egypt. East. Mediterr. Health J. 10, 349–357.

Kapel, C., Saeed, I., 2000. *Echinococcus multilocularis* – a new zoonotic parasite in Denmark. Dan. Veterinaertidsskrift 83, 14–16.

Kapel, C.M.O., Torgerson, P.R., Thompson, R.C.A., Deplazes, P., 2006. Reproductive potential of *Echinococcus multilocularis* in experimentally infected foxes, dogs, raccoon dogs and cats. Int. J. Parasitol. 36, 79–86.

Karamon, J., Sroka, J., Cencek, T., 2012. The first detection of *Echinococcus multilocularis* in slaughtered pigs in Poland. Vet. Parasitol. 185, 327–329.

Karamon, J., Kochanowski, M., Sroka, J., Cencek, T., 2014a. *Echinococcus multilocularis* in Poland – current prevalence in red foxes and first cases in pigs. In: ESCCAP Conference Echinococcus 2014. 8–9th October 2014 (Vilnius, Lithuania), p. 31.

Karamon, J., Kochanowski, M., Sroka, J., Cencek, T., Rozycki, M., Chmurzynska, E., Bilska-Zajac, E., 2014b. The prevalence of *Echinococcus multilocularis* in red foxes in Poland–current results (2009–2013). Parasitol. Res. 113, 317–322.

Karamon, J., Samorek-Pieróg, M., Kochanowski, M., Dabrowska, J., Sroka, J., Golab, E., Umhang, G., Cencek, T., 2016. First detection of *Echinococcus multilocularis* in dogs in a highly endemic area of Poland. Folia Parasitol. 63, 018.

Karim, M.F., Brunetti, E., Rahman, S., Budke, C.M., Ahsan, A.S., Al-Mahtab, M., Zaki, K.M., Alam, M.J., Akbar, S.M., Jalil, M.A., 2015. Abdominal cystic echinococcosis in Bangladesh: a hospital-based study. J. Infect. Dev. Ctries. 9, 70–75.

Kassem, H.H., Abdel-Kader, A.K., Nass, S.A., 2013. Prevalence of hydatid cysts in slaughtered animals in Sirte, Libya. J. Egypt Soc. Parasitol. 43, 33–40.

Katz, A.M., Pan, C.-T., 1958. *Echinococcus* disease in the United States. Am. J. Med. 25, 759–770.

Kedra, A.H., Tkach, V.V., Swiderski, Z., Pawlowski, Z., Emets, A., Pawlowski, J., 2000. Molecular characterisation of *Echinococcus granulosus* from a wild boar. Acta Parasitol. 45, 121–122.

Kern, P., Bardonnet, K., Renner, E., Auer, H., Pawlowski, Z., Ammann, R.W., Vuitton, D.A., 2003. European echinococcosis registry: Human alveolar echinococcosis, Europe, 1982–2000. Emerg. Infect. Dis. 9, 343–349.

Khachatryan, R., 2015. Risk Factors of Alveolar and Cystic Echinococcosis in Patients from Medical Institutions in Yerevan, Republic of Armenia. American University of Armenia, Yerevan, Armenia.

Khalifa, N.O., Khater, H.F., Fahmy, H.A., Radwan, M.E.I., Afify, J.S.A., 2014. Genotyping and phylogenetic analysis of cystic echinococcosis isolated from camels and humans in Egypt. Am. J. Epidemiol. Infect. Dis. 2, 74–82.

Khan, D., Haseeb, M.A., 1984. Hydatidosis of livestock in Pakistan. Folia Parasitol. 31, 288.

Kharchenko, V.A., Kornyushin, V.V., Varodi, E.I., Malega, O.M., 2008. Occurrence of *Echinococcus multilocularis* (Cestoda, Taeniidae) in red foxes (*Vulpes vulpes*) from Western Ukraine. Acta Parasitol. 53, 36–40.

Khurana, S., Das, A., Malla, N., 2007. Increasing trends in seroprevalence of human hydatidosis in North India: a hospital-based study. Trop. Doct. 37, 100–102.

Kia, E.B., Rahimi, H., Sharbatkhori, M., Talebi, A., Fasihi Harandi, M., Mirhendi, H., 2010. Genotype identification of human cystic echinococcosis in Isfahan, central Iran. Parasitol. Res. 107, 757–760.

Kilimcioglu, A.A., Ozkol, M., Bayindir, P., Girginkardesler, N., Ostan, I., Ok, U.Z., 2006. The value of ultrasonography alone in screening surveys of cystic echinococcosis in children in Turkey. Parasitol. Int. 55, 273–275.

Kim, S.J., Kim, J.H., Han, S.Y., Kim, Y.H., Cho, J.H., Chai, J.Y., Jeong, J.S., 2011. Recurrent hepatic alveolar echinococcosis: report of the first case in Korea with unproven infection route. Korean J. Parasit. 49, 413–418.

Kirillova, I., Kirillov, A., 2008. Ecological analysis of cestodes in rodents in the Samara bend. Povolzhskaya Ecol. J. 1, 20–28.

Kirk, K.M., 2010. Sentinels of Arctic Ecosystem Health: Polar Bear and Arctic Fox. University of Alaska. ProQuest Dissertations Publishing, Fairbanks. publication No. 345188, 204 pp.

Klein, C., Massolo, A., 2015. Demonstration that a case of human alveolar echinococcosis in Minnesota in 1977 was caused by the N2 strain. Am. J. Trop. Med. Hyg. 92, 477–478.

Klungsoyr, P., Courtright, P., Hendrikson, T.H., 1993. Hydatid disease in the Hamar of Ethiopia: a public health problem for women. Trans. R. Soc. Trop. Med. Hyg. 87, 254–255.

Knapp, J., Guislain, M.H., Bart, J.M., Raoul, F., Gottstein, B., Giraudoux, P., Piarroux, R., 2008. Genetic diversity of *Echinococcus multilocularis* on a local scale. Infect. Genet. Evol. 8, 367–373.

Knapp, J., Bart, J.M., Giraudoux, P., Glowatzki, M.L., Breyer, I., Raoul, F., Deplazes, P., Duscher, G., Martinek, K., Dubinsky, P., Guislain, M.H., Cliquet, F., Romig, T., Malczewski, A., Gottstein, B., Piarroux, R., 2009. Genetic diversity of the cestode

Echinococcus multilocularis in red foxes at a continental scale in Europe. PLoS Negl. Trop. Dis. 3, e452.

Knapp, J., Staebler, S., Bart, J.M., Stien, A., Yoccoz, N.G., Drogemuller, C., Gottstein, B., Deplazes, P., 2012. *Echinococcus multilocularis* in Svalbard, Norway: microsatellite genotyping to investigate the origin of a highly focal contamination. Infect. Genet. Evol. 12, 1270–1274.

Kokolova, L., Sofronov, V., Platonov, T., Zakharov, E., Verkhovtseva, L., Gavrilyeva, L., 2012. Epizootological situation on zoonosis and parasitic diseases of animals and fish in Yakuia. Ann. M.K. Ammosov North West. Fed. Univ. 9, 86–90.

Kokolova, L., 2007. Epizootiology and Epidemiology of Helminths of Carnivores in Yakutia. Skrabin Helminthology Institute, Yakutia Scinetific Institute of Agricultural Research.

Kolářová, L., Mateju, J., Hrdý, J., Kolářová, H., Hozáková, L., Zampachová, V., Auer, H., Stejskal, F., 2015. Human alveolar echinococcosis, Czech Republic, 2007–2014. Emerg. Infect. Dis. 21, 2263–2265.

Kolářová, L., 1999. *Echinococcus multilocularis*: new epidemiological insights in Central and Eastern Europe. Helminthologia 36, 185–191.

König, A., Romig, T., 2010. Fox tapeworm an underestimated threat – model for estimating risk of contact. Wildl. Biol. 16, 258–267.

Konyaev, S.V., Yanagida, T., Ingovatova, G.M., Shoikhet, Y.N., Nakao, M., Sako, Y., Bondarev, A.Y., Ito, A., 2012a. Molecular identification of human echinococcosis in the Altai region of Russia. Parasitol. Int. 61, 711–714.

Konyaev, S.V., Yanagida, T., Ivanov, M.V., Ruppel, V.V., Sako, Y., Nakao, M., Ito, A., 2012b. The first report on cystic echinococcosis in a cat caused by *Echinococcus granulosus sensu stricto* (G1). J. Helminthol. 86, 391–394.

Konyaev, S., Yanagida, T., Nakao, M., Ingovatova, G., Shoykhet, Y., Bondarev, A., Odnokurtsev, V., Loskutova, K.S., Lukmanova, G.I., Dokuchaev, N.E., Spiridonov, S., Alshinecky, M.V., Sivkova, T.N., Andreyanov, O.N., Abramov, S.A., Krivopalov, A.V., Karpenko, S.V., Lopatina, N.V., Dupal, T.A., Sako, Y., Ito, A., 2013. Genetic diversity of *Echinococcus* spp. In: Russia. Parasitology, vol. 140, pp. 1637–1647.

Korneev, A., Trishin, M., Solovikh, V., Krivulya, Y., Bojenova, I., 2014. Echinococcosis in Orenburg Oblast: epidemiology, immunological and taxonomic aspects. Actual Infect. 4, 46–49.

Kornyushin, V.E., Malega, A., 2011. The Helminths of wild predatory mammals of Ukraine. Cestodes. Vestnik Zoologii e-1.

Kotelnikova, L., Mukhamadieev, I., Budyanskaya, I., 2011. Parasitic diseases of the liver of the residents of Perm Krai. Actual questions about surgery, hepatology, gastointestinology and transfusion medicine. MATE 12, 12–13.

Kouidri, M., Benchaib Khoudja, F., Boulkaboul, A., Selles, M., 2012. Prevalence, fertility and viability of cystic echinococcosis in sheep and cattle of Algeria. Bulg. J. Vet. Med. 15, 191–197.

Kouidri, M., Benchaib Khoudja, F., Boulkaboul, A., Selles, S., 2013. Cystic echinococcosis in small ruminants in Tiaret (Algeria). Glob. Veterinaria 11, 753–758.

Kritsky, D.C., Leiby, P.D., 1978. Studies on sylvatic echinococcosis. V. Factors influencing prevalence of *Echinococcus multilocularis* Leuckart 1863, in red foxes from North Dakota, 1965–1972. J. Parasitol. 64, 625–634.

Kritsky, D.C., Leiby, P.D., Miller, G.E., 1977. The natural occurrence of *Echinococcus multilocularis* in the bushy-tailed woodrat, *Neotoma cinerea rupicola*, in Wyoming. Am. J. Trop. Med. Hyg. 26, 1046–1047.

Kronmann, K.C., Shields, W.W., Sheer, T.A., Crum-Cianflone, N.F., 2008. *Echinococcus* in a U.S. Marine after deployment to Afghanistan. Am. J. Gastroenterol. 103, 2151–2153.

Kumaratilake, L.M., Thompson, R.C.A., Eckert, J., 1986. *Echinococcus granulosus* of equine origin from different countries possess uniform morphological characteristics. Int. J. Parasitol. 16, 529–540.

Kurashvili, B., 1966. Alveolar echinococcosis in Georgia (USSR). In: 1st Int. Congr. Parasitology, Rome, pp. 747–750.

Kuru, B.B., Aypak, S., Aysul, N., 2013. Prevalence of *Echinococcus granulosus* determined with polymerase chain reaction in dogs in Aydin district. Turk. Parazitol. Derg. 37, 78–83.

Kuttubaev, O., Kasimbekov, B., Omorov, R., Karaeva, R., 2004. Echinococcosis in the Kyrgyz Republic. In: Echinococcosis in Central Asia: Problems and Solutions. Dauir Publishing House, Almaty, Kazakhstan, pp. 31–42.

Lahmar, S., Kilani, M., Torgerson, P.R., Gemmell, M.A., 1999. *Echinococcus granulosus* larvae in the livers of sheep in Tunisia: the effects of host age. Ann. Trop. Med. Parasitol. 93, 75–81.

Lahmar, S., Debbek, H., Zhang, L.H., McManus, D.P., Souissi, A., Chelly, S., Torgerson, P.R., 2004. Transmission dynamics of the *Echinococcus granulosus* sheep-dog strain (G1 genotype) in camels in Tunisia. Vet. Parasitol. 121, 151–156.

Lahmar, S., Chehida, F.B., Petavy, A.F., Hammou, A., Lahmar, J., Ghannay, A., Gharbi, H.A., Sarciron, M.E., 2007. Ultrasonographic screening for cystic echinococcosis in sheep in Tunisia. Vet. Parasitol. 143, 42–49.

Lahmar, S., Boufana, B.S., Lahmar, S., Inoubli, S., Guadraoui, M., Dhibi, M., Bradshaw, H., Craig, P.S., 2009a. *Echinococcus* in the wild carnivores and stray dogs of northern Tunisia: the results of a pilot survey. Ann. Trop. Med. Parasitol. 103, 323–331.

Lahmar, S., Rebai, W., Boufana, B.S., Craig, P.S., Ksantini, R., Daghfous, A., Chebbi, F., Fteriche, F., Bedioui, H., Jouini, M., Dhibi, M., Makni, A., Ayadi, M.S., Ammous, A., Kacem, M.J., Ben Safta, Z., 2009b. Cystic echinococcosis in Tunisia: analysis of hydatid cysts that have been surgically removed from patients. Ann. Trop. Med. Parasitol. 103, 593–604.

Lahmar, S., Trifi, M., Ben Naceur, S., Bouchhima, T., Lahouar, N., Lamouchi, I., Maamouri, N., Selmi, R., Dhibi, M., Torgerson, P.R., 2013. Cystic echinococcosis in slaughtered domestic ruminants from Tunisia. J. Helminthol. 87, 318–325.

Lahmar, S., Boufana, B., Ben Boubaker, S., Landolsi, F., 2014a. Intestinal helminths of golden jackals and red foxes from Tunisia. Vet. Parasitol. 204, 297–303.

Lahmar, S., Boufana, B., Jebabli, L., Craig, P.S., Ayari, H., Basti, T., Dhibi, M., Torgerson, P.R., 2014b. Modelling the transmission dynamics of cystic echinococcosis in donkeys of different ages from Tunisia. Vet. Parasitol. 205, 119–124.

Lalošević, D., Lalošević, V., Simin, V., Miljević, M., Čabrilo, B., Bjelić Čabrilo, O., 2016. Spreading of multilocular echinococcosis in Southern Europe: The first record in foxes and jackals in Serbia, Vojvodina Province. Eur. J. Wildl. Res. (in press).

Lamy, A.L., Cameron, B.H., LeBlanc, J.G., Culham, J.A., Blair, G.K., Taylor, G.P., 1993. Giant hydatid lung cysts in the Canadian northwest: outcome of conservative treatment in three children. J. Pediatr. Surg. 28, 1140–1143.

Latif, A.A., Tanveer, A., Maqbool, A., Siddiqi, N., Kyaw-Tanner, M., Traub, R.J., 2010. Morphological and molecular characterisation of *Echinococcus granulosus* in livestock and humans in Punjab, Pakistan. Vet. Parasitol. 170, 44–49.

Laurimaa, L., Suld, K., Moks, E., Valdmann, H., Umhang, G., Knapp, J., Saarma, U., 2015. First report of the zoonotic tapeworm *Echinococcus multilocularis* in raccoon dogs in Estonia, and comparisons with other countries in Europe. Vet. Parasitol. 212, 200–205.

Laurimaa, L., Moks, E., Soe, E., Valdmann, H., Saarma, U., 2016a. *Echinococcus multilocularis* and other zoonotic parasites in red foxes in Estonia. Parasitology 1–9.

Laurimaa, L., Suld, K., Davison, J., Moks, E., Valdmann, H., Saarma, U., 2016b. Alien species and their zoonotic parasites in native and introduced ranges: The raccoon dog example. Vet. Parasitol. 219, 24–33.

Lazarev, G., 2010. Chemotherapy and vaccine prophylaxis against larval cestodosis in the arid zone. Theory Pract. Parasit. Dis. Animals 11, 264–266.
Le Riche, P.D., Soe, A.K., Alemzada, Q., Sharifi, L., 1988. Parasites of dogs in Kabul, Afghanistan. Br. Vet. J. 144, 370–373.
Leiby, P.D., Kritsky, D.C., 1972. *Echinococcus multilocularis*: a possible domestic life cycle in central North America and its public health implications. J. Parasitol. 58, 1213–1215.
Leiby, P.D., Carney, W.P., Woods, C.E., 1970. Studies on sylvatic echinococcosis. 3. Host occurrence and geographic distribution of *Echinococcus multilocularis* in the north central United States. J. Parasitol. 56, 1141–1150.
Li, H.T., Song, T., Duan, X.Y., Qi, X., Feng, X.H., Wang, Y.H., Wen, H., 2013a. Prevalence of human and ovine hepatic hydatid disease diagnosed by ultrasound in Hobukesar Mongolian Autonomous County of Xinjiang. Zhonghua Liu Xing Bing Xue Za Zhi 34, 1176–1178.
Li, J.J., Chen, H.T., Wu, W.P., 2013b. Analysis of larval echinococcosis cases from the national web-based infectious diseases report system in china in 2011. Chin. J. Parasitol. Parasit. Dis. 31, 54–56, 63.
Liccioli, S., Catalano, S., Kutz, S.J., Lejeune, M., Verocai, G.G., Duignan, P.J., Fuentealba, C., Hart, M., Ruckstuhl, K.E., Massolo, A., 2012. Gastrointestinal parasites of coyotes (*Canis latrans*) in the metropolitan area of Calgary, Alberta, Canada. Can. J. Zool. 90, 1023–1030.
Liccioli, S., Duignan, P.J., Lejeune, M., Deunk, J., Majid, S., Massolo, A., 2013. A new intermediate host for *Echinococcus multilocularis*: the southern red-backed vole (*Myodes gapperi*) in urban landscape in Calgary, Canada. Parasitol. Int. 62, 355–357.
Lichtenberger, E., 1957. Equinococosis humana (quiste hidatidico): dos casos. Rev. la Fac. Med. XXV, 119–127.
Lichtenwalner, A., Adhikari, N., Kantar, L., Jenkins, E., Schurer, J., 2014. *Echinococcus granulosus* genotype G8 in maine moose (*Alces alces*). Alces 50, 27–33.
Little, S.E., Johnson, E.M., Lewis, D., Jaklitsch, R.P., Payton, M.E., Blagburn, B.L., Bowman, D.D., Moroff, S., Tams, T., Rich, L., Aucoin, D., 2009. Prevalence of intestinal parasites in pet dogs in the United States. Vet. Parasitol. 166, 144–152.
Litvinenko, O., Pavlokovska, T., 2015. Echinococcosis in human and animals in Ukraine. VetBiotechnol 26, 102–110.
Litvinenko, O., 2012. Distribution of echinococcosis in cattle in Ukraine. Sci. J. NUBiP Ukr.
Litvinenko, O., 2013. Epizootological situation of echinococcosis in pigs in the territory of the Ukraine. Sci. J. NUBiP Ukr. 188.
Litvinenko, O.P., 2015. Echinococcosis in small ruminants. Tvarinnetsvo Ukr. 7, 30–32.
Liu, I.K., Schwabe, C.W., Schantz, P.M., Allison, M.N., 1970. The occurrence of *Echinococcus granulosus* in coyotes (*Canis latrans*) in the central valley of California. J. Parasitol. 56, 1135–1137.
Liu, Q., Cao, L., Zhang, Y., Xu, D., Shang, L., Wang, X., Wei, F., Xiao, L., Ma, R., Cai, J., Zhao, Q., February 2013. Genotypes of *Echinococcus granulosus* in animals from Yushu, Northeastern China. Vector Borne Zoonotic Dis. 13 (2), 134–137. http://dx.doi.org/10.1089/vbz.2012.1050.
Logar, J., Soba, B., Lejko-Zupanc, T., Kotar, T., 2007. Human alveolar echinococcosis in Slovenia. Clin. Microbiol. Infect. 13, 544–546.
Logar, J., Soba, B., Kotar, T., 2008. Serological evidence for human cystic echinococcosis in Slovenia. BMC Infect. Dis. 8, 63.
Loveless, R.M., Andersen, F.L., Ramsay, M.J., Hedelius, R.K., 1978. *Echinococcus granulosus* in dogs and sheep in central Utah, 1971–1976. Am. J. Vet. Res. 39, 499–502.

Lukashenko, N., Zorikhina, V., 1961. Epidemiology of alveococcosis (alveolar echinococcosis) in the central district of the Barab forest-steppe of the Novosibisk Oblast. Med. Parasitol. Parasit. Dis. 2.

Lungu, V., 2013. Optimization of the Epidemiological Surveillance of Echinococcosis in the Republic of Moldova. Chisinau.

Lymbery, A.J., Jenkins, E.J., Schurer, J.M., Thompson, R.C.A., 2015. *Echinococcus canadensis, E. borealis* and *E. intermedius*. What's in a name? Trends Parasitol. 31, 23–29.

Lysenko, M.V., Sakharov, A.I., Astashov, V.L., Samorukov Iu, N., Kuchits, S.S., Semenov, D.P., 2006. Resection of the liver in focal lesions. Voen. Med. Zh 327, 20–22.

Ma, S.M., Maillard, S., Zhao, H.L., Huang, X., Wang, H., Geng, P.L., Bart, J.M., Piarroux, R., May 2008. Assessment of *Echinococcus granulosus* polymorphism in Qinghai province, People's Republic of China. Parasitol. Res. 102 (6), 1201–1206. http://dx.doi.org/10.1007/s00436-008-0894-7.

Ma, X., Liu, C., Yue, J., Cao, C., Ding, J., Mamuti, W., Ito, A., Wen, H., 2009. Genotyping of *Echinococcus granulosus* isolates using microsatellite sequence as the typing marker. Chin. J. Zoonoses 25, 846–849 (Chinese).

Ma, T., Wu, X., Marvin, F., Yun, L., 2014. Infection prevalence study of the main hosts of *Echinococcus* in Ningxia [Chinese]. Ningxia Med. J. 4, 376–378.

Ma, J., Wang, H., Lin, G., Zhao, F., Li, C., Zhang, T., Ma, X., Zhang, Y., Hou, Z., Cai, H., Liu, P., Wang, Y., 2015a. Surveillance of *Echinococcus* isolates from Qinghai, China. Vet. Parasitol. 207, 44–48.

Ma, X., Wang, H., Han, X.M., Zhang, J.X., Liu, Y.F., Zhao, Y.M., Wang, Y.S., Ma, J.Y., Liu, H.Q., Gang, J., 2015b. Survey on Echinococcosis in Maqing County of Qinghai Province. Chin. J. Parasitol. Parasit. Dis. 33, 269–272.

Maas, M., Dam-Deisz, W.D., van Roon, A.M., Takumi, K., van der Giessen, J.W., 2014. Significant increase of *Echinococcus multilocularis* prevalence in foxes, but no increased predicted risk for humans. Vet. Parasitol. 206, 167–172.

Macchioni, G., Lanfranchi, P., Abdullatif Mohamed, A., Testi, F., 1984. Echinococcosis and hydatidosis in Somalia Echinococcosi-idatidosi in Somalia. Bollettino Scientifico della Facolta di Zootecnia e Veterinaria. Univ. Naz. Somala 5, 179–189.

Macpherson, C.N., Craig, P.S., Romig, T., Zeyhle, E., Watschinger, H., 1989. Observations on human echinococcosis (hydatidosis) and evaluation of transmission factors in the Maasai of northern Tanzania. Ann. Trop. Med. Parasitol. 83, 489–497.

Magambo, J.K., Hall, C., Zeyle, E., Wachira, T.M., 1996. Prevalence of human hydatid disease in southern Sudan. Afr. J. Health Sci. 3, 154–156.

Magambo, J.K., Zeyhle, E., Wachira, T., 1998. Hydatid disease in Toposaland, Southern Sudan. Afr. J. Health Sci. 5, 129–132.

Magnino, S., Tamarozzi, F., Casulli, A., Lightowlers, M.W., 2014. Control of cystic echinococcosis in Europe. In: ESCCAP Conference Echinococcus 2014. 8–9th October 2014, Vilnius, Lithuania, p. 19.

Mahdy, O., Maogood, A., Wahab, A., El-Bahy, M., 2014. Epidemiological and molecular characterization of antigens extracted from hydatid cysts of camel, cattle and donkeys in Egypt. Int. J. Basic Appl. Sci. 3, 93–98.

Maillard, S., Benchikh-Elfegoun, M.C., Knapp, J., Bart, J.M., Koskei, P., Gottstein, B., Piarroux, R., 2007. Taxonomic position and geographical distribution of the common sheep G1 and camel G6 strains of *Echinococcus granulosus* in three African countries. Parasitol. Res. 100, 495–503.

Maillard, S., Gottstein, B., Haag, K.L., Ma, S., Colovic, I., Benchikh-Elfegoun, M.C., Knapp, J., Piarroux, R., 2009. The EmsB tandemly repeated multilocus microsatellite:

a new tool to investigate genetic diversity of *Echinococcus granulosus* sensu lato. J. Clin. Microbiol. 47, 3608–3616.

Majorowski, M.M., Carabin, H., Kilani, M., Bensalah, A., 2005. Echinococcosis in Tunisia: a cost analysis. Trans. R. Soc. Trop. Med. Hyg. 99, 268–278.

Makhiyeva, B., Shamkhalov, M., Musaev, Z., Shamkhalov, V., 2010. Distribution of echinococcosis at cattle in Dagestan. Russ. Parasitol. J. 1.

Makhiyeva, B., Shamkhalov, M., Magomedshapiyeva, G., Maksuldova, A., Shamkhalov, V., 2012. Distribution and age dynamics of echinococcosis and bovine cysticercosis in cattle in the central zone of Dagestan and Chechnya. Russ. Parasitol. J. 3, 56–61.

Malczewski, A., Ramisz, A., Rocki, B., Bienko, R., Balicka-Ramisz, A., Eckert, J., 1999. *Echinococcus multilocularis* in red foxes (*Vulpes vulpes*) in Poland: an update of the epidemiological situation. Acta Parasitol. 44, 68–72.

Malczewski, A., 2002. CE and AE in Eastern Europe. In: Proceedings of the NATO Advanced Research Workshop on CEstode Zoonoses: Echinococcosis and Cysticercosis – An emetgent and Global Problem, Poznan, Poland, pp. 81–89.

Maliki, M., Mansouri, F., Bouhamidi, B., Nabih, N., Bernoussi, Z., Mahassini, N., Elhachimi, A., 2004. Hydatidose Alvéolaire Hépatique au maroc. Méd. Trop. 64, 379–380.

Malishev, A., Sobolevskaya, O., 2002. Actual questions on the diagnosis and treatment of human echinococcosis. Vyatka Meical Bull. 1.

Manandhar, S., Horchner, F., Morakote, N., Kyule, M.N., Baumann, M.P., 2006. Occurrence of hydatidosis in slaughter buffaloes (*Bos bubalis*) and helminths in stray dogs in Kathmandu Valley, Nepal. Berl. Munch Tierarztl Wochenschr 119, 308–311.

Manfredi, M.T., Genchi, C., Deplazes, P., Trevisiol, K., Fraquelli, C., 2002. *Echinococcus multilocularis* infection in red foxes in Italy. Vet. Rec. 150, 757.

Manfredi, M.T., Di Cerbo, A.R., Zanzani, S., Moriggia, A., Fattori, D., Siboni, A., Bonazza, V., Filice, C., Brunetti, E., 2011. Prevalence of echinococcosis in humans, livestock and dogs in northern Italy. Helminthologia 48, 59–66.

Mansoorlakooraj, H., Saadati, D., Javadi, R., Heydari, S., Torki, E., Gholami, H., Fard, R.M., 2011. A survey on hydatidosis in livestock in Northern Iran based on data collected from slaughterhouses from 2004 to 2008. Vet. Parasitol. 182, 364–367.

Manterola, C., Benavente, F., Melo, A., Vial, M., Roa, J.C., 2008. Description of *Echinococcus granulosus* genotypes in human hydatidosis in a region of southern Chile. Parasitol. Int. 57, 342–346.

Maravilla, P., Thompson, A.R.C., Palacios-Ruiz, J.A., Estcourt, A., Ramirez-Solis, E., Mondragon-de-la-Pena, C., Moreno-Moller, M., Cardenas-Mejia, A., Mata-Miranda, P., Aguirre-Alcantara, M.T., Bonilla-Rodriguez, C., Flisser, A., 2004. *Echinococcus granulosus* cattle strain identification in an autochthonous case of cystic echinococcosis in central Mexico. Acta Trop. 92, 231–236.

Marcinkutė, A., Šarkūnas, M., Moks, E., Saarma, U., Jokelainen, P., Bagrade, G., Laivacuma, S., Strupas, K., Sokolovas, V., Deplazes, P., 2015. *Echinococcus* infections in the Baltic region. Vet. Parasitol. 213, 121–131.

Mardini, L.B.F., Souza, M.A.T., 1999. Programa de Controle da Hidatidose humana no estado do Rio Grande do Sul. Bol. Hidat. 39–43.

Martínek, K., Kolářová, L., Červený, J., Andreas, M., 1998. *Echinococcus multilocularis* (Cestoda: Taenidae) in the Czech Republic: the first detection of metacestodes in a naturally infected rodent. Folia Parasitol. 45, 332–333.

Martínek, K., Kolářová, L., Červený, J., 2001. *Echinococcus multilocularis* in carnivores from the Klatovy district of the Czech Republic. J. Helminthol. 75, 61–66.

Martinez, G.P., 2011. Human hydatidosis disease: general background and epidemiological situation in Chile, 2001–2009. Rev. Chilena. Infectol. 28, 585–591.

Martinez, P., 2014. Characterization of human hydatidosis mortality: Chile, 2000–2010. Rev. Chilena. Infectol. 31, 7–15.
Martínez-Maya, J.J., Zúñiga, A.I., Jaramillo, A.C.J., Cárdenas, L.J., Navarro, F.R., 1994. Caracterización epidemiológica de la equinococosis/hidatidosis en Zacatecas, México. Vet. Mex. 25, 231–237.
Martin-Hernando, M.P., Gonzalez, L.M., Ruiz-Fons, F., Garate, T., Gortazar, C., 2008. Massive presence of *Echinococcus granulosus* (Cestoda, Taeniidae) cysts in a wild boar (*Sus scrofa*) from Spain. Parasitol. Res. 103, 705–707.
Mas-Coma, S., Valero, M.A., Bargues, M.D., 2008. Effects of climate change on animal and zoonotic helminthiases. Revue Sci. Tech. 27, 443–457.
Massolo, A., Liccioli, S., Budke, C., Klein, C., 2014. *Echinococcus multilocularis* in North America: the great unknown. Parasite 21, 73.
Massolo, A., Preiksaitis, J., Klein, C., Sis, B., Houston, S., Kowalewska-Growchowska, K., 2015. Locally Acquired Alveolar Echinococcosis in an Immunocompromised Patient in Canada: Clinical Presentation and Epidemiologic Investigation. ArcticNet Annual Scientific Meeting, Vancouver, BC, p. 57.
Mastin, A., Brouwer, A., Fox, M., Craig, P., Guitian, J., Li, W., Stevens, K., 2011. Spatial and temporal investigation of *Echinococcus granulosus* coproantigen prevalence in farm dogs in South Powys, Wales. Vet. Parasitol. 178, 100–107.
Mata-Miranda, P., Osnaya-Palma, I., Rodriguez-Prado, U., Gutierrez-Marin, A., Tawil, M., Hernandez-Gonzalez, S., Solano-Ceh, M., Villalvaso, L., Martinez-Maya, J.J., Maravilla, P., Garcia-de-la-Torre, G., Flisser, A., 2007. Epidemiologic and ultrasonographic study of echinococcosis in a community in the state of Mexico. Am. J. Trop. Med. Hyg. 77, 500–503.
Mathers, C.D., Ezzati, M., Lopez, A.D., 2007. Measuring the burden of neglected tropical diseases: the global burden of disease framework. PLoS Negl. Trop. Dis. 1, e114.
Matossian, R.M., Rickard, M.D., Smyth, J.D., 1977. Hydatidosis: a global problem of increasing importance. Bull. World Health Organ. 55, 499–507.
Maurelli, M.P., Rinaldi, L., Pepe, P., Ianniello, D., Amadesi, A., Cringoli, G., 2015. Innovative diagnostic tools to control cystic echinococcosis. In: 9th European Congress on Tropical Medicine and International Health, Basel, Switzerland, p. 48.
Mauti, S., Traore, A., Crump, L., Zinsstag, J., Grimm, F., 2016. First report of *Echinococcus granulosus* (genotype G6) in a dog in Bamako, Mali. Vet. Parasitol. 217, 61–63.
Mazur, O., Fomina, A., 2012. Composition and infection of intermediate and definitive hosts of *Echinococcus multilocularis* (Leuckart, 1858) in basin of lake Baikal (Buryatia). Bull. East Sib. Sci. Centre SO RAMN 5-1, 262–264.
Mazyad, S.A., Mahmoud, L.H., Hegazy, M.M., 2007. *Echinococcosis granulosus* in stray dogs and Echino-IHAT in the hunters in Cairo. Egypt. J. Egypt Soc. Parasitol. 37, 523–532.
Mbaya, H., Magambo, J., Njenga, S., Zeyhle, E., Mbae, C., Mulinge, E., Wassermann, M., Kern, P., Romig, T., 2014. Echinococcus spp. in central Kenya: a different story. Parasitol. Res. 113, 3789–3794.
McManus, D.P., Zhang, L., Castrodale, L.J., Le, T.H., Pearson, M., Blair, D., 2002. Short report: molecular genetic characterization of an unusually severe case of hydatid disease in Alaska caused by the cervid strain of *Echinococcus granulosus*. Am. J. Trop. Med. Hyg. 67, 296–298.
McManus, D.P., 2010. Echinococcosis with particular reference to Southeast Asia. Adv. Parasitol. 72, 267–303.
Meerkhan, A., Abdullah, A., 2012. The epidemiology of hydatidosis in different slaughtered animals in Duhok abattoir, Kurdistan Region of Iraq. In: Second International Conference (Bali (Indonesia)), pp. 45–48.
Mehrabani, D., Oryan, A., Sadjjadi, S.M., 1999. Prevalence of *Echinococcus granulosus* infection in stray dogs and herbivores in Shiraz, Iran. Vet. Parasitol. 86, 217–220.

Melotti, J.R., Muzzall, P.M., O'Brien, D.J., Cooley, T.M., Tsao, J.I., 2015. Low prevalence of *Echinococcus multilocularis* in Michigan, U.S.A.: A survey of coyotes (*Canis latrans*), Red Foxes (*Vulpes vulpes*), and Gray Foxes (*Urocyon cinereoargenteus*), 2009–2012. Comp. Parasitol. 82, 285–290.

Merdivenci, A., 1963. Türkiye'de tilki (*Vulpes vulpes*) lerde ilk helmintolojik araştırma ve ilk *Echinococcus multilocularis* (Leuckart, 1864), Vogel, 1935 olayı. Türk Veteriner Hekim. Derneği Derg. 33, 290–296.

Miller, A.L., Olsson, G.E., Walburg, M.R., Sollenberg, S., Skarin, M., Ley, C., Wahlstrom, H., Hoglund, J., 2016. First identification of *Echinococcus multilocularis* in rodent intermediate hosts in Sweden. Int. J. Parasitol. Parasit. Wildl. 5, 56–63.

Miller, M.J., 1953. Hydatid infection in Canada. Can. Med. Assoc. J. 68, 423–434.

Miman, O., Yazar, S., 2012. Alveolar echinococcosis in Turkey: in the light of the literature. Türkiye Parazitol. Derg. 36, 116–120.

Mirzaei, M., Rezaei, H., Nematollahi, A., 2015. Role of ruminants in the epidemiology of *Echinococcus granulosus* in Tabriz area, Northwest of Iran. Trop. Biomed. 32, 269–275.

Mitchell, J.R., 1977. An abattoir survey of helminths in cattle in Swaziland. J. South Afr. Vet. Assoc. 48, 53–54.

Miterpáková, M., Dubinský, P., 2011. Fox tapeworm (*Echinococcus multilocularis*) in Slovakia – summarizing long-term monitoring. Helminthologia 48, 155–161.

Miterpáková, M., Antolová, D., Ševčíková, Z., Stanko, M., Dinkel, A., Gašpar, V., Dubinský, P., 2006. *Echinococcus multilocularis* in musk rat (*Ondatra zibethicus*): the first finding of the parasite in naturally infected rodent in the Slovak Republic. Helminthologia 43, 76–80.

Mitrea, I.L., Ionita, M., Costin II, Predoi, G., Avram, E., Rinaldi, L., Maurelli, M.P., Cringoli, G., Genchi, C., 2014. Occurrence and genetic characterization of *Echinococcus granulosus* in naturally infected adult sheep and cattle in Romania. Vet. Parasitol. 206, 159–166.

Mobedi, I., Sadighian, A., 1971. *Echinococcus multilocularis* Leuckart, 1863, in red foxes, *Vulpes vulpes* Linn., in Moghan, Azerbaijan Province, northwest of Iran. J. Parasitol. 57, 493.

Mogoye, B.K., Menezes, C.N., Wong, M.L., Stacey, S., von Delft, D., Wahlers, K., Wassermann, M., Romig, T., Kern, P., Grobusch, M.P., Frean, J., 2013. First insights into species and genotypes of *Echinococcus* in South Africa. Vet. Parasitol. 196, 427–432.

Mohamed, R.M., Abdel-Hafeez, E.H., Belal, U.S., Norose, K., Aosai, F., 2014. Human cystic echinococcosis in the Nalut District of Western Libya: A Clinico-epidemiological study. Trop. Med. Health 42, 177–184.

Moks, E., Jogisalu, I., Valdmann, H., Saarma, U., 2008. First report of *Echinococcus granulosus* G8 in Eurasia and a reappraisal of the phylogenetic relationships of 'genotypes' G5-G10. Parasitology 135, 647–654.

Molan, A.L., Baban, M.R., 1992. The prevalence of *Echinococcus granulosus* in stray dogs in Iraq. J. Trop. Med. Hyg. 95, 146–148.

Molan, A.L., 1993. Epidemiology of hydatidosis and echinococcosis in Theqar Province, southern Iraq. Jpn. J. Med. Sci. Biol. 46, 29–35.

Monteiro, D.U., Botton, S.A., Tonin, A.A., Azevedo, M.I., Graichen, D.A., Noal, C.B., de la Rue, M.L., 2014. *Echinococcus canadensis* (G7) and *Echinococcus granulosus* sensu stricto (G1) in swine of southern Brazil. Vet. Parasitol. 202, 335–338.

Moral, I., 2010. In: Ministerio de Salud, G.D.A. (Ed.), Hidatidosis, dagnostico para el equipo de salud (Buenos Aires, Argentina).

Moro, P., Schantz, P.M., 2009. Echinococcosis: a review. Int. J. Infect. Dis. 13, 125–133.

Moro, P.L., McDonald, J., Gilman, R.H., Silva, B., Verastegui, M., Malqui, V., Lescano, G., Falcon, N., Montes, G., Bazalar, H., 1997. Epidemiology of *Echinococcus granulosus* infection in the central Peruvian Andes. Bull. World Health Organ. 75, 553–561.

Moro, P.L., Bonifacio, N., Gilman, R.H., Lopera, L., Silva, B., Takumoto, R., Verastegui, M., Cabrera, L., 1999. Field diagnosis of *Echinococcus granulosus* infection among intermediate and definitive hosts in an endemic focus of human cystic echinococcosis. Trans. R. Soc. Trop. Med. Hyg. 93, 611–615.
Moro, P.L., Garcia, H.H., Gonzales, A.E., Bonilla, J.J., Verastegui, M., Gilman, R.H., 2005. Screening for cystic echinococcosis in an endemic region of Peru using portable ultrasonography and the enzyme-linked immunoelectrotransfer blot (EITB) assay. Parasitol. Res. 96, 242–246.
Moro, P.L., Nakao, M., Ito, A., Schantz, P.M., Cavero, C., Cabrera, L., 2009. Molecular identification of *Echinococcus* isolates from Peru. Parasitol. Int. 58, 184–186.
Morovic, M., 1997. Human hydatidosis in Dalmatia, Croatia. Epidemiol. Infect 119, 271–276.
M'rad, S., Filisetti, D., Oudni, M., Mekki, M., Belguith, M., Nouri, A., Sayadi, T., Lahmar, S., Candolfi, E., Azaiez, R., Mezhoud, H., Babba, H., 2005. Molecular evidence of ovine (G1) and camel (G6) strains of *Echinococcus granulosus* in Tunisia and putative role of cattle in human contamination. Vet. Parasitol. 129, 267–272.
M'rad, S., Oudni-M'rad, M., Filisetti, D., Mekki, M., Nouri, A., Sayadi, T., Candolfi, E., Azaiez, R., Mezhoud, H., Babba, H., 2010. Molecular identification of *Echinococcus granulosus* in Tunisia: First record of the buffalo strain (G3) in human and bovine in the country. The Open Vet. Sci. J 4, 27–30.
M'rad, S., Oudni-M'rad, M., Boubaker, G., Bouazzi, L., Gorcii, M., Nouri, A., Mezhoud, H., Babba, H., 2012. Retrospective study of the distribution and the fertility of hydatid cysts in the child in Tunisia. Pathologie-Biologie 60, 166–169.
Muhtarov, M., 2014. First portable ultrasound based screening study in Bulgaria on the prevalence of cystic echinococcosis in Kardzhali District. TJS 12 (Suppl. 1), 170–174.
Mukbel, R.M., Torgerson, P.R., Abo-Shehada, M.N., 2000. Prevalence of hydatidosis among donkeys in northern Jordan. Vet. Parasitol. 88, 35–42.
Muminov, A., Mahmadov, F., Shodmonov, I., 2004. Echinococcosis of people and animals of Tadjikistan. In: Echinococcosis in Central Asia: Problems and Solutions. Dauir Publishing House, Almaty, Kazakhstan, pp. 20–24.
Mwambete, K.D., Ponce-Gordo, F., Cuesta-Bandera, C., 2004. Genetic identification and host range of the Spanish strains of *Echinococcus granulosus*. Acta Trop. 91, 87–93.
Nabokov, S., Vasiliev, R., 1978. Clinical picture and classification of alveococcosis. Vestn. Khir Im. I I Grek. 120, 31–35.
Nahmias, J., Goldsmith, R., Schantz, P., Siman, M., el-On, J., 1991. High prevalence of human hydatid disease (echinococcosis) in communities in northern Israel: epidemiologic studies in the town of Yirka. Acta Trop. 50, 1–10.
Nahorski, W.L., Knap, J.P., Pawlowski, Z.S., Krawczyk, M., Polanski, J., Stefaniak, J., Patkowski, W., Szostakowska, B., Pietkiewicz, H., Grzeszczuk, A., Felczak-Korzybska, I., Golab, E., Wnukowska, N., Paul, M., Kacprzak, E., Sokolewicz-Bobrowska, E., Niscigorska-Olsen, J., Czyrznikowska, A., Chomicz, L., Cielecka, D., Myjak, P., 2013. Human alveolar echinococcosis in Poland: 1990–2011. PLoS Negl. Trop. Dis. 7, e1986.
Nakamura, K., Ito, A., Yara, S., Haranaga, S., Hibiya, K., Hirayasu, T., Sako, Y., Fujita, J., 2011. A case of pulmonary and hepatic cystic Echinococcosis of CE1 stage in a healthy Japanese female that was suspected to have been acquired during her stay in the United Kingdom. Am. J. Trop. Med. Hyg. 85, 456–459.
Nakao, M., Xiao, N., Okamoto, M., Yanagida, T., Sako, Y., Ito, A., 2009. Geographic pattern of genetic variation in the fox tapeworm *Echinococcus multilocularis*. Parasitol. Int. 58, 384–389.
Nasrieh, M.A., Abdel-Hafez, S.K., Kamhawi, S.A., Craig, P.S., Schantz, P.M., 2003. Cystic echinococcosis in Jordan: socioeconomic evaluation and risk factors. Parasitol. Res. 90, 456–466.

Naughton, D., 2012. The Natural History of Canadian Mammals. Canadian Museum of Nature and University of Toronto Press, 784 p.
Nazirov, F.G., Ilkhamov, I.L., Ambekov, N.C., 2002. Echinococcosis in Uzbekistan: types of problems and methods to improve treatment [article in Russian]. Med. J. Uzb. 2, 2–5.
Neghina, R., Neghina, A.M., Marincu, I., Iacobiciu, I., 2010. Epidemiology and epizootology of cystic echinococcosis in Romania 1862–2007. Foodborne Pathog. Dis. 7, 613–618.
Niu, M., An, Y., 2012. The epidemiological investigation and prevention measures of echinococcosis in Sunan County, Gansu Province [Chinese]. Chin. J. Clin. Ration. Drug Use 5, 45–48.
Nusilaiti, M., Yan Hao, Z., 2016. Echinococcosis (hydatid disease) infection in livestock in 14 areas in Xinjiang, 2014. Chin. J. Zoonoses 32, 202–205.
Obwaller, A., Schneider, R., Walochnik, J., Gollackner, B., Deutz, A., Janitschke, K., Aspock, H., Auer, H., 2004. *Echinococcus granulosus* strain differentiation based on sequence heterogeneity in mitochondrial genes of cytochrome c oxidase-1 and NADH dehydrogenase-1. Parasitology 128, 569–575.
Oguz, B., Deger, S., 2013. Cystic echinococcosis and cysticerci of *Taenia hydatigena* in cattle and sheep slaughtered in a Van Local Slaughterhouse. Turk. Parazitol. Derg. 37, 186–189.
O'Hern, J.A., Cooley, L., 2013. A description of human hydatid disease in Tasmania in the post-eradication era. Med. J. Aust. 199, 117–120.
Ok, U.Z., Ozkol, M., Kilimcioglu, A.A., Dinc, G., Bayindir, P., Ostan, I., Pabuscu, Y., Ozcan, C., Korkmaz, M., Coskun, S., Yuksel, H., Girginkardesler, N., 2007. A province-based study using sampling method to investigate the prevalence of cystic echinococcosis among primary school children in Manisa, Turkey. Acta Trop. 103, 116–122.
Okolo, M.I., 1986. Prevalence and public health implications of *Echinococcus granulosus* in rural dogs in Eastern Nigeria. Int. J. Zoonoses 13, 19–24.
Oksanen, A., Lavikainen, A., 2015. *Echinococcus canadensis* transmission in the North. Vet. Parasitol. 213, 182–186.
Omar, M., Sultan, K., Haridy, M., Omran, A., 2013. Prevalence of cystic echinococcosis in slaughtered ruminants in different abattoirs, Upper Egypt. Am. J. Animal Vet. Sci. 8, 117–121.
Omer, R.A., Dinkel, A., Romig, T., Mackenstedt, U., Elnahas, A.A., Aradaib, I.E., Ahmed, M.E., Elmalik, K.H., Adam, A., 2010. A molecular survey of cystic echinococcosis in Sudan. Vet. Parasitol. 169, 340–346.
Omer, R.A., Daugschies, A., Romig, T., 2011. Cystic echinococcosis in Sudan and South Sudan: research history of a neglected zoonosis. Berl. Munch Tierarztl Wochenschr 124, 521–527.
Onac, D., Gyorke, A., Oltean, M., Gavrea, R., Cozma, V., 2013. First detection of *Echinococcus granulosus* G1 and G7 in wild boars (*Sus scrofa*) and red deer (*Cervus elaphus*) in Romania using PCR and PCR-RFLP techniques. Vet. Parasitol. 193, 289–291.
Orea-Martinez, J.G., Perez-Corro, M.A., Contreras-Vera, R.A., Breton-Marquez, J.H., 2013. A case of hydatid cyst caused by *Echinococcus granulosus* in Puebla, Mexico, that resulted in successful surgical treatment. Rev. Gastroenterol. Mex. 78, 116–119.
Ortblad, K.F., Lozano, R., Murray, C.J., 2013. The burden of HIV: insights from the Global Burden of Disease Study 2010. AIDS 27, 2003–2017.
Ortlepp, J.R., 1937. South African Helminths—Part I. Onderstepoort J. Vet. Sci. Anim. Ind. 9, 311–336.
Osman, Y., Hou, Y.Y., Zhao, J.S., Mamtimin, Y., 2014. Retrospective analysis of echinococcosis surgerical cases in Xinjiang from 2005 to 2013. Chin. J. Parasitol. Parasit. Dis. 32, 334–338.

Otero-Abad, B., Torgerson, P.R., 2013. A systematic review of the epidemiology of echinococcosis in domestic and wild animals. PLoS Negl. Trop. Dis. 7, e2249.

Ouchène, N., Bitman, I., Zéroual, F., Ouchène-Khélifi, N., 2014. Cystic echinococcosis in wild boars (*Sus scrofa*) and slaughtered domestic ruminants in Algeria. Asian J. Anim. Vet. Adv. 9, 767–774.

Oudni-M'rad, M., M'rad, S., Ksia, A., Lamiri, R., Mekki, M., Nouri, A., Mezhoud, H., Babba, H., 2016. First molecular evidence of the simultaneous human infection with two species of *Echinococcus granulosus* sensu lato: *Echinococcus granulosus* sensu stricto and *Echinococcus canadensis*. Parasitol. Res. 115, 1065–1069.

Owor, R., Bitakaramire, P.K., 1975. Hydatid disease in Uganda. East Afr. Med. J. 52, 700–704.

PAHO, 2015. Informe Epidemiológico en la Región de América del Sur – 2009–2014. In: Echinococcosis. Panamerican Health Organization.

Palmer, S.R., Biffin, A.H., 1987. The changing incidence of human hydatid disease in England and Wales. Epidemiol. Infect 99, 693–700.

Palmer, S.R., Biffin, A.H., Craig, P.S., Walters, T.M., 1996. Control of hydatid disease in Wales. BMJ 312, 674–675.

Palmer, C.S., Traub, R.J., Robertson, I.D., Devlin, G., Rees, R., Thompson, R.C.A., 2008. Determining the zoonotic significance of *Giardia* and *Cryptosporidium* in Australian dogs and cats. Vet. Parasitol. 154, 142–147.

Palmieri, J.R., Schantz, P., Masbar, S., Purnomo, B., 1984. Absence of *Echinococcus granulosus* in dogs from Central Sulawesi (Celebes) Indonesia. J. Parasitol. 70, 180–181.

Pantieleiev, V., Nartailakov, M., Mustafin, A., Gabdrakhimov, S., Musharapov, D., 2011. Lazer treatment of the liver for hepatic alveolar echinococcosis. Med. Rep. Bashkorstan 6, 46–49.

Pappaioanou, M., Schwabe, C.W., Sard, D.M., 1977. An evolving pattern of human hydatid disease transmission in the United States. Am. J. Trop. Med. Hyg. 26, 732–742.

Parsa, F., Fasihi Harandi, M., Rostami, S., Sharbatkhori, M., 2012. Genotyping *Echinococcus granulosus* from dogs from Western Iran. Exp. Parasitol. 132, 308–312.

Pednekar, R.P., Gatne, M.L., Thompson, R.C.A., Traub, R.J., 2009. Molecular and morphological characterisation of *Echinococcus* from food producing animals in India. Vet. Parasitol. 165, 58–65.

Peregrine, A.S., Jenkins, E.J., Barnes, B., Johson, S., Polley, L., Barker, I.K., 2012. Alveolar hydatid disease (*Echinococcus multilocularis*) in the liver of a Canadian dog in British Columbia, a newly endemic region. Can. Vet. J. 53.

Peregrine, A.S., 2015. Alveolar echinococcosis in dogs: an emerging issue? Vet. Rec. 177, 567.

Perez Fontana, V., 1949. Boletin de la Oficina Sanitaria Panamericana. Pan Am. Sanit. Bureau 28, 124–156.

Pezeshki, A., Akhlaghi, L., Sharbatkhori, M., Razmjou, E., Oormazdi, H., Mohebali, M., Meamar, A.R., 2013. Genotyping of *Echinococcus granulosus* from domestic animals and humans from Ardabil Province, northwest Iran. J. Helminthol. 87, 387–391.

Piarroux, M., Piarroux, R., Giorgi, R., Knapp, J., Bardonnet, K., Sudre, B., Watelet, J., Dumortier, J., Gerard, A., Beytout, J., Abergel, A., Mantion, G., Vuitton, D.A., Bresson-Hadni, S., 2011. Clinical features and evolution of alveolar echinococcosis in France from 1982 to 2007: results of a survey in 387 patients. J. Hepatol. 55, 1025–1033.

Piarroux, M., Piarroux, R., Knapp, J., Bardonnet, K., Dumortier, J., Watelet, J., Gerard, A., Beytout, J., Abergel, A., Bresson-Hadni, S., Gaudart, J., 2013. Populations at risk for alveolar echinococcosis, France. Emerg. Infect. Dis. 19, 721–728.

Piarroux, M., Gaudart, J., Bresson-Hadni, S., Bardonnet, K., Faucher, B., Grenouillet, F., Knapp, J., Dumortier, J., Watelet, J., Gerard, A., Beytout, J., Abergel, A., Wallon, M., Vuitton, D.A., Piarroux, R., 2015. Landscape and climatic characteristics

associated with human alveolar echinococcosis in France, 1982 to 2007. Euro Surveill. 20 pii:21118.
Piccoli, L., Bazzocchi, C., Brunetti, E., Mihailescu, P., Bandi, C., Mastalier, B., Cordos, I., Beuran, M., Popa, L.G., Meroni, V., Genco, F., Cretu, C., 2013. Molecular characterization of *Echinococcus granulosus* in south-eastern Romania: evidence of G1-G3 and G6-G10 complexes in humans. Clin. Microbiol. Infect. 19, 578–582.
Pierangeli, N.B., Soriano, S.V., Roccia, I., Bergagna, H.F., Lazzarini, L.E., Celescinco, A., Kossman, A.V., Saiz, M.S., Basualdo, J.A., 2010. Usefulness and validation of a coproantigen test for dog echinococcosis screening in the consolidation phase of hydatid control in Neuquen, Argentina. Parasitol. Int. 59, 394–399.
Pijáček, M., 2011. The occurrence of the tapeworms of *Echinococcus* spp. in animals in the Czech Republic. Spec. Seminar Infect. Echinococcosis 7–8.
Pilaca, A., Vyshka, G., Pepa, A., Shytaj, K., Shtjefni, V., Bocari, A., Beqiri, A., Kraja, D., 2014. A Neglected Zoonosis in Albania: why echinococcosis is becoming a surgeon's exclusivity? Mediterr. J. Hematol. Infect. Dis. 6, e2014013.
Plesker, R., Nadler, T., Dinkel, A., Romig, T., 2009. A case of an *Echinococcus ortleppi* infestation in a red-shanked douc langur (*Pygathrix nemaeus*) in northern Vietnam. Vietnam. J. Primatol. 3, 75–81.
Pliyeva, A., Uspenskii, A., 2006. Epizootiology processes of echinococcosis in carnivores and agricultural animals. Vet. Pathol. 2, 120–122.
Polydorou, K., 1983. Stray-Dog Control in Cyprus – Primitive and Humane Methods. Int. J. Stud. Anim. Probl. 4, 146–151.
Pomamarev, N., Kostykov, M., 2012. The Study of Helminths of Wild Foxes of the Altay District, vol. 1. Bulletin of the Altay State Agrarian University, pp. 57–59.
Pomamarev, N., Tikhaya, N., Kostykov, M., Nekrasov, V., 2011. Helminth Fauna of Wild Carnivores in Different Ecological Zones of the Altay District, vol. 5. Bulletin of the Altay State Agrarian University, pp. 64–67.
Postnova, V., Shendo, G., Bazeltseva, L., Slavina, A., Postnov, A., Kruglova, N., 2010. Seroepidemiological assessment of the situation of echinococcosis in Astrakhan region 2007–2009. Theory Pract. Parasit. Dis. Animals 11, 365–367.
Pour, A.A., Hosseini, S.H., Shayan, P., 2011. Comparative genotyping of *Echinococcus granulosus* infecting buffalo in Iran using cox1 gene. Parasitol. Res. 108, 1229–1234.
Rafiei, A., Hemadi, A., Maraghi, S., Kaikhaei, B., Craig, P.S., 2007. Human cystic echinococcosis in nomads of south-west Islamic Republic of Iran. East. Mediterr. Health J. 13, 41–48.
Raimkylov, K.M., Kuttubaev, O.T., Toigombaeva, V.S., 2015. Epidemiological analysis of the distribution of cystic and alveolar echinococcosis in Osh Oblast in the Kyrgyz Republic, 2000–2013. J. Helminthol. 89, 651–654.
Rainova, I., Marinova, I., Harizanov, R., Jordanova, D., Kaftandjiev, I., Bikov, I., Tsvetkova, N., 2014. Parasitic diseases in Bulgaria in 2012. Probl. Inf. Parasit. Dis. 42, 29–38.
Raisolsadat, M., 2010. Alveolar hydatid cyst: A case report. Zahedan Med. Res. J. 12, 47–50.
Rajabloo, M., Hosseini, S.H., Jalousian, F., 2012. Morphological and molecular characterisation of *Echinococcus granulosus* from goat isolates in Iran. Acta Trop. 123, 67–71.
Rajkovic-Janje, R., Bosnic, S., Rimac, D., Dragicevic, P., Vinkovic, B., 2002. Prevalence of helminths in wild boar from hunting grounds in eastern Croatia. Z. für Jagdwiss. 48, 261–270.
Raoul, F., Quéré, J.P., Rieffel, D., Bernard, N., Takahashi, K., Scheifler, R., Wang, Q., Qiu, J., Yang, W., Craig, P.S., Ito, A., Giraudoux, P., 2006. Distribution of small mammals in a pastoral landscape of the Tibetan plateaus (Western Sichuan, China) and relationship with grazing practices. Mammalia 3-4, 214–225.

Raoul, F., Hegglin, D., Giraudoux, P., 2015. Trophic ecology, behaviour and host population dynamics in *Echinococcus multilocularis* transmission. Vet. Parasitol. 213, 1262, 171.

Rausch, R.L., Fay, F.H., 2002. Epidemiology of alveolar Echinococcosis, with reference to St Lawrence Island, Bering Sea. In: Craig, P., Pawlowski, Z. (Eds.), Cestode Zoonoses: Echinococcosis and Cysticercosis – An emergent and global problem. IOS Press, pp. 309–325.

Rausch, R.L., Richards, S.H., 1971. Observations on parasite-host relationships of *Echinococcus multilocularis* Leuckart, 1863, in North Dakota. Can. J. Zool. 49, 1317–1330.

Rausch, R., Schiller, E.L., 1951. Hydatid disease (Echinococcosis) in Alaska and the importance of rodent intermediate hosts. Science 113, 57–58.

Rausch, R., Schiller, E.L., 1956. Studies on the helminth fauna of Alaska. XXV. The ecology and public health significance of *Echinococcus sibiricensis* Rausch & Schiller, 1954, on St. Lawrence Island. Parasitology 46, 395–419.

Rausch, R., Williamson, F.S.L., 1959. Studies on the Helminth Fauna of Alaska. The Parasites of Wolves, *Canis lupus L.* J. Parasitol. 45, 395–403.

Rausch, R.L., Fay, F.H., Williamson, F.S., 1990. The ecology of *Echinococcus multilocularis* (Cestoda: Taeniidae) on St. Lawrence Island, Alaska. II. Helminth populations in the definitive host. Ann. Parasitol. Hum. Comp. 65, 131–140.

Rausch, R.A., 1959. Notes on the prevalence of hydatid disease in Alaskan Moose. J. Wildl. Manage. 23, 122.

Rausch, R.L., 1960. Recent studies on hydatid disease in Alaska. Parassitologia 2, 391–398.

Rausch, R.L., 1967. On the ecology and distribution of *Echinococcus* spp. (Cestoda: Taeniidae), and characteristics of their development in the intermediate host. Ann. Parasitol. Hum. Comp. 42, 19–63.

Rausch, R.L., 1995. Life cycle patterns and geographic distribution of *Echinococcus* species. In: Thompson, R.C.A., Lymbery, A.J. (Eds.), *Echinococcus* and Hydatid Disease. Cab International, Wallingford, pp. 89–134.

Rausch, R.L., 2003. Cystic echinococcosis in the Arctic and Sub-Arctic. Parasitology 127 (Suppl.), S73–S85.

Razikov, S., Adilova, M., 2011. Prevalence of cestodes in dogs before going to the summer pastures of sheep. Theory Pract. Parasit. Dis. Animals 12, 416–417.

Razmi, G., Sardari, K., Kamrani, A., 2006. Prevalence of *Echinococcus granulosus* and other intestinal helminths of stray dogs in Mashhad area. Iran. Archives Razi 61, 143–148.

Reyes, M.M., Taramona, C.P., Saire-Mendoza, M., Gavidia, C.M., Barron, E., Boufana, B., Craig, P.S., Tello, L., Garcia, H.H., Santivanez, S.J., 2012. Human and canine echinococcosis infection in informal, unlicensed abattoirs in Lima, Peru. PLoS Negl. Trop. Dis. 6, e1462.

Richter, J., Orhun, A., Grüner, B., Müller-Stöver, I., Reuter, S., Romig, T., Häussinger, D., Kern, P., 2009. Autochthonous cystic echinococcosis in patients who grew up in Germany. Euro Surveill. 14, 39–45.

Riley, W.A., 1933. Reservoirs of *Echinococcus* in Minnesota. Minn. Med. 16, 744–745.

Rinaldi, L., Maurelli, M.P., Capuano, F., Perugini, A.G., Veneziano, V., Cringoli, S., 2008a. Molecular update on cystic echinococcosis in cattle and water buffaloes of southern Italy. Zoonoses Public Health 55, 119–123.

Rinaldi, L., Maurelli, M.P., Veneziano, V., Capuano, F., Perugini, A.G., Cringoli, S., 2008b. The role of cattle in the epidemiology of *Echinococcus granulosus* in an endemic area of southern Italy. Parasitol. Res. 103, 175–179.

Risse, G.B., 2005. Glimpses of a hidden burden: hydatid disease in eighteenth-century Scotland. Bull. Hist. Med. 79, 534–543.

Rodas, J.D., Pizurno, O., Rodríguez, A., 2011. Quiste hidatídico hepático– Puesta al día y presentación de un caso clínico autóctono. Tendencias Med.

Rodriguez-Prado, U., Jimenez-Gonzalez, D.E., Avila, G., Gonzalez, A.E., Martinez-Flores, W.A., Mondragon de la Pena, C., Hernandez-Castro, R., Romero-Valdovinos, M., Flisser, A., Martinez-Hernandez, F., Maravilla, P., Martinez-Maya, J.J., 2014. Short report: Genetic variation of *Echinococcus canadensis* (G7) in Mexico. Am. J. Trop. Med. Hyg. 91, 1149–1153.

Roinioti, E., Papathanassopoulou, A., Theodoropoulou, I., Simsek, S., Theodoropoulos, G., 2016. Molecular identification of *Echinococcus granulosus* isolates from ruminants in Greece. Vet. Parasitol. 226, 138–144.

Rokni, M.B., 2009. Echinococcosis/hydatidosis in Iran. Iran. J. Parasitol. 4, 1–16.

Romano, M.N., Brunetti, O.A., Schwabe, C.W., Rosen, M.N., 1974. Probable transmission of *Echinococcus granulosus* between deer and coyotes in California. J. Wildl. Dis. 10, 225–227.

Romanov, E., Mishonkova, A., 2010. Geoinformation Prediction Centers Parasitic Infestation *Sus scrofa domestica* (Linnaeus, 1758) in the Ulyanovsk Region, 2. Bulletin of the Altai State Agrarian University, pp. 50–55.

Romig, T., Dinkel, A., Mackenstedt, U., 2006. The present situation of echinococcosis in Europe. Parasitol. Int. 55 (Suppl.), 187–191.

Romig, T., Omer, R.A., Zeyhle, E., Huttner, M., Dinkel, A., Siefert, L., Elmahdi, I.E., Magambo, J., Ocaido, M., Menezes, C.N., Ahmed, M.E., Mbae, C., Grobusch, M.P., Kern, P., 2011. Echinococcosis in sub-Saharan Africa: emerging complexity. Vet. Parasitol. 181, 43–47.

Romig, T., Ebi, D., Wassermann, M., 2015. Taxonomy and molecular epidemiology of *Echinococcus granulosus* sensu lato. Vet. Parasitol. 213, 76–84.

Romig, T., 2002. Spread of *Echinococcus multilocularis* in Europe? In: Craig, P.S., Pawłowski, Z. (Eds.), Cestode Zoonoses: Echinococcosis and Cysticercosis – an Emergent and Global Problem, pp. 65–80.

Romig, T., Deplazes, P., Jenkins, D., Giraudoux, P., Massolo, A., Craig, P.S., Wassermann, M., Takahashi, K., de la Rue, M., 2017. Ecology and life cycle patterns of *Echinococcus* species. In: Thompson, R.C.A., Deplazes, P., Lymbery, A.J. (Eds.), *Echinococcus* and Echinococcosis, Part A, vol. 95, pp. 213–314.

Rosenzvit, M.C., Zhang, L.H., Kamenetzky, L., Canova, S.G., Guarnera, E.A., McManus, D.P., 1999. Genetic variation and epidemiology of *Echinococcus granulosus* in Argentina. Parasitology 118, 523–530.

Rospotrebnadzor, 2016. Federal Service for the Supervision of Consumer Protection and Welfare. http://rospotrebnadzor.ru.

Rossi, P., Tamarozzi, F., Galati, F., Pozio, E., Akhan, O., Cretu, C.M., Vutova, K., Siles-Lucas, M., Brunetti, E., Casulli, A., network, H.e, 2016. The first meeting of the European Register of Cystic Echinococcosis (ERCE). Parasit. Vectors 9, 243.

Rossouw, G.J., Knott-Craig, C.J., Erasmus, P.E., 1992. Cardiac echinococcosis: cyst removal in a beating heart. Ann. Thorac. Surg. 53, 328–329.

Rostami, S., Shariat Torbaghan, S., Dabiri, S., Babaei, Z., Ali Mohammadi, M., Sharbatkhori, M., Fasihi Harandi, M., 2015. Genetic characterization of *Echinococcus granulosus* from a large number of formalin-fixed, paraffin-embedded tissue samples of human isolates in Iran. Am. J. Trop. Med. Hyg. 92, 588–594.

Rostami Nejad, M., Hoseinkhan, N., Nazemalhosseini, E., Cheraghipour, K., Abdinia, E., Zali, M., 2007. An analysis of hydatid cyst surgeries in patients referred to hospitals in Khorram-Abad, Lorestan during 2002–06. Iran. J. Parasitol. 2, 29–33.

Rukopisi, N., Zaitsiev, I., 2015. Repeat Operations for Alveolar Echinococcosis (Tomsk).

Sa, M.J.S.d., Souza, C.F.A.E., Teixeira, A.C.P., Oliveira, A.D.L., Souza, R.M.D., Ribeiro, R.M.P., 1998. Hidatidose bovina: frequencia no estado de Minas Gerais: 1996. Hig. Aliment. 12, 19–22.

Saad, M.B., Magzoub, M., 1986. *Echinococcus granulosus* infection in dogs in Tambool. Sudan. J. Helminthol. 60, 299–300.

Saberi-Firouzi, M., Kaffashian, F., Hayati, E., Ghaderi, A., Keshavarz, H., Arshadi, S., Arshadi, C., Sotoudehmaram, M., Massarrat, M., Ghalambor, M., 1998. Prevalence of hydatidosis in nomadic tribes of southern Iran. Med. J. Islam. Repub. Iran 12.

Sadikov, V., 1963. Distribution of the agents of echinococcosis and alveococcosis in dogs and foxes in Uzbekistan. In: Helminths in Man, Animals and Plants and their Control K. I. Skryabin 85th Anniversary, Volume, 61–62.

Sadjjadi, S.M., 2006. Present situation of echinococcosis in the Middle East and Arabic North Africa. Parasitol. Int. 55 (Suppl.), S197–S202.

Saeed, I., Kapel, C., Saida, L.A., Willingham, L., Nansen, P., 2000. Epidemiology of *Echinococcus granulosus* in Arbil province, northern Iraq, 1990–1998. J. Helminthol. 74, 83–88.

SAG, 2015. Informe beneficio y hallazgos patológicos en mataderos nacionales 2014. In: Subdepartamento de sanidad animal, u.d.v.d.e., división de protección pecuaria, Servicio Agrícola y Ganadero (SAG).

Salem, C.O., Schneegans, F., Chollet, J., Jemli, M.E., 2011. Epidemiological studies on echinococcosis and characterization of human and livestock hydatid cysts in Mauritania. Iran. J. Parasitol. 6, 49–57.

Salinas-López, N.-, Jiménez-Guzmán, F., Cruz-Reyes, A., 1996. Presence of *Echinococcus oligarthrus* (Diesing, 1863) Lühe, 1910 in *Lynx rufus* texensis Allen, 1895 from San Fernando, Tamaulipas state, in north-east Mexico. Int. J. Parasitol. 26 (7), 793–796.

Samavatian, A., Valilou, M.R., Lotfi, A., Khani, M.Y., Mirzaei, H., 2009. Study of the incidence rate of liver and lung hydatidosis in slaughtered cattle and buffaloes, at Ahar Abattoir (Arasbaran Region – Northwestern Iran) during 2007–2008. Buffalo Bull. 28, 218–222.

Samuel, W.M., Ramalingam, S., Carbyn, L.N., 1978. Helminths in coyotes (*Canis latrans Say*), wolves (*Canis lupus L.*), and red foxes (*Vulpes vulpes L.*) of southwestern Manitoba. Can. J. Zool. 56, 2614–2617.

Sanchez, G.A., De Bernard, C., Sousa, O.E., 1992. Hepatic polycystic hydatidosis. Clinical and histopathologic report of the second native case of echinococcosis in the Republic of Panama. Rev. Med. Panama 17, 3–11.

Sánchez, E., Cáceres, O., Náquira, C., Garcia, D., Patiño, G., Silvia, H., Volotão, A.C., Fernandes, O., September 2010. Molecular characterization of *Echinococcus granulosus* from Peru by sequencing of the mitochondrial cytochrome C oxidase subunit 1 gene. Mem. Inst. Oswaldo Cruz 105 (6), 806–810.

Sanchez, E., Caceres, O., Naquira, C., Miranda, E., Samudio, F., Fernandes, O., 2012. *Echinococcus granulosus* genotypes circulating in alpacas (*Lama pacos*) and pigs (*Sus scrofa*) from an endemic region in Peru. Mem. Inst. Oswaldo Cruz 107, 275–278.

Sánchez-González, J., Rivera, C.A., V.M., Cruz Luna, A., F.M., Andrade, Q.M., 1997. Anticuerpos anti-*Echinococcus* (hidatidosis) mediante hemaglutinación pasiva, en sujetos expuestos a riesgo. Rev. Mex. Patol. Clin. 44.

Santivanez, S.J., Gutierrez, A.M., Rosenzvit, M.C., Muzulin, P.M., Rodriguez, M.L., Vasquez, J.C., Rodriguez, S., Gonzalez, A.E., Gilman, R.H., Garcia, H.H., Peru, C.W.G., 2008. Human hydatid disease in Peru is basically restricted to *Echinococcus granulosus* genotype G1. Am. J. Trop. Med. Hyg. 79, 89–92.

Sariozkan, S., Yalcin, C., 2009. Estimating the production losses due to cystic echinococcosis in ruminants in Turkey. Vet. Parasitol. 163, 330–334.

Sarkari, B., Mansouri, M., Khabisi, S.A., Mowlavi, G., 2015. Molecular characterization and seroprevalence of *Echinococcus granulosus* in wild boars (*Sus scrofa*) in south-western Iran. Ann. Parasitol. 61, 269–273.

Sawyer, J.C., Schantz, P.M., Schwabe, C.W., Newbold, M.W., 1969. Identification of transmission foci of hydatid disease in California. Public Health Rep. 84, 531–541.

Scala, A., Varcasia, A., Pipia, A.P., Pilo, C., Garippa, G., 2006. First molecular isolation of *Echinococcus granulosus* horse strain (G4) in Sardinia (Italy). Parassitologia 48, 344.

Schantz, P.M., Alstine, C.V., Blacksheep, A., Sinclair, S., 1977a. Prevalence of *Echinococcus granulosus* and other cestodes in dogs on the Navajo reservation in Arizona and New Mexico. Am. J. Vet. Res. 38, 669–670.

Schantz, P.M., von Reyn, C.F., Welty, T., Andersen, F.L., Schultz, M.G., Kagan, I.G., 1977b. Epidemiologic investigation of echinococcosis in American Indians living in Arizona and New Mexico. Am. J. Trop. Med. Hyg. 26, 121–126.

Schantz, P.M., Chai, J., Craig, P.S., Eckert, J., Jenkins, D.J., Macpherson, C.N.L., Thakur, A., 1995. Epidemiology and control of hydatid disease. In: Thompson, R.C.A., Lymbery, A.J. (Eds.), Echinococcus and Hydatid Disease. CAB International, Australia, pp. 233–331.

Schantz, P.M., Wang, H., Qiu, J., Liu, F.J., Saito, E., Emshoff, A., Ito, A., Roberts, J.M., Delker, C., 2003. Echinococcosis on the Tibetan Plateau: prevalence and risk factors for cystic and alveolar echinococcosis in Tibetan populations in Qinghai Province, China. Parasitology 127, S109–S120.

Schiller, E.L., 1955. Studies on the helminth fauna of Alaska. XXVI. Some observations on the cold-resistance of eggs of *Echinococcus sibiricensis* Rausch and Schiller, 1954. J. Parasitol. 41, 578–582.

Schmidt, D., Tiendrebeogo, H., Roudaut, M., Nozais, J.P., Coulibaly, A.O., Ouattara, K., Metras, D., 1978. Epidemiological study of pulmonary hydatid cyst in the Ivory Coast. 3 cases. Bull. Soc. Pathol. Exot. Fil. 71, 154–157.

Schneider, R., Gollackner, B., Schindl, M., Tucek, G., Auer, H., 2010. *Echinococcus canadensis* G7 (pig strain): an underestimated cause of cystic echinococcosis in Austria. Am. J. Trop. Med. Hyg. 82, 871–874.

Schneider, R., Aspöck, H., Auer, H., 2011. Unexpected increase of alveolar echincoccosis, Austria, 2011. Emerg. Infect. Dis. 19, 475–477.

Schurer, J., Shury, T., Leighton, F., Jenkins, E., 2013. Surveillance for *Echinococcus canadensis* genotypes in Canadian ungulates. Int. J. Parasitol. Parasit. Wildl. 2, 97–101.

Schurer, J.M., Gesy, K.M., Elkin, B.T., Jenkins, E.J., 2014a. *Echinococcus multilocularis* and *Echinococcus canadensis* in wolves from western Canada. Parasitology 141, 159–163.

Schurer, J.M., Ndao, M., Quewezance, H., Elmore, S.A., Jenkins, E.J., 2014b. People, pets, and parasites: one health surveillance in southeastern Saskatchewan. Am. J. Trop. Med. Hyg. 90, 1184–1190.

Schurer, J.M., Rafferty, E., Farag, M., Zeng, W., Jenkins, E.J., 2015. Echinococcosis: An economic evaluation of a veterinary public health intervention in rural Canada. PLoS Negl. Trop. Dis. 9, e0003883.

Schurer, J.M., Pawlik, M., Huber, A., Elkin, B., Cluff, D., Pongracz, J., Gesy, K., Wagner, B., Dixon, B., Merks, H., Bal, M.S., Jenkins, E.J., 2016. Intestinal parasites of gray wolves (*Canis lupus*) in northern and western Canada. Can. J. Zool. 94, 643–650.

Schweiger, A., Ammann, R.W., Candinas, D., Clavien, P.A., Eckert, J., Gottstein, B., Halkic, N., Muellhaupt, B., Prinz, B.M., Reichen, J., Tarr, P.E., Torgerson, P.R., Deplazes, P., 2007. Human alveolar echinococcosis after fox population increase, Switzerland. Emerg. Infect. Dis. 13, 878–882.

Scrimgeour, E.M., Mehta, F.R., Suleiman, A.J., 1999. Infectious and tropical diseases in Oman: a review. Am. J. Trop. Med. Hyg. 61, 920–925.

Seesee, F.M., Sterner, M., Worley, D.E., 1983. Helminths of the Coyote (*Canis latrans* Say) in Montana. J. Wildl. Dis. 19, 54–55.

Seimenis, A., 2003. Overview of the epidemiological situation on echinococcosis in the Mediterranean region. Acta Trop. 85, 191–195.

Shafiq, M., Tanveer, A., Athar, M., 2005. Epidemiology and Economical Aspects of Hydatidosis/Echinococcosis in Different Animals and Man (Dissertation). University of the Punjab, Lahore, p. 415.
Shahnazi, M., Hejazi, H., Salehi, M., Andalib, A.R., 2011. Molecular characterization of human and animal *Echinococcus granulosus* isolates in Isfahan, Iran. Acta Trop. 117, 47–50.
Shaikenov, B.S., Torgerson, P.R., Usenbayev, A.E., Baitursynov, K.K., Rysmukhambetova, A.T., Abdybekova, A.M., Karamendin, K.O., 2003. The changing epidemiology of echinococcosis in Kazakhstan due to transformation of farming practices. Acta Trop. 85, 287–293.
Shaikenov, B.S., Rysmukhambetova, A.T., Massenov, B., Deplazes, P., Mathis, A., Torgerson, P.R., 2004. The use of a polymerase chain reaction to detect *Echinococcus granulosus* (G1 strain) eggs in soil samples. Am. J. Trop. Med. Hyg. 71, 441–443.
Shaikenov, B.S., 2006. Distribution and ecology of *Echinococcus multilocularis* in Central Asia. Parasitol. Int. 55 (Suppl.), S213–S219.
Shambesh, M.A., Craig, P.S., Macpherson, C.N., Rogan, M.T., Gusbi, A.M., Echtuish, E.F., 1999. An extensive ultrasound and serologic study to investigate the prevalence of human cystic echinococcosis in northern Libya. Am. J. Trop. Med. Hyg. 60, 462–468.
Sharbatkhori, M., Mirhendi, H., Jex, A.R., Pangasa, A., Campbell, B.E., Kia, E.B., Eshraghian, M.R., Harandi, M.F., Gasser, R.B., 2009. Genetic categorization of *Echinococcus granulosus* from humans and herbivorous hosts in Iran using an integrated mutation scanning-phylogenetic approach. Electrophoresis 30, 2648–2655.
Sharbatkhori, M., Mirhendi, H., Harandi, M.F., Rezaeian, M., Mohebali, M., Eshraghian, M., Rahimi, H., Kia, E.B., 2010. *Echinococcus granulosus* genotypes in livestock of Iran indicating high frequency of G1 genotype in camels. Exp. Parasitol. 124, 373–379.
Sharbatkhori, M., Fasihi Harandi, M., Mirhendi, H., Hajialilo, E., Kia, E.B., 2011. Sequence analysis of cox1 and nad1 genes in *Echinococcus granulosus* G3 genotype in camels (*Camelus dromedarius*) from central Iran. Parasitol. Res. 108, 521–527.
Shariatzadeh, S.A., Spotin, A., Gholami, S., Fallah, E., Hazratian, T., Mahami-Oskouei, M., Montazeri, F., Moslemzadeh, H.R., Shahbazi, A., 2015. The first morphometric and phylogenetic perspective on molecular epidemiology of *Echinococcus granulosus* sensu lato in stray dogs in a hyperendemic Middle East focus, northwestern Iran. Parasit. Vectors 8, 409.
Sharifi, I., Tasbiti, A., 1994. Helminth parasites of stray dogs in Kerman. Iran. J. Publ. Health 23, 13–23.
Sharifi, I., Daneshvar, H., Ziaali, N., Fasihi Harandi, M., Nikian, Y., Ebrahimi, A., Keshavarz, H., Masoud, J., 1996. Evaluation of a control program on hydatid cyst in the city of Kerman. J. Kerman Univ. Med. Sci. 3, 168–174.
Sharma, M., Sehgal, R., Fomda, B.A., Malhotra, A., Malla, N., 2013. Molecular characterization of *Echinococcus granulosus* cysts in north Indian patients: identification of G1, G3, G5 and G6 genotypes. PLoS Negl. Trop. Dis. 7, e2262.
Sharrif, L., Al-Rawashdeh, O., Al-Qudah, K., Al-Ani, F., 1998. Prevalence of gastrointestinal helminthes, hydatid cysts and nasal myiasis in camels in Jordan. In: Proceedings of the Third Annual Meeting for Animal Production under Arid Condition. United Arab Emirates University, pp. 108–114.
Shehatha, J., Alward, M., Saxena, P., Konstantinov, I.E., 2009. Surgical management of cardiac hydatidosis. Tex. Heart Inst. J. 36, 72–73.
Sherifi, K., Rexhepi, A., Hamidi, A., Behluli, B., Zessin, K.H., Mathis, A., Deplazes, P., 2011. Detection of patent infections of *Echinococcus granulosus* ("sheep-strain", G1) in naturally infected dogs in Kosovo. Berl. Munch Tierarztl Wochenschr 124, 518–521.
Shevkoplyas, V., Lopatin, V., 2009. Key aspects of antiparasitic activities in the Krasnodar region. Agric. Gaz. Urals 7, 86–88.

Shi, D., 1995. The first report of *Echinococcus multilocularis* in domestic dogs in Zhang County of Gansu Province. Chin. Med. J. 108, 615–617.

Shimalov, V.V., Shimalov, V.T., 2000. Helminth fauna of the wolf (*Canis lupus* Linnaeus, 1758) in Belorussian Polesie. Parasitol. Res. 86, 163–164.

Shimalov, V., 2011. The prevalence of *Echinococcus multilocularis* (cestoda, taeniidae) in the southwestern part of Belarus. News Natl. Acad. Sci. Ser. Biol. Sci. 4, 108–112.

Shodmonov, I., Razikov, S., 2015. Epidemiological situation of echinococcosis. Modern problems of science and education. Mod. Prob. Sci. Educ. 2.

Sigurdarson, S., 2010. Dogs and echinococcosis in Iceland. Acta Vet. Scand. 52, 6.

Sikó, B., Bokor, E., Fekeás, E., Nemes, I., Murai, E., Gubányi, A., 1995. Occurrence and epidemiology of *Echinococcus granulosus* and *E. multilocularis* in the Covasna county, East Carpathian Mountains, Romania. Parasitol. Hung 28, 43–56.

Siko, S.B., Deplazes, P., Ceica, C., Tivadar, C.S., Bogolin, I., Popescu, S., Cozma, V., 2011. *Echinococcus multilocularis* in south-eastern Europe (Romania). Parasitol. Res. 108, 1093–1097.

Simsek, S., Cevik, A., 2014. First detection and molecular characterization of *Echinococcus equinus* in a mule in Turkey. Acta Parasitol. 59, 773–777.

Simsek, S., Eroksuz, Y., 2009. Occurrence and molecular characterization of *Echinococcus granulosus* in Turkish mouflon (*Ovis gmelinii anatolica*). Acta Trop. 109, 167–169.

Simsek, S., Balkaya, I., Koroglu, E., 2010. Epidemiological survey and molecular characterization of *Echinococcus granulosus* in cattle in an endemic area of eastern Turkey. Vet. Parasitol. 172, 347–349.

Simsek, S., Kaplan, M., Ozercan, I.H., 2011. A comprehensive molecular survey of *Echinococcus granulosus* in formalin-fixed paraffin-embedded tissues in human isolates in Turkey. Parasitol. Res. 109, 411–416.

Simsek, S., Roinioti, E., Eroksuz, H., 2015. First Report of *Echinococcus equinus* in a Donkey in Turkey. Korean J. Parasitol. 53, 731–735.

Singh, B.B., Dhand, N.K., Ghatak, S., Gill, J.P., 2014a. Economic losses due to cystic echinococcosis in India: Need for urgent action to control the disease. Prev. Vet. Med. 113, 1–12.

Singh, B.B., Sharma, R., Sharma, J.K., Gill, J.P.S., 2014b. Molecular detection of *E. granulosus* sheep strain (G1) infections in naturally infected dogs in Punjab (India). Helminthologia 51, 269–272.

Singh, B.B., 2011. Molecular Epidemiology of Echinococcosis in Northern India and Public Health Significance (Dissertation). University Ludhiana, Indiana.

Sirol, J., Lefevre, M., 1971. [Human echinococcosis in Chad, a breeding country (3 cases at Fort-Lamy, Chad)]. Bull. Soc. Pathol. Exot. Fil. 64, 887–890.

Skelding, A., Brooks, A., Stalker, M., Mercer, N., Villa, E., Gottstein, B., 2014. Hepatic alveolar hydatic disease (*Echinococcus multilocularis*) in a boxer dog for southern Ontario. Can. Vet. J. 55.

Snabel, V., Altintas, N., D'Amelio, S., Nakao, M., Romig, T., Yolasigmaz, A., Gunes, K., Turk, M., Busi, M., Huttner, M., Sevcova, D., Ito, A., Altintas, N., Dubinsky, P., 2009. Cystic echinococcosis in Turkey: genetic variability and first record of the pig strain (G7) in the country. Parasitol. Res. 105, 145–154.

Sobrino, R., Gonzalez, L.M., Vicente, J., Fernandez de Luco, D., Garate, T., Gortazar, C., 2006. *Echinococcus granulosus* (Cestoda, Taeniidae) in the Iberian wolf. Parasitol. Res. 99, 753–756.

Somily, A., Robinson, J.L., Miedzinski, L.J., Bhargava, R., Marrie, T.J., 2005. Echinococcal disease in Alberta, Canada: more than a calcified opacity. BMC Infect. Dis. 5, 34.

Soriano, S.V., Pierangeli, N.B., Pianciola, L., Mazzeo, M., Lazzarini, L.E., Saiz, M.S., Kossman, A.V., Bergagna, H.F., Chartier, K., Basualdo, J.A., 2010. Molecular

characterization of *Echinococcus* isolates indicates goats as reservoir for *Echinococcus canadensis* G6 genotype in Neuquen, Patagonia Argentina. Parasitol. Int. 59, 626–628.
Sorochenko, E., 1972. The distribution of echinococcosis amongst the indiginous population of the Nenets National District. Questions of Medical Geography of the North European part of the USSR. In: Proceedings of a Scientific Conference (Almaty), pp. 49–52.
Sotiraki, S., Chaligiannis, I., 2010. Cystic echinococcosis in Greece. Past and present. Parasite 17, 205–210.
Sotiraki, S., Himonas, C., Korkoliakou, P., 2003. Hydatidosis-echinococcosis in Greece. Acta Trop. 85, 197–201.
Soule, C., Fabien, J.F., Maillot, E., 1995. Animal echinococcosis in France: national survey in slaughterhouses in 1989. Epidemiol. et Sante Animale 28, 131–141.
Sousa, O.E., Lombardo Ayala, J.D., 1965. Informe de un caso de hidatidosis den sujeto nativo Panameno. Primer caso autoctono. Arch. Medicos Panamenos 14, 79–86.
Staubach, C., Hoffmann, L., Schmid, V.J., Ziller, M., Tackmann, K., Conraths, F.J., 2011. Bayesian space-time analysis of *Echinococcus multilocularis* infections in foxes. Vet. Parasitol. 179, 77–83.
Stefanic, S., Shaikenov, B.S., Deplazes, P., Dinkel, A., Torgerson, P.R., Mathis, A., 2004. Polymerase chain reaction for detection of patent infections of *Echinococcus granulosus* ("sheep strain") in naturally infected dogs. Parasitol. Res. 92, 347–351.
Stehr-Green, J.K., Stehr-Green, P.A., Schantz, P.M., Wilson, J.F., Lanier, A., 1988. Risk factors for infection with *Echinococcus multilocularis* in Alaska. Am. J. Trop. Med. Hyg. 38, 380–385.
Stejskal, F., Trojánek, M., Oliverius, M., Kolbeková, P., Kolářová, L., 2014. Cystic and alveolar echinococcosis in the Czech Republic: diagnostics and follow up. In: International Symposium. Innovation for the Management of Echinococcosis (Besançon), p. 70.
Steta, J., Torre, A., 2009. Mexican-native human echinococcosis: case report of an underestimated disease. Ann. Hepatol. 8, 251–254.
Storandt, S.T., Kazacos, K.R., 1993. *Echinococcus multilocularis* identified in Indiana, Ohio, and east-central Illinois. J. Parasitol. 79, 301–305.
Storandt, S.T., Kazacos, K.R., 2012. *Echinococcus multilocularis* identified in Michigan with additional records from Ohio. J. Parasitol. 98, 891–893.
Storandt, S.T., Virchow, D.R., Dryden, M.W., Hygnstrom, S.E., Kazacos, K.R., 2002. Distribution and prevalence of *Echinococcus multilocularis* in wild predators in Nebraska, Kansas, and Wyoming. J. Parasitol. 88, 420–422.
Suarez, J.P., Rivera, E., Rico, F., Ramirez, P., Lopez de Lara, A., 1995. Paquipleuritis Granulomatosa de Origen Parasitario, 8, p. 54.
Suleimanova, K., Shinkina, E., 2015. Questions about echinococcosis in cattle. In: Materials of an International Scientific Practical Conference in Celebration of 85 years of the Ural Government Academy of Veterinary Medicine and 100th Anniversary of the Birth of Professor Vasiliniya Grigoryevicha Martinova. Tronitsk, pp. 70–73.
Sultanov, A., Abdybekova, A., Abdibaeva, A., Shapiyeva, Z., Yeshmuratov, T., Torgerson, P.R., 2014. Epidemiology of fishborne trematodiasis in Kazakhstan. Acta Trop. 138, 60–66.
Svobodová, V., 2014. The occurrence of the tapeworms of *Echinococcus* spp. in animals in the Czech Republic. In: Special Seminar, Infections of Echinococcosis, Prague, pp. 7–8.
Sweatman, G.K., Williams, R.J., 1963. Survival of *Echinococcus granulosus* and *Taenia hydatigena* eggs in two extreme climatic regions of New Zealand. Res. Vet. Sci. 4, 199–216.
Sweatman, G.K., 1952. Distribution and incidence of *Echinococcus granulosus* in man and other animals with special reference to Canada. Can. J. Public Health 43, 480–486.
Sydler, T., Mathis, A., Deplazes, P., 1998. *Echinococcus multilocularis* lesions in the livers of pigs kept outdoors in Switzerland. Eur. J. Vet. Pathol. 4, 43–46.

Tabain, I., Sviben, M., Ljubin-Sternak, S., Vilibic-Cavlek, T., Mlinaric-Galinovic, G., 2011. Seroprevalence of *Echinococcus granulosus* infection in Croatian patients with cystic liver disease. J. Helminthol. 85, 192–195.

Takacs, A., Szabo, L., Juhasz, L., Takacs, A.A., Lanszki, J., Takacs, P.T., Heltai, M., 2014. Data on the parasitological status of golden jackal (*Canis aureus* L., 1758) in Hungary. Acta Vet. Hung. 62, 33–41.

Tamayo Meneses, L., Pacheco Llerena, R., Fernandez Hinojosa, R., Chungara Montaño, J., 2004. Hidatidosis. Experiencia institucional. Rev. la Soc. Boliv. Pediatría 43, 149–154.

Tang, C.T., Quian, Y.C., Kang, Y.M., Cui, G.W., Lu, H.C., Shu, L.M., Wang, Y.H., Tang, L., 2004. Study on the ecological distribution of alveolar *Echinococcus* in Hulunbeier Pasture of Inner Mongolia, China. Parasitology 128, 187–194.

Tappe, D., Brehm, K., Frosch, M., Blankenburg, A., Schrod, A., Kaup, F.J., Matz-Rensing, K., 2007. *Echinococcus multilocularis* infection of several Old World monkey species in a breeding enclosure. Am. J. Trop. Med. Hyg. 77, 504–506.

Tarish, J.H., Al-Saqur, I.M., Al-Abbassy, S.N., Kadhim, F.S., 1986. The prevalence of parasitic helminths in stray dogs in the Baghdad area, Iraq. Ann. Trop. Med. Parasitol. 80, 329–331.

Tashani, O.A., Zhang, L.H., Boufana, B., Jegi, A., McManus, D.P., 2002. Epidemiology and strain characteristics of *Echinococcus granulosus* in the Benghazi area of eastern Libya. Ann. Trop. Med. Parasitol. 96, 369–381.

Tavakoli, H., Bayat, M., Kousha, A., 2008. Hydatidosis infection study in human and livestock populations during 2002–2007. Am. Eur. J. Agric. Environ. Sci. 4, 473–477.

Terentyeva, Z., 2007. Distribution of echinococcosis in the Oreburg Region. Vet. Pathol. 2, 164–168.

Thapa, N.K., Armua-Fernandez, M.T., Kinzang, D., Gurung, R.B., Wangdi, P., Deplazes, P., 2017. Detection of *Echinococcous granulosus* and *Echinococcus ortleppi* in Bhutan. Parasitol. Int. (in press).

Thompson, R.C.A., Jenkins, D.J., 2014. *Echinococcus* as a model system: biology and epidemiology. Int. J. Parasitol. 44, 865–877.

Thompson, R.C.A., McManus, D.P., 2002. Towards a taxonomic revision of the genus *Echinococcus*. Trends Parasitol. 18, 452–457.

Thompson, R.C.A., Smyth, J.D., 1975. Equine hydatidosis: a review of the current status in Great Britain and the results of an epidemiological survey. Vet. Parasitol. 1, 107–127.

Thompson, R.C.A., Boxell, A.C., Ralston, B.J., Constantine, C.C., Hobbs, R.P., Shury, T., Olson, M.E., 2006. Molecular and morphological characterization of *Echinococcus* in cervids from North America. Parasitology 132, 439–447.

Thompson, R.C.A., Kutz, S.J., Smith, A., 2009. Parasite zoonoses and wildlife: emerging issues. Int. J. Environ. Res. Public Health 6, 678–693.

Thompson, R.C.A., 2013. Parasite zoonoses and wildlife: One Health, spillover and human activity. Int. J. Parasitol. 43, 1079–1088.

Thompson, R.C.A., 2017. Biology and systematics of *Echinococcus*. In: Thompson, R.C.A., Deplazes, P., Lymbery, A.J. (Eds.), *Echinococcus* and Echinococcosis, Part A, vol. 95, pp. 65–110.

Tiaoying, L., Jiamin, Q., Wen, Y., Craig, P.S., Xingwang, C., Ning, X., Ito, A., Giraudoux, P., Wulamu, M., Wen, Y., Schantz, P.M., 2005. Echinococcosis in Tibetan populations, western Sichuan Province. China Emerg. Infect. Dis. 11, 1866–1873.

Tigre, W., Deresa, B., Haile, A., Gabriel, S., Victor, B., Pelt, J.V., Devleesschauwer, B., Vercruysse, J., Dorny, P., 2016. Molecular characterization of *Echinococcus granulosus* s.l. cysts from cattle, camels, goats and pigs in Ethiopia. Vet. Parasitol. 215, 17–21.

Todorov, T., Boeva, V., 1999. Human echinococcosis in Bulgaria: a comparative epidemiological analysis. Bull. World Health Organ. 77, 110–118.

Tolnai, Z., Szell, Z., Sreter, T., 2013. Environmental determinants of the spatial distribution of *Echinococcus multilocularis* in Hungary. Vet. Parasitol. 198, 292–297.

Torgerson, P.R., Budke, C.M., 2003. Echinococcosis—an international public health challenge. Res. Vet. Sci. 74, 191—202.
Torgerson, P.R., Craig, P.S., 2009. Risk assessment of importation of dogs infected with *Echinococcus multilocularis* into the UK. Vet. Rec. 165, 366—368.
Torgerson, P.R., Macpherson, C.N., 2011. The socioeconomic burden of parasitic zoonoses: global trends. Vet. Parasitol. 182, 79—95.
Torgerson, P.R., Shaikenov, B.S., Baitursinov, K.K., Abdybekova, A.M., 2002. The emerging epidemic of echinococcosis in Kazakhstan. Trans. R. Soc. Trop. Med. Hyg. 96, 124—128.
Torgerson, P.R., Burtisurnov, K.K., Shaikenov, B.S., Rysmukhambetova, A.T., Abdybekova, A.M., Ussenbayev, A.E., 2003. Modelling the transmission dynamics of *Echinococcus granulosus* in sheep and cattle in Kazakhstan. Vet. Parasitol. 114, 143—153.
Torgerson, P.R., Oguljahan, B., Muminov, A.E., Karaeva, R.R., Kuttubaev, O.T., Aminjanov, M., Shaikenov, B., 2006. Present situation of cystic echinococcosis in Central Asia. Parasitol. Int. 55 (Suppl.), S207—S212.
Torgerson, P.R., Schweiger, A., Deplazes, P., Pohar, M., Reichen, J., Ammann, R.W., Tarr, P.E., Halkik, N., Mullhaupt, B., 2008. Alveolar echinococcosis: from a deadly disease to a well-controlled infection. Relative survival and economic analysis in Switzerland over the last 35 years. J. Hepatol. 49, 72—77.
Torgerson, P.R., Ziadinov, I., Aknazarov, D., Nurgaziev, R., Deplazes, P., 2009. Modelling the age variation of larval protoscoleces of *Echinococcus granulosus* in sheep. Int. J. Parasitol. 39, 1031—1035.
Torgerson, P.R., Keller, K., Magnotta, M., Ragland, N., 2010. The global burden of alveolar echinococcosis. PLoS Negl. Trop. Dis. 4, e722. http://dx.doi.org/10.1371/journal.pntd.0000722.
Torgerson, P.R., de Silva, N.R., Fevre, E.M., Kasuga, F., Rokni, M.B., Zhou, X.N., Sripa, B., Gargouri, N., Willingham, A.L., Stein, C., 2014. The global burden of foodborne parasitic diseases: an update. Trends Parasitol. 30, 20—26.
Torgerson, P.R., Devleesschauwer, B., Praet, N., Speybroeck, N., Willingham, A.L., Kasuga, F., Rokni, M.B., Zhou, X.-N., Fèvre, E.M., Sripa, B., Gargouri, N., Fürst, T., Budke, C.M., Carabin, H., Kirk, M.D., Angulo, F.J., Havelaar, A., de Silva, N., 2015. World Health Organization Estimates of the Global and Regional Disease Burden of 11 Foodborne Parasitic Diseases, 2010: A Data Synthesis. PLoS Med. 12, e1001920. http://dx.doi.org/10.1371/journal.pmed.1001920.
Torgerson, P.R., 2013. The emergence of echinococcosis in central Asia. Parasitology 140, 1667—1673.
Torgerson, P., 2017. Echinococcosis in Western Europe, Eastern Europe and Central Asia. In: Hotez, P. (Ed.), Neglected Tropical Diseases — Europe and Central Asia. Springer.
Torres, F., 2012. Identificación de la presencia de hidatidosis en el Camal Municipal de la ciudad de Puyo, Provincia de Pastaza. Univ. Cent. del Ecuad.
Torrez Salazar, J., Torrez Salazar, J., Peñaloza Chej, A.M., Iriarte La Fuente, D., 2009. Quiste hidatídico hepático caso clínico en la caja nacional de salud - Cochabamba. Gac. Médica Boliv. 32, 48—51.
Trachsel, D., Deplazes, P., Mathis, A., 2007. Identification of taeniid eggs in the faeces from carnivores based on multiplex PCR using targets in mitochondrial DNA. Parasitology 134, 911—920.
Tranbenkova, N.A., 1992. The ecology of *Echinococcus multilocularis* (Leuckart, 1863) and *E. granulosus* (Batsch, 1786) on the Kamchatka Peninsula. Med. Parasitol. Mosk. 45—47.
Traub, R.J., Robertson, I.D., Irwin, P.J., Mencke, N., Thompson, R.C.A., 2005. Canine gastrointestinal parasitic zoonoses in India. Trends Parasitol. 21, 42—48.
Trezeguet, M.A., Torres, P., Kistermann, J.C., NBernasconi, G., Lacunza, J., Ramos, M., 2011. In: Alimentaria, S.N.d.S.y.C. (Ed.), Sistema de vigilancia epidemiologica de la

hidatidosis animal, en establecimientos con bovinos, ovinos o porcinos, por medio de la faena en la republica argentina. SENASA.
Troncy, P., Graber, M., 1969. Echinococcosis-hydatidosis in Central Africa. 3. Carnivore teniasis due to *Echinococcus granulosus* (Batsch 1786-Rudolphi 1801). Revue d'elevage et de medecine veterinaire des pays tropicaux 22, 75–84.
Tserennadmid, T., Enkhjargal, T., 2000. Current status of clinical laboratory in Mongolia. Rinsho Byori 48, 200–204.
Tsukada, H., Li, W., Duo, H., Guo, Z., Fu, Y., Peng, M., Shen, X., Jing, J., Yuan, A., Ni, M., He, S., Huang, F., Feng, K., Ishikawa, K., Kobayashi, I., Shinohara, A., Nonaka, N., 2014. A comparison of the diet and fine-scale distribution of sympatric Tibetan and red foxes in Qinghai, PR China. Wildl. Biol. 20, 356–361.
Tulin, A.I., Ribenieks, R., Pogodina, E.N., Stutska, R., Shavlovskis, I., Gardovskis, I., 2012. Diagnostics and surgical treatment of liver echinococcosis in Latvia. Vestn. Khir Im. I I Grek. 171, 38–44.
Turčeková, Ľ., Šnábel, V., Dudiňák, V., Gašpar, V., Dubinský, P., 2009. Prevalence of cystic echinococcosis in pigs from Slovakia, with evaluation of size, fertility and number of hydatid cysts. Helminthologia 46, 151–158.
Umhang, G., Lahoreau, J., Nicolier, A., Boue, F., 2013a. *Echinococcus multilocularis* infection of a ring-tailed lemur (*Lemur catta*) and a nutria (*Myocastor coypus*) in a French zoo. Parasitol. Int. 62, 561–563.
Umhang, G., Richomme, C., Boucher, J.M., Hormaz, V., Boue, F., 2013b. Prevalence survey and first molecular characterization of *Echinococcus granulosus* in France. Parasitol. Res. 112, 1809–1812.
Umhang, G., Chihai, O., Boue, F., 2014a. Molecular characterization of *Echinococcus granulosus* in a hyperendemic European focus, the Republic of Moldova. Parasitol. Res. 113, 4371–4376.
Umhang, G., Comte, S., Raton, V., Hormaz, V., Boucher, J.M., Favier, S., Combes, B., Boue, F., 2014b. *Echinococcus multilocularis* infections in dogs from urban and peri-urban areas in France. Parasitol. Res. 113, 2219–2222.
Umhang, G., Grech-Angelini, S., Maestriani, O., Hormaz, V., Boucher, J.-M., Peytavin, C., Richomme, C., Boue, F., 2014c. Identification of lifecycle of *Echinococcus canadensis* in Corsica. In: ESCAAP Conference Echinococcus, Vilnius, Lithuania, p. 45.
Umhang, G., Richomme, C., Hormaz, V., Boucher, J.M., Boue, F., 2014d. Pigs and wild boar in Corsica harbor *Echinococcus canadensis* G6/7 at levels of concern for public health and local economy. Acta Trop. 133, 64–68.
Umur, S., 2003. Prevalence and economic importance of cystic echinococcosis in slaughtered ruminants in Burdur, Turkey. J. Vet. Med. B Infect. Dis. Vet. Public Health 50, 247–252.
Uspenskii, A., Malkhova, E., Ershovova, T., 2013. Performance of coordination plans of scientific researches in veterinary parasitology. Russ. Parasitol. J. 2, 48–53.
Usubalieva, J., Minbaeva, G., Ziadinov, I., Deplazes, P., Torgerson, P.R., 2013. Human alveolar echinococcosis in Kyrgyzstan. Emerg. Infect. Dis. 19, 1095–1097.
Utuk, A.E., Simsek, S., 2013. Molecular characterization of the horse isolate of *Echinococcus granulosus* in Turkey. J. Helminthol. 87, 305–308.
Utuk, A.E., Simsek, S., Koroglu, E., McManus, D.P., 2008. Molecular genetic characterization of different isolates of *Echinococcus granulosus* in east and southeast regions of Turkey. Acta Trop. 107, 192–194.
Uysal, V., Paksoy, N., 1986. *Echinococcosis multilocularis* in Turkey. J. Trop. Med. Hyg. 89, 249–255.
Uzunlar, A.K., Yilmaz, F., Bitiren, M., 2003. *Echinococcosis multilocularis* in south-eastern Anatolia, Turkey. East Afr. Med. J. 80, 395–397.

Vaiyeva, J., Khairulin, A., Shalmenov, M., Yastreb, V., 2012. Experience of struggle with echinococcosis on an example of one facilities of the West-Kazakhstan area. Russ. Parasitol. J. 4, 100–102.

Valenzuela Ramos, M.C., Navarrete Arellano, N., Soto Fernández, A., 2010. Quiste hidatídico. Presentación de un caso. Rev. Mex Cirugía Pediátrica 17, 38–43.

van Asch, B., Zhang, A-b., Oskarsson, M.C.R., Klütsch, C.F.C., Amorim, A., Savolainen, P., 2013. Pre-Columbian origins of Native American dog breeds, with only limited replacement by European dogs, confirmed by mtDNA analysis. Proc. R. Soc. Lond Biol. 280.

van der Giessen, J.B.W., de Vries, A., Chu, M.L., Mulder, J., Teunis, P., Takumi, K., 2005. *Echinococcus multilocularis*: increasing presence in the southern border area in the Netherlands. Bull. Scand. Baltic Soc. Parasitol. 14, 156.

Van Kesteren, F., Mastin, A., Mytynova, B., Ziadinov, I., Boufana, B., Torgerson, P.R., Rogan, M.T., Craig, P.S., 2013. Dog ownership, dog behaviour and transmission of *Echinococcus* spp. in the Alay Valley, southern Kyrgyzstan. Parasitology 140, 1674–1684.

Vaniscotte, A., Raoul, F., Poulle, M.L., Romig, T., Dinkel, A., Takahashi, K., Guislain, M.H., Moss, J., Tiaoying, L., Wang, Q., Qiu, J., Craig, P.S., Giraudoux, P., 2011. Role of dog behaviour and environmental fecal contamination in transmission of *Echinococcus multilocularis* in Tibetan communities. Parasitology 138, 1316–1329.

Varcasia, A., Canu, S., Lightowlers, M.W., Scala, A., Garippa, G., 2006. Molecular characterization of *Echinococcus granulosus* strains in Sardinia. Parasitol. Res. 98, 273–277.

Varcasia, A., Canu, S., Kogkos, A., Pipia, A.P., Scala, A., Garippa, G., Seimenis, A., 2007. Molecular characterization of *Echinococcus granulosus* in sheep and goats of Peloponnesus, Greece. Parasitol. Res. 101, 1135–1139.

Varcasia, A., Garippa, G., Pipia, A.P., Scala, A., Brianti, E., Giannetto, S., Battelli, G., Poglayen, G., Micagni, G., 2008a. Cystic echinococcosis in equids in Italy. Parasitol. Res. 102, 815–818.

Varcasia, A., Piseddu, T., Pipia, A.P., Schianchi, G., Marongiu, A., Petruzzi, V., Scala, A., Garippa, G., 2008b. Epidemiological and biomolecular updates on cystic echinococcosis in pigs and wild boars of Sardinia (Italy). Lucr. Stiintifice Med. Veterinara 41, 385–387.

Varcasia, A., Tanda, B., Giobbe, M., Solinas, C., Pipia, A.P., Malgor, R., Carmona, C., Garippa, G., Scala, A., 2011. Cystic echinococcosis in Sardinia: farmers' knowledge and dog infection in sheep farms. Vet. Parasitol. 181, 335–340.

Vargas Rivera, I., Martinez Maya, J.J., Jaramillo Arango, C.J., 1995. Caracterizacion de la hidatidosis porcina en el rastro frigorifico Los Reyes La Paz, Estado de Mexico, Mexico. Veterinaria Méx. 26, 365–368.

Varodi, E.I., Kornyushin, V.V., Kharchenko, V.A., Yemets, A.M., 2007. Distribution of *Echinococcus granulosus* in Ukraine. In: Abstracts of Tenth International Helmithological Symposium (Stará Lesná), p. 40.

Veit, P., Bilger, B., Schad, V., Schafer, J., Frank, W., Lucius, R., 1995. Influence of environmental factors on the infectivity of *Echinococcus multilocularis* eggs. Parasitology 110, 79–86.

Vejdani, M., Vejdani, S., Lotfi, S., Najafi, F., Nazari, N., Hamzavi, Y., 2013. Study of operated primary and secondary (recurrence) hydatidosis in hospitals of Kermanshah, west of Islamic Republic of Iran. East. Mediterr. Health J. 19, 671–675.

Vera, O., Fernández, J., López, R., Caihuara, D., 2004. El quiste hidatidico: Un problema de Salud Publica en La Paz. Rev. Méd. 10.

Vera, O., Pinilla, H., Soliz, M., 2006. Quistes hidatídicos pulmonares gigantes: A propósito de tres casos. Rev. Cuad. 51.

Vergles Rataj, A., Bidovec, A., Žele, D., Vengušt, G., 2010. *Echinococcus multilocularis* in the red fox (*Vulpes vulpes*) in Slovenia. Eur. J. Wildl. Res. 56, 819–822.

Verster, A., Collins, M., 1966. The incidence of hydatidosis in the Republic of South Africa. Onderstepoort J. Vet. Res. 33, 49–72.

Villalobos, N., Gonzalez, L.M., Morales, J., de Aluja, A.S., Jimenez, M.I., Blanco, M.A., Harrison, L.J., Parkhouse, R.M., Garate, T., 2007. Molecular identification of *Echinococcus granulosus* genotypes (G1 and G7) isolated from pigs in Mexico. Vet. Parasitol. 147, 185–189.

Villarreal Jiménez, A., Padua Gabriel, A., Lezama Urtecho, C., Galindo Rujana, M.E., Argüero Sánchez, R., 1995. Quiste hidatídico pulmonar con ruptura mixta: informe de un caso. Rev. Méd IMSS 33.

Villena, E., Uzqueda, F., 2011. Fortalecimiento cooperacion tecnica para mejorar el control de la hidatidosis en Bolivia.

Villeneuve, A., Polley, L., Jenkins, E., Schurer, J., Gilleard, J., Kutz, S., Conboy, G., Benoit, D., Seewald, W., Gagne, F., 2015. Parasite prevalence in fecal samples from shelter dogs and cats across the Canadian provinces. Parasit. Vectors 8, 281.

Vlad, D.C., Neghina, A.M., Dumitrascu, V., Marincu, I., Neghina, R., Calma, C.L., 2013. Cystic echinococcosis in children and adults: a seven-year comparative study in western Romania. Foodborne Pathog. Dis. 10, 189–195.

Vuitton, D.A., Zhou, H., Bresson-Hadni, $, et al., 2003. Epidemiology of alveolar echinococcosis with particular reference to China and Europe. Parasitology 127, $87–107.

Vuitton, D.A., Wang, Q., Zhou, H.X., Raoul, F., Knapp, J., Bresson-Hadni, S., Wen, H., Giraudoux, P., 2011. A historical view of alveolar echinococcosis, 160 years after the discovery of the first case in humans: part 1. What have we learnt on the distribution of the disease and on its parasitic agent? Chin. Med. J. 124, 2943–2953.

Vuitton, D.A., Demonmerot, F., Knapp, J., Richou, C., Grenouillet, F., Chauchet, A., Vuitton, L., Bresson-Hadni, S., Millon, L., 2015. Clinical epidemiology of human AE in Europe. Vet. Parasitol. 213, 110–120.

Vural, G., Baca, A.U., Gauci, C.G., Bagci, O., Gicik, Y., Lightowlers, M.W., 2008. Variability in the *Echinococcus granulosus* cytochrome C oxidase 1 mitochondrial gene sequence from livestock in Turkey and a re-appraisal of the G1-3 genotype cluster. Vet. Parasitol. 154, 347–350.

Wachira, T.M., Bowles, J., Zeyhle, E., McManus, D.P., 1993. Molecular examination of the sympatry and distribution of sheep and camel strains of *Echinococcus granulosus* in Kenya. Am. J. Trop. Med. Hyg. 48, 473–479.

Wagner, S., 2014. Diversität von *Echinococcus* spp. Proben (M.Sc. thesis). University of Hohenheim, Stuttgart, Germany.

Wahlers, K., Menezes, C.N., Wong, M., Mogoye, B., Frean, J., Romig, T., Kern, P., Grobusch, M.P., 2011. Human cystic echinococcosis in South Africa. Acta Trop. 120, 179–184.

Wahlers, K., Menezes, C.N., Wong, M.L., Zeyhle, E., Ahmed, M.E., Ocaido, M., Stijnis, C., Romig, T., Kern, P., Grobusch, M.P., 2012. Cystic echinococcosis in sub-Saharan Africa. Lancet. Infect. Dise 12, 871–880.

Wahlstrom, H., Enemark, H.L., Davidson, R.K., Oksanen, A., 2015. Present status, actions taken and future considerations due to the findings of *E. multilocularis* in two Scandinavian countries. Vet. Parasitol. 213, 172–181.

Waloch, M., 2012. Cystic echinococcosis in Poland in 2010. Przegl. Epidemiol. 66, 311–313.

Wang, Q., Vuitton, D.A., Qiu, J.M., Giraudoux, P., Xiao, Y.F., Schantz, P.M., Raoul, F., Li, T.Y., Yang, W., Craig, P.S., 2004. Fenced pasture: a possible risk factor for human alveolar echinococcosis in Tibetan pastoralist communities of Sichuan, China. Acta Trop. 90, 285–293.

Wang, Y., He, T., Wen, X., Li, T., Waili, T.T., Zhang, W., Zhou, H., Zheng, H., Wen, H., Davaadorj, N., Gambolt, L., Mukhar, T., Rogan, M.T., Craig, P.S., 2005. Human cystic echinococcosis in two Mongolian communities in Hobukesar (China) and Bulgan (Mongolia). Trans. R. Soc. Trop. Med. Hyg. 99, 692–698.

Wang, Q., Qiu, J., Yang, W., Schantz, P.M., Raoul, F., Craig, P.S., Giraudoux, P., Vuitton, D.A., 2006. Socioeconomic and behavior risk factors of human alveolar echinococcosis in Tibetan communities in Sichuan, People's Republic of China. Am. J. Trop. Med. Hyg. 74, 856–862.

Wang, Z., Wang, X., Liu, X., 2008. Echinococcosis in China, a review of the epidemiology of *Echinococcus* spp. EcoHealth 5, 115–126.

Wang, C., He, H., Li, M., Lei, F., Root, J.J., Wu, Y., Qin, J., 2009. Parasite species associated with wild plateau pika (*Ochotona curzoniae*) in southeastern Qinghai Province, China. J. Wildl. Dis. 45, 288–294.

Wang, L., Wu, W., Zhu, X., 2010. The endemic status of hydatidosis in China from 2004 to 2008. Chin. J. Zoonoses 26, 699–702.

Wang, H., Lu, C., Liu, X., Zhang, W., 2015a. Metastatic and prognostic factors in patients with alveolar echinococcosis. Int. J. Clin. Exp. Pathol. 8, 11192–11198.

Wang, N., Wang, J., Hu, D., Zhong, X., Jiang, Z., Yang, A., Deng, S., Guo, L., Tsering, D., Wang, S., Gu, X., Peng, X., Yang, G., 2015b. Genetic variability of *Echinococcus granulosus* based on the mitochondrial 16S ribosomal RNA gene. Mitochondrial DNA 26, 396–401.

Wang, Q.H., Shang, W.J., Zhao, C.T., Zhang, S.W., Lu, S.L., Liu, X.D., 2015c. Epidemic Status of Echinococcosis in Gannan Tibetan Autonomous Prefecture of Gansu Province during 2007–2011. Chin. J. Parasitol. Parasit. Dis. 33, 45–48.

Ward, J.W., Bradshaw, R.C., 1956. New records of the occurrence of the hydatid tapeworm, *Echinococcus granulosus*, in Central and South Mississippi. J. Parasitol. 42, 35.

Ward, J.W., 1965. Additional records of *Echinococcus granulosus* from dogs in the lower Mississippi region. J. Parasitol. 51, 552–553.

Wassermann, M., Aschenborn, O., Aschenborn, J., Mackenstedt, U., Romig, T., 2015. A sylvatic lifecycle of *Echinococcus equinus* in the Etosha National Park, Namibia. Int. J. Parasitol. Parasites Wildl. 4, 97–103.

Wei, W., Huang, Y., Huang, L., Yi, Y., De, Y., 2014. 2007–2012 Sichuan Province hydatid disease endemic area of direct network reporting analysis [Chinese]. Chin. J. Pathog. Biol. 9, 68–70.

Welburn, S.C., Beange, I., Ducrotoy, M.J., Okello, A.L., 2015. The neglected zoonoses—the case for integrated control and advocacy. Clin. Microbiol. Infect. 21, 433–443.

Wenker, C., Hoby, S., 2011. Alveolar echinococcosis in lowland gorillas: A threat for the European captive population?. In: Proceedings of the American Association of Zoo Veterinarians 2011 Annual Conference, p. 196.

WHO, 2010. Working to Overcome the Global Impact of Neglected Tropical Diseases. First WHO Report on Neglected Tropical Diseases. WHO press, Geneva, Switzerland.

WHO, 2015. European Hospital Morbidity Database, World Health Organization Regional Office for Europe. Available at: http://data.euro.who.int/hmdb/.

Wilson, J.F., Rausch, R.L., Wilson, F.R., 1995. Alveolar hydatid disease. Review of the surgical experience in 42 cases of active disease among Alaskan Eskimos. Ann. Surg. 221, 315–323.

Wiwanitki, V., 2004. A summary of hydatid disease in Thailand. Inter J. Trop. Med. 2.

Wnukowska, N., Salamatin, R., Golab, E., 2011. Human echinococcosis in Poland in 2003–2010 according to the serological tests results of NIPH-NIH. Przegl. Epidemiol. 65, 455–458.

Wobeser, G., 1971. The occurrence of *Echinococcus multilocularis* (Leukart, 1863) in cats near Saskatoon, Saskatchewan. Can. Vet. J. 12, 65–68.

Wu, X.H., Wang, H., Zhang, J.X., Ma, X., Liu, Y.F., Han, X.M., Liu, H.Q., Cai, H.X., Zhao, Y.M., Ma, J.Y., Liu, P.Y., Zeng, C., 2007. An epidemiological survey on echinococcosis in Zhiduo County of Qinghai Province. Chin. J. Parasitol. Parasit. Dis. 25, 229–231.

Wu, X., 2015. Investigation of Ningxia livestock hydatid infection status in 2012. Ningxia Med. J. 37, 277–278.

Xhaxhiu, D., Kusi, I., Rapti, D., Kondi, E., Postoli, R., Rinaldi, L., Dimitrova, Z.M., Visser, M., Knaus, M., Rehbein, S., 2011. Principal intestinal parasites of dogs in Tirana, Albania. Parasitol. Res. 108, 341–353.

Yamamoto, N., Kishi, R., Katakura, Y., Miyake, H., 2001. Risk factors for human alveolar echinococcosis: a case-control study in Hokkaido, Japan. Ann. Trop. Med. Parasitol. 95, 689–696.

Yamasaki, H., Nakao, M., Nakaya, K., Schantz, P.M., Ito, A., 2008. Genetic analysis of *Echinococcus multilocularis* originating from a patient with alveolar echinococcosis occurring in Minnesota in 1977. Am. J. Trop. Med. Hyg. 79, 245–247.

Yamov, V., Antropov, V., 2008. Epizootiology of helminths of pigs in South Tyumen. Agrar. Vestn. Urala 5, 70–71.

Yan, N., Nie, H.M., Jiang, Z.R., Yang, A.G., Deng, S.J., Guo, L., Yu, H., Yan, Y.B., Tsering, D., Kong, W.S., Wang, N., Wang, J.H., Xie, Y., Fu, Y., Yang, D.Y., Wang, S.X., Gu, X.B., Peng, X.R., Yang, G.Y., 2013. Genetic variability of *Echinococcus granulosus* from the Tibetan plateau inferred by mitochondrial DNA sequences. Vet. Parasitol. 196, 179–183.

Yanagida, T., Mohammadzadeh, T., Kamhawi, S., Nakao, M., Sadjjadi, S.M., Hijjawi, N., Abdel-Hafez, S.K., Sako, Y., Okamoto, M., Ito, A., 2012. Genetic polymorphisms of *Echinococcus granulosus* sensu stricto in the Middle East. Parasitol. Int. 61, 599–603.

Yang, Y.R., Rosenzvit, M.C., Zhang, L.H., Zhang, J.Z., McManus, D.P., 2005. Molecular study of *Echinococcus* in west-central China. Parasitology 131, 547–555.

Yang, Y.R., Ellis, M., Sun, T., Li, Z., Liu, X., Vuitton, D.A., Bartholomot, B., Giraudoux, P., Craig, P.S., Boufana, B., Wang, Y., Feng, X., Wen, H., Ito, A., McManus, D.P., 2006. Unique family clustering of human echinococcosis cases in a chinese community. Am. J. Trop. Med. Hyg. 74, 487–494.

Yang, Y.R., Craig, P.S., Ito, A., Vuitton, D.A., Giraudoux, P., Sun, T., Williams, G.M., Huang, Z., Li, Z., Wang, Y., Teng, J., Li, Y., Huang, L., Wen, H., Jones, M.K., McManus, D.P., 2007. A correlative study of ultrasound with serology in an area in China co-endemic for human alveolar and cystic echinococcosis. Trop. Med. Int. Health 12, 637–646.

Yang, Y.R., Craig, P.S., Sun, T., Vuitton, D.A., Giraudoux, P., Jones, M.K., Williams, G.M., McManus, D.P., 2008. Echinococcosis in Ningxia Hui Autonomous Region, northwest China. Trans. R. Soc. Trop. Med. Hyg. 102, 319–328.

Yang, S-j., Wang, Q-y., Wang, J., Tian, T., Jiang-shan, Z., Kang, C., Wu, W-p., 2015. Surveillance for hydatidosis in Emin, Xinjiang, 2007–2013. Dis. Surveill. 30, 130–133.

Yassien, M., Ahmed, A., Soliman, S., Youssef, A., 2013. Prevalence of parasitic infection of slaughtered cattle and buffalo in municipal abattoir at Ismailia, Egypt. In: Tokora, M., Uga, S. (Eds.), Parasitic Zoonoses in Asian-Pacific Region. Sankeisha Publishing Agent, pp. 19–24.

Yazar, S., Yaman, O., Cetinkaya, F., Sahin, I., 2006. Cystic echinococcosis in Central Anatolia, Turkey. Saudi Med. J. 27, 205–209.

Yemets, O.M., 2013. *Echinococcus granulosus* in wild canidae and ungulates. In: XV. Conference of Ukrainian Scientific Society of Parasitologists. (Chernivtsi, Ukraine.).

Yena, S., Sanogo, Z.Z., Keita, A., Sangare, D., Sidibe, S., Delaye, A., Doumbia, D., Diallo, A., Soumare, S., 2002. Surgery for pulmonary hydatid cyst in Mali. Ann. Chir. 127, 350–355.

Young, E., 1975. Echinococcosis (hydatidosis) in wild animals of the Kruger National Park. J. South Afr. Vet. Assoc. 46, 285–286.

Yu, S.H., Wang, H., Wu, X.H., Ma, X., Liu, P.Y., Liu, Y.F., Zhao, Y.M., Morishima, Y., Kawanaka, M., 2008. Cystic and alveolar echinococcosis: an epidemiological survey in a Tibetan population in southeast Qinghai, China. Jpn. J. Infect. Dis. 61, 242–246.

Yudin, V., 2012. Helminth infestation carnivorous mammals of Primorye and the Amur Region. Modern Problems of Wildlife management. Game Manag. Fur Farming 1, 609–610.

Zanaj, S., 1997. Frekuenca e ekinokokozës në ripërtypësit e rretheve Tiranë, Korçë, Vlorë, Elbasan dhe Lezhë. Bul. i Shk. 2, 113–117.

Zariffard, M., Massoud, J., 1998. Study of *Echinococcus granulosus* and *Echinococcus multilocularis* infections in Canidiae in Ardabile province of Iran. Arch. Inst. Razi. 48/49, 47–52.

Zaytsev, I., 2013. Indications for reoperation with liver alveococcosis. Sib. Med. J. 8, 110–112.

Zeder, M.A., 2008. Domestication and early agriculture in the Mediterranean Basin: Origins, diffusion, and impact. Proc. Natl. Acad. Sci. U.S.A. 105, 11597–11604.

Zeyhle, E., Abel, M., Frank, W., 1990. Epidemiologische Untersuchungen zum Vorkommen von *Echinococcus multilocularis* bei End- und Zwischenwirten in der Bundesrepublik Deutschland. Mittl. Öster Gesel. Trop. Parasitol. 12, 221–232.

Zhang, L., Eslami, A., Hosseini, S.H., McManus, D.P., 1998. Indication of the presence of two distinct strains of *Echinococcus granulosus* in Iran by mitochondrial DNA markers. Am. J. Trop. Med. Hyg. 59, 171–174.

Zhang, L.H., Joshi, D.D., McManus, D.P., 2000. Three genotypes of *Echinococcus granulosus* identified in Nepal using mitochondrial DNA markers. Trans. R. Soc. Trop. Med. Hyg. 94, 258–260.

Zhang, Y., Bart, J.M., Giraudoux, P., Craig, P., Vuitton, D., Wen, H., 2006. Morphological and molecular characteristics of *Echinococcus multilocularis* and *Echinococcus granulosus* mixed infection in a dog from Xinjiang, China. Vet. Parasitol. 139, 244–248.

Zhao, Y.M., Tong, S.X., Jing, T., Chong, S.G., Cai, X.P., Jing, Z.Z., Han, J., 2009. Investigation on echinococcosis in animals in Gannan Tibetan Autonomous Prefecture. Chin. J. Parasitol. Parasit. Dis. 27, 27–30.

Zhekamukhova, M., Maksidova, Z., Mantaeva, S., Sarbasheva, M., Shikhalieva, M., AM., B, 2012. The Echinococcosis Epizootology Peculiarities of Meat and Wool Breed Sheep in the Farm Enterprises of the Kabardino-Balkarian Foothills Zone, 12. Bulletin of the Krasnoyarsk State Agrarian University.

Zheng, J., Du, F., Wang, D., 2014. The experience of 526 patients with hepatic echinococcosis in Tibet. J. Hepatobil. Surg. 22 (6), 459–461 (Chinese).

Zhong, X., Wang, N., Hu, D., Wang, J., Liu, T., Gu, X., Wang, S., Peng, X., Yang, G., 2014. Sequence analysis of cytb gene in *Echinococcus granulosus* from Western China. Korean J. Parasitol. 52, 205–209.

Ziadinov, I., Mathis, A., Trachsel, D., Rysmukhambetova, A., Abdyjaparov, T.A., Kuttubaev, O.T., Deplazes, P., Torgerson, P.R., 2008. Canine echinococcosis in Kyrgyzstan: using prevalence data adjusted for measurement error to develop transmission dynamics models. Int. J. Parasitol. 38, 1179–1190.

Ziadinov, I., Deplazes, P., Mathis, A., Mutunova, B., Abdykerimov, K., Nurgaziev, R., Torgerson, P.R., 2010. Frequency distribution of *Echinococcus multilocularis* and other helminths of foxes in Kyrgyzstan. Vet. Parasitol. 171, 286–292.

Ziaei, H., Fakhar, M., Armat, S., 2011. Epidemiological aspects of cystic echinococcosis in slaughtered herbivores in Sari abattoir, North of Iran. J. Parasit. Dis. 35, 215–218.

Zuko, A., Obradović, Z., 2014. Cystic echinococcosis in Bosnia and Herzegovina – an overview. ESCCAP. *Echinococcus* 2014 Abstract Booklet.

INDEX

'*Note:* Page numbers followed by "f" indicate figures and "t" indicate tables.'

D

E

F

G

H

I

K

L

M

N

O

P

R

S

T

CONTENTS OF VOLUMES IN THIS SERIES

Volume 46

Volume 47

Volume 48

Volume 49

Volume 50

Volume 51

Volume 52

Volume 53

Volume 54

Volume 55

Volume 61

Volume 62

Volume 63

Volume 64

Volume 65

Volume 66

Volume 67

Volume 68

Volume 69

Volume 70

Volume 71

Volume 72

Volume 73

Volume 74

Volume 75

Volume 76

Volume 77

Volume 78

Volume 79

Volume 80

Volume 81

Volume 82

Volume 87

Volume 92

FSC
www.fsc.org
MIX
Papier | Fördert gute Waldnutzung
FSC® C083411

Kolibri 360 GmbH
Ferdinand-Jühlke-Straße 7
99095 Erfurt, Deutschland
produktsicherheit@kolibri360.de